PLANETARY VOLCANISM
A Study of Volcanic Activity in the Solar System
Second edition

WILEY-PRAXIS SERIES IN ASTRONOMY AND ASTROPHYSICS
Series Editor: John Mason, B.Sc., Ph.D.

Few subjects have been at the centre of such important developments or seen such a wealth of new and exciting, if sometimes controversial, data as modern astronomy, astrophysics and cosmology. This series reflects the very rapid and significant progress being made in current research, as a consequence of new instrumentation and observing techniques, applied right across the electromagnetic spectrum, computer modelling and modern theoretical methods.

The crucial links between observation and theory are emphasised, putting into perspective the latest results from the new generations of astronomical detectors, telescopes and space-borne instruments. Complex topics are logically developed and fully explained and, where mathematics is used, the physical concepts behind the equations are clearly summarised.

These books are written principally for professional astronomers, astrophysicists, cosmologists, physicists and space scientists, together with post-graduate and undergraduate students in these fields. Certain books in the series will appeal to amateur astronomers, high-flying 'A'-level students, and non-scientists with a keen interest in astronomy and astrophysics.

ROBOTIC OBSERVATORIES
Michael F. Bode, Professor of Astrophysics and Assistant Provost for Research, Liverpool John Moores University, UK

THE AURORA: Sun–Earth Interactions
Neil Bone, School of Biological Sciences, University of Sussex, Brighton, UK

PLANETARY VOLCANISM: A Study of Volcanic Activity in the Solar System, Second edition
Peter Cattermole, formerly Lecturer in Geology, Department of Geology, Sheffield University, UK, now Principal Investigator with NASA's Planetary Geology and Geophysics Programme

DIVIDING THE CIRCLE: The Development of Critical Angular Measurement in Astronomy 1500–1850
Second edition
Allan Chapman, Wadham College, University of Oxford, UK

THE DUSTY UNIVERSE
Aneurin Evans, Department of Physics, University of Keele, UK

COMET HALLEY - Investigations, Results, Interpretations
Volume 1: Organization, Plasma, Gas
Volume 2: Dust, Nucleus, Evolution
Editor: John Mason, B.Sc., Ph.D.

ELECTRONIC AND COMPUTER-AIDED ASTRONOMY: From Eyes to Electronic Sensors
Ian. S. McLean, Department of Astronomy, University of California at Los Angeles, California, USA

URANUS: The Planet, Rings and Satellites
Ellis D. Miner, Cassini Project Science Manager, NASA Jet Propulsion Laboratory, Pasadena, California, USA

THE PLANET NEPTUNE: An Historical Survey Before Voyager, Second edition
Patrick Moore, CBE, D.Sc.(Hon.)

THE HIDDEN UNIVERSE
Roger J. Tayler, Astronomy Centre, University of Sussex, Brighton, UK

Forthcoming titles in the series are listed at the back of the book.

PLANETARY VOLCANISM

A Study of Volcanic Activity in the Solar System
Second edition

Peter Cattermole
Department of Earth Sciences
University of Sheffield

JOHN WILEY & SONS
Chichester • New York • Brisbane • Toronto • Singapore

Published in association with
PRAXIS PUBLISHING
Chichester

Wiley Editorial Offices

John Wiley & Sons Ltd, Baffins Lane,
Chichester, West Sussex PO19 1UD, England

John Wiley & Sons, Inc., 605 Third Avenue,
New York, NY 10158-0012, USA

Jacaranda Wiley Ltd, G.P.O. Box 859, Brisbane
Queensland 4001, Australia

John Wiley & Sons (Canada) Ltd, 22 Worcester Road,
Rexdale, Ontario M9W 1L1, Canada

John Wiley & Sons (SEA) Pte Ltd, 37 Jalan Pemimpin 05-04,
Block B, Union Industrial Building, Singapore 2057

A catalogue record for this book is available from the British Library

ISBN 0-471-96051-9

Printed and bound in Great Britain by Hartnolls Ltd, Bodmin

Table of Contents

Author's Preface xi
Acknowledgements xiii
List of Plates xv
List of Tables xxiii

1   **Introduction**   **1**
    1.1   The Sun's family 1
    1.2   Planetary crusts 1
    1.3   Of planets and plates 2
    1.4   What is volcanism? 3
    1.5   The study of volcanoes 4
    1.6   Comparative volcanology 5
    1.7   Remote-sensing and geological exploration 6

    Colour plate Section comes between 1 and 2

2   **The making of planets**   **9**
    2.1   The solar nebula 9
    2.2   Primordial materials 12
    2.3   Planetary accretion 15
    2.4   Planetary heating 16
    2.5   Mantle segregation and convection 21
    2.6   Lithospheres 22

3   **The generation and evolution of magmas**   **24**
    3.1   The production of magma 24
    3.2   Mantle source rocks and partial melts 28
    3.3   Magmatic evolution 31
        3.3.1   Fractional crystallization 32
        3.3.2   Major element behaviour 33
        3.3.3   Trace elements and isotopes 36
    3.4   Fractionation in the solar system 38
    3.5   Epilogue 41

4   **Magma ascent and eruption**   **43**
    4.1   Melt formation at depth 43
    4.2   Magma ascent 44

4.3   Types of volcanic eruption 48
4.4   Controls on the rise and effusion rates of magmas 50
4.5   The role of volatiles 52
      4.5.1   Eruption clouds 54
      4.5.2   Phreatomagmatic eruptions 56
4.6   Rates and amounts of magma generation on planets 58

5   The rheological properties and behaviour of volcanic flows   60
5.1   Rheological concepts 61
5.2   Volcanic flows as Bingham liquids 62
5.3   Viscosity 65
5.4   Flow behaviour 67
5.5   Magma discharge rates 73
5.6   The cooling of lava flows 74
5.7   Latent heat of crystallization effects 76
5.8   Rheological analysis of volcanic flows 77
      5.8.1   Quantitative models 78
      5.8.2   Yield strength models 79
      5.8.3   The derivation of effusion rate 80
5.9   Estimates of Viscosity 82
5.10 Rheological data 83
      5.10.1   Yield strength and viscosity 83
      5.10.2   Effusion rate and flow duration 83
      5.10.3   Analysis of quantitative data 86
      5.10.4   Rheology and lava composition 91
5.11 Pyroclastic flows 93
5.12 Lahars (mudflows) 95
5.13 Sulphur flows 95

6   Volcanic landforms   97
6.1   The diversity of volcanic landforms 97
6.2   Flood lavas 100
6.3   Shield volcanoes 105
6.4   Paterae 107
6.5   Volcanic cones 109
6.6   Volcanic domes and tholoids 112
6.7   Pyroclastic flows 112
6.8   Coronae, novae and arachnoids 116
6.9   Maars 116

7   Volcanic plains and their development   117
7.1   Basalts of the Earth's oceanic plains 117
7.2   Terrestrial continental flood basalts 120
      7.2.1   Iceland 121
      7.2.2   Greenland and the North Atlantic Province 123

	7.2.3	The Columbia River Plateau	125
	7.2.4	Other flood basalt provinces	128
	7.2.5	Characteristics of terrestrial flood lava sequences	130
7.3	Continental Rifts		130
	7.3.1	The East African Rift	130
	7.3.2	The Rio Grande Rift	132
	7.3.3	Continental rifts – the wider context	133
7.4	Mercurian plains volcanism		135
7.5	Volcanic plains on Venus		137
	7.5.1	Venus: general physiography	138
	7.5.2	Local surface composition and characteristics	141
	7.5.3	Volcanic nature of the Venusian plains	141
	7.5.4	Age of Venusian plains	148
	7.5.5	Eruption characteristics on Venus	148
	7.5.6	Flowfields and lava channel systems	149
	7.5.7	Distribution and characteristics of small volcanic structures	155
7.6	Lunar volcanic plains		159
7.7	Volcanic plains on Mars		160
	7.7.1	Noachian plateau plains – the cratering record	162
	7.7.2	Plateau plains – morphological characteristics	163
	7.7.3	Hesperian-age ridged plains	164
	7.7.4	Significance of plains ridges	166
	7.7.5	Hesperian-age flow plains	171
	7.7.6	Tempe Terra volcanic province	174
	7.7.7	Volcanic plains of Amazonis, Memnonia and Aeolis	176
	7.7.8	Amazonian volcanic plains	177
	7.7.9	The northern plains of Mars	177
	7.7.10	Summary	177

8	**Lunar volcanism**		**179**
8.1	Highland volcanism		179
	8.1.1	KREEP	180
	8.1.2	Fra Mauro Basalts	181
	8.1.3	The KREEP-rich Apollo 15 rocks	183
	8.1.4	Pre-Imbrium KREEP-rich flood volcanism	184
	8.1.5	Age and origin of KREEP	186
8.2	Highland volcanic features		188
	8.2.1	The Cayley and Descartes Formations	189
	8.2.2	Other highland volcanic structures	192
8.3	Mare volcanism		195
	8.3.1	Thickness of mare deposits	197
	8.3.2	Mapping and subdivision of mare units	200
	8.3.3	Galileo multispectral data (Clementine mission)	203
	8.3.4	Dark mantling deposits	209
	8.3.5	Sequence of lava emplacement	209

8.3.6   Petrology and geochemistry of mare basalts 211
8.3.7   Petrogenesis of mare basalts 214
8.4   Emplacement of mare lavas: morphological evidence 215
8.4.1   The morphology of lunar flows 216
8.4.2   Sinuous rilles 222
8.4.3   Domes, low shields and other volcanic complexes 225
8.4.4.   Mare arches and ridges (wrinkle ridges) 231
8.4.5   Dark mantling materials 236
8.4.6   Mare filling as exemplified by the Orientale Basin 237

9   Shield volcanoes and terrestrial examples   243
9.1   Introduction 243
9.2   General characteristics of terrestrial shields 244
9.3   The Hawaiian shields 246
9.3.1   Growth cycle of a typical shield 249
9.3.2   Petrological history of Hawaiian volcanoes 254
9.3.3   Genesis of Hawaiian-type shields 256
9.4   Caldera formation and other summit activity 257

10   Martian central volcanism   261
10.1 Distribution of central volcanoes 261
10.2 Classification of volcano types 263
10.3 Ages of central volcanoes on Mars 264
10.4 Tharsis and Elysium – gravity and tectonics 265
10.5 The Tharsis volcanic rise 267
10.5.1   The Tharsis Montes 268
10.5.2   Olympus Mons 275
10.5.3   Older Tharsis volcanoes 280
10.5.4   Alba Patera 285
10.6 The Elysium volcanic rise 294
10.6.1   The Elysium shields 294
10.6.2   Apollinaris Patera 297
10.7 Highland Paterae of the Hellas Region 298
10.7.1   Amphitrites Patera 299
10.7.2   Hadriaca Patera 299
10.7.3   Tyrrhena Patera 302
10.7.4   Implications for early patera volcanism 304
10.8 Temporal sequence of Martian central volcanism 305
10.9 Controls on Martian volcanism 306

11   Central volcanism on Venus   309
11.1 Distribution of large volcanic structures on Venus 310
11.2 Morphology of Venusian shield volcanoes 310
11.2.1   Intermediate-sized volcanoes 310
11.2.2   Large volcanic shields 311

| 11.2.3 Coronae and related structures .... 315 |
| 11.2.4 Pancake domes and related features .... 323 |

11.3 Volcanic rises 325
    11.3.1 The volcanic rise of Beta Regio 325
    11.3.2 The volcanic rise of Atla Regio 328
    11.3.3 The volcanic rises of Eistla Regio 331
    11.3.4 The volcanic rises of Bell and Phoebe Regiones 335
    11.3.5 Western Ishtar Terra 338
11.4 Controls on centralized volcanism 339
    11.4.1 Structure of interior and lithosphere of Venus 340
    11.4.2 Magma uprise and plume activity on Venus 343
    11.4.3 Magma reservoirs and Neutral Buoyancy Zones 347
    11.4.4 Age of volcanism and resurfacing on Venus 348
11.5 Epilogue 350

12 Volcanism on Io   **351**
12.1 Surface composition of Io 351
12.2 Surface features on Io 353
    12.2.1 Mountain massifs 354
    12.2.2 Plains and intervent flows 355
    12.2.3 Vent materials 357
12.3 Volcanic plumes 362
12.4 The role of silicates and sulphur on Io 362
12.5 Interior structure of Io 366

13 Volcanism on icy satellites   **367**
13.1 Ice and rock Lithospheres and asthenospheres 367
13.2 Cryovolcanism 368
13.3 The Jovian satellites 369
    13.3.1 Europa 370
    13.3.2 Ganymede 371
    13.3.3 Callisto 372
13.4 The satellites of Saturn and Uranus 372
    13.4.1 Dione and Tethys 373
    13.4.2 Ariel and Titania 375
    13.4.3 Miranda 376
13.5 Satellites of Neptune 378
    13.5.1 Triton 378

Appendix 1 382
Appendix 2 387
References 388
Index 411

Author's Preface

An opportunity to revise and update the original edition of *Planetary Volcanism* could not be missed. The first edition was written some six years ago and, not surprisingly, is now somewhat out-of-date. This is particularly true of our knowledge of Venus but also applies to work on the outer planet moons and to the Moon. Then, of course, the planetary community has been beavering away at research into the geology of Mars, thus, all-in-all, there is much new work to discuss.

The broad plan of the original edition has been followed but since there is so much new information about Venus and Mars, in particular, it was found expedient to separate discussion of centralized volcanism into individual chapters. I also considered it wise to bring the information on the rheological aspects of flows under one roof, so to speak, the result being what I believe to be a more coherent treatment. This will be found in chapter 5.

Over the past few years it has given me pleasure to receive many favourable, encouraging, comments about the first edition which, it was noted in the first edition, I wrote with some trepidation. Not surprisingly there also have been criticisms, mostly constructive, and the new edition provides a vehicle for incorporating some of these into a new text. I have included, therefore, a little more chemistry in this edition and also more weight has been given to some petrogenetic aspects of the subject. It remains nonetheless essentially a work based on photogeological interpretation and analysis of remotely-acquired spectral and thermal data rather than laboratory experiment. It does not delve into the complex problems of phase relations in magmas, for instance, or the segregation of elements within the planets and their moons, although one could argue that these are both a part of volcanism. It is in no sense a definitive text which covers every aspect of planetary volcanism, rather it is an offering which seeks to inform and stimulate those who have an interest in this fascinating topic.

While my own views and ideas colour this book, of necessity it is founded on the painstaking research of a multitude of scientists with a variety of backgrounds, and holding often quite diverse views about the nature and origin of the features revealed by spacecraft imagery. I have tried to abstract from the immense literature impartially and synthesize what I have found. I hope I have done this fairly and accurately, and that I have achnowledged all whose work I have represented. I hope that the endeavour will prove worthwhile, and both stimulating and useful to a range of scientists in many countries.

I would like to re-iterate my thanks to Ron Greeley and John Guest for their continued interest and support, and to those at both NASA Headquarters and NSSDC who have given me continuing assistance and encouragement. As with the first edition, this book is dedicated to Patrick Moore who started me on the road to geological exploration of the Moon many years ago and whose undying enthusiasm is a role model to all who have an interest in the Universe at large.

Peter Cattermole, Sheffield, October 1995.

Acknowledgements

I wish to acknowledge the many scientists whose work forms the basis of this book. Although too numerous to mention individually, I owe them all a great debt and hope that I have represented their views accurately and fairly. I specifically thank David Okerson of NASA headquarters who had an important part to play in supplying the excellent Magellan images for the new Venus chapter, Ron Greeley, John Guest and Jim Head for their continuing willingness to keep me up-to-date with planetary research by sending new publications as they become available, and Dave Rothery for helping me with photos of outer planet moons.

Several of the diagrams used herein have been published in the *Journal of Geophysical Research* and are copyright by the American Geophysical Union. Others have been reproduced by courtesy of *Icarus* and *Space Science Letters*. Plate 5 is copyright of *National Geographic* and was taken in September 1993 by Robert Sissons. A large number of figures and several plates are reproduced by kind permission of the National Aeronautics and Space Administration and I would like to take this opportunity to thank that organization for its continued assistance.

List of Plates

Chapter 1

1.1   Galileo image of the cratered asteroid, Ida 7

Chapter 5

5.4   Long sinuous rille crossing western Mare Imbrium near the
        crater Brayley 68

5.5   Upper surface of fresh pahoehoe flow showing development
        of ropy texture 69

5.6   Section through a typical aa flow, showing upper clinkery part
        and denser inner region with vesicle trains 70

5.7   Pressure ridges on the surface of a rhyolite obsidian flow, Big
        Glass Mountain, California 71

5.16   Blocky texture of massive mudflow associated with caldera
        collapse of Mount Meru, N. Tanzania 94

Chapter 6

6.1   Sequence of flood basalts with vertical dykes exposed in the
        cliffs of Isla La Gomera, Canary Islands 98

6.2   Flood lavas on the western side of Mare Imbrium 99

6.4   Synoptic photomosaic of the King's Bowl lava field, East Snake
        River Plain 102

6.5   The interior of a large collapsed lava tube in the Snake River
        Plains, Idaho 103

6.6   Region of tube- and channel-fed flows on the west flank of Alba
        Patera, Mars 104

6.7   (a) Shuttle radar image of the volcanic calderas atop three
        approximately 1000-m-high shield volcanoes of the western
        Galápagos Islands, Ecuador 106

6.8   Photomosaic of the Martian patera volcano, Alba Patera, situated
        in northern Tharsis 108

6.9   The Martian highland volcano, Tyrrhena Patera 109

6.10   Group of aligned monogenetic cones on the north flank of Meru
        volcano, Tanzania 110

6.11   Large phonolite dome, approximately 3.5 km in diameter, intruded
        into the lower flanks of Meru volcano, northern Tanzania 112

6.12   Circular steep-sided volcanic domes in Apha Regio, Venus 113

6.13 Aerial view of large mudflow on eastern side of Meru volcano, showing distinctive hummocky topography superimposed on broad lobate flow 113
6.14 The corona, Aramaiti 115

Chapter 7
7.12 Mercurian smooth plains southeast of the Caloris Basin 136
7.15 Geometrically-transformed image of Venusian surface at the Venera 14 landing site 142
7.16 High-resolution Magellan image of the Bereghinya Planitia region of Venus 143
7.17 Magellan image of the plains region of Lavinia Planitia 144
7.18 (a) Reticulate plains, with impact crater just above centre and shield field towards bottom right. (b) Gridded plains with intersecting lineaments 146
7.19 (a) Lobate plains with intermediate radar-backscattering characteristic. (b) Wrinkle ridges on a part of Lavinia Planitia 147
7.20 200-km-long segment of a sinuous Venusian lava channel in Lavinia Planitia 150
7.21 Lava flow fields on the eastern flank of Sapas Mons shield volcano 151
7.22 The region of Mylitta Fluctus, in northern Lada Terra 152
7.25 The association of shield fields and fracture belts on the surface of Niobe Planitia 157
7.26 Synoptic view of a part of Atla Regio 158
7.27 Magellan image of three flat-topped "pancake" domes in Eistla Regio 159
7.28 Scalloped dome in Alpha Regio 160
7.30 Noachian-age plateau plains south of Elysium Planitia 166
7.32 Noachian-age smooth facies ridged plains northeast of Arabia 166
7.33 Wrinkle ridges crossing the plains near Hesperia Planum. Note the asymmetric profile of the ridges and their rather sinuous courses 167
7.34 Plains units on the borders of Isidis, showing possible volcanic flows and spatter ridges or exhumed *en echelon* dikes 168
7.35 50-km-diameter caldera structure on the ridged plains of Syrtis major Planitia 169
7.37 Hesperian-age sheet flows northeast of Uranius Patera 172
7.38 Extensive flood lavas on the southern flanks of Tharsis 173
7.39 Plains-style volcanic features in the Tempe terra volcanic province 175
7.40 (b) Diverse morphology of three volcanic units near 5˚S, 175˚W 176

Chapter 8
8.1 Oblique view of Fra Mauro, with Bonpland and Parry 182
8.2 Photomicrograph of crystallized KREEP, showing laths of calcic plagioclase and tabular orthopyroxene set in a brown glass 184
8.3 (a) View of the lunar surface around the Apollo 15 landing site 185

8.3   (c) Detailed view of boundary between dark basalts of
        Palus Putredinis and more heavily cratered, higher albedo
        Apennine Bench Formation 186

8.5   The Descartes region of the Moon, showing the lineated and
        furrowed deposits and prominent domes of the Descartes
        Formation and the smoother *light plains* of the Cayley 189

8.6   Groups of smooth-textured domes typify the Cayley plains to the
        northeast of the crater, Ritchey 190

8.7   (b) High oblique view of the same region as 8.3(a), showing also
        the long sinuous rille, Rima Bradley 192

8.8   Oblique view of Gruithuisen γ and δ, on the edge of Mare Imbrium 194

8.9   The region northeast of Prinz, showing the upland massifs of the
        Harbinger Montes, surrounded and embayed by Imbrium lavas 195

8.10   Enlarged view of two Harbinger massifs, showing the typical
        rounded topography, tree-bark texture and basal talus aprons 196

8.11   Prominent volcanic dome or low cone within the light plains
        of the floor of the farside ring, Mendeleev 197

8.12   Dome-like landforms within the ponded units on the floor of the
        farside ring, Waterman 198

8.13   Oblique view looking north across Mare Crisium, showing the
        smooth, dark mare lavas and the rugged, higher-albedo highland
        massifs marking the rim of the Crisium basin 199

8.14   Imprint of Surveyor III footpad into the lunar regolith, as
        photographed by the Apollo 12 LM crew 200

8.16   (Left) Unfiltered photograph of part of the lunar nearside, showing
        Mare Crisium towards the right-hand side of the image, with Mare
        Foecunditatis to its south, then Maria Tranquillitatis and Serenitat
        is further west. (Right) Colour difference photograph of the lunar
        nearside obtained by Ewen Whitaker. Note the redder (lighter)
        interior flows of Mare Serenitatis, as well as the details revealed
        within both Mare Imbrium and Oceanus Procellarum 204

8.18   Cryptomare on the floor of the elongate structure Schiller 207

8.19   The Grimaldi - Riccioli region of the Moon, showing the extensive
        low albedo mare units within Grimaldi and the less extensive ones
        inside Grimaldi 208

8.20   (a) Lunar Orbiter photograph of Mare Serenitatis 210

8.21   Distribution of Stage 1 and 3 basalts on the southern border of Mare
        Serenitatis 211

8.23   Eratosthenian-age lobate flows on the southwestern surface of
        Mare Imbrium 216

8.24   Northeastward continuation of Eratosthenian-age lobate flows
        depicted in 8.24 217

8.25   Sinuous leveed flow east of Lansberg 218

8.26   Sinuous rilles north of the flooded crater Prinz 219

8.28   Sinuous rilles and other volcanic features in the vicinity of Prinz
and Krieger 222
8.30   Series of aligned depressions near Gruithuisen, Mare Imbrium 224
8.31   Rough-textured domes, cones, sinuous and linear rilles in the Marius
Hills region 226
8.32   (a) Prominent 9-km-wide domes near Milichius. (b) Smooth 8 ¥ 4 km
dome with summit depression on Mare Nubium 227
8.33   Low shields near Wichmann, Herigonius region, Oceanus Procellarum 229
8.34   (Top) Small cinder cone in Mare Nubium, west of Lassell 230
8.35   Mare ridges mimicking the ramparts of the buried ring structure,
Flamsteed P 231
8.36   The tortuous, narrow spine of a wrinkle ridge lies on the western edge
of abroad mare arch in the southern two-thirds of this image, but
crosses to the opposite side towards the north 232
8.38   Features of mare arch/ridge system near Harbinger Montes,
Mare Imbrium 234
8.40   Dark mantling deposits here are seen covering the Aristarchus Plateau,
occupying the bottom right-hand part of this image 237
8.41   Synoptic view of the Orientale Basin showing the distribution of
mare plains units and areas shown in following figures 238
8.42   Northern part of Mare Orientale 239
8.43   Southern part of Mare Orientale 240
8.44   Coalescing low shields on the floor of Lacus Veris 241

Chapter 9
9.1   Oblique view across South Pit and Mokuaweoweo on the plateau-like
summit of the Hawaiian shield, Mauna Loa 244
9.7   Oblique view of cinder and spatter cones developed on the
Southwest Rift Zone of Mauna Loa 253
9.13   Pit craters along Kilauea's East Rift Zone, as photographed in 1954 260

Chapter 10
10.3   Mosaic showing the summit region of Arsia Mons 267
10.4   Enlarged view of southwest flank of Arsia Mons, showing prominent
embayment composed of coalescent pits and circumferential grabens 268
10.6   Summit region of Ascraeus Mons, showing the nested summit caldera
complex, radial flows and flank embayments 271
10.7   High-resolution mosaic of the south rim of Ascraeus Mons' summit
caldera 272
10.8   The summit of Pavonis Mons, showing the 45-km-diameter caldera
within its larger, shallow depression 273
10.9   Mosaic showing landslide to the northwest of Arsia Mons 274
10.10   Synoptic view of Olympus Mons 276
10.11   The summit region of Olympus Mons, showing the nested caldera pits,
radiating flow texture and broad flow terraces 277

10.12   Olympus Mons aureole texture 279
10.14   Group of three volcanic shields situated east of Ceraunius Fossae 281
10.15   JPL enhanced image of Ceraunius Tholus, showing the finely
         striated shield flanks and the prominent channels on the north side 282
10.16   Biblis Patera, a smaller shield west of Pavonis Mons 283
10.17   The 110 km ¥ 83 km nested caldera complex at the summit of
         Uranius Patera 284
10.18   Flow types characteristic of the southeast flanks of Alba Patera 286
10.19   The younger summit caldera of Alba Patera 287
10.22   Anastomosing channel networks on the northern flank of Alba Patera 290
10.24   Volcanic spatter ridge on the eastern flank of Alba Patera 292
10.26   (above) General view of Elysium Mons. (below) Higher-resolution
         mosaic showing the hummocky texture typical of the shield flanks,
         together with several prominent channels north of the caldera and
         a long scarp on the eastern flank 295
10.27   Summit region of Hecates Tholus showing the nested caldera
         complex, aligned pit rows, channels and generally rather smooth
         appearance of the circum-summit area 296
10.28   View of the isolated central volcano Apollinaris Patera, with its
         100-km-diameter caldera 298
10.29   Ancient volcanic ring structures close to the south rim of Hellas 299
10.30   Hadriaca Patera, an ancient volcanic structure on the northeast
         rim of the Hellas basin 300
10.31   Channels on the flanks of Hadriaca Patera 301
10.32   The summit region of the eroded volcanic structure, Tyrrhena Patera 302
10.33   High-resolution image of a part of the summit shield margin of
         Tyrrhena Patera, northwest of the volcano summit 303

Chapter 11

11.3   Intermediate-sized volcanic structures in Atla Regio 314
11.4   Radar-bright lobate volcanic flows originate in a volcanic
         depression developed along a NE-SW fracture line in Phoebe Regio 315
11.5   The prominent chain of coronae running along Parga Chasma
         and Themis Regio 316
11.6   The 400-km-diameter concentric corona, Tamfana 317
11.7   (a) The corona structure, Selu, situated in Alpha Regio 318
11.9   Artemis Chasma 321
11.10   The circular depression, Aramaiti, located south of Aphrodite Terra
         in Aino Planitia 322
11.11   (a) Steep-sided pancake domes in Tinatin Planitia. (b) Unusual
         steep-sided dome located east of Beta Regio 324
11.13   The central part of Beta Regio 327
11.14   (a) The volcanic rise of Atla Regio 329
11.16   Guor Linea Rift and Gula Mons 332

11.18   Shield volcano straddling the rift system extending southeastward
         from Bell Regio 336
11.19   Ishtar Terra 337
11.20   The boundary between Danu Montes and the radar-dark plateau
         of Lakshmi Planum 338

Chapter 12

12.2   The mountain massif of Boosaule Montes, showing the adjacent
         large vent structure and surrounding layered and intervent plains
         deposits with low shields, calderas and rifts 354
12.3   The region of Nemea Planum, showing layered plains units with
         linear rifts and eroded scarps 355
12.4   Prominent white fumarolic deposit at base of Iopolis Planum
         layered plains unit 356
12.5   The prominent Ionian low shield, Ra Patera is seen towards the
         lower margin of this mosaic 358
12.7   The two disk-shaped shield volcanoes, Apis and Inachis Tholi,
         100 and 140 km in diameter respectively 360
12.8   A prominent umbrella-shaped plume may be seen rising from the
         limb of Io, silhoueeted aginst the dark sky background 361

Chapter 13

13.3   Voyager image of Ganymede showing the two principal terrain types
         (i) dark cratered terrain, and (ii) grooved, light terrain 371
13.4   Voyager-1 mosaic covering the Saturn-facing partion of the leading
         hemisphere of Dione 373
13.5   Voyager-2 synoptic view of one hemisphere of Tethys 374
13.6   (a) Voyager-2 mosaic of Ariel, showing prominent grooves floored by
         relatively smooth plains units 375
13.7   A Voyager-2 image of a part of Brownie Chasma, showing the
         3-km-wide "rille" of Sprite Vallis 376
13.8   Voyager-2 mosaic of Miranda, showing Arden Corona,
         Inverness Corona and Elsinore Corona 377
13.10   Triton: smooth plains, showing a major depression and overlapping
         layers towards the bottom of the image 379
13.11   Triton: hummocky terrain, showing a smooth-floored depression
         with central rough unit and several narrow channels, probably
         evacuated volcanic tubes or fissures 380
13.12   Canteloupe terrain on Neptune's largest moon, Triton 381

Colour plate section (between pages 8 and 9)

Plate 1   View of the volcanic shield of Mauna Loa from Mauna Kea

Plate 2   The smouldering cone of the dangerous stratovolcano, Gunung Merapi, central Java

Plate 3   Unwelded ignimbrite deposit, Isla La Gomera, Canary Islands

Plate 4   Huge block of intrusive feldspar-porphyry transported by laharic flows from Meru caldera and dumped about 8 km from source

Plate 5   Havoc wreaked by mudflows during the paroxysmal eruption of Gunung Agung, Bali, during May 1963

Plate 6   False colour Landsat image of Big Island, Hawaii

Plate 7   Stratigraphic map of Kilauea shield, Hawaii, showing pattern of flows erupted since 10,000 B.P.

Plate 8   Oblique colour view of the large shield volcano, Sapas Mons

Plate 9   Spectral classes for mare basalts of the lunar nearside

Plate 10   SSI composite mosaic used to distinguish the principal units in the lunar highlands and maria

Plate 11   Mosaic of the Tharsis region of Mars, including the western end of the equatorial canyon system

Plate 12   A Magellan altimetry and radar image composite of a part of Aino Planitia

Plate 13   The two shield volcanoes Sif and Gula Montes

Plate 14   Gravity and topography for the region between Atla and Beta Regiones

Plate 15   Space Shuttle view of the erupting Rabaul volcano, Papua New Guinea, on September 19, 1994

Plate 16   Voyager mosaic of Babbar Patera and environs, a low profile shield volcano on Io

Plate 17   Active flow-field of yellow sulphur with leveed margins at Kawah Mas, Gunung Papandayan, Java

Plate 18   The fractured disk of Jupiter's moon, Europa, acquired from a distance of 241,000 km by Voyager-2

Plate 19   Voyager mosaic of a part of Neptune's moon, Triton

List of tables

Chapter 2
2.1 Condensation temperatures of solar nebula materials at 10^{-3} atm 10
2.2 Elemental abundances in the solar photosphere and corona 13
2.3 Chemical compositions of carbonaceous chondrites 14

Chapter 3
3.1 Composition of mantle partial melts at differing temperatures and
    pressures 31
3.2 Major element chemical variation in an upward sequence through the
    differentiated Rhiw Igneous Complex 34
3.3 Distribution coefficients of selected trace elements in basaltic rocks 37

Chapter 4
4.1 Planetary magma production rates 59

Chapter 5
5.1 Yield strength and viscosity estimates for lava flows on the Earth, Moon
    and Mars 84
5.2 Estimated average effusion rates for terrestrial, lunar and martian flows 85

Chapter 7
7.1 Temporal and dimensional data for terrestrial continental flood basalt
    provinces 120
7.2 Chemical anayses of different basalt types 122
7.3 Chemical analyses for a variety of rift valley volcanic rocks 132
7.4 Composition of surface materials at Venera and Vega sites compared
    with terrestrial oceanic basalt 141
7.5 Size, distribution and area of volcanic features on Venus 149

Chapter 8
8.1 The lunar stratigraphic column 180
8.2 Major and selected trace element composition of terra anorthosite,
    KREEP and mare basalt 181
8.3 Classification of sampled basalt types, showing major chemical and/or
    mineralogical characteristics that distinguish each group 213

Chapter 9

9.1   K-Ar ages for seven Hawaiian shield volcanoes 249

Chapter 10

10.1 Number of craters greater than or equal to 1 km diameter per 10^6 km^2 for
individual martian volcanoes, compared with absolute ages derived from
chronologies of Neukum and Wise (1976) and Soderblom (1977) 264

1

Introduction

1.1 THE SUN'S FAMILY

Nine planets, their twenty seven major satellites, a host of smaller rock and ice bodies, meteorites and comets make up the Sun's family. Planet Earth may be unique in that it has both a hydrosphere and a breathable atmosphere, conditions which currently support life. The four *inner planets*: Mercury, Venus, Earth and Mars, form a group of dense, rocky, silicate worlds often named the terrestrial group. The next group of four – Jupiter, Saturn, Uranus and Neptune – constitute the *outer planets* (also called the Jovian worlds). These are of much larger size, lower density and consist predominantly of gas, or gas and ice. In fact, two pairs can be identified: Jupiter and Saturn are gas giants which have an essentially hydrogen-helium composition, while Uranus and Neptune are ice giants that are made from compounds of carbon, nitrogen and oxygen. The outermost planet, Pluto, is a maverick, being roughly the same size as the Moon and having a smaller companion, Charon, which is just over half its diameter. Comets, asteroids and comets complete the Sun's family.

1.2 PLANETARY CRUSTS

The outer skin of a planetary body is known as its crust. Crusts of the inner planets are composed of silicate rocks, as are those of asteroids. Stony meteorites – believed to have originated in the asteroid belt – represent samples of the most primitive silicate material we have to date. Outer planet moons have crusts that may be partly silicate and partly ice. Io is unique in having extensive surface and near-surface deposits of sulphur and its compounds, as well as silicate rocks.

All of the planets accreted from the Solar Nebula, an inhomogeneous mixture of gas, dust and ice. The heat of accretion, plus that generated by decay of short-lived radioactive nuclides, generated sufficient thermal energy to bring about extensive melting during the first 100 Ma of Solar System history. Core formation – known to have taken place early on in planetary evolution – also generated heat by supplying enormous amounts of gravitational potential energy. The response to all this heat production was that the Earth experienced a major geochemical upheaval. During this stage the core formed, the mantle was emplaced and underwent at least some partial melting, and massive quantities of volatiles were degassed. It was only when radiative cooling became sufficiently advanced that a solid outer skin was able to form. A similar pattern may have been followed on the other rocky planets.

The most primitive planetary crusts may well have been comparable to a crystallized scum on the top of a kind of magma "ocean". Because the Moon has not recycled its outer layer, we can observe what its early crust was like. The plagioclase-rich highland crust, known to be at least 4 Ga old, is enriched in calcium and aluminium and is readily explicable as having formed by flotation of plagioclase feldspar crystals in a vast magma ocean. A similar notion has been suggested for the Earth; however Earth has recycled its crust several times, therefore evidence for an early magma ocean is lacking and it is cited purely from argument. Whether similar magma oceans existed on the other inner worlds is not known. Regardless of the details of this early stage, it is clear that primitive silicate crusts were crystallized by cooling of magmas at or near planetary surfaces. This was a part of the long process of chemical fractionation.

The chilled outer skin of the Earth would have been thin early on. It would have been continually impacted by incoming planetesimals and injected from below by plumes of hot molten material rising above mantle plumes. Evidence from Archaean terrains suggests that primitive magmas were komatiitic, as may also have been the case on Mars. The development of Earth's unique continental crust is still contentious but appears to rely on the development of plate tectonics, itself attendant upon the cycling of water within and at the surface of the Earth, and upon vigorous convection within the mantle layer.

1.3   OF PLANETS AND PLATES

One of the most fundamental characteristics of the Earth's surface is the linear concentration of intense geological activity. In recent years the refining of the model of plate tectonics has allowed geologists to recognize that concentration of volcanic and seismic activity along global-scale oceanic rift/ridge systems is a response to upwelling of hot mantle material at divergent plate margins. Similar but chemically distinct volcanism, seismicity and subduction is also focussed along marginal oceanic locations where lithospheric plates are diverging, as around the Pacific, and where deep subduction trenches characterize the ocean floors. In intraplate settings active volcanism occurs above long-lived hot spots (mantle plumes), as at Hawaii.

The segmented nature of the Earth's lithosphere may be unique amongst the planets. Certainly there is no evidence for such a structure on either the Moon or Mercury. Mars is characterized by massive shield volcanoes and major lithospheric bulges that are believed related to long-lived mantle plumes and hot spots that have been active beneath an overlying crust that is immobile tangentially. However that is not to say that plate activity may not have occurred in the past. Recently (Sleep, 1994) the suggestion has been made that plate activity did once occur. Venus remains somewhat contentious in this respect. There are global-scale deformed belts and distinct zones of major volcanic structures to be sure, however the pattern and nature of these is not, by general agreement, analogous with terrestrial subduction-divergence activity. During the last 500 Ma, geological activity on this planet appears to have been dominated by widespread plume activity; however, it is not beyond the realms of possibility that earlier in its history some form of lateral plate movement may have occurred. It is a moot point as to whether or not Earth-type subductive movements could ever have taken place on Venus, bearing in mind the very high ambient temperatures and consequent buoyant nature of the Venusian lithosphere.

1.4   WHAT IS VOLCANISM?

Volcanism is a major surface manifestation of the outward transfer of energy from the interior of a planet. The extremely widespread distribution of volcanic structures and their attendant products on the surfaces of all of the inner planets, and Jupiter's innermost Galilean satellite, Io, implies that at some period in their evolution they had hot, largely molten, interiors. Thus there is direct evidence from both the Moon and the Earth that at an early stage they experienced at least one phase of general melting, the evidence for this primarily being provided by a diversity of crustal rock types which can be explained only by the slow process of chemical fractionation (differentiation). Unmanned missions to Mercury, Venus and Mars have all revealed considerable diversity of crustal materials here too, confirming that these planets also must have undergone some degree of differentiation. Implicit in this fact is an acceptance that magmatic materials were generated early during planet evolution. Volcanism is an inevitable consequence of planetary differentiation and crustal generation by magmatic processes.

The heat which accumulated inside the planets early in solar system history was predominantly generated by accretional energy and the decay of short-lived radioactive nuclides. The slow build-up of thermal energy eventually brought about melting. In the course of the same process, hydrous and volatile elements were mobilized and their gases moved outwards in the general direction of space. Elements such as sodium, potassium, argon and nitrogen would have been involved in the degassing. The controlling influences on this would have been the molecular weights of the elements and gases concerned, their condensation temperatures, chemical reactivity, the mass of each parent body, and the extent to which melting released and circulated the various chemical components. The inner planets have suffered quite different fates in this respect. The Moon and Mercury have virtually completed their degassing and now are built primarily from anhydrous refractory silicates. Mars is rather different in having retained some of its volatile constituents as an atmosphere, and indeed may still be actively degassing. Its carbon dioxide mantle has been of the utmost importance in the planet's development, even though it exerts a pressure only about 0.008 that at the Earth's surface. The terrestrial atmosphere, on the other hand, is quite unlike that of Mars, for here oxygen is an abundant constituent and is generally believed to have originated in the photosynthetic activity of plants, which have existed on our world for at least 10^9 years. Earth's atmospheric nitrogen and argon both appear to have come from the degassing of the interior, the latter by the radioactive decay of potassium ions within the outer crust. Venus, while almost identical in size to the Earth, has a dense, choking mantle of carbon dioxide. This planet's position, much closer to the Sun than Earth, apparently has played a significant role in producing the very high temperatures which prevail at the surface and which have forced most of the lighter elements to escape completely. The interplay of these processes during the formative stages of planetary evolution eventually manufactured the planets' solid crusts which were, therefore, formed essentially by volcanism, while the continued injection of magma into and outpouring of magma onto this thin outer skin, gradually thickened and strengthened it.

1.5   THE STUDY OF VOLCANOES

Volcanic landforms are varied and depend largely on the bulk chemistry of the related parent magmas, the nature of the fractures or feeders which tap them and convey them towards the surface, and the modifying effects of the presence or absence of a hydrosphere and atmosphere. Earth is geologically complex and consequently reflects this in a rich diversity of volcanic landforms. Mars has some of the largest constructional volcanic landforms, as well as the most widespread lava flows, so far discovered, many being orders of magnitude larger than their terrestrial counterparts. There is, however, less diversity than on Earth. Venus has widespread lava plains, massive flowfields, large shield volcanoes – some of which may still be active (Anderson, 1984) – steep-sided domes and unique volcano-tectonic landforms called coronae. Mercury is less well imaged but does have volcanic units somewhat similar to the Moon's mare-filling lavas. Jupiter's Galilean moon Io is the most volcanically active world of all and apparently is resurfacing itself at the amazingly rapid rate of 100 m per million years (Johnson *et al.*, 1980). Several outer planet moons show evidence for a style of volcanism involving nitrogen compounds and ices.

The geological study of volcanoes, their products and the hazards they pose is a specialist discipline but is based on principles that are well known and fundamental to geology. It would be customary for an Earth-based geologist, on entering new territory, to make a series of careful field observations, noting the geomorphological characteristics of the study area, the nature of the rock types at outcrop, and superposition-transection relations between individual rock units. As a basis for this work he or she would normally expect to have a topographic map, probably aerial photographs, the wherewithal to collect rock (and possibly gas) samples and, if funds allowed, to conduct selected geophysical experiments. Subsequently rock samples would be subjected to a variety of petrological and geochemical tests and the data synthesized into a coherent picture of the geological evolution of the region under scrutiny .

Such information is usually readily obtainable in the terrestrial environment, if with some physical hardship, but the situation is vastly different elsewhere in the solar system. While in some ways the study of volcanism on other planets is a simpler task, due largely to the apparent absence of plate tectonics and its complicating effects, other factors combine to make the planetary volcanologist's task less than easy, not the least of these being the factor of remoteness. Largely due to this, only the Moon has been blessed with a series of human visits and both topographical and geological mapping of other planets has been derived by a variety of remote-sensing methods which are both highly technical and extremely costly. Nevertheless in the space of less than a quarter of a century high-resolution images of all of the inner planets and the solid surfaces of some of the satellites of the

Jovian group have been obtained, and an array of successful seismic, geochemical, spectroscopic, meteorological and infrared thermal experiments have been conducted. In addition, lander probes have descended onto the Moon, Venus and Mars, sampled the surface materials and sent back a variety of scientific data which has been utilized in extending geological knowledge. Missions planned for the 1990's seek to put down landers onto the surface of Mars, study in detail one of its tiny satellites, Phobos, and obtain very high resolution images from Venus.

1.6 COMPARATIVE VOLCANOLOGY

Current understanding of terrestrial volcanology is derived from both field and experimental data and from theoretical modelling. Its objectives are not simply to map out the worldwide distribution of volcanoes or the different occurrences of the variety of volcanic materials they have generated, but also to comprehend which of a number of physical factors govern the shape of volcanic landforms, the styles and patterns of eruptions and the distribution of rock types in different sectors of individual volcanic complexes; also how series of volcanic rocks are genetically related. Furthermore volcanologists seek to establish whether volcanism has followed much the same pattern with time, or whether there have been long-term changes, and, if the latter, why such changes may have occurred.

Since planetary exploration became sophisticated enough to allow for a variety of imaging, non-imaging experiments and sample returns from the surfaces of distant bodies, it has been of burning interest to know what the similarities and differences are between our own planet and the rest. This is particularly so with respect to volcanic activity. On the Earth the array of volcanic processes is very great, ranging from quiet effusion of magma at huge mid-oceanic volcanoes to violently explosive ejections of magma and ash during continental Plinian outbreaks; from the noisy expulsion of internal gases on the deep ocean floor to the catastrophic slurrying of hot magmatic materials and mud in lahars; from the formation of diamond-bearing diatremes to the building of massive strato-volcanoes along those continental margins beneath which subduction of the Earth's oceanic crust is taking place. Nearly all of this activity is concentrated along boundaries between lithospheric plates.

In accepting the view that all of the planets and their satellites were accreted from the same solar nebula, and in the knowledge that volcanism is the surface expression of inexorable internal planetary processes common to all, the geologist is committed to a comparative approach. The Earth has served the scientific community well as the first volcanic case history, but volcanic rocks are evident also upon Mercury, Venus, the Moon, Mars and Io. Since each of these bodies evolved at a different position within the primordial nebula, not surprisingly each has a different mass, volume, internal structure, surface temperature and atmospheric make-up, and it would be strange indeed if each of these bodies had evolved in an identical way. That they have not is at once apparent when the geomorphology of their surfaces is observed, and no more so is this emphasized than when the products of volcanism are studied. Comparisons made between volcanic features on the inner planets and the Earth allow geologists to arrive at certain reasoned judgements concerning their origin, based on the wealth of terrestrial field, laboratory and theoretical data that has accrued. In this way it has become possible not only to learn more about them and what they represent, but also to view our own planet in a more enlightened way.

1.7   REMOTE-SENSING AND GEOLOGICAL EXPLORATION

Visual imagery has formed the backbone of solar system geological exploration. These images have been obtained by a variety of methods, including conventional photography, vidicon systems and radar. The first high-resolution images of the Moon were procured by Ranger 7 in the summer of 1964; between 1964 and 1971 more than 8000 images of Mars were sent back from a series of US Mariner spacecraft. Mercury experienced three fly-bys of Mariner 10 in 1973, and the first surface photographs of Venus were transmitted in June 1975 by the Soviet probe Venera 9. Much more recently, high-resolution radar imagery has been transmitted from the Soviet Venera 15 and 16 probes orbiting Venus, and from the highly successful U.S. Magellan probe. Data have also been obtained with the use of Earth-based radar facilities, such as those at Arecibo. Most of this data has been transmitted in digital form.

The high degree of manipulability of digitized imagery renders it amenable to enhancement, geometric correction and multispectral classification on sophisticated image-processing systems, a brief résumé of which may be found in Greeley (1985) and a more complete synthesis in Colwell (1983). Suffice it to say that the qualitative and quantitative analysis of photographic and TV hard-copy has provided a springboard for the more sophisticated experimental work, laboratory simulation studies and modelling procedures that subsequently have been completed and published. While remotely sensed photographic data is extremely important, it alone cannot solve all planetological problems. Volcanic features always repay first-hand study, ideally by a trained geologist, but much may also be learned at close range via unmanned devices.

Geological exploration of the planets really commenced with the Soviet lander Luna 9, which in January 1966 landed on the Moon's Oceanus Procellarum, unfolded its communications antennae and transmitted the first close-up pictures to Earth. Later that same year, a similar probe (Luna 13) used a mechanical arm to investigate the physical attributes of the lunar regolith, its onboard instruments recording a relatively low level of natural radioactivity, suggestive of the mare material being basaltic in composition. Four months after the first of the two Soviet probes landed, the American craft Surveyor 1 set down close by the lunar crater Flamsteed, sending back a plethora of close-up pictures and conducting a series of most valuable on-surface experiments over the next eight months. The six successful Apollo missions sent back not only photographs but also lunar samples, so that by the time Apollo 17 re-entered the Earth's atmosphere at the close of 1972, over 381 kg of lunar rocks became available for study. Added to this must be the small but significant sample return from the Soviet Luna programme (0.31 kg). Over 35 500 individual samples exist and theoretically are available to the scientific community. Absolute ages derived for some of these rocks have enabled geologists to put a timescale on lunar events, while laboratory geochemical and petrological analyses have provided a working knowledge of lunar rocks and their genesis.

Apollo not only took photographs and returned samples, the command-module pilots

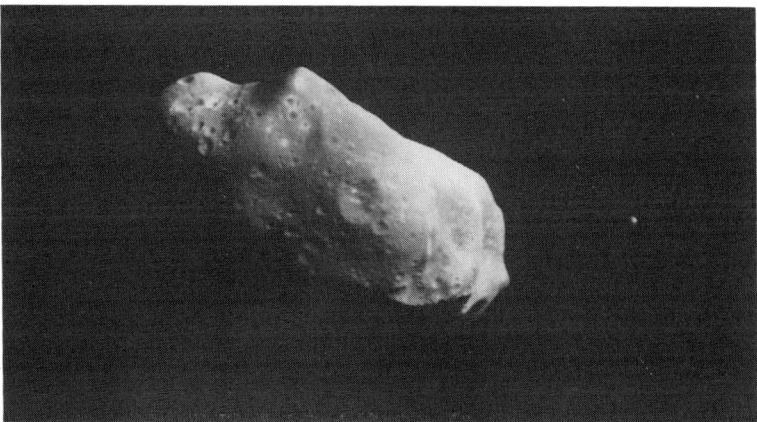

Fig. 1.1: Galileo image of the cratered asteroid Ida.

on the later missions were in control of an array of instruments which produced data from orbit, via techniques such as XRF-spectrometry, gamma-ray spectrometry, magnetometry, gravimetry and bi-static radar. Thus a wealth of supporting data is available to supplement the imagery and returned samples. Much of this information is of inestimable value to the planetary volcanologist, and a useful résumé of the methods used and some of the Apollo results may be found in Moore *et al.*, (1980).

Imaging systems were nevertheless the principal experiments onboard some early missions; thus our data set for the study of volcanic materials on Mercury is the TV photography of about half of its surface obtained by Mariner 10. There is little doubt that visual data provides the most spectacular evidence for planetary volcanism, but it needs supporting information. The two Mars Viking Orbiters, launched in 1975, carried not only high-resolution vidicons, but also hardware to collect data regarding atmospheric water vapour content (an infrared spectrometer), thermal and albedo characteristics (IR radiometers) and various radio and radar sensors aimed at studying celestial mechanics and atmospheric properties, and carrying out topographic profiling. The Viking Landers, beside carrying out their own program of surface imaging, conducted molecular analysis (gas chromatography), a biology experiment, inorganic analysis of the soil (XRF-spectrometer) and meteorological experiments, and recorded both the physical and magnetic properties of the soils at the two chosen landing sites in Chryse and Utopia. The scientific results of this highly successful project, plus the earlier data from Mariner 9, form the basis for volcanic studies on Mars. The data bank for Mars is nevertheless inferior to the lunar data set for two reasons: (i) neither lander set down a trained geologist, and (ii) there is no extensive returned sample collection.

In response to much more recent developments our geological perception of Venus is rapidly changing, largely as a result of the stream of new radar imaging data released as a result of the highly successful Magellan mission. This is being added to the information gleaned by earlier Venera probes and the US Pioneer Venus mission and is rapidly revealing both the impact, volcanic and tectonic history of that planet, and providing insights to its internal structure with high-resolution gravity data. With respect to the only other world which we know to exhibit volcanic materials, namely Io, the data is derived solely from the highly successful Voyager spacecraft, which passed by Jupiter during 1979.

Recently the Galileo spacecraft passed by asteroids Ida and Gaspra on its way to Jupiter, revealing them to be irregular bodies with heavily cratered surfaces (Fig. 1.1). During a 71-day-long orbit of the Moon, the tiny spacecraft Clementine systematically mapped the Moon's surface in eleven different colours in the visible and near infrared parts of the spectrum. In addition, it took tens of thousands of high resolution and mid-infrared thermal images. Currently plans are well advanced for a Russian/International mission to Mars in 1996. It remains to be seen whether or not this becomes a reality. The regrettable failure of the U.S. Mars Observer spacecraft on August 21 1993, has spurred NASA into considerable Mars-oriented activity: it is hoped to launch Mars Global Surveyor towards the planet during November 1996. This is a part of a long-term plan to launch low-mass spacecraft towards Mars every 26 months through the year 2005.

Plate 1: View of the volcanic shield of
Mauna Loa from Mauna Kea.
Photo: Patrick Moore.

Plate 2: The smoulding cone of the
dangerous stratovolcano, Gunung Merapi
(2914 m), located in Central Java. A
major eruption in A.D. 1006 partially
buried the temple of Borobudur in ash.
Merapi was active again during 1994.
Photo: Peter Cattermole.

Plate 3: Unwelded ignimbrite deposit,
Isla de Gomera, Canary Islands. The
ash-flow is cut by a dyke and overlain
by air-fall pyroclastics.
Photo: Peter Cattermole.

Plate 4: Huge block of intrusive feldspar-
porphyry transported by laharic flows
from Meru caldera and dumped about
8km from source.
Photo: Charles Downie.

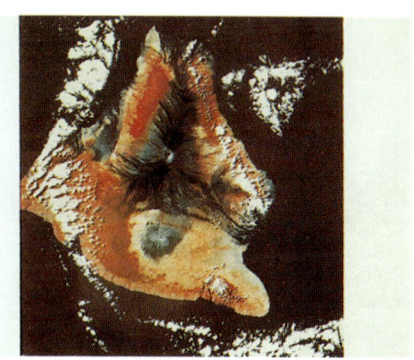

Plate 5: Havoc wreaked by mudflows during the paroxysmal eruption of Gunung Agung, Bali, during May 1963. Photograph taken in September 1963 by Robert Sissions. Courtesy of National Geographic.

Plate 6: False colour Landsat image of Big Island, Hawaii. Mauna Kea lies towards the top of the image, while the extensive lava flow–fields associated with the huge shield of Mauna Loa are the most prominent features of the region. Kilaeua is located towards the right of Mauna Loa's Northeast Rift Zone.

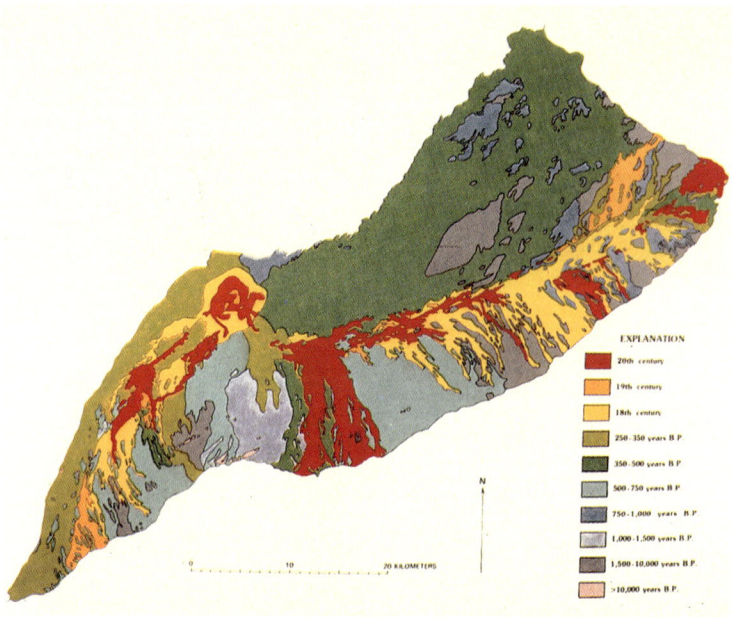

EXPLANATION

20th century
19th century
18th century
250-350 years B.P.
350-500 years B.P.
500-750 years B.P.
750-1,000 years B.P.
1,000-1,500 years B.P.
1,500-10,000 years B.P.
>10,000 years B.P.

Plate 7: Stratigraphic map of Kilaeua shield, Hawaii, showing pattern of flows since 10,000 years B.P. After Holcomb 1980.

Plate 8: Oblique colour view of the large shield volcano, Sapas Mons. Lava flows extend for hundreds of kilometres around the summit which rises 1.5 km above the adjacent plains. Another shield, Maat Mons. can be seen in the background. Magellan image P–40176.

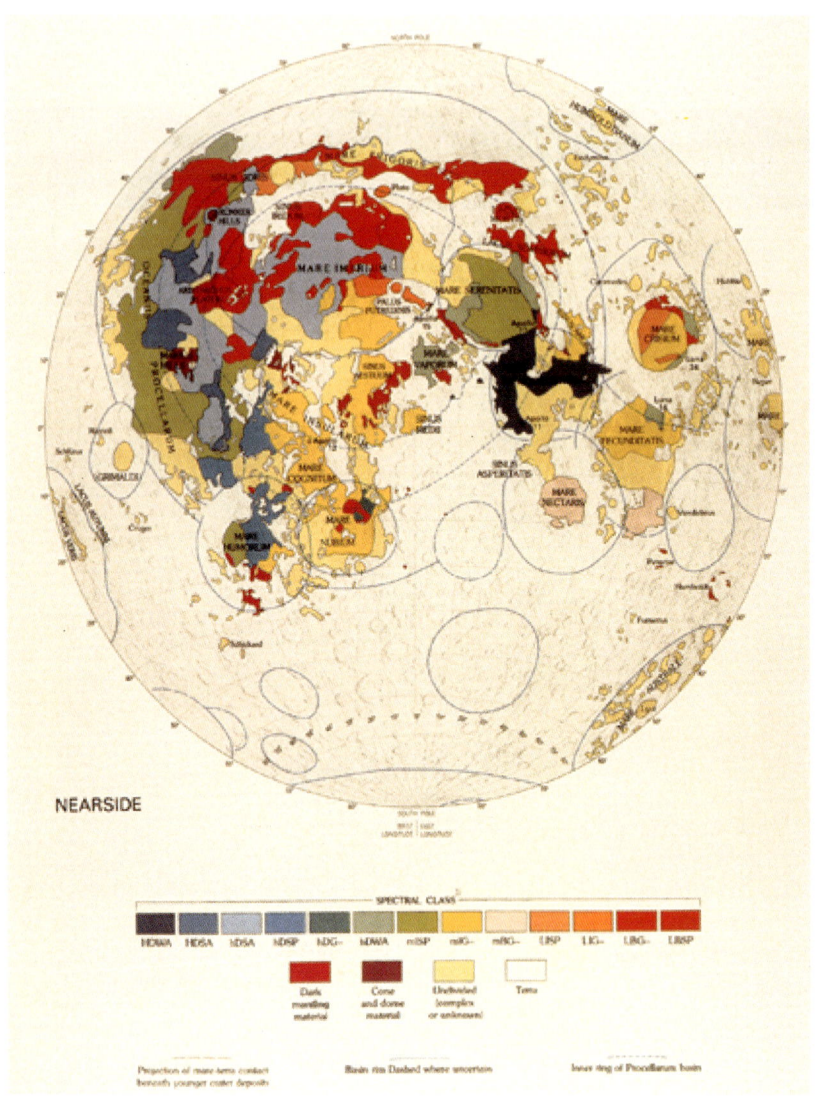

Plate 9: Spectral classes for mare basalts of the lunar nearside. From Wilhelms 1987, after Pieters 1978.

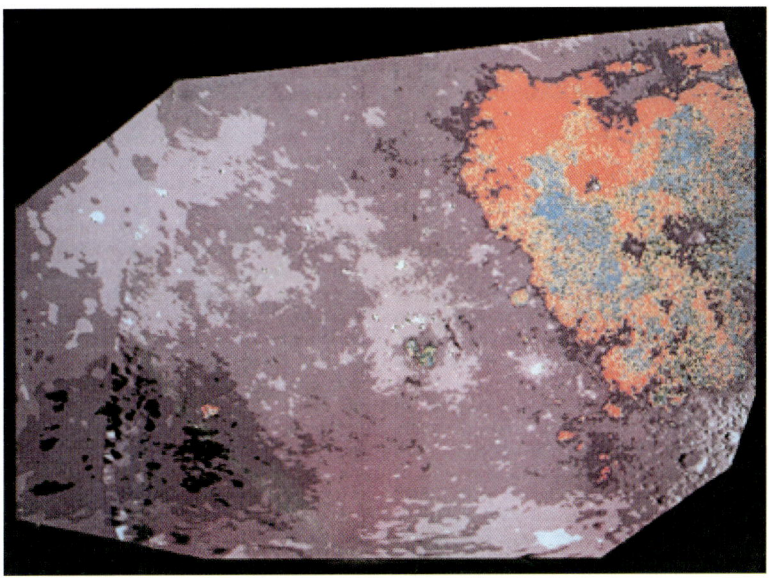

Plate 10: SSI colour composite mosaic used to distinguish the principal units in the lunar highlands and maria. Highland units are colour-coded by brightness from low to high albedo as black, dark purple, purple, pink and white. The unusually low albedo highland units at lower left outline much of the South Pole–Aitken basin. All mare materials fall in the lowest albedo slice, but are colour-coded on the basis of UV/VIS to represent increasing TiO_2 content (red represents low TiO_2, through yellow, green and blue, which represents high TiO_2 content. From Pieters et al, 1993.

Plate 11: Colour mosaic of the Tharsis region of Mars, including the western end of the equatorial canyon system. Courtesy of Alfred McEwen, USGS Flagstaff.

Plate 12: A Magellan altimetry and radar image composite of a part of Aino Planitia. Colours represent elevations of the surface: red and magenta being the highest and blue the lowest. From north to south, the volcanic shields of Ushas, Innini and Hathor Montes are depicted. Ushas rises slightly less than 2 km above the adjacent plains, Innini rises 2.8 km, and Hathor 2.6km. The chain of three volcanoes is thought to be the result of activity above a rising mantle plume. Magellan image MGN–116 P–42385.

Plate 13: The two shield volcanoes Sif and Gula Mons are shown on this image. The superimposed colours represent altimetric data. Both volcanoes are believed to consist of thick accumulations of lava. The Sif flow–fields are especially bright in the radar image because of the roughness of the material exposed. A rift runs from the southeastern flank of Gul Mons. Magellan image P–38346AC.

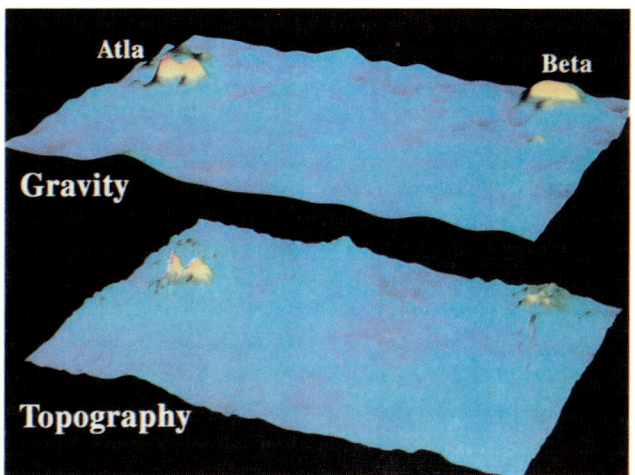

Plate 14: Gravity and topography for the region between Atla and Beta Regions. The high gravity and topography values are typical of sites of rifting and volcanism on Venus. Positive (light blue) and negative (dark blue) anomalies correspond almost perfectly with topography. Magellan image MGN–113, P–42356AC.

Plate 15: Space Shuttle view of erupting Rabaul Volcano, Papua New Guinea, on September 19, 1994. Taken near the peak of activity from the 6–km–wide crater, the 18,000–metre–high plume is being carried westwards by the prevailing winds. At lower levels, where wind speeds are low, a symmetrical blanket of yellowish ash was deposited over a region 20 km across. Over 50,000 people were forced to evacuate their homes as a result of this eruption. Photo courtesy of NASA.

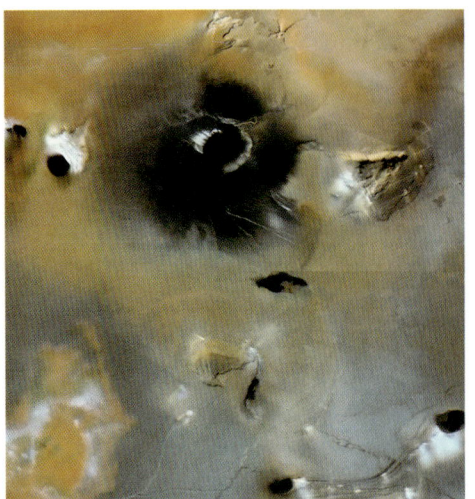

Plate 16: Voyager mosaic of Babbar Patera and environs, a low profile shield volcano on Jupiter's Galilean moon, Io. The volcano is surrounded by volcanic plains generated by fallout from geyser–like plumes. Image courtesy of Alfred McEwen, USGS Flagstaff.

Plate 17: Active flow of yellow sulphur with leveed margins at Kawah Mas, Gunung Papandayán, Western Java. Molten brown sulphur at fumarolic vents gives way to orange sulphur and then yellow towards the distal margin. Photo: Peter Cattermole.

Plate 18: The fractured disk of Jupiter's moon, Europa, acquired from a distance of 241,000 km by Voyager 2. The complex pattern of fractures is well seen as is the paucity of impact craters. Image P–21764C, courtesy of NASA.

Plate 19: Voyager mosaic of a part of Neptune's moon, Triton. The large polar "cap" of nitrogen ice is in the lower half of the image, is highly reflective and pinkish in hue. Dark plumes from suspected volcanic vents can be seen in this region. The region to the north, redder in colour, is of canteloupe terrain which is traversed by ridge/ trough features. Hummocky terrain separates the two regions. JPL mosaic P–34764, courtesy of NASA.

2

The making of planets

2.1 THE SOLAR NEBULA

Astronomers believe they understand quite well how stars are born and evolve, but are far less sure about the formation of planetary systems. In large part this is because they have but one that they can study, our own, whereas astronomers have millions of stars upon which to base their research. Not surprisingly, therefore, no single picture of planet formation is accepted by all planetary scientists (see, for instance: Alfvén and Arrhenius 1976; Clark *et al.* 1972; Elsasser 1963; Hartmann 1978; Hoyle 1946; Ringwood 1979; Safronov 1972). This is not to say that the Solar System is unique, for the probability is high that other such systems are attached to Sun-like stars. Whether or not Earth as an abode of life is unique, is a totally different question which remains unanswered and is likely to for many years to come.

As one eminent planetary scientist has noted, (Ross Taylor, 1992) there are two obvious truths: (1) the Solar System is isolated, the distance of the nearest stars exceeding the diameter of the system by more than 50 000 times, and (2) the whole system effectively lies in one plane, with most of the members orbiting the Sun and rotating in the same direction. Together these provide evidence for a common origin for all bodies within the system: such a configuration is hardly likely to arise from a random accumulation of such a diverse family of objects. The notion of a common origin has been the general consensus since the time of Laplace, some 200 years ago.

Most planetary scientists today entertain a modified version of Laplace's original solar nebula idea, preferring either that the Sun formed first from one cloud of matter, while the planets condensed from a second cloud which was captured by the Sun, or that Sun and planets had the same source, a vast rotating nebula (Cameron, 1978; Prentice, 1978; Reeves, 1978; Ross Taylor, 1992). The modern consensus is that the Sun and planets were formed from a single nebula which was injected by material from a nearby exploding supernova. The nebula would have been but a fragment from a huge molecular cloud, its mass and angular momentum having determined much of its subsequent evolution. As the fragment collapsed into a disk, dust settled towards the mid-plane. The Sun evolved by gravitational contraction and the central condensation of the cloud, while the planets and other objects

Table 2.1: Condensation temperatures of solar nebula materials at 10^{-3} atm. (After Grossman and Larimer, 1974.)

Mineral phase	Composition	Temperature (°C)
Corundum	Al_2O_3	1410
Melilite	$Ca_2Al_2SiO_7$-$Ca_2MgSi_2O_7$	1205
Perovskite	$CaTiO_3$	1200
Spinel	$MgAl_2O_4$	1150
Metallic iron	Fe (Ni)	1130
Forsterite (olivine)	Mg_2SiO_4	1120
Diopside (pyroxene)	$CaMgSi_2O_6$	
Enstatite (pyroxene)	$MgSiO_3$	1100
Anorthite (plagioclase)	$CaAlSi_2O_8$	
Alkali feldspar	$(Na,K)AlSi_3O_8$	980
Troilite	FeS	430
Magnetite	Fe_3O_4	135
Ice, methane, etc.	H_2O, CH_4, CO_2, H_2, etc.	<0

were created from a cocoon nebula that enveloped the proto-Sun. The mass of the nebula is believed to have been about 0.15 solar masses, with only 0.03 solar masses being retained by the planets and the rest blown away by the solar wind (Hartmann, 1983).

Star formation is quite well understood, largely because stars can be studied at most stages in their evolution. They differ from planets, however, in that they condense from fragments of giant molecular clouds and therefore exhibit considerable uniformity. Planets, on the other hand, are accreted, piece-by-piece, from solid fragments in circumstellar disks. Should the planetesimals growing within these reach sufficient size before the gaseous components within the nebula are dissipated by T-Tauri type activity, the gas may collapse by gravitational attraction, giving rise to Jovian-type gas giants. On the other hand, if the gas is blown away before this can occur, rocky, gas-impoverished planets would result, giving rise to a very different type of system. Each potential planetary system could have followed a very different evolutionary path.

Once the nebula started to clear due to T-Tauri activity, there was depletion of H and He in the inner nebula. Once the disk had formed, collisions between finely dispersed material tended to cancel out noncircular motions until a system of parallel circular coplanar orbits was established and hydrostatic equilibrium would have been attained. Its rate of contraction would diminish and it would have cooled predominantly by infrared radiation, from temperatures >2000 K. As cooling took place and condensation temperatures were reached, microscopic dust grains would have begun to form. These would have been in equilibrium with the enveloping gas and other minerals at a particular temperature and pressure. The earliest minerals to form would have been refractory compounds, e.g. those containing tungsten, osmium, zirconium, and oxides of calcium, titanium and aluminium. Between 1625 and 1125 K, materials would have condensed such as are now found in carbonaceous chondrites. As the temperature continued to fall, iron and nickel condensed, forming an alloy; Mg-silicates condensed as dust particles at temperatures of between

1400 and 1300 K. These were followed by sodium and potassium silicates (feldspars). At still lower temperatures, the grains would have reacted with the gas, and by the time the temperature had fallen to around 500 K, the dust would have consisted of olivine, feldspar and oxidized Fe minerals. This assemblage is represented in 4.56 Ga-old meteorites. With even lower temperatures, carbon (in the form of graphite), carbon-rich silicate minerals, and complex organic molecules would also have appeared. From 500 to 200 K water vapour in the nebula would have reacted with some of the dust grains to form hydrated minerals (e.g. serpentine, tremolite and talc). Where temperatures hovered around 200 K in the cloud, many more hydrated minerals condensed, together with ice crystals. In the outer regions of the nebula, where temperatures were even lower, ammonia and methane formed ices, some of which mixed with water to form hydrated ices. As the grains aggregated in bodies having metre dimensions and the nebular cloud dispersed, sunlight would have been able to sublimate any unshielded ice in the inner regions, and water-ice became restricted to the outer reaches of the nebula, beyond the present asteroid belt.

On this basis, the gross composition of each planet can be explained as though, at some stage of its cooling history, nebular gas was dispersed by the solar wind and each planet retained residual condensates at temperatures corresponding to its heliocentric distance (Table 2.1). Mercury lies in the refractory-rich zone, Venus and Earth in the lower density silica-rich zone, Mars in a still lower density oxidized-iron zone, while the outer worlds lay in a zone rich in ices. While such a notion is a useful principle, research suggests things may not have been quite that simple.

Additionally, the way in which the dust was redistributed may well have varied from region to region: far from the Sun's gravitational pull the nebular disk may have split off rings of matter from its outer edge; these may have subdivided into several large gaseous protoplanets (Jovian worlds). Closer to the Sun, where the dust would have settled more rapidly into the mid-plane, asteroid-sized bodies would accrete; in a few million years these would have coalesced into moon-sized planetesimals. Later again, perhaps within 100 m.y., perturbations by Jupiter's gravity and multiple collisions would induce them to accrete into the terrestrial planets. Since the material would be dominated by local dust – which would attain chemical equilibrium with the surrounding medium – it would be very sensitive to temperature. Thus the observed compositions must reflect the maximum temperature experienced at each locality within the nebula.

Undoubtedly this is a gross oversimplification. For a start, tiny grains would continue to react with the enveloping gas as it cooled, long after the formation of the larger grains. This would have the effect of impregnating the smaller grains with the more volatile constituents (e.g. water, carbon compounds and halogens). That this occurred is evidenced by the relatively high proportion of volatiles in the matrix of carbonaceous meteorites. Then again, turbulence within the nebula could transport smaller grains far from the central plane. Alternatively it could mix compositionally distinct grains formed at quite different distances from the Sun. Then again, should equilibrium conditions not be attained (which is a very strong likelihood) chemical distribution would be different again. It is an extremely complex topic!

The compositions of the outer planets are a function of their time of formation, of the clearing of the nebular gases and volatiles during early T-Tauri activity, and the more rapid accretion of the two gas giants. It is envisioned that Uranus and Neptune both grew in a nebula that already had lost much of its H and He. That factors other than heliocentric distance controlled their chemistry is supported by the observed marked zonation within the Asteroid Belt, a development which took place at a late stage in the nebula's evolution. The inner planets formed from material already depleted in elements volatile at around 1100 K which is why, for example, the noble gases (Ne, Ar, Kr and Xe) are severely depleted in the Earth.

Many meteorite constituents were formed in a high O/H environment so that O was enriched with respect to H by at least two orders of magnitude. Since these formed $<10^6$ years after the birth of the cloud, the H and He must have been lost very early on. The present solar wind has only 10 atoms/cm^3 at the location of the Earth's orbit – compared to 10^{15} atoms/cm^3 in the original nebula. This gives a measure of how effectively the gaseous envelope has been removed.

2.2   PRIMORDIAL MATERIALS

There is a natural curiosity regarding the nature of the primitive stuff from which the Sun and planets were formed. A number of candidates exist. These include the Sun (which predated all of the planets), Jupiter (which predated the terrestrial planets), comets and certain kinds of chondritic meteorites.

The Sun is composed predominantly of H and He, a fact not realized until as recently as 1925 (Payne, 1925). The composition of the solar photosphere is given in Table 2.2. These data remain the source of information for the elements H, C, N and O in the primitive solar nebula, since they are depleted in meteorites. Fe has posed problems and still does; however, recent research (Holweger, 1988) involving analysis of Fe II lines, confirms equivalence between meteoritic and solar photospheric values. That there has been chemical fractionation within the Sun is evidenced by comparing the above data with those for the solar corona (Table 2.2), the main difference being that elements with high ionization potentials are depleted in the corona.

The analysis of planetary crustal rocks and of spectroscopic data tells us little about the bulk composition of the planets since they are inhomogeneous due to chemical fractionation. If we are to learn anything about the primitive stuff from which the solar nebula was made, finding very ancient material is clearly a necessity, since only this will have the slightest likelihood of having escaped chemical fractionation or alteration. We know that Jupiter formed early, could not it be representative of the pristine stuff? The answer is that while it formed early on it contains far more heavy elements than the Sun and does not exhibit solar ratios of H/He. As to comets, recent analyses, including those from Comet Halley, indicate that these ancient bodies show wide differences compared to the solar photosphere, particularly with respect to Si and Fe. On this basis we have to reject both candidates.

Meteorites are samples of ancient material and, of the various groups, the CI chondrites appear to be the most primitive. Thus a common exercise is to compare the composition of the CI chondrites with solar abundances, many workers considering the former to be representative of nebular composition at an early stage. The reasons for this are twofold: (i) the odd-mass nuclei fall on a smooth curve of abundances versus mass ratio, and so retain a memory of nucleosynthesis; and (ii) there is quite a good match for non-gaseous elements

Table 2.2: Elemental abundances in the solar photosphere and corona ($\log N_H = 12.00$). Data from Anders, E. and Grevesse, N. (1989) and Holweger, H. (1990).

Photosphere		Corona	
H	12.00	H	
He	10.99	He	10.14
Li	1.16	C	7.90
Be	1.15	N	7.40
B	2.6	O	8.30
C	8.56	F	4.00
N	8.05	Ne	7.46
O	8.93	Na	6.38
F	4.56	Mg	7.59
Ne	8.09	Al	6.47
Na	6.33	Si	7.55
Mg	7.58	P	5.24
Al	6.47	S	6.93
Si	7.55	Cl	4.93
P	5.45	Ar	5.89
S	7.21	K	5.14
Cl	5.5	Ca	6.46
Ar	6.56	Sc	4.04
K	5.12	Ti	5.24
Ca	6.36	V	4.23
Sc	3.10	Cr	5.81
Ti	4.99	Mn	5.38
V	4.00	Fe	7.65
Cr	5.67	Ni	6.22
Mn	5.39	Cu	4.31
Fe	7.51	Zn	4.76
Co	4.92		
Ni	6.25		
Cu	4.21		
Zn	6.25		

between CIs and the solar photosphere. Many anomalies exist, however, and there is clear evidence that these meteorites are not pristine, for instance, they have been subjected to aqueous alteration with the consequent formation of hydrated minerals. Also, compared to other chondrite groups, CIs are enriched in volatiles, siderophile and chalcophile elements, and there is some variation between different CI samples too. Despite this, study of their chemistry has considerable value in the quest for primordial matter, their elemental abundances most closely resembling the solar photosphere data for the non-gaseous elements (Table 2.2). Meteoritic data are therefore taken as the best estimate of solar nebula composition that we have to date.

Table 2.3: Chemical compositions of carbonaceous chondrites. (After Henderson 1982.)

	Carbonaceous chondrite type I (Orgueil)	Carbonaceous chondrite type II (Murray)
FeS	15.07	7.67
SiO_2	22.56	28.69
TiO_2	0.07	0.09
Al_2O_3	1.65	2.19
MnO	0.19	0.21
FeO	11.39	21.08
MgO	15.81	19.77
CaO	1.22	1.92
Na_2O	0.74	0.22
K_2O	0.07	0.04
P_2O_5	0.28	0.32
H_2O	19.89	12.42
Cr_2O_3	0.36	0.44
NiO	1.23	1.50
CoO	0.06	0.08
C	3.10	2.78
Others	6.96	0.62
Total	100.65	100.04

The CI chondrites have an unusual chemistry which is unique among small bodies in that it closely resembles that of the Sun (Table 2.3). They are distinct from other meteorites, not least because they alone contain curious spherical particles (chondrules) which are dispersed within a fine-grained matrix. The chondrules themselves are composed of olivine and pyroxene, together with minor amounts of Fe/Ni alloy and troilite (FeS). The co-existent Fe-bearing silicates and Fe metallic phases imply an oxygen-depleted environment for that part of the solar nebula in which they formed. Their chemistry contrasts strongly with the terrestrial environment, where in both the crust and mantle, all iron is combined with oxygen and no independent metallic phase is found. The chondrites also hold abundant inclusions of refractory Ca-Al phases, i.e. perovskite, melilite, spinel and garnet, which indicate elevated condensation temperatures. The modern view is that the chondrules formed above the midplane of the nebula, where nebular flares caused localized intense heating of dust. In contrast, the matrix is characterized by low-temperature condensates such as the hydrated silicate, serpentine and clay minerals, together with about 5% volatile organic compounds. The presence of both high-and low-temperature phases discloses that prior to accretion the primordial nebula consisted of a mixture of refractory and non-refractory phases.

2.3   PLANETARY ACCRETION

The first problem attached to understanding any accretional process is to find an acceptable mechanism whereby small cosmic particles came together. The accretion of freely moving particulate materials can occur only where some force acts upon them which exceeds that of the rebound after particle-particle collision. Both magnetic and electrical dipoles could provide such a force, and while the former is restricted to magnetic minerals, the latter affects all materials and it may have been the most significant in causing adhesion and clustering of small particles. Safronov (1972) has shown that in a swarm of co-orbiting dust-sized particles that have just condensed from the solar nebula, travel velocities would be of the order of only 50 m s^{-1} for kilometre-sized fragments. Regolith-free targets would experience mass-gain only within a narrow range of velocities where the impact velocity was just greater than the escape velocity, for below this critical level rebound would occur, while above it shattering and crater formation would take place. If even a thin regolith were present this would give the body a granular surface texture, resulting in increased cohesion and as a result the velocity band for mass-gain would widen considerably. Ip (1974) and Alfvén and Arrhenius (1976) have suggested that mutual collisions between grains have a tendency to make the orbits of colliding grains similar, which gives to any collection of grains a kind of viscosity which may entrain them into a "jet stream". If grains can be focussed in this way, many of the problems attendant upon accretion may be solved. The presence of gas in the nebula would also have played a very significant part because it would tend to decelerate dust-sized grains to a level more conducive to accretion. Additionally it would have effectively damped random chaotic motions within the cloud, coercing the tiny particles to rotate in near-coplanar, circular orbits with small relative velocities (Cameron, 1973; Hartmann, 1972; Safronov, 1972; Urey, 1952). The time taken for the particles to assemble in the equatorial plane could have been as short as 1000 years (Huang, 1973).

Regardless of the precise manner in which the initial accretion of planetesimals took place, it seems clear that a number of distinct phases would be passed through as the cloud material rotated about the proto-Sun. The initial stage would have been the actual condensation of silicate and ice grains from the cloud, which would have been followed by aggregation into larger particles, either by gravitational growth or by particle-particle collisions. Eventually a stage would be reached when the radii of the larger bodies had increased to between 2 and 5 km, by which time impact velocities would have risen to between 2 and 50 m s^{-1}. Such bodies probably would have developed a regolith which effectively inhibited rebound of infalling particles and promoted growth by collision. In contrast, smaller bodies would tend to shatter and fragment as these greater infall velocities were attained. As a consequence the larger objects in part would grow at the expense of the smaller ones, and because they were being struck at velocities only slightly above their own escape velocities, the larger bodies would maximize on their own accretion by minimizing on mass-loss and in growth terms would "run away" from the others. In so doing they would sweep up much of the more finely comminuted debris generated by shattering of the smaller fragments and would eventually attain planetary dimensions.

The rate at which accretion occurred can be constrained by exotic isotopic anomalies which arise in certain types of chondritic meteorite. These are due to the decay products of short-lived radionuclides such as ^{26}Al (half-life 10^6 years), ^{29}I (half-life 16×10^6 years) and ^{244}Pu (half-life 82×10^6 years) which are believed to have been generated in a nearby supernova outburst which took place during the formative years of the solar system. The parent radionuclides cannot have been indigenous to the solar cloud since if they had been, they should be uniformly distributed among all meteorites, which they are not. Evidently they were expelled by the supernova explosion at high velocities and captured by the solar nebula. After capture they were locked into some chondrites, a process which, because of the very short half-lives of the isotopes concerned, can only have occupied a few millions of years. This constrains the accretion time of the parent bodies of the chondritic meteorites.

2.4   PLANETARY HEATING

One consequence of the accretionary process is that, each time there is an impact, the kinetic energy of the impacting grain is almost entirely converted into heat. Since the thermal energy per unit surface is dM_{em}/dt, where M_{em} is the mass of the embryonic body and t is the temperature, the average formation temperature of any celestial body is proportional to M_{sc}/t_{inf}, where M_{sc} is the mass of the final secondary body and t_{inf} is the infall time (Alfvén and Arrhenius, 1976). Thus the proto-planets gained thermal energy during accretion as infalling planetesimals imparted their kinetic energy. Assuming that the infall time was similar for all bodies, the formation temperature of the proto-planets must have been proportional to their present masses. Ip (1974) effected numerical integrations of the equations for thermal power per unit surface area delivered by an impacting mass, and on the basis of his data some general conclusions can be drawn concerning the terrestrial planets. Firstly, Mercury should have a temperature maximum closer to its centre than the Earth but, because of its small size, that temperature should be lower. Secondly, Venus's heat structure should be very similar to Earth's but with the melted region closer to the centre. Thirdly, the heating capacity of Mars should be at least one order of magnitude lower than Earth's and the temperature maximum rather close to the surface, such that if a melted region ever existed it must have been located at about 0.9 of the planetary radius.

The impact record fossilized on the solid surfaces of all of the planets and their satellites indicates that during the first 500 million years of solar system history, all were bombarded by meteoroids, some of very large dimensions. These would have contributed substantially to their energy budgets, and it has been calculated that they may have raised the surface temperature of the Earth as high as 10 000°C; however, it is unclear exactly how much of this heat was retained and how much was lost to Space. Certainly if the majority of infalling particles were of dust size, accretion would necessarily have had to be completed within about 10 000 years for core-forming temperatures to be achieved (i.e. 2000-4000°C). On the other hand, for bodies of larger dimensions, say 0.1% Earth mass, their kinetic energy would have penetrated deeper, been more effectively implanted, and consequently accretion might have taken several million years.

Since there is strong geochemical evidence that Earth's core grew both during and after accretion, apparently from a relatively homogeneous, chondritic parent material, the

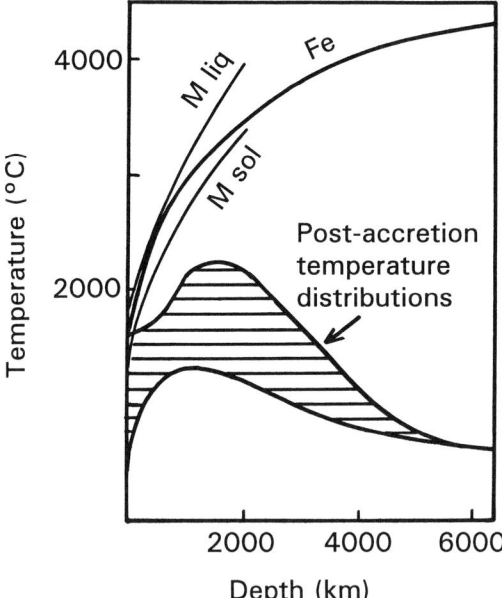

Fig. 2.1: Melting temperature curves for iron (Fe) and melting interval of mantle solidus and liquidus (Higgins and Kennedy, 1971; Kennedy and Higgins, 1973) together with estimated post-accretional temperature distribution within the Earth prior to core formation (Safronov, 1972; Ringwood, 1975).

energy required for core generation must have been supplied predominantly by impact energy. Although accretional energy was a most important source of planetary heat, much of this would have been lost to space, especially if it was implanted within a relatively thin surface layer. Radiative and conductive processes – the two which would most readily reduce the heat budget of a proto-planet – are both a function of surface area, so in terms of heat losses, the smaller planets would have suffered more than their larger companions because of their greater surface-area/volume ratio. Deeper sources of heat would, however, be more effectively retained and should only be lost to the surface by convection.

Other sources of heat existed early in planetary development. Tidal interactions may have generated energy, but little is known of their effects. More is known about two other heating mechanisms, namely adiabatic heating and the decay of long-lived radionuclides. With respect to the former, the hydrostatic pressure exerted by an overlying mass of accreted materials causes the interior temperature of a planet to rise. In the Earth's case, immediately after accretion adiabatic heating had the potential to raise the core temperature by as much as 900°C. Had there been no other source of energy the core temperature must have fallen almost at once, due to conduction. It may be shown that the terrestrial core could not have been generated simply via adiabatic heating, since the projected increase in melting point with depth for potential core-forming materials is either greater or at least as great as the adiabatic gradient. Estimates of the interior temperature of the Earth at a time immediately after accretion but before core formation show that interior temperatures may have been too low to allow the melting of Mg-rich silicates; nevertheless, metallic iron particles in chondritic parent material may have melted because the melting temperature of the pure metal is somewhat less than that of the silicate (Fig. 2.1).

If this was the case, then droplets of metallic iron may have segregated and begun their descent toward the region of the core. Ringwood (1979) argues that core segregation cannot have taken longer than a few hundred million years to accomplish, since the present overall Pb/U ratio of the upper mantle/crust system must have been established at around 4.55×10^9 years – the "age" of the Earth derived from these two elements. The fractionation of lead from uranium could only have been effected at this time by the segregation of a metallic iron/nickel core, which would carry down with it lead but not uranium (Ringwood 1960). Measurements of the partition coefficients for lead between iron alloys and basaltic melt subsequently allowed Oversby and Ringwood (1971) to confirm that the time interval separating accretion from core formation was probably less than 10^8 years.

Most thermal models agree that a proto-Earth of approximately chondritic composition with a major part of its radioactivity reservoir locked inside it will heat up gradually with time. Thermal energy would be generated by the slow decay of long-lived radionuclides such as ^{238}U, ^{235}U, ^{232}Th and ^{40}K. The build up of radioactive heat would have arisen largely by virtue of the insulating properties of the proto-lithosphere. The segregation of a core is the direct result of such a process and it has been estimated that the thermal energy released as the metal phase sank and lost gravitational potential energy, was sufficient to have heated the whole Earth to over 2000°C (Birch, 1965; Tozer, 1965). At this point the solid mantle may have begun to convect in response to the rising interior temperature, which would have given a more positive slope to the geothermal gradient. The consequence of the rising geotherm was that it may well have eventually intersected the silicate solvus, whereupon partial melting of at least the outer few hundred kilometres occurred. The settling rate of core-forming elements would have been enhanced once the internal temperature had come close to the melting point for the silicates, since at that point the silicate matrix would have become softened sufficiently to permit droplets of a dense metallic phase to sink downwards through it. The precipitation of metallic droplets would certainly have preceded the general melting of the silicate phase and there seems little doubt that core formation was a fairly rapid process compared to the generation of the silicate mantle. This would, of course, apply to all the inner planet group. Core infall must have provided a major source of heat during the early stages of terrestrial planet evolution.

The presence of sizeable amounts of FeO and FeS in the Earth's core is now generally accepted and is indicated by the density of the outer core which is significantly lower than an Fe/Ni alloy under the appropriate temperature/pressure conditions. This would have played a significant role in depressing the melting point of the metal phase but not the silicate, which would permit core formation prior to any general melting of the mantle. The bulk oxygen content would have played a major role in determining the size of siderophile cores within the terrestrial planets. Thus if the Earth had been of carbonaceous chondrite initial composition, and no volatile loss had occurred, then it should be almost entirely lithophile with just a small chalcophile core and certainly no free metal. Because

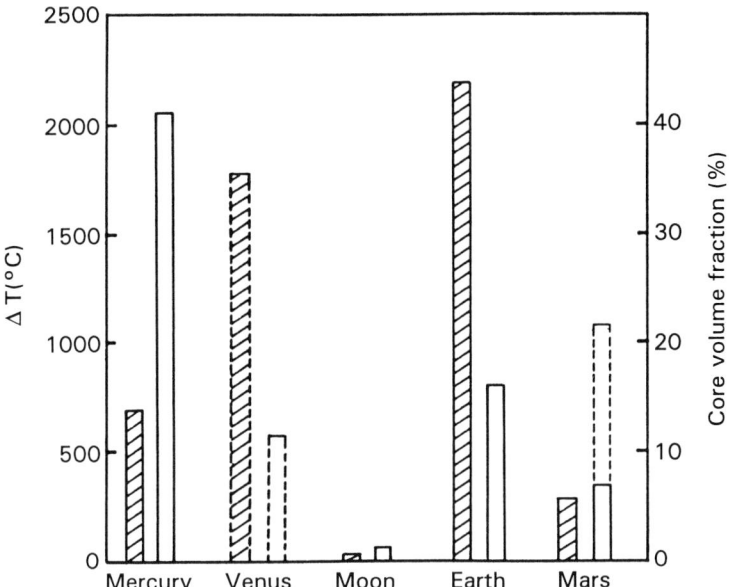

Fig. 2.2: Mean temperature rise due to core segregation for the terrestrial planets (hatched bars) (Solomon, 1979) compared with planetary volume fraction occupied by dense metallic cores (open bars) (Solomon, 1980).

the Earth actually has a 3485 km radius metallic core which accounts for 32% of the planetary mass (16.4% of the volume), it must be assumed that free iron remained after all of the available O_2 and S had combined with other lithophile and chalcophile elements, and therefore that the Earth must have lost a few per cent of its original oxygen quota.

The only other planet for which there is significant geochemical data is the Moon, which is less compressed than the Earth due to its smaller size. If they are viewed as having accreted in the same region of the primordial nebula, together the Earth and Moon demonstrate a bulk chemistry equivalent to chondritic meteorites, but with depletion in all volatiles, including the alkali elements. Magnetic data suggests that the Moon probably did form a core early on, probably about 4×10^9 years ago (Runcorn, 1980). If it does have a core its Fe/Si ratio must be lower than the Earth's and it is unlikely to have a radius greater than 400 km which means it accounts for less than 1% of the planet's volume (Solomon, 1980). The generation of such a small core would have incremented the lunar heat budget by only about 10°C (Solomon, 1979); thus the heat necessary to bring about such melting as was required to produce the early differentiation of its anorthositic crust must have been provided by accretional energy (Johnston and Toksöz, 1977).

Photogeological evidence reveals that small outpourings of basalt-like extrusions occurred over the highland regions about 4.2×10^9 years ago from which it can be deduced that sufficient heat had been generated within this rather small, "dry" planet during the first 0.3×10^9 years of lunar history for some volcanism to take place. The subsequent interior heating which eventually produced a melted source region for the basin-filling mare basalts between 3.9 and 3.1×10^9 years ago, must have been of later radiogenic origin (Solomon and Chaiken, 1976). Apollo data has established that the source region for the mare basalts, which, it should be remembered, account for only 0.2% of the total crustal volume, was enriched in refractory lithophile elements at the expense of volatile and siderophile elements compared with the terrestrial mantle, suggesting it concentrated more refractory condensates than Earth. Since geological evidence indicates that there has been little if any overall expansion or contraction of the lunar volume since the primordial crust was formed, the heat balance inside the Moon must have been maintained by the outer regions cooling while the central parts heated up. This implies that our satellite developed a gradually thickening lithosphere which, as is known from Apollo data, is at least 70 km thick. Even though the differentiated mare basalts are of relatively minor importance in terms of their volume, they do illustrate that partial melting is a widespread planetary process, while the degree of chemical variation they exhibit also reveals some heterogeneity in the lunar mantle.

Mars also is less compressed than the Earth and therefore of lower density. Its moment of inertia indicates that there is a core but this is either smaller or significantly less dense than the Earth's. Johnston and Toksöz (1977) and Solomon (1979) calculate the Martian core to have a radius of between 1400 and 2000 km, accounting for between 7 and 21% of the total volume. Geophysical and geochemical constraints indicate that the Martian mantle must be relatively enriched in FeO compared with the Earth, perhaps by as much as a factor of three (McGetchin et al., 1981). The long history of volcanism and extensional stress which this planet has experienced suggests that both core infall and mantle segregation may have occurred somewhat later on Mars than on the other terrestrial planets (Schubert, 1979; Solomon and Chaiken, 1976). Furthermore, the strongly developed extensional faulting implies that significant cooling cannot have affected Mars during the last $1\text{-}2 \times 10^9$ years of its history.

Since Venus does not have a natural satellite, its moment of inertia cannot be determined. It is, however, only slightly smaller and less dense than Earth so it seems reasonable to assume that its internal structure and composition are similar. There is probably a metallic core, somewhat smaller than Earth's, which accounts for about 23% of the planetary mass and there is likely to be more oxidized iron in its mantle. It certainly cannot have a higher Fe/Si ratio than the Earth's. The massive CO_2 atmosphere must have been outgassed, like the Earth's, but to outgas an oxidized atmosphere of such volume requires substantial Fe to have resided in the planet's interior (Ringwood and Anderson, 1977). The close similarity in mass and density between Venus and the Earth points probably to a similar early thermal history, with the separation of the core controlling it.

The last member of the inner planet group, Mercury, is approximately 15% too dense to have been derived from the same parent material as the other terrestrial planets and even if it assumed that all the more volatile elements were removed during accretion, this is still true. Mercury's relatively elevated density implies that there must be a higher proportion of Fe-rich core material and less mantle-forming silicates than in the other planets. Urey (1952) calculated that an Fe-rich core with a diameter of about 1800 km and a further 600 km of refractory-rich mantle materials are required to satisfy the physical constraints. Since it was formed very close to the proto-Sun, where accretional temperatures were very high, it has been suggested that Mercury accreted from heterogeneous material in which most of the Fe has condensed and soon after Mg-silicates had started to form. Since core formation would lead to a significant increase in the planet's radius, extensional fractures ought to be visible in the ancient cratered terrain, believed to have formed about 4×10^9 years ago. On the contrary, widespread lobate fault scarps have been interpreted as reversed (compressional) faults which implies a shrinkage of 1-2 km in the Mercurian radius prior to the close of the Great Bombardment, i.e. about 3.9×10^9 years ago (Solomon, 1977). The only reasonable conclusion to be drawn is that core formation must have preceded the generation of the earliest preserved crust, following which the thermal evolution was dominated by slow cooling. The measured or estimated core dimensions and calculated mean temperature rises generated by planetary core formation are shown in Fig. 2.2.

2.5 MANTLE SEGREGATION AND CONVECTION

The formation of a silicate phase may initially have been only localized within the Earth; however, gradually such pockets as did form must have risen, taking with them entrapped radioactive materials having lithophile affinities. These trapped elements would have slowly heated them up until melting occurred, with the result that small sialic bodies were formed. Exactly how long it took for the Earth's outer layers to take on their present aspect and, by implication, a temperature distribution resembling its present pattern is unknown, but, on account of the widespread evidence for compressive stresses in the Earth's continental crust, it generally has been assumed that convective motions were eventually set up within an Earth-encompassing, silicate-rich, mantle layer. The oldest authenticated terrestrial rocks date back to 3.8×10^9 years ago, when it is generally believed that the crust would have been very thin. Younger sequences of deformed rocks, including supracrustal and magmatic rocks, are volumetrically more significant around 2.8×10^9 years ago, while extensive basaltic intrusions have been dated between 2.5 and 1.2×10^9 years. Radiometric data thus suggests that a significant volume of magmatic crustal material was being generated during later Archaean times, and it therefore seems reasonable to presume that mantle segregation was complete $1-1.5 \times 10^9$ years after the Earth had accreted and that planet-wide volcanism had by this time become important. Once this had been accomplished, convective transport of thermal energy became very important and consequently there would have been a slow reduction in the rate at which the Earth's interior was heated.

Diverse evidence from the Moon, meteorites and Mars indicates that on these bodies too, chemical fractionation was well underway during Archaean times. Radiometric dating of lunar rocks indicates that its anorthositic crust formed between 4.5 and 4×10^9 years ago. Basin-filling basaltic lavas were being generated from the lunar mantle at 3.9×10^9 years. Extensive volcanic plains developed on Mars at an early stage (about 3.5×10^9 years), assuming that crater dates derived for Martian surface units may safely be calibrated with lunar data. Thus magmatic activity was well established on both worlds 10^9 years after accretion was completed, and, as will be shown in a later section, there is evidence from Mars that enormous amounts of heat were being lost from the planet during the early stages of patera volcanism, tentatively dated at around 3×10^9 years. The situation on Mercury and Venus is less clear, but if intercrater plains units on the former are volcanic in origin, the same also is true of Mercury. Venus probably underwent a similar early thermal evolution to the Earth. Mantle melting and segregation was, however, a much slower process than core formation.

It is now generally accepted that mantle convection has played a major part in the subsequent development of the larger terrestrial planets (Schubert, 1979; Schubert *et al.*, 1979; Sharpe and Peltier, 1979). Following on from early core formation it is clear that cooling has dominated the thermal histories of these worlds. The gravitational potential energy released during core infall on the Earth (and possibly Venus) would have been adequate to melt it completely, but this did not happen, and some process must, therefore, have inhibited this. Such a process could have been convection within a molten medium such as a silicate mantle layer, which would have the potential to remove thermal energy quite rapidly. Subsequent cooling would have lowered internal temperatures to a level at which a mantle solidified, whereupon subsolidus convection would have commenced and indeed continues on the Earth to the present day. Although their thermal histories are not the same, the rate of cooling for all of the terrestrial planets would have fallen with time, as mantle viscosity increased and convective processes became less vigorous. From the time when the first near-surface internal melting occurred, volcanism, in one guise or another, has played a part in planetary development, and in particular, in surface modification. On Mercury and the Moon volcanism commenced early and also terminated early because thermal energy generation either quickly declined or energy was lost to space; by implication such mantle activity as developed must also have declined relatively rapidly. In contrast, an extended history of volcanism characterizes both the Earth and Mars and appears to have typified Venus. Therefore more vigorous mantle activity is indicated for these bodies, particularly the Earth and, in the latter case at least, has continued to the present day. Careful study of the chemistry (including isotope chemistry), mineralogy and morphology of volcanic rocks preserved at the surfaces of these solid bodies provides important clues to the fundamental internal processes of which volcanic rocks are the surface manifestation. During the course of this study it will be shown how considerable heterogeneity must have existed within the silicate mantles of the terrestrial planets. Magmatic activity on Jupiter's moon Io involves both silicate magmas and sulphur compounds. The peculiar nature of volcanism on this small moon demands a separate approach and its study is reserved for a separate chapter.

2.6   LITHOSPHERES

The outermost layer of a planet is termed its lithosphere. On the Earth this includes the crust and the upper part of the mantle layer, above the Low Velocity Zone. A similar situation applies also to Mars and Venus. Planetary lithospheres concentrate silicate materials and most developed very early in their history by chilling of basalt-like magmas that rose from their underlying mantles. The Moon's ancient silicate skin – composed predominantly of anorthosite – is known to have existed 4.4 Ga ago; it is likely that this also was the case for the other inner planets. As time progressed there has been a tendency for planetary lithospheres to thicken. This thickening process may not have been uniform and it is known that there are, at least on Mars, Venus and the Earth, regions of relatively thick and thin lithosphere. Mass inequalities within the terrestrial lithosphere are balanced by isostasy; thus the less dense and relatively thick continental crust "floats" on the denser oceanic-type crust below. Considerable debate continues regarding other planets' lithospheres, generally-speaking this being fuelled by the considerable degree of latitude available in the interpretation of gravity data. Suffice it to note at this point, that planetary lithospheres developed early in planetary history and are subject to considerable change in response to both internal and external processes.

3

The generation and evolution of magmas

3.1 THE PRODUCTION OF MAGMA

The principles which govern the production of magma from solid rock are extremely complex and have been summarized by Best (1982), Cox *et al.* (1979) and Presnall (1969). In general terms there are two principal modes by which melting may be accomplished: (i) equilibrium fusion, in which the melt produced continually reacts and equilibrates with the crystalline residue, and (ii) fractional fusion, where the melt which is produced is immediately isolated from the system, whereupon no reaction between it and the crystalline residue is possible. The former is the reverse of fractional crystallization and the melts generated give a continuous but limited compositional range; in contrast, the latter produces individual melts which span a much greater compositional and temperature range.

On the Earth, and there is every reason to suppose this also is true of the other terrestrial planets, most magmas are silicate liquids. However, the Earth's outer core is the only proven liquid region inside the planet and this is composed largely of metallic iron; consequently it cannot be the source region for magmas which approach the Earth's surface. Silicate magmas, therefore, must have their origin either in the crust or mantle layers, both of which are essentially solid. If molten liquid is to be generated from such materials it is necessary for some degree of partial melting to occur.

The principal current source of thermal energy within the Earth is long-lived radioactivity but it may be shown that the mantle, where modern research indicates that most magmas originate, has a relatively low radioactive content which would be inadequate to produce large-scale partial melting, except over unacceptably lengthy periods of time (Oxburgh and Turcotte, 1978; Schubert, 1979). Radioactives tend to be concentrated in surface rocks because radioactive elements are preferentially partitioned into partial melts of silicate materials which tend to migrate towards the surface. The distribution of radioactive elements in the mantles of the other terrestrial planets is unknown; nevertheless those processes which lead to their concentration in crustal rocks on Earth may be expected to have operated there also and it is reasonable to anticipate that there would be a similar upward concentration of radioactive heat sources. It is worth noting, however, that if whole-mantle convection had characterized any of these planets it would have tended to homogenize the heat sources present, assuming the convective system had sufficient time to stabilize.

Melt production may occur only where there is a change in either the temperature (T), pressure (P) or bulk chemistry (X) of the source material. If the Earth's mantle was essentially immobile, then, by whatever means changes in P, T or X were brought about, the distribution pattern for melting sites and, by inference, surface volcanism, should be random. This is, of course, patently not the case, since between 90 and 95% of current volcanic activity is concentrated along mid-oceanic ridge zones, within circum-Pacific-type island arcs or continental rift valleys. Furthermore at different times in geological history, the distribution pattern for volcanic eruptions has changed. This non-randomness is a direct result of the convective escape of the Earth's interior heat which is manifested in plate tectonics. Such localization of volcanism is not uniquely terrestrial, as is witnessed by the concentration of shield volcanicity in the Tharsis and Elysium regions of Mars, the confinement of lunar basalt volcanism largely to the mare basins and the association of major volcanic structures on Venus with crustal rift zones. Plate tectonics need not be imputed in these cases but some regional mobility of the mantle layer is implied.

Magmas are ephemeral and produced by localized partial melting followed by an outward escape of any liquid phase generated towards a planet's surface. Because incongruent melting governs most magmatic processes and since this occurs over a wide range of temperatures, reaction relationships ensure that all partial melts are relatively enriched in the more soluble, lower melting point components of systems, leaving behind a more refractory residue. Most silicate liquids are between 5 and 10% less dense than their crystallized equivalents, but for partial melts being generated within the Earth's mantle the density contrast may be as great as 20% (i.e. for basaltic liquid the density is approximately 2.65 g cm^{-3}, while for solid mantle, it approximates to 3.3 g cm^{-3}). This density contrast aids in their ascent towards the surface since they will be gravitationally unstable with respect to the encompassing rocks. The velocity of uprise will be dependent upon other factors, including composition, apparent viscosity, tectonic setting and conduit geometry.

The most reasonable interpretations of recent geophysical studies of the Earth's lithosphere, particularly of its seismic and electrical properties, suggest that between 1 and 3% partial melt may reside in that part of the upper mantle known as the low velocity zone (LVZ), which forms a continuous layer beneath the crust (Birch, 1969; Lambert and Wyllie, 1972; Yoder, 1976). On a worldwide scale this layer may not constitute a major magma source – an inference based on the observation that there is little magmatic activity in intraplate settings – ; however, it may be the dominant source region where plate movements disturb either T, P or X, giving rise to local partial melting and magma generation.

The simplest way to produce a partial melt is to raise the temperature of the source rock, but large amounts of energy are required to produce even a small volume of melt, a result of the great contrast between the specific heat of silicate rocks (i.e. the amount of heat required to change the temperature of the liquid or solid) and the heat of fusion (the amount of heat which must be added to melt a unit mass that is at the temperature where the liquid and crystals co-exist). For silicates, the specific heat ranges from $0.8 - 1.3$ J g^{-1} °C^{-1}, while the heat of fusion lies between 270 and 420 J g^{-1} °C^{-1} (Williams and McBirney, 1979) meaning that the latter is around 300 times greater than the former. Hence 100 J of thermal energy may raise the temperature of 1 g of solid rock by about 100°C, but will produce only a few

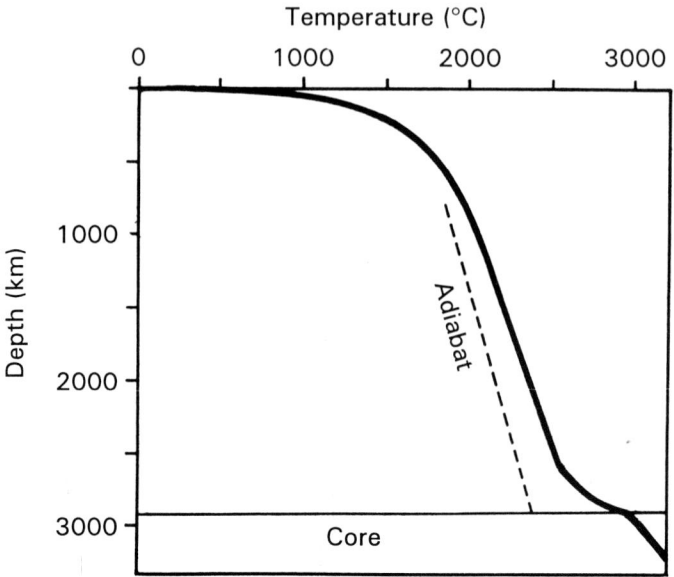

Fig. 3.1: A possible temperature profile inside the Earth. Steep gradients characterize the crust, but within the convecting part of the mantle the gradient cannot be much steeper than that of the adiabat (dashed line).

hundredths of a gram of silicate melt from a solid silicate material already close to its melting point. Bearing this in mind, it is instructive to look at the ways in which partial melts of mantle rocks may be produced at lithospheric plate boundaries.

The temperature for the onset of melting of mantle materials increases by about $10°C$ for every 10^8 Pa rise in pressure (which corresponds to a depth increase of 3 km) – the geothermal gradient for the Earth at these depths (Fig. 3.1).

One hundred kilometres below the surface of the Earth, mantle materials may experience a temperature significantly above that necessary to produce some melt at zero pressure, but because the effect of increasing pressure is to elevate the melting point, they will remain solid. If sufficient heat is supplied, however, the temperature may be raised above that required to produce some melting, or alternatively, if the pressure is reduced sufficiently, the temperature necessary for fusion may fall below the local temperature (Fig. 3.1). In a convecting mantle system of the type believed to give rise to terrestrial plate tectonics, both possibilities exist. The existence of small amounts of hydrated substances within the source rock will also have an effect, since they will depress the melting point of silicate materials below those typical of anhydrous systems.

The Earth's rigid plates are about 100 km thick. At mid-oceanic ridges, individual plates grow by the addition of mantle-derived material at their following margins while destruction of the opposite margins is achieved along subduction zones (Fig. 3.2).

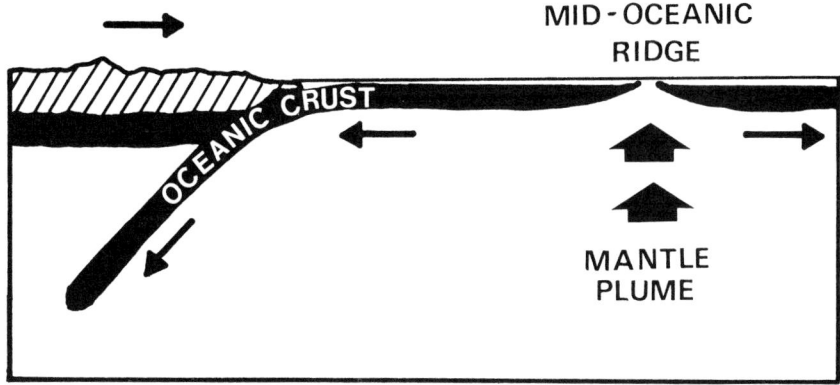

Fig. 3.2: General plate structure of the Earth. At divergent plate margins, plates grow by the accretion of mafic material from rising mantle plumes. At convergent margins, energy is released during foundering of the subducted plate, generating seismic and magmatic activity as the lithosphere is recycled.

Where plate divergence occurs beneath mid-oceanic ridges, a plume of slowly rising mantle material will be hotter and therefore less dense than the mantle on either side; as a result during its ascent the rising plume will undergo a degree of partial melting, resulting in an expansion in volume. This, incidentally, promotes a 1-2 km elevation of the mid-oceanic ridge above the surrounding ocean floor. Along the flanks of the ridge system the adjacent lithospheric slabs slither inexorably away at about 1 cm per year, the slab material becoming older as distance from the ridge crest increases, until, at the opposing (destructive) margin, a slab may be around 200 million years old. During the period in which the newly generated crust slides slowly away from the axial ridge, at least the upper portion of the slab will cool and so the preceding cold plate margin will necessarily become denser than the engulfing mantle. As it founders, phase transformations will enhance the density contrast, while the gravitational energy of the subducted slab will be utilized in overcoming the frictional drag of the opposing plate. This will be released partly as seismic and partly as thermal energy. The slab also will be heated by the engulfing, hotter mantle, until eventually the temperature of the thin basaltic layer may exceed its solidus, a point which would be reached sooner if hydrated materials (e.g. metasomatized lavas, ocean floor sediments) were attached to the descending slab.

3.2   MANTLE SOURCE ROCKS AND PARTIAL MELTS

The chemical composition of any partial melt is not only governed by the type of melting, but also by the constitution of the source rock and the degree of partial melting involved. The nature of the source rock will be a function of several factors, including the pre-accretional history of the planetary material from which it was derived, the planetary mass and its early evolution, therefore magmas derived by partial metling of planetary interiors must reflect their chemistry and evolutionary history. The Earth's mantle is not available for direct analysis, but representative samples of old oceanic floor and mantle are believed to have been found in ophiolite sequences, while relatively unmodified mantle materials are also found within kimberlite pipes and in xenoliths brought up in some basaltic lavas (Carmichael *et al.*, 1974; Dawson, 1971). In broad terms the Earth's mantle has been found to have a composition of 90% (SiO_2+MgO+FeO) + 10% (Al_2O_3+CaO+Na_2O), implying that it is ultramafic. The abundant presence of harzburgite (80% olivine + 20% orthopyroxene) in nodules and within ophiolite complexes suggests that olivine and its various polymorphs probably constitute the upper mantle, which is confirmed by seismic evidence. Partial melting of such material must have produced the wide variety of volcanic rocks seen at the Earth's surface and there is much evidence supportive of the view that this applies also to the other terrestrial planets.

The predominance on Earth of basaltic lava at both divergent plate margins and intraplate oceanic settings, and its relative abundance in continental rift valleys and convergent plate tectonic environments, together with its primitive, mantle-like isotopic signature, has led to a plethora of experiments involving the generation of different kinds of basaltic melt from a variety of potential mantle materials under controlled conditions. The main purpose of these has been to establish which particular starting material most closely satisfies the available petrological, physical and geochemical data. One of the most extensive series of experiments involved a synthetic variety of mantle peridotite termed pyrolite , i.e. *pyr*oxene-*oli*vine rock (Ringwood, 1966; Green and Ringwood, 1970). This indicated that transitions between the Earth's internal layers could be accounted for by phase changes undergone by pyrolite under appropriate P-T conditions (Fig. 3.3).

Subsequent work has confirmed that partial melting of peridotite produces basalt under upper mantle conditions such as are believed to pertain in the Earth's LVZ. Peridotite thus has supplanted the other favoured candidate for a mantle source rock, eclogite, principally since if the latter is invoked it (i) implies a phase change at the continental Moho and (ii) necessitates 100% partial melting to generate chemically identical basalt. Since at both the oceanic and continental Moho a compositional change is encountered, the former is not justified. With respect to the latter, 100% partial melting would produce a totally liquid layer which would completely attenuate seismic S-waves; this does not occur.

Since the P- and S-wave velocities of silicates like olivine, enstatite and garnet are very similar, it is impossible to arrive at an estimate of mantle composition from the absolute seismic wave velocities; however an estimate of mantle mineralogy can be derived from the seismic discontinuities which occur at depths of 420 and 670 km (Fig. 3.3). These are due to solid state phase changes which are temperature dependent and recent experimental work has indicated they reflect the presence of different polymorphs of the mineral olivine (Kawada, 1977; Ito and Yamada, 1982) which, to manifest themselves in major

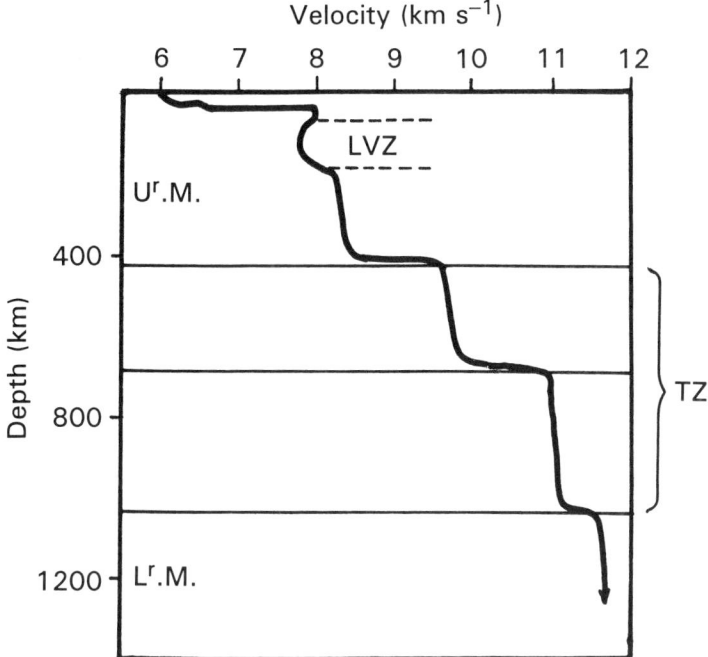

Fig. 3.3: Variations in seismic P-wave velocities within the Earth's mantle. Ur.M., upper mantle; Lr.M., lower mantle; LVZ, low velocity zone; TZ, transition zone.

discontinuities, must be present in substantial amounts.

The 420 km discontinuity can be shown to correspond to a change from a to b olivine which occurs at 1800°C and 1.515×10^{10} Pa, while the 670 km discontinuity appears due to the decomposition of b-olivine into a perovskite-wustite mixture at a temperature of around 2000°C and a pressure of 2.4×10^{10} Pa. Estimates suggest that olivine may be present in the upper mantle to the extent of 70%, which is very similar to that in lherzolite (Maalöe and Aoki, 1977) which currently is the most favoured candidate for the mantle source material. Interestingly, the nature of these changes implies that material ascending from the lower into the upper mantle, as it would in a plume, would undergo a temperature increase, since the transformation entails an exothermic reaction.

Experiments involving dry peridotite systems indicate that any initial basaltic liquid is generated along a curve with a positive temperature-pressure gradient, beginning at 1200°C and 10^5 Pa pressure (Fig. 3.5); however, since the field of anydrous basalt does not intersect the geotherms, melting could not occur without the intervention of some outside influence. Such an influence could be a rising plume of hot mantle material, either beneath an oceanic ridge or in an intraplate location, whereupon the increased thermal regime could perturb the geothermal gradient sufficiently for it to intersect the basalt dry melting curve, as shown in Fig. 3.4.

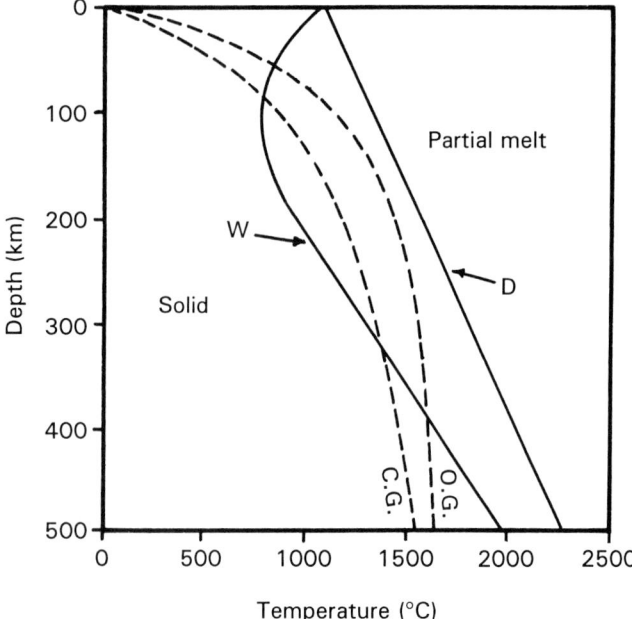

Fig. 3.4: Geothermal gradients within the Earth's upper mantle and their relationship to hydrous and anhydrous melting curves for peridotite. O.G., oceanic geotherm; C.G., continental geotherm; D, dry melting curve; W, wet melting curve.

A further factor which must be of at least local importance, is the presence of volatiles. Uncombined water is capable of lowering the melting temperature by as much as 500°C, the melting curve for the water-saturated system having a negative P-T gradient at low pressures, but becoming positive with increasing depth. It is known that only small amounts of water are locked up in mantle materials (less than 1%), but even 0.1% water would enable some partial melt to be formed, whereas it would be impossible in the dry system. Importantly, the same amount of water would allow greater amounts of partial melting with increasing depth. Therefore water may be a more significant control on the degree of partial melting than lithostatic pressure.

The composition of liquids formed by the partial melting of mantle peridotite (1 in Table 3.1) are dependent upon the degree of partial melting (which is largely determined by the temperature) and by the local liquid/mineral relationships (which are pressure-dependent). At low degrees of mantle fusion, i.e. within the range 1200-1250°C at 10^5 Pa pressure, an upper mantle partial melt would have a composition similar to the tholeiitic basalt (2 in Table 3.1). At higher temperatures (and greater degrees of partial melting) liquids become increasingly more enriched in magnesium. If the pressure is increased but remains below the 10^9 Pa level, a small degree of melting produces a similar silica-oversaturated basaltic liquid. At greater pressures, corresponding to deeper melting levels, liquids become successively more impoverished in silica, with the result that at pressures

Table 3.1: Composition of mantle partial melts at differing temperatures and pressures.

	(1) lherzolite	(2) tholeiite	(3) alkali basalt
SiO_2	45	49	45
TiO_2	0.1	1.5	2.3
Al_2O_3	3	18	16
Fe_2O_3	1	1.6	6.5
FeO	7	6	8
MgO	40	8	7.5
CaO	3	11	9
Na_2O	0.2	3	3.2
K_2O	0.03	0.1	0.8

of above 2×10^9, silica-deficient minerals like nepheline may crystallize, giving rise to alkali-basalts (analysis 3, Table 3.1).

The validity of such experimental results is confirmed by various pieces of evidence, one of which is the distribution of different basalt types in the vicinity of mid-oceanic ridges. Results from the FAMOUS project (see summary in Ehlers and Blatt, 1980) show how in the vicinity of the median ridges which sit above the hottest zones of the sub-oceanic mantle, lavas are dominated by tholeiites and other silica-saturated lavas appropriate to conditions of relatively shallow melting. In contrast, towards the margins of the rift zones the typical lavas are alkali-basalts and other silica-undersaturated types which appear, therefore, to have been generated at greater depths. Such basaltic melts are generated from relatively anhydrous mantle (less than 0.1% water). Where higher water contents prevail the phase relationships change completely, giving initial melts with a significantly greater proportion of SiO_2 and lesser amounts of MgO than are typical of basalt. Hydrated mantle material would be available where oceanic plates are being ingested at subduction zones, for instance, in the circum-Pacific belt. The consumption of the small amounts of water in ocean-floor basalt and the greater amounts locked up in the veneer of marine sediments attached to the preceding edge of the plunging slab, would be quite sufficient to explain the andesitic volcanism so typical of this region. That hydration is apparently important in andesite formation is indicated by the frequent relics of hydrated minerals found in their phenocryst assemblages.

3.3 MAGMATIC EVOLUTION

The extremely widespread occurrence of basalt-like rocks on the terrestrial planets suggests that basaltic partial melts developed widely in the inner regions of the solar system. This implies that pyrolite-like source rocks must have been present at upper mantle levels, which suggests at least a degree of homogeneity with respect to the primordial matter from which these planets were made. The subsequent evolution of these melts may have followed quite different paths since the chemical and physical environments in which they crystallized and their temporal evolution were not necessarily the same. For example, lunar basalts crystallized under highly reducing conditions, with the result that the rocks lack Fe^{3+} but may hold metallic iron. Terrestrial and Martian lavas, on the other hand, must have been generated under oxidizing conditions, with the result that Fe^{3+} and (OH) compounds typically are a part of the mineral assemblage.

Magma generated at depth within a planet and transported rapidly enough to the surface environment, there to be erupted as lava, may complete its freezing so quickly that its pristine chemistry becomes little modified. Most magmas, however, do not fulfil these requirements and show various degrees of chemical modification due to fractional crystallization. Indeed, any volume of magma which has cooled slowly will have completed its crystallization over a long period and is bound to have experienced fractionation; thus deep-seated plutons all show some chemical changes, while the great majority of extrusive bodies will have experienced some degree of modification. Any field geologist who has studied igneous rocks will know that compositional variations may occur even across a single outcrop and certainly will be observable on a scale of hundreds of metres. It is not surprising, therefore, that many years ago petrologists began enquiring as to whether the variable rocks of a single region had some genetic origin, say in a common parent magma or by the mixing of two source magmas. One way of confirming or refuting a suspected common origin of a suite of igneous rocks is to study the mineralogy, noting the pervasive occurrence of similar mineral species, e.g. alkali pyroxenes or amphiboles, but often a more effective method is to study rock chemistry, in particular the trends shown by both major and trace elements, stable isotopes or the ratios of pairs of elements, any or all of which may indicate a common genetic link.

3.3.1   Fractional crystallization

It is now well-established that most mantle-derived melts suffer some degree of differentiation as they rise upward through the crust. The most important method by which this is achieved is fractional crystallization; however, gas transfer, liquid immiscibility and thermo-gravitational fusion may also be involved. The process of magmatic differentiation was first thoroughly investigated by N. L. Bowen whose classic studies were published in 1928 as *The evolution of the igneous rocks*. Bowen studied igneous rocks in the field extensively and conducted a series of laboratory experiments which formed the basis of later work on silicate systems. He stressed that in rock melts the various mineral phases precipitating do not usually crystallize simultaneously. This largely is a function of the different melting temperatures of the common rock-forming silicates. As a consequence the melt and the crystals separating from it are not of the same composition, the former gradually changing composition during crystallization and following a liquid line of descent towards the invariant point for the system concerned. In most cases, there is a continual interaction between melt and crystals, in particular there being free interchange of the ions which enter into those solid solution series which epitomize the common silicate species.

The simplest means whereby a magma body becomes fractionated is where early-crystallized minerals, being generally denser than the enclosing melt, settle downwards under the influence of gravity. As a result of this physical separation of phases, the liquid residue becomes progressively impoverished in the more refractory components and

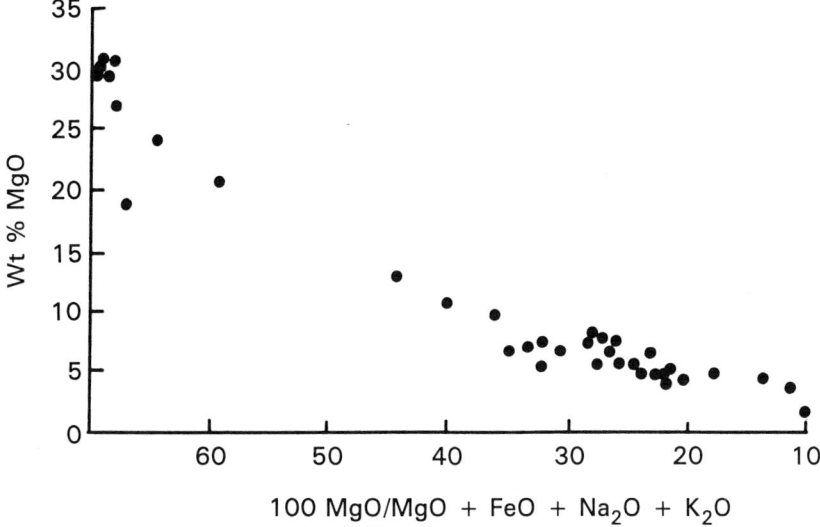

Fig. 3.5: Plot of MgO versus MgO + FeO + Na$_2$O + K$_2$O for layered rocks of the Rhiw Complex, North Wales. The importance of the withdrawal of MgO via crystallization and settling of Mg-olivine crystals is reflected in the observed variation.

relatively enriched in the more fusible ones. The chemical fractionation process has two aspects. Firstly, most common silicates show ionic substition (solid solution), such as Ca for Na in the plagioclases and Mg for Fe^{2+} in the olivines. Fractional crystallization therefore impoverishes any liquid precipitating these phases in Mg and Ca with respect to Fe^{2+} and Na, with the result that the ratios: Mg/(Mg + Fe^{2+}) and Ca/(Ca + Na) exhibit a systematic decrease throughout a series of volcanic or plutonic rocks derived from a common parent magma (Fig. 3.5).

Secondly, there is a trend towards a minimum melting point composition (eutectic). For the range of volcanic rocks encountered on the Earth there are two eutectics: (a) that between silica and alkali feldspar, and (b) that between alkali feldspar and feldspathoid. Overall, fractional crystallization gradually eliminates Mg and Fe silicates (e.g. olivine and pyroxene), leaving a liquid residual dominated by Na and K alumino-silicates (e.g. orthoclase and albite).

3.3.2 Major element behaviour

Clearly the efficiency of the fractionation process varies from one suite of rocks to another, while the chemical constitution of the parent magma may also vary; consequently the two trends may be present to differing degrees in different suites. Nonetheless the dominant trend in major element distribution during the fractional crystallization of basalt-like magmas is for depletion (via Mg-olivine precipitation) of the residual melt in Mg at an early stage, with the result that the residual liquid becomes enriched in Fe, Si, Al, Ti, Ca,

Table 3.2: Major element variation in an upward sequence through the differentiated Rhiw igneous complex. (After Cattermole, 1976).

	Chilled margin	Picrite	Leuco gabbro	Magnetite gabbro	Granophyre
SiO_2	44.10	38.70	44.50	38.50	58.60
Al_2O_3	12.20	5.50	18.00	16.00	16.40
Fe_2O_3	3.78	4.48	1.44	5.18	1.56
FeO	9.25	9.05	4.47	14.53	4.70
MgO	12.80	30.06	7.00	8.00	1.50
CaO	9.40	6.10	15.20	13.70	9.40
Na_2O	2.50	0.40	5.20	1.25	6.40
K_2O	0.18	0.09	1.85	0.32	2.40
TiO_2	0.93	0.43	0.13	0.29	0.29

Na and K. With the removal of more ferriferous olivines, early removal of some Fe will reduce the degree of iron enrichment in the melt and may have a pronounced effect on the path of fractionation. Subsequently, with the precipitation of Ca-plagioclase and Ca/Mg/Fe pyroxenes due to the cooling of the system and the slowly changing chemical composition of the magma, the residual melt fraction becomes progressively enriched in the more fusible components, i.e. Na,K, Si and (OH). In this way, during the slow cooling and fractional crystallization of plutonic basaltic magma, a wide variety of rock compositions may be generated and ultimately a small volume of 'granitic' residue may be produced (Table 3.2).

Some of the minor and trace elements can substitute for major ions (e.g. Ni^{2+} and Co^{2+} substitute for Mg^{2+}, Cr^{3+} substitutes for Fe^{3+}), but others may have ionic dimensions which either are too large or too small, or whose bond type and valency are too different, to effect substitution with a major cation. Such elements as K, Rb, Ba, U, Th, Zr and Nb fall into this class, as do gases like CO_2, H_2O and HCl. Such elements will tend either to be completely retained or gradually enriched in a basaltic melt until such time as their concentration in the melt became high enough for their own species to precipitate out. They have become known as the *incompatible elements* and many volcanic rocks can be shown to have concentrations of the incompatible elements which are 200 times those characteristic of their mantle source rocks. Such concentrations can be explained by small degrees of partial melting and liquid extraction (i.e. less than 0.5%) or by extreme liquid fractionation.

The crystal-liquid fractionation process outlined above is frequently ascribed to simple gravitational segregation as witnessed in that classic petrological example, the Palisades Sill (Jaeger, 1968; Walker, 1969) but recent work suggests that the repetitive rhythmic layering so typical of intrusions such as those of Skaergaard and Bushveld, hitherto accepted as classic cases of gravity segregation, may in fact be due either to deposition of crystals from density currents or to *in situ* bottom crystallization (Irvine, 1979; McBirney and Noyes, 1979). Either way, the process of crystal fractionation has the same effect upon the liquid line of descent.

Another explanation for fractionated silicate systems is liquid immiscibility, which has had a chequered career among geologists as a viable fractionation process, but which has received limited support from experimental work (Roedder, 1951). Immiscibility within silicate magmas appears to be of restricted importance, but evidently has played a significant role in the fractionation of some high Fe-rich basaltic magmas and highly alkaline melts (Roedder, 1979), furthermore it has been shown to have operated in silicate-carbonate systems; which may have important connotations for nephelinite-carbonatite genesis. Some of the most striking evidence in favour of the natural coexistence of conjugate liquids is afforded by some lunar ferrobasalts in which immiscible globules of brown glass are found within clearer siliceous glass that crystallized between early-formed phenocrysts.

The ability of high-temperature volatiles to transport more fusible constituents, e.g. K, Na and Si, is well-known, but there is little actual evidence from natural occurrences that a magma body has been enriched by this method. Fenitization, so common in alkaline nephelinite-carbonatite ring complexes, is, however, one manifestation of volatile transport. Of greater importance in large-scale magma diversification is the process of thermogravitational diffusion. This has been invoked to explain the compositional zoning of rhyolitic pyroclastic flow deposits and their associated magma chambers and was first proposed by Hildreth (1979) to account for the Bishop ash-flow tuff of eastern California. In the magma chamber parent to this ash-flow sequence there is a significant upward enrichment in SiO_2 and H_2O, while minor and trace element concentrations show a tenfold variation between the earliest and latest magmatic fraction erupted. The cap of the Bishop chamber is strongly stratified, while the lower core is more uniform, crystal-rich and of intermediate silica content. The style of zonation observed here, and in many other similar suites, cannot be ascribed to simple gravity segregation and according to Hildreth (1979) it is most realistically explained by diffusion within an essentially liquid system, whereby the effect of gravity on the melt together with a vertical thermal gradient is sufficient to generate a stratiform convective system. This has the effect of enhancing the diffusive movement of elements and allowing a degree of chemical transport well above that possible in static magmatic systems. Under such conditions zonation may develop for each chemical element in response to thermal gradient, the gravity field, and gradients in atomic structure of the magma. Since the uppermost, highly hydrated layers are less polymerized, this has a major influence on the bonding characteristics of different metallic cations present in the magma. In the silica-rich roof of the Bishop chamber, those elements showing a marked concentration, apart from SiO_2 and H_2O, are Na, Li, Be, Rb, Cs, Nb, Mo, F etc. (the same elements as are enriched in pegmatites and molybdenum-porphyry deposits), while there is downward enrichment into the intermediate "core" magma of elements such as Cu and Au (which are major components of copper-porphyry deposits). This particular process may as a consequence be of the utmost importance in the generation of major ore deposits.

In volcanic centres with protracted lives, there is potential for two or more unlike magmas to mix and blend. Such a process would necessarily lead to further magmatic diversity. That this occurs may be shown by the development of hybrid rocks and net-veined complexes, a characteristic, for example, of the Hebridean Province of Tertiary age in western Britain (see Harker, 1904; Richey *et al*, 1930; Daly, 1933). In Iceland, too, tholeiitic fissure eruptions have generated mixed rocks, in which decimetre- to centimetre-sized mafic inclusions account for up to 10% of rhyolite lava, which occur alongside truly hybrid rocks which exhibit thoroughly homogenized fabric (Blake, 1984). Then again, the assimilation of country rocks during the forceful injection of plutonic bodies may produce contaminated rocks which further add to the variety of igneous products (see, for instance, Bowen, 1928; Nockolds, 1934).

3.3.3   Trace elements and isotopes

Eight major elements comprise 98% by weight of the Earth's crust; these are oxygen, silicon, aluminium, iron, magnesium, calcium, sodium and potassium. During fractionation of partial melts derived either by mantle fusion or fusion of pre-existing crustal rocks, these major constituents will be concentrated either in the early-crystallized material or the residual melt, according to the principles outlined above. They are, however, not the only ones since the minor elements (manganese, titanium and phosphorus) and the less abundant trace elements are also involved. The way in which trace elements become distributed in a crystallizing magma may be predicted on the basis of crystal field theory, as was shown by Curtis (1964), and the fractionation trends shown by such elements provide important information concerning differentiation processes. Those elements which find it easiest to gain sites in a crystallizing phase are those with ions of similar size and electrical charge to one or other of the major elements in the crystal structure (Mason, 1966). Thus, for instance, Sr^{2+} substitutes for Ca^{2+}, and Cr^{3+} for Fe^{3+} in many silicate minerals. The ease with which a particular element can enter into a mineral is reflected in its partition coefficient KD, where

$$KD = \frac{\text{concentration of element in phase 1}}{\text{concentration of element in phase 2}}$$

and is constant for any particular equilibrium. It is customary for partition coefficients to be written so that the concentration of the element in the mineral is divided by that in the liquid. In this way the size of KD reflects the facility with which an element can enter the structure of a particular mineral. The distribution coefficients of a number of trace elements which occur in basaltic rocks are given in Table 3.3.

Elements which preferentially are concentrated in the liquid phase during either melting or crystallization are termed incompatible, while those which preferentially are retained of extracted in the residual or crystallizing solid phases, are called compatible. Naturally, as different minerals are consumed during melting or new minerals appear during crystallization, the equilibrium assemblage will change, and therefore certain elements may change status, becoming compatible instead of incompatible or vice versa (e.g. Sr is incompatible in ultramafic systems but would become compatible in the presence of significant plagioclase). In terms of magma generation from mantle material, K, Rb, Sr, Ba, Zr, Th and light rare earth elements (REE) are incompatible with respect to the commonly crystallizing minerals in such systems, i.e. olivines, pyroxenes, spinels and garnet, and have become known as large ion lithophile (LIL) elements.

The importance of these less abundant or "dispersed" elements cannot be too strongly stressed, since they have far-reaching implications for all petrogenetic models. The reasons are not hard to appreciate and have already been implied: their relative concentrations in

Table 3.3: Distribution coefficients of selected trace elements in basaltic rocks. (After Cox *et al.*, 1979).

Mineral	K	Rb	Sr	Sm	Eu	Ni	Cr
Olivine	0.001	0.001	0.001	0.002	0.002	10	0.2
Orthopyroxene	0.001	0.001	0.01	0.010	0.013	4	2
Clinopyroxene	0.002	0.001	0.07	0.26	0.20	2	10
Garnet	0.001	0.001	0.001	0.22	0.32	0.4	2
Spinel	0.01	0.01	0.01	0.05	0.03	5	10
Plagioclase	0.2	0.07	2.2	0.07	0.3	0.01	0.01
Amphibole	1	0.3	0.5	0.52	0.59	3	12
Fe-mica	2.7	3.1	0.08	0.03	0.03	3.5	7

the various minerals which have crystallized from a magma can be very disparate. Furthermore, while major element variation within a rock suite may range over an order of magnitude, that for the trace elements may be three times this. The trace element constitution of a magma, therefore, may often impose constraints on the character of a potential source rock, in particular the minerals which were present but were then fused to generate a partial melt. It may also indicate which mineral species were involved in subsequent fractional crystallization.

While there is no simple overall pattern to the geochemical behaviour of the elements, the following broad groupings may be noted for igneous rocks derived by crystal-liquid fractionation processes:

1. Many of the transition elements (Sc to Cu) are enriched in the mafic fractions and show fair to strong inter-element correlations.
2. Many elements with atomic number $\geq$ (Y), but excluding the platinum group elements, tend to be preferentially concentrated in the melt phase during the fractionation of basaltic magma.
3, The elements Cu, Fe, Ni, Co, Ag, Au, Se, Te, In, Tl and Re tend to occur where a magmatic sulphide fraction develops.
4. Immiscible aqueous phases tend to be enriched in Na, K, Ca, Mg, Cl, S (as sulphate), Li, B, P, C (as bicarbonate or carbonate), Zn and W.

In addition to the dispersed elements described above, modern petrology has made increasing use of radioactive and stable isotopes as petrogenetic indicators. Isotopes of O, Rb, Sr, Nd, Sm, Pb, U and Th, in particular, have been much studied in a variety of terrestrial and lunar rocks (see Faure, 1977). The importance of isotope variations is that they frequently survive the kind of fractionation which is concomitant upon the formation and evolution of magmas. Thus where trace element concentrations and ratios may have been severely modified, say, by alteration, isotope ratios may be utilized as geochemical "tracers" of magmatic origins. Herein is not the place to delve into isotope systematics; the reader is referred to Cox *et al.*, 1979 or Maalöe, 1985), however, a number of important connotations of isotopic data may be noted at this point. Since isotopes of heavy trace elements are not fractionated during normal chemical processes, estimates of present day $^{87}Sr/^{86}Sr$ ratios for terrestrial mantle material (considered to have behaved as a closed system) may be obtained by studying the isotope ratios of those rocks likely to have been derived from them, which, in the case of the Earth, means the oceanic basalts. Many recent tholeiitic basalts have $^{87}Sr/^{86}Sr$ values of around 0.704 (Faure and Powell, 1972), which is in accord with the general observation that most mantle-derived volcanics have ratios less than 0.706, implying that this is the characteristic isotopic signature of the Earth's upper mantle. (In contrast, magmatic rocks derived from, or contaminated by, long-lived sources with high Rb/Sr values, e.g. sialic crust, should have higher $^{87}Sr/^{86}Sr$ ratios, i.e. above 0.706.)

Isotopic ratios for meteorites have provided invaluable clues to the early chemistry of the Earth and other terrestrial planets since they condensed out of the same primordial material at the same time. Initial isotopic ratios for the most primitive kinds of meteorites provide a reference against which to compare other planetary data. While meteorites comprise a very complex family of rocks, the ratios for all types fall within the range: 0.698-0.700, while their ages vary between 4.5 and 4.6×10^9 years. The primitive *basaltic achondrites* are particularly useful, since they have very low initial Rb content, which means that their initial ratio (about 0.699) is virtually independent of age (Papanastassiou and Wasserburg, 1969). This value has been widely used, therefore, for comparative purposes and gives an indication of the initial ratio of the Earth and the Moon.

While Rb/Sr isotopic data show that terrestrial basaltic rocks have mantle signatures, this does not mean to say that variations do not occur; the same applies to the lunar data. Mid-oceanic ridge basalts (MORB) which are tholeiitic, have $^{87}Sr/^{86}Sr$ ratios which average 0.7028, while oceanic island basalts (OIB), also known as *plume basalts*, average 0.7034 (Hofmann and Hart, 1975; Tatsumoto *et al.*, 1965). Now such rocks are considered to have been derived from partial melting of lherzolitic mantle material which is composed of different minerals (e.g. spinel, olivine, pyroxene, etc.). Clearly the various species within lherzolite will have different $^{87}Rb/^{87}Sr$ ratios, but at high temperature, because diffusion rates are high, the $^{87}Sr/^{86}Sr$ ratios will nonetheless be identical for all. At lower temperatures (below 1100°C), however, diffusion rates become too low for homogenization to occur and the various species acquire different $^{87}Sr/^{86}Sr$ values. Partial melting of mantle material takes place at temperatures of around 1500°C; consequently the isotopic ratios of the primary melts will be the same as the average ratio for the source material. The important implications of the different isotopic ratios observed for MORB and OIB, therefore, are that their sources must have had different ratios and been characterized by different chemistry. Since the only obvious potential source of such variation – contamination by seawater, is ruled out by Nd-Sm isotopic data, a degree of upper mantle heterogeneity is implied.

3.4   FRACTIONATION IN THE SOLAR SYSTEM

Apart from the Earth, the only large geochemical data sets available are those for the Moon and meteorites. One group of the latter, in particular, are interesting in this context; these are the eucrites, a much-studied family of differentiated basaltic meteorites. The

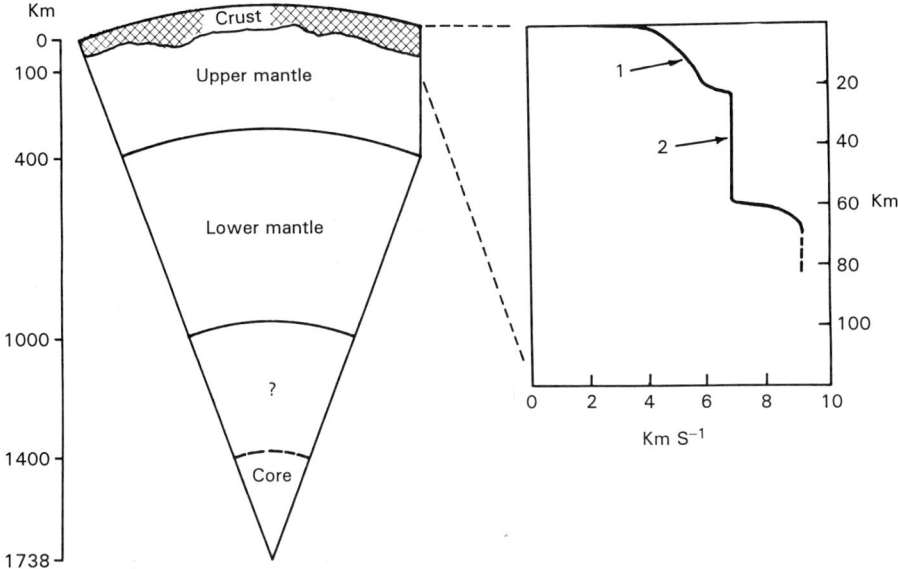

Fig. 3.6: (Left) Internal structure of the Moon, based on seismic data. (After Goins, N. R. *et al.*, (1977), *Proc.Lunar Planet.Sci.Conf.,8th*, 471-486.) (Right) Seismic velocity profile for outermost 150 km of the Moon. 1, lunar mare basalts, 2 lunar anorthositic gabbros. (After Toksöz, M.N. *et al.*, (1973) *Proc.Lunar Sci.Conf.,4th*, 2529-2547.)

more contentious SNC meteorites may represent samples from Mars. Study of all of these samples sheds light on the early differentiation of the silicate planets and, coupled with lunar seismic data, constrains models for the Moon's interior.

Seismic data reveal a layered structure for the Moon, and it has been proposed that beneath the outer skin is a thick layer of olivine-pyroxene rock which may extend downwards to depths of about 300 km (Fig. 3.6).

The seismic velocity profile for the uppermost 150 km of the Moon, beneath the eastern part of Oceanus Procellarum, indicates there to be an approximately 1 km thick, loosely consolidated regolith layer, underlain by material probably corresponding to mare basalt. At a depth of 20 km, seismic velocities increase and, down to depths of between 60-75 km, are consistent with a layer of anorthositic gabbro composition. Below this level, the seismic velocities are more appropriate to denser pyroxenites (Goins *et al.*, 1977; Latham *et al.*, 1978). The 60-75 km discontinuity appears to mark the crust-mantle boundary.

The highland rocks of the Moon are highly shocked anorthosites, which are largely composed of what is, by terrestrial standards, very calcic feldspar (calcic bytownite/anorthite). Less abundant highland rock types include norites, olivine-gabbros and (rare) dunites. Currently there is much support for the view that the plagioclase-rich rocks represent an accumulation of crystals which rose towards the top of an ancient lunar "magma ocean". Whether this hypothesis is accepted or not, the source for the melt out of which the feldspar precipitated must have been the underlying mantle, which, according to the magma ocean theory, would represent a downward-settled crystal accumulate complementing the anorthosites. As radiometric dating gives us an age of between 4.6 and 4.4×10^9 years for the highland crust, this segregation must have occurred very early in lunar history. Since melting of the whole Moon is unlikely to have occurred (Ringwood, 1979), partial melting processes are more likely to have played the major role in differentiating the outer layers. Partial fusion of an ultramafic source rock could not itself have produced anorthosite, since it would contain insufficient Ca and Al to allow significant precipitation of calcic plagioclase, consequently the generation of anorthosite can best be ascribed to the segregation of calcic plagioclase as a near-liquidus phase from a basalt-like partial melt.

Basaltic rocks are common on the Moon; however, the very extensive basin-filling mare basalts, which were extruded between $4.0-3.2 \times 10^9$ years ago, differ not only among themselves but also from terrestrial basalts in a number of ways. Firstly, they contain strongly zoned Ti-clinopyroxene and very calcic plagioclase, while late-crystallizing tridymite and cristobalite are common; secondly, ilmenite is usually abundant and may become a major constituent; and thirdly, they all lack Fe^{3+} and may contain metallic iron. The latter implies they must have been produced in a highly reducing environment and, since they have a low concentration of volatile species also, this environment must have been one of low oxygen fugacity. Finally, and very significantly, in contrast to the highland anorthosites which are relatively enriched in europium, all the mare basalts show a strong depletion in Eu^{2+}. This fact provides a vital clue regarding early lunar history: because Eu^{2+} ions partition strongly into plagioclase crystals in equilibrium with a silicate melt, the latter must become depleted in europium, thus if the olivine-pyroxene layer represents a complementary crystal differentiate, it should be characterized by Eu^{2+} depletion. While no direct sampling of mantle rock could be undertaken, it is the consensus that the mare basalts were derived from it by partial melting. This being so, the differentiated basalts should retain the trace element and isotopic signature of their source. Their characteristic europium depletion provides confirmation of this, thereby giving the argument consistency and showing that the mare source material must have fractionated from the material of the highland crust *prior* to mare basalt extrusion

The majority of primitive lunar mare basalts have lower Al_2O_3 than the most primitive terrestrial MORB glasses. With respect to this element,however, lunar mare lavas may be divided into two groups: an aluminous group represented by high-Al Apollo 12 basalts and the Apollo 14, and Lunae 16 and 24 very low titanium group (VLT basalts) which have similar Al content to the eucrites and a lower-Al group including most Apollo 11, 15, 17 and the majority of Apollo 12 lavas (Ridley, 1975). This disparity between the two groups cannot be due to fractional crystallization and appears to be a manifestation either of source heterogeneity or differences in the degree of partial melting. The systematically lower Al signature of most lunar lavas compared with terrestrial basalts could be due either to an inherently lower Al_2O_3 content for the lunar basalt source, or to the retention of some Al-rich phase in the source material (see Wood *et al.*, 1970).

Mare basalts and eucrites all have a pronounced depletion in Na_2O relative to terrestrial basalts. This indicates a similarity between the source rocks for the lunar basalts and the eucrites and implies that there was significant pre-planetary loss of volatile elements from the material from which the Moon and eucrite parent body formed. The only exception to this is the sample of Shergotty (one of the SNC group). Should the SNC meteorites be confirmed as having an origin in Mars, this would indicate a retention of significant volatiles in the matter which accreted in Mars.

Another element which shows a marked variation is chromium. Compared with lunar mare basalts and eucrites, terrestrial MORBs show a pronounced negative Cr anomaly. Since there is no compelling evidence in support of the view that there were intrinsic differences between the three bodies in respect of this element (Papike and Bence, 1978), the anomaly appears to be a function of the higher oxygen fugacity under which the MORBs crystallized, this having the effect of stabilizing Cr-spinel and the Cr^{3+} in pyroxenes (Sato, 1978).

Lastly, the sources for the different planetary basalts have intrinsically different FeO values and $Mg/(Mg + Fe^{2+})$ ratios. These important chemical indicators are signatures acquired from first-order chemical differences between the primordial matter from which they eventually grew and also from early protoplanet differentiation. The least fractionated terrestrial MORB glasses are considered to be the most primitive basalt magmas; these have $Mg/(Mg + Fe^{2+}) = 0.72$. The low-Ti (VLT) lunar mare basalts, in contrast, have ratios between 0.60 and 0.64, while the eucrites nearly all fall within the range 0.35-0.45. If the latter are to be considered as primitive partial melts, as many workers believe them to be (Walker *et al.*, 1979), then their chemistry should reflect the chemical constitution and residual mineralogy of their sources, which must, therefore, have had significantly lower $Mg/(Mg + Fe^{2+})$ ratios than most terrestrial basalts, and by implication, the Earth's mantle.

Little has thus far been said of Mars. The surface sample set is not very helpful in the present context; however, the group of eucrite-type bodies known as the SNC meteorites may shed some light on the primitive chemistry of this planet. These olivine-pyroxene samples have very different oxygen isotope and volatile contents from the other eucrites. The volatile contents are much higher in the SNCs. This indicates an origin in a body which must have been large enough to retain volatiles. Furthermore the 1.3 million year formational ages for this group imply that the body was also large enough to retain sufficient internal heat such that internal melting occurred over a period at least as long as 3.0×10^9 years. Such a body must have been larger than the Moon (Wasson, 1985) which adds some circumstantial weight to the proposed Mars origin.

Analysed Venusian rocks indicate a basaltic chemistry for plains-forming lavas, which trace element values for U, Th and K suggest somewhat more evolved types basalt, akin perhaps to certain terrestrial continental rocks. Regrettably the sample bank is very small and little more can be said at this stage.

3.5 EPILOGUE

The lunar mare basalts cover about 17% of the lunar surface. Similar, generally low-lying, plains units cover approximately 20% of those parts of Mercury imaged by Mariner 10, and occupy 30% of the more northerly hemisphere of Mars. Radar imagery reveals roughly 85% of the Venusian surface is given to volcanically-generated plains. The widespread occurrence of such surfaces on all of the inner planets points to a general episode of basalt-like volcanism relatively early during the history of the solar system. It is clear that volcanic activity on the Moon was finished very early (i.e. by 3.1×10^9 ago) and this appears to be true also for Mercury. The volcanic resurfacing of the northern part of Mars may have commenced at the same time but extensive plains- and shield-building activity appears to have continued for much longer there. For Venus there is considerable uncertainty about the oldest volcanic rocks, but much evidence to support the notion that plains and shield-building volcanism continued into quite recent times.

Lowman (1976, 1978) has termed this period of plains formation and shield-building the "second differentiation", which may have occurred at quite different times on the inner worlds in response to such factors as their differing masses and condensation temperatures. This appears to be a not inappropriate name for the phase of partial melting and fractional crystallization which was the natural development after the primary chemical differentiation was completed. Subsequent internal activity may or may not have given rise to a phase when plate tectonic activity occurred. Earth certainly reached this stage of geological evolution and it is possible that Venus too developed this way. There is still some debate as to whether or not plate movements commenced on Mars; today Mars appears to be relatively inert tectonically, but Kaula (1975) opines that a stage of plate movement may have taken place on Mars until lithospheric thickening prevented further lateral plate activity.

4

Magma ascent and eruption

Apart from the Earth, the only other worlds upon which an active eruption has been seen is Jupiter's third largest satellite, Io, which was imaged by the two Voyager spacecraft in 1979-1980 and Neptune's moon, Triton. Consequently our firsthand knowledge of volcanic eruptions comes almost entirely from the Earth and any survey of eruption styles and mechanisms necessarily must be based on terrestrial experience. Yet despite our ignorance concerning active volcanism on other worlds, it is possible to speculate with a fair degree of confidence about the processes which have generated the various volcanic landforms revealed in spacecraft images. Naturally there is also a dearth of firsthand information regarding how magmas form and where, but our knowledge of the laws of physics, coupled with modern geophysical data relating to the Earth's interior, allows us to make certain informed predictions.

4.1   MELT FORMATION AT DEPTH

Exactly how magmatic liquids form, collect and then rise up through planetary mantles and crusts is unknown; however the laws of physics dictate that in a gravitational field any rock melt which is less dense than its surroundings will tend to rise towards the surface. How far and at what rate such uprise may be accomplished is a function of a complex interplay of factors which include properties of the magma itself, such as viscosity, yield strength, density, volatile content and bulk chemistry, and the dimensions of the conduit which are a function of planetary gravity, the nature of the enclosing rocks, the depth of the magma source region and the lithostatic pressure gradient. A comprehensive appraisal of the principles of hydrostatic and hydrodynamic equilibria which govern the rise of magmas from deep source regions has been undertaken by Williams and McBirney (1979, pp. 43-63).

At deep levels in the Earth's crust or in the upper regions of the mantle, rock melts may commence their lives in "zones of melting" where they start to collect along intergranular boundaries. With only a few per cent partial melting, any magmatic liquid that is generated probably will exist as rather tenuously connected intergranular films which, because of capillary attraction, can draw away from the encompassing solid matrix only with considerable difficulty. As melting proceeds, however, and a greater volume of liquid accumulates, eventually the pockets may join and form dendritic channel networks composed of liquid-crystal "slush". Modelling experiments conducted by Scott and Stevenson (1980) suggest that, in the zone of melting, regions of locally high porosity material are able to ascend through regions of lower porosity, giving rise to what have been termed magmons The porous flow of this buoyant liquid through partially molten mantle rock is believed to be the initial process that leads to magma segregation in the asthenosphere. At such deep levels the matrix material would be able to deform, therefore it could compact to fill the void left behind as the more buoyant liquid flows through it and ascends (Turcotte and Ahern, 1978; Waff and Bulau, 1979). Since the porous flow mechanism sees matrix material being left behind, it differs from diapiric rise, where there is combined upward movement of both melt and matrix.

4.2   MAGMA ASCENT

Subsequent ascent of any melt so created either could be a function of the pressure resulting from the volumetric expansion associated with melting or of the inherent buoyancy of the liquid generated. Williams and McBirney (1979) have shown, however, that the former is insignificant compared with the latter, and evidently it is the hydrostatic instability of a lower-density melt overlain by higher-density rocks which causes it to rise. This buoyant force is in turn counteracted by both the frictional and viscous resistance of the enclosing material, and only if it exceeds the total of all the resistive forces operating against it will the magma be able to rise towards a planet's surface.

During ascent, a magma will tend to equilibrate both thermally and chemically with the enclosing rocks and with the crystals it holds in suspension. The very slow rise of a melt through rocks at temperatures below their melting points would soon lead to its crystallization and the cessation of upward progress; however, should there be repeated influx of successive batches of newly generated magma, then this gradually would lead to the warming of the overlying rocks, until eventually the rising melt need only equilibrate to a temperature and composition appropriate to its depth. Thereafter it might continue to rise without excessive further loss of heat. Such repetitious activity is to be expected because a rising body of hot magma will leave behind it a thermally disturbed and volumetrically expanded zone which successive batches of melt would exploit in preference to pursuing virgin courses in unheated rocks. This effect may at least partly explain the persistence of volcanism at given loci over prolonged periods.

The general absence of deep seismic activity beneath major volcanoes suggests that in most regions of the Earth rock melts do not rise directly from the base of the lithosphere. Indeed the consensus view appears to be that beneath most large volcanoes, batches of new magma rise occasionally to fill relatively shallow storage reservoirs. This suspicion is supported by numerous field observations. For instance, in Hawaii it has been observed often that while effusive eruptions are taking place from the summit of Mauna Loa, an active lava lake exists within the nearby, but 1000-metre-lower, caldera of Kilauea. Clearly two contemporaneous columns of magma are here operating as quite separate systems.

Geophysical evidence indicates that the density of the Earth's LVZ may be less than that of the overlying mantle. In consequence, sinusoidal bulges (Rayleigh-Taylor

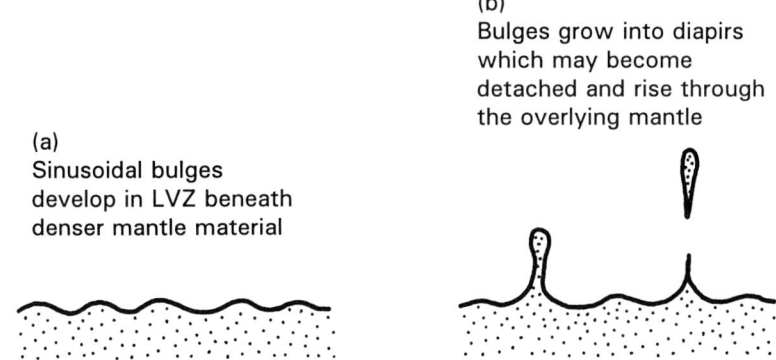

(b)
Bulges grow into diapirs
which may become
detached and rise through
the overlying mantle

(a)
Sinusoidal bulges
develop in LVZ beneath
denser mantle material

Fig. 4.1: Development of Raleigh-Taylor instabilities and generation of diapirs.

instabilities) may be generated in the LVZ as less-dense material collects in localized bulges beneath the denser mantle above (Fig. 4.1). Smaller bulges cool, thereby losing their density contrast relatively quickly and dying away. In contrast, larger bulges slowly grow upwards, their rate of progress gradually accelerating until they form steep-sided diapirs or plumes. The wavelength of these will be a function of the thickness and viscosity contrast between the contrasting layers. Marsh (1978) notes that in the Aleutian-Cascade andesitic chain the spacing of individual volcanoes is about 70 km, while calculations by Fyfe (1973) predict diapir wavelengths on the scale of tens of kilometres for granitic plutons.

The ascent of diapiric bodies is frequently achieved by forceful shouldering aside of the country rocks. At relatively shallow levels (i.e. in the upper crust) where the enclosing rocks are relatively brittle, magma emplacement may be accomplished largely by engulfment of blocks (stoping) from the walls and roof surrounding the ascending diapir and/or uplift of the roof rocks. Much exploitation of existing weaknesses such as faults, joints and bedding planes would be expected. These effects are aided by thermal stresses set up in the country rocks as intrusion proceeds, accordingly, wherever temperature gradients are steep, differential expansion occurs. Because trapped groundwater becomes heated to magmatic temperatures and experiences pressures of hundreds of gigapascals in such an environment, assuming it is free to expand, the vapour will open incipient cracks and detach wedges of wall rock which fall into the magma chamber. Even at depths of several kilometres, the pressures induced may be sufficient to propagate fractures well beyond the extremities of the rising pluton (Spence and Turcotte, 1985), especially if the enclosing rocks are already being subjected to differential tectonic stresses.

Considerable experimental and modelling work has been completed upon the diapiric rise of mantle melts (Grout, 1945; Ramberg, 1967,1970; Fyfe, 1973; Marsh and Carmichael, 1974; Elder, 1976; Fedotov 1975, 1977; Carmichael *et al.*, 1977; Marsh, 1978; Marsh and Kantha, 1978). A diversity of opinions exists regarding how the process actually operates. Some workers consider that magmas rise through the asthenosphere as slender columns which remain attached to their source, while others assert they become detached from their source and rise solely in response to their own buoyancy. From theoretical considerations alone, the latter appears the more acceptable; there is little likelihood that any reasonable density contrast in the Earth's mantle would be able to provide the pressure differential necessary to drive a slender column upwards. Detachment also is in closer accord with the observations: (a) that terrestrial volcanism is episodic, (b) that discrete batches of magma are of limited volume, and (c) that they are chemically distinct.

One important characteristic of diapiric activity is that it may involve only very limited mechanical segregation of liquid and crystals within the region of melting, instead there being production of a mush of crystals and liquid with a bulk density below that of the enclosing rocks. The continued upward progress of such a diapir could lead to further melting but it is to be anticipated that only at relatively shallow levels would such a process produce an effective mechanical separation of the magmatic liquid from the solid matrix.

An alternative mode of ascent may be provided by a kind of solution stoping termed zone melting. This relies upon the notion that, if a body of relatively anhydrous magma has an appreciable vertical extent, there is no way that it can approach equilibrium with the same crystals at its top and near its bottom, assuming uniform temperature. In turn, this depends on the premise that there must be a significant increase in the temperature of the solidus with depth. As a result, a pod of liquid can rise by melting its roof while at the same time crystallizing near its floor. Such a process would allow a zone of melting to pass upwards with limited physical transfer of melt, and, incidentally, has been advocated by Harris (1957) as one way in which incompatible elements may be enhanced in partial melts of mantle peridotite.

The occurrence of block-filled pipes or narrow funnel-shaped bodies (diatremes) of kimberlite in various terrestrial continental locations (see for instance, Boyd and Meyer, 1979; Nixon, 1973) is generally taken to imply that there has been rapid rise of highly gas-charged magmas, probably from considerable depth. It appears that on the way towards the surface some volatile-rich melts lose sufficient dissolved gas due to boiling that a highly expanded, fluidized column of melt and solids may rise with sufficient power to cause significant upwarping and even disruption of the overlying cap rock. Should the latter occur, explosive activity is typically manifested in maars or tuff rings. McGetchin and Ulrich (1973) have discussed the inferences for the terrestrial planets from the lithostatic rise of gas and magma in a diatreme, while Pai *et al* (1978) modelled a lunar example.

Regions of the Earth such as the Columbia River Plateau, the British Hebridean Tertiary province and the state of Northern Queensland in Australia, are the locations of intensively studied and very widespread dyke swarms. The Tertiary dykes of northern Great Britain nicely illustrate the scale of such swarms, outcropping over a region which is approximately 400 km long and 280 km wide (Fig. 4.2). These and similar swarms elsewhere were associated with eruptions of large-volume basaltic flows from fissures which must have tapped extensive magma reservoirs that lay somewhere between the base of the Earth's crust and the top of the asthenosphere. On some of the other terrestrial planets, fissure volcanism appears to have been even more extensive than on the Earth and it might be expected that future generations of planetary geologists may discover enormous dyke families on these worlds too, if sufficient erosion has occurred. Parallel or subparallel dyke swarms and related fissure eruptives typify regions of brittle lithosphere subjected to tectonic rifting. Here magma has invaded more-or-less vertical dilatational cracks that have been propagated in a direction perpendicular to the principal tensional stress. Radial

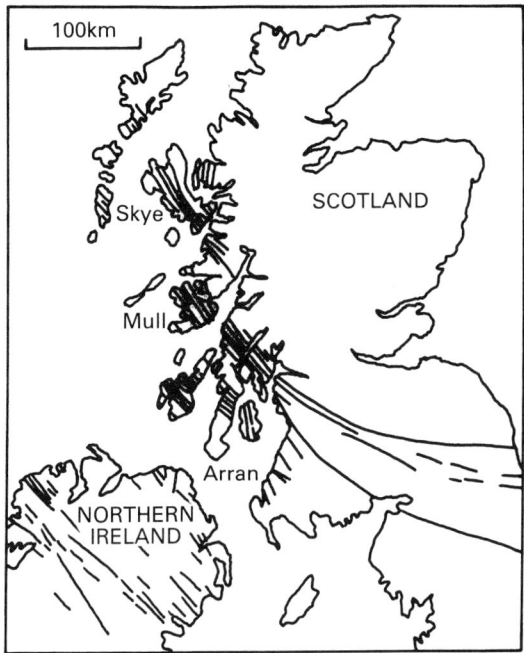

Fig. 4.2: Tertiary basic dyke complex of northwestern Britain.

and annular dyke swarms also develop, usually being associated with high-level central intrusions where magma has risen either along dilatant fissures generated normal to the direction of least stress, and thus orientated tangentially with respect to the perimeters of subvolcanic plutons, or along ring-faults produced by cauldron subsidence.

The ascent of magma on Mars is likely to follow a similar course, however, there is evidence to suggest that Martian partial melts may be iron-enriched or ultrabasic (Wilson and Head, 1993). Additionally, because of the lower surface gravity, diapiric ascent rates are anticipated to be lower than on Earth; the result of this is that larger diapiric bodies would rise to shallower depths on Mars. This favours the intrusion of dykes that would be wider by a factor of two compared to those of the Earth which, in turn, enables much higher mass effusion rates to be achieved that would reach a factor of five times those typical of our own world. For the same style of cooling-controlled effusive activity, volcanic flows might be expected to be about six times as long as those typical of the Earth.

In contrast to the above, on Venus the high atmospheric pressure would inhibit the formation of neutral buoyancy zones and shallow magma reservoirs (Head and Wilson, 1992). The latter would only begin to appear when the magma gas content increases significantly. Since magmas would tend to retain their volatiles for much longer than on either the Earth or Mars, Venusian magmas of similar (basalt-like) chemistry might exhibit quite a wide range of rheology, allowing quite viscous flows to generate domes or be intruded as dykes, sills and high-level intrusions.

4.3   TYPES OF VOLCANIC ERUPTION

On Earth there are two principal styles of volcanic activity: there are eruptions which emanate from a central vent (central eruptions) and those which are associated with fissures (fissure eruptions). The former are related to the rise of magma up a single conduit which taps a magma source at a much deeper level and the erupted products are distributed in a roughly radial pattern around the vent. Fissure eruptions occur in regions suffering crustal extension and magma is able to force its way towards the surface along one or more pathways situated along fractures. At the surface, lava may erupt from open fissures, while magma coagulating in the subsurface pathways forms dykes. Because of the rather different geometry involved, the distribution pattern for volcanic materials associated with fissure eruptions generally is quite unlike that of central episodes (Fig. 4.3). The extreme case of terrestrial fissure eruption is where very large volumes of fluid basaltic magma is erupted near continental margins during the early stages of continental divergence. During such major extensional movements, low viscosity basaltic lavas may be erupted from laterally extensive linear vents or fissures. These provide pathways for flood basalts which may cover vast areas, giving rise to thick sequences of lavas such as those of the Columbia River Plateau region of the western USA, where extensive Quaternary basalt flows cover an area of 130 000 km^2 and attain a thickness of over 1 km.

The state of the materials which emerge from a vent, whether it be of the central or fissure type, is a function of several factors. Because these often change with time, even during the course of a single eruptive outbreak, the amount and kind of material being erupted may alter, and the height and lateral distance it may travel can change. These variations give rise to an extensive array of volcanic landforms whose diversity is a function of variable eruption style. Hawaiian eruptions, for instance, are characterized by relatively quiet effusion of hot, fluid, basaltic lavas and are manifested in the kind of activity so frequently observed on the active volcanic islands of the Hawaiian chain. This particular style of eruption is brought about because the high-pressure gases within the rising magma are able to escape both readily and steadily, thereby preventing large-scale coalescence of gas bubbles within the magma and the possibility that they may suddenly be released. The facility with which gases can escape often leads to spectacular fire fountains in which molten basalt may rise as high as 400 m into the atmosphere before falling back to the ground around the vent, coalescing into molten pools which drain away as lava flows. Fire fountains may play for hours on end, and a series of particularly long-lived and spectacular ones were captured on film during the 1959 eruption of Kilauea, Hawaii.

The island of Stromboli, situated between the Italian mainland and Sicily, gives its name to a more explosive type of eruption typified by the spasmodic escape of gases trapped within basic lavas, giving rise to a series of either regular or irregular explosions separated from one another on a timescale measurable in minutes. Such Strombolian eruptions are common on the Earth, typifiying activity at centres as widely separated as the much-studied Sicilian volcano Etna, Mount Erebus in Antarctica and numerous stratovolcanoes high in the Andes. The products thrown out during typical Strombolian eruptions are mainly lava which forms bombs and blocks generally of small size that do not travel far from the vent area. Many Martian eruptions may have been of this kind, the main difference between Martian equivalents of this style and those of Earth being that,

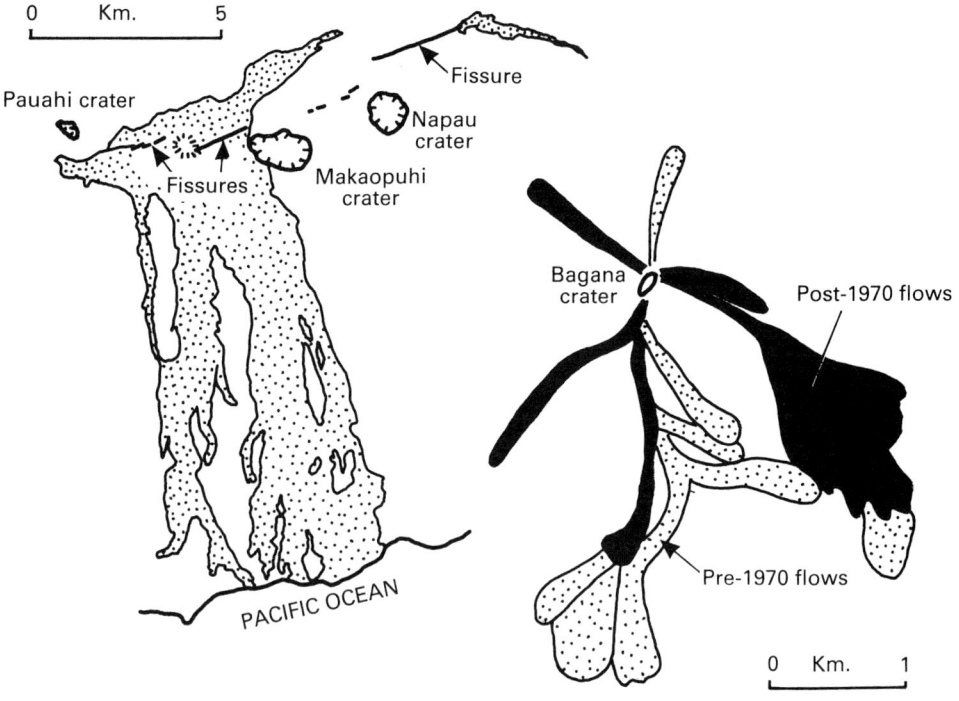

Fig. 4.3: Contrasting distribution of volcanic flows associated (left) with 1969-71 fissure eruption of Mauna Ulu, Hawaii (Swanson, D. A. *et al.*, 1979) and (right) central vent eruption of Bagana volcano, Papua New Guinea (Bultitude, R. J., 1976).

although the larger clasts will remain near the vent, the finer materials will be more broadly disseminated on Mars, the finest debris being carried high into the Martian atmosphere in convecting ash plumes. Also, there would be a greater degree of comminution of debris on Mars, generating much more finer material.

While lava involved in Strombolian activity is generally slightly more viscous than that characteristic of Hawaiian eruptions, its viscosity is not so great that really violent explosive activity develops. This occurs where more viscous magmas are involved, if there is a major build-up of dissolved gases beneath the vent area. When this occurs, as it has at Vulcano in the Aeolian Islands (and also at Stromboli), widely separated but highly explosive outbreaks send into the atmosphere clouds of ash as well as molten lava to give Vulcanian eruptions. In response to the spasmodic but very violent escape of superheated volatiles that is their trademark, huge blocks may be tumbled around in the vicinity of the vent, while large portions of the volcanic superstructure may be destroyed and the morphology of the volcanic mountain dramatically transformed. Where the escape of magmatic gases is more continuous, massive, optically dense, eruption plumes composed of fine-grained ash give rise to what sometimes are termed Vesuvian eruptions, after the famous volcano that forms such a dramatic backdrop to the Italian port of Naples. More often, however, volcanologists classify eruptions of this kind as sub-Plinian.

True Plinian eruptions are amongst the most violent seen on the Earth and were responsible for the much publicized events at Mount St. Helens in 1980. There have been even more catastrophic explosions, for instance those of the Indonesian island, Krakatau in 1883, and that at Bezymianny (Kamchatka) in 1955. At Krakatau, the ejection of an enormous pumice cloud was followed by outflow of fast-moving glowing avalanches, also made from pumice. Their eruption caused the draining of the subterranean magma chamber whose roof collapsed. This set up a chain of tsunamis which swept the densely populated coasts of both Java and Sumatra, causing great loss of life. When Bezymianny blew up, over 4 km^3 of volcanic ash was blasted out to cover an area of 57 000 km^2. In all of these cases major changes took place in the shape of the volcanic superstructure and, characteristically, huge volumes of material were swept up into the eruption cloud. Where similarly violent eruptions generate ground-hugging "glowing avalanches" or nuées ardentes, as was the case at Krakatau as well as at Mont Pelée in 1902, the term Peléan eruption is often used. These violent and excessively dangerous outbreaks are usually associated with the growth of a lava dome above slowly-rising, viscous magma. Similar nuées burst out of the vent during the 1955 Bezymianny eruption too, but whereas at Mont Pelée they composed the main part of the eruption, in Kamchatka there was a massive vertical eruption cloud too.

Plinian eruptions on Mars would see eruption clouds rising roughly five times higher than their terrestrial equivalents, for the same mass eruption rate. One important consequence of the lower atmospheric pressure would be that Plinian eruptive deposits on Mars would be about 100 times finer than those of Earth, being almost entirely sub-centimetre in grade. It would also be more probable that basaltic magmas would give rise to Plinian style eruptions. This is partly a function of the greater degree of fragmentation on that planet, but also the apparent frequent interaction of basic magma types with groundwater reservoirs.

4.4   CONTROLS ON THE RISE AND EFFUSION RATES OF MAGMAS

Volcanic processes on the silicate planets (among which, incidentally, it is necessary to include Io, since its sulphur volcanism is believed to be driven by a silicate mantle) are dependent upon planetary gravity, surface temperature and atmospheric characteristics. Because there are significant differences between one planet and the next in terms of at least one of these controls, it is entirely natural to anticipate that there will be variations: (a) in the way magmas reach the surface, (b) in the way they behave in the vicinity of vents and (c) in the morphologies of the volcanic structures themselves and in the form and extent of their eruptive products. To give an example, on the Moon and Io, which effectively are atmosphereless, volatile expansion in the near-surface environment would be extreme when compared with either the Earth or Venus. In consequence, lunar and Ionian pyroclasts would be predicted to have much higher ejection velocities and wider distribution than comparable terrestrial or Venusian eruptives.

In the immediate vicinity of a volcanic vent the eruptive products which accumulate may include molten lava, solid pyroclasts and blocks of country rock (xenoliths) which may or may not be volcanic in origin. The nature, grain size, proportions and lateral extent of these various materials will be a function of several different factors to which a brief consideration will now be given. Should the reader require a more detailed treatment, then the following papers will be found particularly informative: Fedotov (1978), Head and Wilson (1986), Head et al. (1981), Mouginis-Mark et al. (1982), Shaw (1980), Shaw and Swanson (1970), Solomon (1975), Weertman (1971), Wilson and Head (1981, 1983), Wilson et al. (1980).

The first point to be made is that in the deeper parts of the Earth's crust rocks can deform plastically, while in the near-surface layers they may be partly unconsolidated. For this reason the tensile strength of crustal materials will tend to reach a maximum at some intermediate level. As a consequence, once a steady flow of magma towards a planet's surface is established, there is little likelihood that the pressure within the melt in the near-surface environment will be very different from the local lithostatic pressure. The rise velocity of a magma (u_d) at a depth below that at which volatile exsolution occurs, is a function of the magma density (ρ), viscosity (η) and yield strength (y), together with conduit diameter/fissure width (w), planetary gravity (g) and the effective density difference between the rising magma and the enclosing rocks ($\delta\rho$). It may be represented by the equation:

$$u_d = (w^2 g \delta\rho - 2yw)/12\eta$$

while the corresponding mass eruption rate, M, may be shown by

$$M = wL\rho u_d = w^2 L\rho(wg\delta\rho - 2y)/12\eta$$

where L is the fissure length measured in the horizontal plane. It will be noticed that the eruption rate is strongly dependent upon the fissure width (or conduit size). Because differences in tectonic regime will largely control the upper limits on fissure width and length via strain accumulation and the depth at which a magma originates will control the lower limit on fissure width by the cooling constraints it imposes, both of these factors will be significantly more important than gravity differences between planets. Wilson and Head (1983) suggest that they may be as least as important as the effect that the different rheologies of varying magma types may have on an individual planet. Fedotov (1978) in modelling the expected parameters for hydraulic fractures on the Earth, arrived at a width to length ratio in the range 10^{-2} to 10^{-3} for near-surface layers, and 10^{-3} to 10^{-4} for the deep crust.

The width of a tectonically produced fissure will depend upon the amount of stress which has to be alleviated and the area of the strained region, while its length is dependent on the stress distribution prior to the fracturing. When a fissure initially opened in response to a large excess (nonhydrostatic) pressure in the source region, if the pressure is relieved during eruption, then it may close and further progress of the magma be prevented. On the other hand, when during the course of an eruption there is wall erosion, as often occurs, the width of a fissure may be enlarged. Because wall erosion is likely to be at a maximum where the magma rise velocity is at its peak, there will be a propensity for the rising sheet of magma to concentrate into a small number of discrete, vertical zones. This tendency would be expected to be most pronounced during the early stages of an eruption, as was observed during the 1973 Heimay eruption, in Iceland (Thorarinsson *et al.*, 1973).

Magma cannot attain the surface of a planet should a fissure be too narrow or a conduit of a diameter too small; the limits to both will be imposed by the effects of magmatic cooling or the presence of a yield strength in the melt. For an eruption to be maintained it is necessary that the mass flow rate in a fissure system is at least equal to the mass of magma erupted per unit time from the surface vent. Since the rise and effusion rates of a magma are predominantly a function of the vent geometry, it follows that the greatest rise velocity for a given mass flow rate will be that appropriate to a circular conduit, the rise velocity decreasing as the ratio of the length to the width of a vent or fissure increases. For terrestrial basalts with viscosities greater than 10^2 Pa s^{-1} and very low yield strengths, fissure widths in the range 0.2 to 0.6 m are necessary in order for eruptions from depths of between 500 m and 200 km to occur. Very high volume terrestrial flood eruptions like those which produced the Columbia River basalts are estimated to have had average output rates of between 10^2 and 10^7 kg s^{-1} m^{-1}, (Swanson *et al.*, 1975) and would require fissures 4 m wide to accommodate such elevated effusion rates. Mare-filling basaltic eruptions on the Moon are estimated to have had even higher mass eruption rates, i.e. 10^6-10^9 kg s^{-1}, which would necessitate fissure widths at least three times those typically found on the Earth, and it is possible that they may have attained widths of about 10 metres (Wilson and Head, 1981). As far as Mars is concerned, it can be shown from a study of lava flow parameters that mass eruption rates often must have exceeded 10^5 kg s^{-1} (Baloga and Pieri, 1985; Cattermole, 1986) which would imply that fissure widths were probably at least twice those of Earth. That they were appears to be confirmed by the dimensions of suspected spatter ridges revealed on some Viking images (Cattermole, 1987). Due to the enhanced surface temperatures and the much smaller ratio of subsurface to surface pressure on Venus, the minimum fissure width required to permit magma to reach its surface before significant cooling occurred would be smaller than on the Earth. The magma would also be hotter on entering a fissure in the first place. For an effective density difference approaching 100 kg m^{-3}, and a magma source depth of between 1 and 10 km, Head and Wilson (1986) calculate that a minimum fissure width of between 0.3 and 1.1 m would be required. If the source were as deep as 100 km, the minimum width would increase to around 1.9 m. Such a range of values, interestingly, is not dissimilar to values calculated for terrestrial basalts by the same workers in 1981.

4.5   THE ROLE OF VOLATILES

As it approaches a planetary surface, a rising body of magma will begin to exsolve gas bubbles in response to the decreasing external pressure. Exsolution releases energy which in turn increases the ascent velocity. The general geometry of a magma conduit or fissure for three different cases is shown in Fig. 4.4. In the first (a) the rising magma is gas-free, therefore no volatile exsolution occurs and the magma rises through the vent prior to

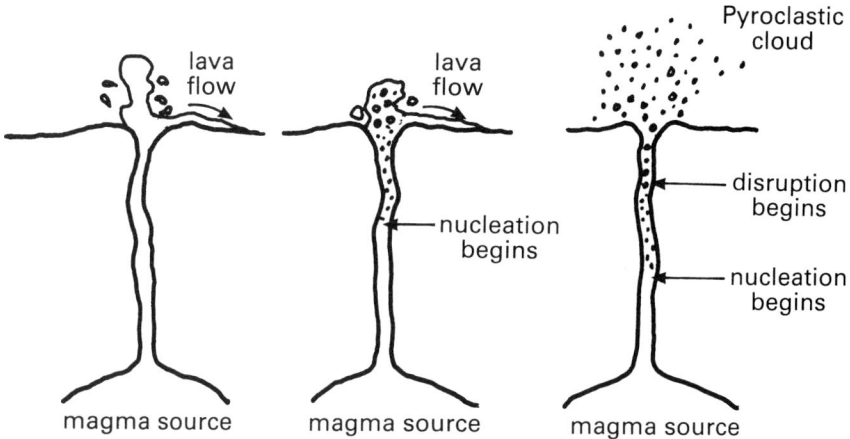

Fig. 4.4: General geometry of a volcanic conduit for three different cases: (a) gas-free eruption; (b) eruption where gas is exsolved but volume never approaches value sufficient to disrupt magma into pyroclasts; (c) eruption where gas volume is large enough for disruption of magma into pyroclasts. (After Wilson, L. and Head, J. W., (1981).)

flowing away down the volcano's flanks in a relatively quiescent fashion. The driving force for ascent would be the magma density contrast, aided probably by a build-up of stresses around the magma chamber prior to eruption, a function perhaps of an injection of fresh magma from below, or some measure of tectonic readjustment. It is the release of stress in the near-surface rocks which initially drives the eruption and supplies the replacement volume necessary to counterbalance the volume of the erupted melt (Wadge, 1981). For melts containing a gas component the conditions are more complex. At some depth beneath the surface the rising magma will begin to vesiculate but the exsolved gas volume may fail to attain a value sufficient to disrupt the magma into pyroclasts (b). In this situation, fire fountaining will occur but the molten clots will fall back around the vent to coalesce into a flow. This would be the position with a typical Hawaiian eruption. Thirdly (c), the growing population of exsolved bubbles may exceed a critical ratio with respect to the magma (which on Earth is about 0.75), whereupon the magma disrupts into pyroclasts which are ejected explosively as an ascending cloud of gas and solids. Depending on the rate at which volatile exsolution occurs and the atmospheric density, in some circumstances very high convecting clouds may develop, distributing fine material over a very large region, while in others the pyroclast-laden cloud may collapse to form, particularly in the case of viscous silicic melts, a ground-hugging pyroclastic flow. Where fluid basic magma is involved, the cloud may simply consist of molten lava droplets, forming a huge fire fountain which eventually feeds a lava flow (Sparks et al., 1978).

On Mars, because eruption cloud instability takes place at lower mass eruption rates than on Earth (for the same volatile content), pyroclastic flow formation is inherently more likely for basic magma types than on the Earth. For the same magma chemistry and volatile content, eruption velocities on Mars would be at least one and a half times those of Earth, fountains feeding pyroclastic flows rising over twice the height of their terrestrial equivalents (Wilson and Head, 1993).

4.5.1   Eruption clouds

For the disruption of magma to take place in the first place, thereby disassembling it into pyroclasts, the volatile content, assuming the predominant gas is H_2O, must exceed about 0.07 wt% on the Earth. On Mars this critical value is less at about 0.01 wt%, and the low atmospheric pressure would facilitate pyroclast formation. Unless martian magmas were unexpectedly volatile-deficient, low discharge rate effusive eruptions should be characterized by Strombolian style activity. The abundant presence of CO_2 and H_2O ice on Mars strongly suggests that most Martian eruptions would have been explosive, regardless of their chemistry, therefore even basaltic eruptions would have been characterized by severe fire fountaining.

On the atmosphereless Moon the situation would have been even more extreme and the disruption of magma into pyroclasts must have been extremely common. Very high mass eruption rates generally are believed to have pertained on the Moon during the period of mare volcanism, implying that optically dense eruption clouds would have fountained above vent areas, and predictions suggest that landing zones could have been up to 4 km wide along active fissures and up to 6 km across around central vents. Enormous fire fountains are believed to have accompanied the formation of lunar sinuous rilles (Head and Wilson, 1979) and these may have caused substantial thermal erosion of the adjacent lunar surface rocks.

In complete contrast, the atmospheric pressure in the Venusian lowlands is so extreme (about 9×10^6 Pa) that at least 2 wt% H_2O (or 5 wt% CO_2) would need to be dissolved in the magma before pyroclasts could begin to form on Venus. In the highlands, where the pressure is less, the figures would be 1 wt% and 3 wt% respectively. Pyroclastic deposits are therefore less likely to be generated on Venus than on either the Earth, Mars or the Moon. Thornhill (1993) notes that the stability of eruption plumes on Venus depends strongly on initial eruption temperature, as well as the atmospheric pressure at a vent site. Convecting plumes could form for both H_2O- and CO_2-driven eruptions; these, however, would require temperatures >730°C, would be favoured by high altitude vents and a considerable amounts of a highly soluble volatile such as H_2O. Under these conditions plume heights of around 40 km could be achieved. From what we know of Venus the likelihood of sufficient volatiles being present militates against such activity and major plumes are unlikely to form. Should very high eruption rates be involved, for the same conditions, pyroclastic flows would tend to form since plumes collapse at an early stage. Whatever the exact conditions turn out to be, plume heights on Venus would be significantly lower than those of the Earth.

When a convecting eruption cloud is able to remain in a stable condition, its maximum elevation is proportional to the fourth root of the mass eruption rate (Settle, 1978). This can be very important in terms of the distribution of pyroclastic materials, as can be illustrated by noting that, for the same mass eruption rate, an eruption plume could rise five times higher on Mars than on the Earth, yet could attain only a third the height on

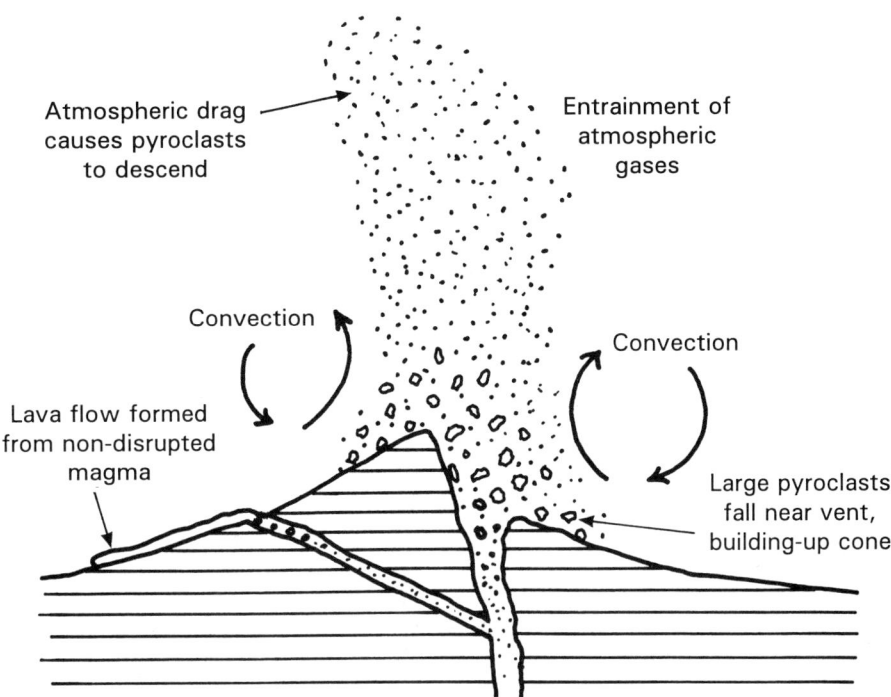

Atmospheric drag
causes pyroclasts
to descend

Entrainment of
atmospheric
gases

Convection

Convection

Lava flow formed
from non-disrupted
magma

Large pyroclasts
fall near vent,
building-up cone

Fig. 4.5: Schematic diagram illustrating the more important processes which may occur in the neighbourhood of a volcanic vent.

Venus. However, this proportionality is a function of the physical properties of each planet's atmosphere. On any planet with an appreciable atmosphere, gas/pyroclast clouds generated in this way will strongly interact with the atmospheric gases which become entrained in the eruption cloud and heated by the hot tephra. During terrestrial eruptions, atmospheric gases begin to have an affect when rock fragments within a cloud cease to rise and their horizontal velocity carries them further from the central column (Plate 15). At this stage their motion becomes dominated by the now much larger mass of the atmosphere and the final part of the trajectory for small pyroclasts is significantly affected by atmospheric drag, which causes them to fall (Fig. 4.5). On Mars, where the atmospheric pressure at ground level is only 0.006 times that of the Earth, the initial expansion of an erupting cloud will be much greater and subsequent atmospheric drag far less. Furthermore, since the Martian gravity is only about one-third the Earth's, for a given rise velocity, an eruption plume will rise to roughly three times the height it would on Earth. On this basis, air-fall pyroclastic blankets on Mars are expected not only to be thinner but significantly more dispersed than on Earth.

Where a planet has no appreciable atmosphere, as is the case with the Moon, at very modest heights above the vent area individual pyroclasts will become widely separated and as a result the expanding gas tends to lose good thermal contact with the hot fragments it entrains. Where the magma ascent velocity was too rapid to allow gas bubbles to coalesce and grow, steady lunar eruptions would result. Wilson and Head (1983), in modelling lunar eruption characteristics, assumed that carbon monoxide was the dominant volatile present (Housley, 1978). On this basis they calculated that, for a 0.07 wt% volatile content, steady lunar eruptions involving pyroclasts largely of sub-millimetre size, would be characterized by ejection velocities of about 40 m s^{-1} and would be capable of ejecting pyroclasts out to a distance of around 1 kilometre. Where magma ascent was relatively tardy, giving time for bubbles to coalesce and grow to larger size, then non-steady (Strombolian) style eruptions would occur since the gas fraction in the escaping magma may have risen to as much as 10%, leading to ejection velocities of the order of 500 m s^{-1} and emplacement of ejecta out to distances measurable in tens of kilometres. Thus there is a strong likelihood that large areas of extremely thin pyroclastics outcrop on the Moon but are irresolvable in currently available imagery.

Eruptions on the Jovian moon Io appear to involve either SO_2 or sulphur (Smith *et al.* 1972) as the volatile constituent. The observed 300 km-high eruption plumes imply ejection velocities of about 1 km s^{-1}, which in turn suggests that vent-exiting materials actually are in the gaseous form. In complete contrast, the very high temperatures prevailing in the atmosphere of Venus (650-750 K) and the elevated atmospheric gas pressures (4-10 MPa) will ensure that pyroclastic deposits on that planet are very much less extensive than on the other terrestrial planets. Indeed, it is conceivable that very little pyroclastic material exists there.

On Earth, eruption cloud heights observed between 1970-74 show that the maximum height attained by a cloud is positively correlated with the rate at which pyroclastic material is generated by an explosive eruption. By applying scaling relationships developed for convective thermal plumes to the maximum height a cloud reaches, estimates of the amount of thermal power released have been obtained (Settle, 1978). Thus the scaling relationship between height and thermal flux observed in industrial hot gas plumes, i.e.

$$dH \propto QH^{1/4}$$

appears to be broadly representative of volcanic eruption clouds. This being so, the effective thermal power output of an explosive eruption is approximately linearly related to an eruption's pyroclastic mass flux. As an example, the paroxysmal eruption of Gunung Agung on 16th May 1963, with its maximum attained altitude of 10 km, and pyroclastic volume of 3×10^7 m^3, generated 10^{13} cal s^{-1}.

4.5.2   Phreatomagmatic eruptions

Thus far the volatiles involved in eruptions have been assumed to be magmatic in origin. However, on the Earth, groundwater may gain access to a vent, or disrupted magma may be blasted out through standing water, giving rise to phreatic eruptions. Should this take place, there will be much more extensive disruption of the pyroclasts, which may become very finely divided and then ejected either steadily, to give phreato-Plinian events, or episodically, generating Surtseyan eruptions (Self and Sparks, 1978). Maar craters are aften formed under

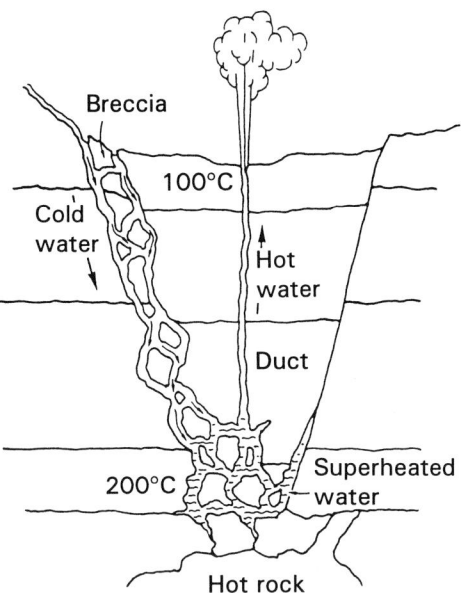

Fig. 4.6: Cross-section through a geyser. This may form where magma is sufficiently close to the surface and where the caprock is fractured. Water percolating downwards becomes superheated, expands and the entire water column is transformed into steam.

such circumstances. Because of the suspected extensive subsurface volatile reservoir on Mars, the likelihood of phreatic eruptions there must be very high and some preliminary modelling of this style of eruption has been undertaken by Mouginis-Mark *et al.*, 1982). The observation of blanketed and channelled flanks to certain Martian volcanoes (e.g. Ceraunius Tholus, Uranius Patera, Alba Patera and Hecates Tholus) has led some workers to speculate they were the products of volcanic density currents. Such eruptions might be expected where rising magma came into contact with a thick permafrost layer, generating a base surge (Reimers and Kolmar, 1979). Such activity may have been common early in Martian history. Interestingly, base surges may also have been generated within outflow deposits associated with Venusian impact events. Pyroclastic flows would be expected to generate long run-out deposits on planets with low atmospheric pressures.

Geysers are a common terrestrial phenomenon and involve the interaction of shallow intrusions with groundwater. A similar situation, not necessarily involving silicic magma, has been proposed for violent eruptions believed to typify certain outer planet moons; such eruptions may have been driven by either nitrogen or methane. On Earth, geyser activity is found in such widely dispersed regions as New Zealand, Iceland and Yellowstone National Park, Western U.S.A. Geysers effectively are boiling springs which periodically spew out a column of steam into the atmosphere as the magmatic conduit is intermittently emptied (Fig. 4.6). A similar style of activity may carry nitrogen compounds high into the atmosphere above moons such as Triton.

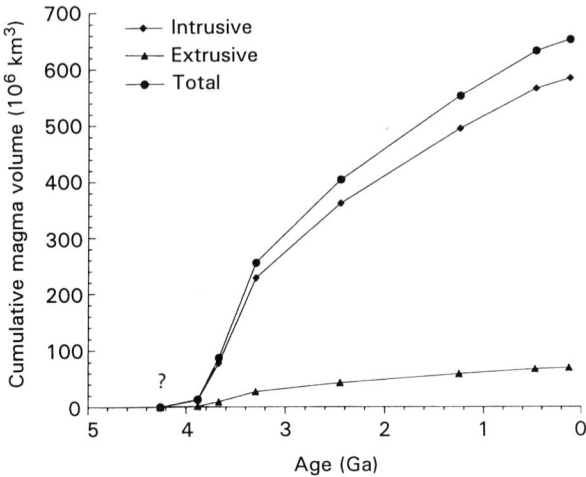

Fig. 4.7: Cumulative magma production on Mars as a function of time. (After Greeley and Schneid, 1991)

Finally, the build-up of gas pressure beneath congealed lava plugs, such as occurred within the superstructure of Mount St. Helens during 1980 (Foxworthy and Hill, 1982), is typical of low effusion rate eruptions of high viscosity magma. The resultant Vulcanian explosions are highly dangerous and, on Earth, often have caused great loss of life and extensive damage to property. While such magmas may not be widespread beyond the Earth (see, for instance, Francis and Wood, 1982), Head and Wilson (1979) have invoked a similar phenomenon to explain certain lunar dark halo deposits.

4.6   RATES AND AMOUNTS OF MAGMA GENERATION ON PLANETS

An estimate of the amounts of magma generated within the terrestrial planets, particularly Mars, has been made by Greeley and Schneid (1991). By mapping the extent of volcanic materials, estimating their thickness, assigning ages by crater counting and superposition relations, they were able to determine volumes of material generated at different periods in Martian history. Rates of magma production were then arrived at after making reasonable assumptions concerning the ratio of extrusive to intrusive activity and were then compared with rates estimated for Venus, Earth and the Moon. The results of their calculations are presented in Table 4.1 and Fig. 4.7. The data suggests that Martian magma production peaked during the Early Hesperian (circa 3.3 Ga), when 168×10^6 km^3 of magma was generated. This period of activity manifested itself in the plateau-forming flood lavas and (inferred) intrusions into the ancient cratered crust. The average thickness of volcanic plains arrived at in the above work is 320 m in the Late Noachian to 320 m during the Middle Amazonian. The average is roughly half that for the lunar maria. The trend they note for Mars is that the thickness of volcanic deposits increased with time. Such a trend could signify gradual changes in eruptive style, composition or viscosity.

Table 4.1: Planetary magma production rates. Rates for Venus are for the last 1 Ga; those for Earth are for the last 180×10^6 y, those for the Moon, over the last 3.85 Ga, and those for Mars, for the last 3.9 Ga.

Planet	Extrusive production (km³/yr)	Total magma production (km³/yr)	Scaled extrusive *production† (km³/yr)	Scaled total production‡ (km³/yr)
Venus	≤2.0	≤19	≤0.63	≤0.78
Earth	3.7-4.1	26-34	1.0	1.0
Moon	0.0024	0.025	0.052	0.069
Mars	0.018	0.17	0.042	0.052

*Assumes intrusive/extrusive ratio of 8.5:1 for Venus, Moon and Mars.

†Scaled to Earth mass and Earth extrusive production rates (~30 km³/y) over the last 180×10^6 y.

‡Scaled to Earth mass and total magma production rates (~30 km³/y) over the last 180×10^6 y.

Comparison of magma production rates on Mars, Earth and Venus indicates that extrusive production rates appear to be a function of planetary mass; Mars has the lowest rate and Earth the highest. Results for total magma production rates also scale with planetary mass, but, even when scaled to the terrestrial mass and production rate, magma generation on Mars is significantly lower than for the other inner planets. Lunar rates appear anomalously high; thus, despite having a lower mass than Mars, both extrusive and total magma production are greater than for Mars. This could reflect the greater importance of tidal stresses in lunar evolution.

5

The rheological properties and behaviour of volcanic flows

Lava flows move downslope and have a natural predilection for finding topographic lows. In this respect their behaviour is akin to that of water under the influence of a gravitational field; however, in most other respects they behave rather differently. For instance, regardless of whether on a steep or shallow slope, the front of a moving lava flow will cease to move when the rheological properties of the lava prevent its further progress.

On the Earth it has been possible, albeit with considerable difficulty (and often at great personal risk to volcanologists!) to obtain relatively accurate *in situ* measurements of such important physical properties of flowing lavas as temperature, viscosity and gas content and to measure, usually from a more comfortable distance, the height and optical density of eruption clouds, and the temporal and spatial progress of a variety of volcanic bodies, including active lava flows, viscous domes, pyroclastic flows and lahars. The collection of this data has enabled us to relate the rheology of erupted materials to the morphologies of the landforms they produce and to gain insights into which particular physical properties have been paramount in dictating their behaviour. Those physical processes which have the more pronounced effects include radiative cooling, vesiculation, crystallization, and non-Newtonian flow.

Studies of volcanism elsewhere in the solar system have, in the main, suffered from the severe disadvantage of having only the solidified, compacted and often weathered products of volcanism available for study, therefore any statements concerning the rheological behaviour of non-terrestrial volcanic flows must necessarily be highly speculative, since the most we can do is to use our terrestrial experience as a model. Since rheological modelling of even closely studied lavas such as those erupted on Hawaii, in Japan and from Mount Etna has proved to be fraught with difficulties and is still full of uncertainties, it is essential to recognize that any hypotheses regarding the rheological properties and eruption rates of lavas beyond the Earth must be developed cautiously, in the knowledge that they can, at the very best, only be reasonable approximations to the truth. A brief review of quantitative rheological models which have been invoked to explain flow behaviour both on and beyond the Earth has recently been presented by Baloga (1988).

5.1 RHEOLOGICAL CONCEPTS

Rheology has become a study in its own right and a large number of largely conceptual or semi-empirical models have been produced which deal with a wide variety of materials. Most of these depend in some measure upon classical ideas concerning viscous, plastic and elastic behaviour, a full discussion of which would be beyond the scope of this book. Should the reader require a more rigorous treatment then is given here, then Chapter 5 in *Mount Etna: the anatomy of a volcano* (Chester *et al.*, 1985) can be recommended wholeheartedly.

An ideal *viscous fluid* will deform irreversibly when subject to stresses of any magnitude as a result of flow, while the degree of deformation will change continually as a function of time. In contrast, an ideal *elastic solid* when subjected to the same stresses, will suffer instantaneous deformation but not flowage, the amount of deformation being directly proportional to the applied stress. *Plastic* materials bridge the gap between these two extremes, since if the stresses applied do not exceed some critical value they behave as elastic solids; however, once this value is exceeded the material will flow and thus respond in the manner of a fluid. Because the magma erupted from a volcanic vent is essentially a solidifying fluid, lavas are best described by the viscous and plastic models.

The way in which fluids behave was studied by Newton who discovered the law of liquid behaviour. In simplest terms this can be written:

$$stress = viscosity \times rate\ of\ strain$$

and is applicable to what have become known as Newtonian liquids. For any given temperature, the viscosity is different for every liquid and is best visualized as the amount of stress required to produce a given rate of strain. The behaviour of a Newtonian liquid may be plotted on a graph of shear stress versus rate of strain where it is represented by a straight line passing through the origin (Fig. 5.1).

Because they are at temperatures below their liquidus when erupted, most lavas contain crystals and also gas bubbles; therefore they do not behave exactly like Newtonian liquids. In 1968, Shaw and his co-workers investigated the rheology of flowing lava during a Hawaiian eruption and showed that it had non-Newtonian characteristics. Similar work on the 1971 eruption of Mount Etna by Walker (1967), Robson (1967) and Hulme (1974) confirmed that a typical flow exhibited non-Newtonian behaviour throughout most of its length.

Natural lavas in fact appear to behave like *Bingham liquids*, in that they possess a linear stress versus rate of strain characteristic once the applied stress exceeds a certain finite stress value – the *yield strength*. Below the yield strength the strain rate becomes zero and the liquid does not flow. On a plot of shear stress versus rate of strain, a Bingham liquid, therefore, also may be represented by a straight line but not passing through the origin, instead intersecting the y-axis at a value appropriate to its yield strength (Fig. 5.1).

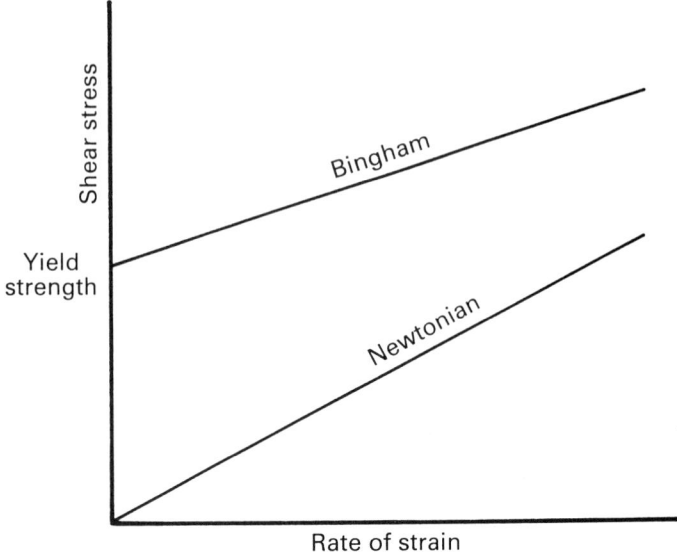

Fig.5.1: Plot of rate of strain versus shear stress, showing the difference between Newtonian and Bingham fluids.

5.2   VOLCANIC FLOWS AS BINGHAM LIQUIDS

During the early 1970's, several workers addressed the problem of lava flow behaviour. Johnson (1970) compared lava flows with the flow of glaciers and rock debris, which behave as Bingham liquids within confining channels. Hulme (1974) also argued that lava can best be modelled as an isothermal Bingham liquid in laminar motion and that the morphology and dimensions of lava flows primarily are a function of the non-Newtonian character of the magma, rather than the effects of its cooling. He demonstrated that when unconfined Bingham liquids flow downslope, they will assume a predetermined constant width and depth. The extent to which a lava flow will expand laterally is governed by the pressure gradient created in response to the variations in lava depth across it. In other words, the driving force for lateral expansion is provided by the excess hydrostatic pressure at the centre of the flow; therefore lateral spreading will cease when the shear strength at the base of the flow becomes equal to the yield stress, at which point the transverse profile becomes fixed. Motion downslope can then only continue if the shear stress at the base of the flow in that direction exceeds the yield strength. Because the shear strength is proportional to the depth of the fluid, there is a critical depth which has to be exceeded before movement can take place. However, it can be appreciated that only a small increment in flow depth may be needed to instigate a large increase in flow rate.

    Not surprisingly, along each margin of a flow there must be a zone where the lava

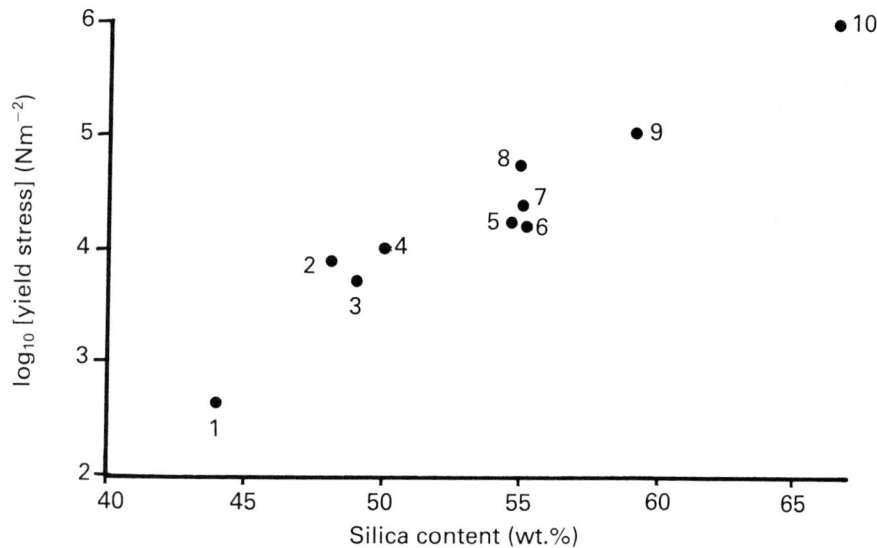

Fig. 5.2: Plot of SiO$_2$ versus yield stress for a variety of silicate rocks.
1, Mare Imbrium basalt; 2, Mount Etna (1910); 3, O-shima basalt; 4, Mauna Loa (1942);
5, Tristan da Cunha 1; 6, Hekla (1947); 7, Paricutin; 8, Tristan da Cunha 2; 9, Teide;
10, Chao diorite. (After Hulme, G. (1976).)

depth is less than this critical value; as a consequence marginal banks of chilled lava will
accrue and between these solid *levees* the mobile fluid flows. The natural development of
such banks was exploited by Hulme (1974, 1976) who devised a method by which estimates
of yield strength could be obtained, assuming that data for the local gravity, levee width,
lava density and local gradient were available. Calculation of the yield strength is achieved
by using the equation

$$\text{yield strength } (S_y) = 2w_b g \rho \alpha^2$$

where w_b = width of levee, ρ = density of the lava, g = local gravity and α = the local
slope.

This method of deriving estimates of yield strength has been used widely for flows on
the Earth, Moon and Mars (see Cattermole, 1987; Hulme, 1976; Moore *et al.*, 1978;
Schonfeld, 1979; Zimbelman, 1985); however there are numerous problems associated
with it, as there usually are with models that assume "ideal" conditions. For instance,
during flow emplacement degassing usually occurs; then again, crystals may separate,
either settling towards the flow base or floating towards its top. Because the chemical
interactivity of liquid magma, crystals and gas bubbles complicates the rheological
behaviour, it is almost impossible to characterize the rheology of a flowing lava over the
complete course of its emplacement. Thus while the viscosity of the liquid magma
predominantly is a function of its chemistry and temperature alone, the *apparent viscosity*

of the bulk lava is dependent not only upon the liquid viscosity but also the concentration of crystals and gas bubbles, and the stress conditions under which the lava is flowing.

Direct measurements of molten, volatile-rich basalts in the Makaopuhi lava lake, Mauna Loa, Hawaii, indicated yield strength values near 10^2 N m^{-2} (Shaw *et al.*, 1968). Remote measurements made by Moore *et al.* (1978) on flows 600 m from the rim of the same lake indicated a yield strength near 4×10^3 N m^{-2}, while several kilometres from the rim, values had increased to 2×10^4 N m^{-2}. While there is a strong dependence of yield stress upon both temperature and silica content (Fig. 5.2), these differences cannot be explained by changes in silica content since this is fairly constant at approximately 51% and only a small range in temperature.

It is likely, therefore, that the variation is a function of changes in volatile content, which should decrease rapidly with distance from the source region, and the proportion of entrained solids, which would be expected to increase with distance from the vent due to cooling.

Yield strength values calculated for non-terrestrial flows, using the method of Hulme mentioned above, show a wide range. Values of 2×10^2 N m^{-2} were derived for lunar Mare Imbrium flows by Booth and Self (1973). Martian lavas appear generally to have somewhat greater yield strengths, the range being from 8.8×10^3 to 4.5×10^4 N m^{-2} for flank flows on Olympus Mons (Hulme, 1976), from 1.2×10^3 to 2.8×10^4 N m^{-2} on Ascraeus Mons (Zimbelman, 1985) and from 1.9×10^3 to 2.8×10^4 N m^{-2} on Alba Patera (Cattermole, 1987).

Another study of the Mokaopuki lava lake flows by Settle (1979a) suggested that while basaltic flows in many ways behaved as Bingham liquids, they could also be modelled as crystal suspensions in Newtonian liquids and it is interesting to note that his calculated suspension viscosities for high shear rate lava flowage – which might be anticipated to occur over a large part of a flow's emplacement history – agreed with measured viscosities to within a factor of two and were also in agreement with estimates of Bingham fluid plastic viscosity published by Moore and Schaber in 1974.

Bingham models have also been used to study the rheology of mudflows. Excellent conditions were provided in 1980, during the catastrophic eruption of Mount St. Helens, for Fink and co-workers (1981) to calculate yield strengths and plastic viscosities for three flows, based on a geometry model. They obtained yield strengths within the range 400-110 Pa and viscosities between 20 and 320 Pa s^{-1}. These figures agreed well with both measured and estimated values previously quoted in the literature. The rheological characteristics of pyroclastic flows are less well known, in large part due to the dangers involved in collecting field data. Clearly their behaviour is non-Newtonian and both laminar and turbulent flow regimes occur, the latter being typical of the upper gas-rich layers and the former of the denser basal layers, where the effective viscosity is higher.

The difficulties of deriving rheological data by morphological measurements was reinforced by the work of Fink and Zimbelman (1985) who showed after *in situ* measurements of cross-flow profiles along lavas erupted during 1983 and 1984 at Kilauea, Hawaii, that incorrect assumptions about pre-flow topography could also introduce errors into the estimation of yield strength from flow profiles. Despite all of these shortcomings, however, attempts to characterize flow rheology on the basis of Hulme-type methodology must be considered useful, if only in providing order of magnitude estimates.

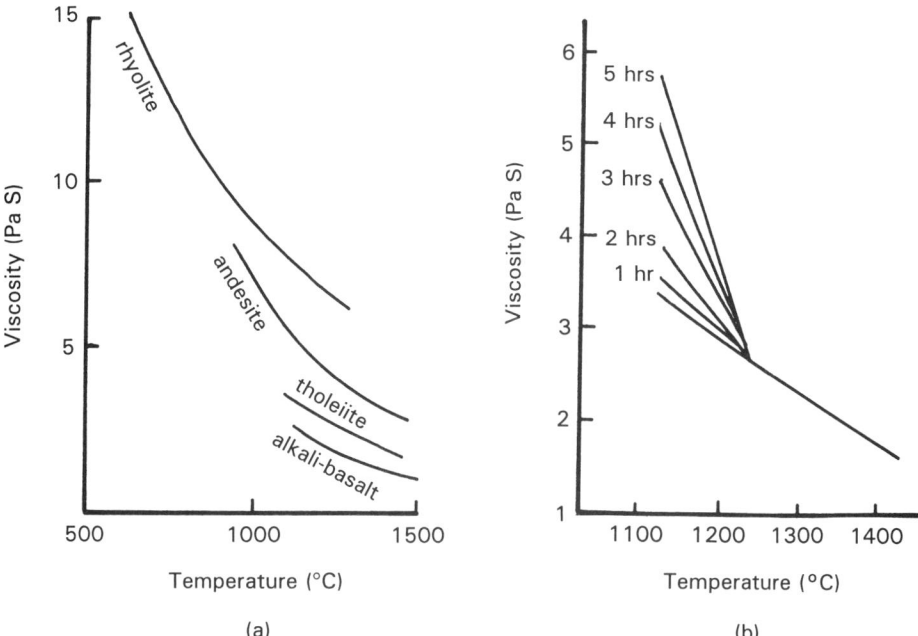

Fig. 5.3: (a) Viscosity versus temperature for a variety of common terrestrial volcanic rocks. (After Murase and MacBirney, 1973.) (b) Plot of viscosity versus temperature for a basalt allowed to cool without disturbance, showing how the viscosity increases with time. (After Murase and MacBirney, 1973)

5.3 VISCOSITY

The most important single physical property of a magma is its viscosity. This may be defined as the ratio of shear stress to the rate of shear strain and is a measure of the internal resistance of a fluid to flow. On the assumption that lavas behave as non-Newtonian fluids, the *effective viscosity* of a non-Newtonian fluid is that quantity which may be substituted for the Newtonian viscosity in a mathematical expression for Newtonian flow under geometrically similar conditions. The main effect of viscosity is to determine the lateral extent and thickness of flows. The ratio of the horizontal extent (H) to the thickness (V) is known as the *aspect ratio* and in natural flows has been found to range from less than 2 to more than 1000.

The viscosities of molten silicate systems are strongly dependent upon the proportion of silicon they contain, a fact which can be illustrated by noting that a silicate liquid with a Si/O ratio of 1:4 has a viscosity which is over ten times less than one where the ratio is 1:2. The variation in viscosities of some common igneous rocks within and above their melting range, as established in the laboratory by Murase and McBirney (1973), is illustrated in Fig. 5.3(a).

Field measurements of lavas and laboratory data from both natural and synthetic compositions indicate that viscosities fall not only with decreasing silica content but also with increasing temperature. At temperatures below the onset of crystallization the viscosity tends to increase with time and, assuming the flow remains undisturbed, may continue to increase for several hours prior to attaining a steady state (Fig. 5.3(b)).

It comes as no surprise, therefore, to learn that viscosity increases by several orders of magnitude between the vent and the terminus of basaltic flows (Shaw *et al.*, 1968; Pinkerton and Sparks, 1978; McBirney and Murase, 1984). One of the most complete sets of observations on a natural lava was that obtained by Minakami (1951) during the 1950/1 eruption of Mihara volcano on O-shima Island, Japan. He recorded the temperature, flow velocity, viscosity and flow dimensions at four stations along the course of a flow as it descended the volcano's slopes. As it reached the lower slopes, the flow became wider and thicker, while its velocity decreased by one order of magnitude. The viscosity increased with increasing distance from the vent and was nearly two orders of magnitude greater at a point 450 m from its source (Sakuma, 1954).

The effects of increasing amounts of volatiles upon the viscosity of silicate magmas are rather difficult to study in detail; however, there is little doubt that the addition of H_2O to silica-rich melts has the effect of lowering the viscosity, largely because water disrupts the strongly bonded framework of linked silica and alumina in such highly polymerized melts. In low-silication magmas, such as basalts, the incorporation of volatiles appears to have a lesser effect because of the relatively restricted presence of appropriate silicate linkages in these relatively weakly polymerized magmas (Scarfe, 1973). Pressure also has an effect upon viscosity, experimental work indicating that the effect of increasing pressure on basaltic and andesitic melts is to decrease their viscosities.

The above conditions will apply in general to simple cases, for instance, a melt above its crystallization temperature which has not commenced to exsolve its volatiles. As soon as either gas bubbles or crystals are generated, however, more complex conditions apply. The formation of crystals within a lava render it more viscous because they impede the shearing flow of the magma, while entrained bubbles slow down shear flow due to both the effects of the gas pressure within the bubbles and also the surface tension at the bubble-liquid interfaces. Finally, as we have noted above, real magmas have a yield strength which increases dramatically as they begin to crystallize. All of these factors together have a fundamental effect upon the viscosities of magmatic systems and therefore play important roles in determining both the character of flow motion and the morphology of lava flows and associated structures.

A further complexity regarding flow behaviour is introduced where rapid rates of shearing motion increase the thermal energy of the viscous lava faster than heat can be dissipated. In consequence, the temperature rises and has the effect of reducing the apparent viscosity not only in its own right but because crystals may go back into solution and therefore reduce their concentration. Sometimes the strong localization of this thermal feedback mechanism along specific internal flow surfaces means that subsequent shearing is confined to narrow zones, with the result that continued application of stress causes flow along these restricted zones rather than through the whole body.

Estimates of the viscosity of lavas on other planets have been made using the method first outlined by Hulme in 1974. This is based on the measurement of certain physical parameters of flows appearing on medium to high resolution imagery. The parameters required are the flow length, total width, width of marginal levees and slope of the ground. Of these, data for slope and levee width are the most prone to error. The relationship explored by Hulme is the following

$$\eta = F/E(g\rho)^3(S_y)^4/\alpha$$

where η is the viscosity, E is the effusion rate, g is the acceleration due to gravity, ρ is the density, S_y is the yield strength, α the slope and F a dimensionless parameter that is a function of the total flow width and the width of the levees. The input of a reasonable density is not a problem, but the effusion rate can only be estimated from the flow length according to the empirical method of Walker (1973). Notwithstanding the inherent shortcomings, Schonfeld (1979), using an estimated effusion rate of 3.5×10^7 m^3 s^{-1} and a lava density of 2.7 g cm^{-3} in connection with data obtained by Moore et $al.$ (1978), derived viscosities for flows on the Martian volcano Arsia Mons, of between 0.3 and 6 Pa s.

Sulphur flows would be anticipated to behave in a totally different way to silicate ones, largely because the apparent viscosity of sulphur displays a local minimum of around 430 K. Consequently the subject is treated separately in Chapter 11.

5.4   FLOW BEHAVIOUR

Having established the fundamental rheological properties, it is possible to apply these to the principles of fluid dynamics and come up with some predictions about flow behaviour in extrusive magmas. The way in which fluids flow can be related to the dimensionless *Reynold's number* (*Re*) which is a measure of the relative magnitude of the inertial and viscous forces affecting a mobile fluid. It can be expressed by the ratio

$$Re = 2\rho_m HV/\eta$$

where ρ_m is the average density of the magma, V the velocity, η the viscosity and H the uniform depth of the flowing body. Empirical calculations indicate that where Re is less than about 2000, fluids move by laminar flow, whereas if Re is greater than about 2000, flow tends to be turbulent.

Under most conditions, because of the generally considerable viscosity of terrestrial silicate magmas (i.e. greater than 10^2 Pa s) and their correspondingly high yield strengths, flow will be laminar except close to vents. Lunar lavas are generally less viscous and have lower yield strength, therefore turbulent flow may be achieved much more readily. Head and Wilson (1983) have shown that on slopes of between 5° and 6°, if a critical value for mass eruption rate of 3×10^7 kg s^{-1} is exceeded, then turbulent flow will occur in both lunar and terrestrial basalts. Terrestrial basalts rarely are extruded with mass eruption rates that exceed 3×10^6 kg s^{-1} and therefore flow characteristically is laminar. However, lunar basalts seem to have been erupted quite commonly at rates of at least 10^8 kg s^{-1} (Hulme and Fielder, 1977; Head and Wilson, 1981), particularly those associated with sinuous rilles, and these bear obvious witness to the turbulent nature of these rapidly erupted, high-volume, flows and their capacity to effect thermal erosion of the lunar surface

Fig. 5.4: Long sinuous rille crossing western Mare Imbrium near the crater Brayley. Such rilles are believed to have been incised into the mare surface by the turbulent flow of high-volume fluid basalt lavas. Frame width about 35 km. Part of Apollo 17 panoramic camera frame 3130.

Fig. 5.5: Upper surface of fresh pahoehoe flow showing development of ropy texture. La Mancha, Isla de la Palma. Photo: Peter Cattermole.

Fig. 5.6: Section through a typical aa flow, showing upper clinkery part and denser inner region with vesicle trains. Bonito flow, Sunset Crater, Arizona. Photo: Peter Cattermole.

(Fig. 5.4). A similarly turbulent flow regime will almost certainly have characterized Martian rilles of this type.

The actual velocity of flow is determined by the interplay of a complex array of factors which include viscosity, volume, density, slope, rate of effusion and nature of the lava channel. The rate of flow diminishes with increasing distance from the source region; furthermore there will be pronounced velocity gradients across the flow at any given point along its course. Magma flowing via an open channel flows most rapidly at the top and middle, but once it crusts over and an insulating roof forms, the hottest and most rapidly moving region drops beneath the surface.

Among terrestrial lavas the ocean island basalts of Hawaii are among the most fluid. Measurements made during the 1855 eruption, indicated a rate of flow of 64 km h^{-1} (0.29 m s^{-1}) over slopes of between 10° and 25°, while lavas erupted from Mauna Loa in 1940, travelled at velocities of up to 40 km h^{-1} (0.18 m s^{-1}) over very low slopes (Williams and McBirney, 1979). The initial flow rate (first few minutes) during the 1805 eruption of Vesuvius approached 75 km h^{-1} (0.35 m s^{-1}), but, generally speaking, rates are somewhat lower than those of Hawaii. More siliceous magmas flow at much slower rates, and, although actual field data are sparse, rates of tens of metres per hour appear to be typical.

The fluidity of the lava also controls the type of flow which develops. In Hawaii, for instance, two types of flow predominate: *pahoehoe* and *aa*. Pahoehoe lavas have glassy, smooth, billowy surfaces which occasionally have ropy tops and often contain widespread smooth areas (Fig. 5.5).

More fluid pahoehoe flows generally advance with a rolling motion, where the upper part of the flow, moving faster than the lower, is rolled over and buried beneath the advancing flow front. Less fluid flows move more slowly, tending to advance by the successive generation of bulbous "toes" at the flow front. Aa flows are characterized by rough, hackly and fragmented surfaces. The upper part of a typical aa flow is vesicular and the surface resembles furnace clinkers, individual clinkers being up to 15 cm across. Below the surface the fragments become welded together and the flow interior is denser as a result (Fig. 5.6).

Of the two, pahoehoe flows are the less viscous (above 10^2 Pa s), the ropy surfaces developing in response to the differential motion between the less viscous interior and more viscous skin. Typically terrestrial pahoehoe flows are less than 15 m thick, occur in both distal and proximal locations, cover areas of between 1 and 1000 km^2 and advance at rates measurable in metres per minute. Aa lavas, with their somewhat higher viscosities (greater than $10^3 - 10^9$ Pa s), move more slowly (few metres per hour), form thicker units (10 – 100 m), occur in central or proximal locations and cover areas of between 1 and 100 km^2. Commonly, channels develop in aa flows but tubes are rare. In contrast, pahoehoe often show the development of tubes. It is common on Hawaiian volcanoes to follow a pahoehoe flow downstream from the vent area and see it changing into aa. This change occurs in response to decreasing fluidity, produced either by cooling, degassing or crystallization.

A third type of flow, *block lava*, is rare in Hawaii, but may be widespread in other locations. Such lavas are composed of large fragments, usually smooth, with irregular surfaces that give a relief of up to several metres. It is common to find ridges developed which lie perpendicular to the flow direction. These flows proceed both slowly and irregularly, so that there is a tendency for some layers to shear over others. Of the three types mentioned, they are also the most viscous (above 10^4 Pa s) and it is the combination of elevated viscosity and their tendency for occasional rapid flow which causes the lava to break into angular clasts which subsequently become rounded or fragmented. Flows of this kind advance at rates of only a few metres per day and may be found in central or proximal locations.

Another morphological feature which reflects the internal behaviour of volcanic flows is the development of surface ridges, often developed on the surfaces of terrestrial silicic lavas (Fig. 5.7).

Fink (1980), after studying the obsidian flows of Big and Little Glass Mountains in northern California, concluded that the regular spacing and anticlinal profile of surface ridges was a manifestation of the contrast in viscosity between the hotter interior of the flow and its cooler upper surface. Since similar structures are found in dacites and andesites and are somewhat comparable with ropy pahoehoe basalt, it is clear this is a fairly widespread phenomenon. Fink was also able to calculate minimum viscosities of many flows for which such data were otherwise unavailable, and applied his method to somewhat controversial lava features in the Arcadia region of Mars which were suggested to be of silicic composition. However application of the same method to a wider spectrum of

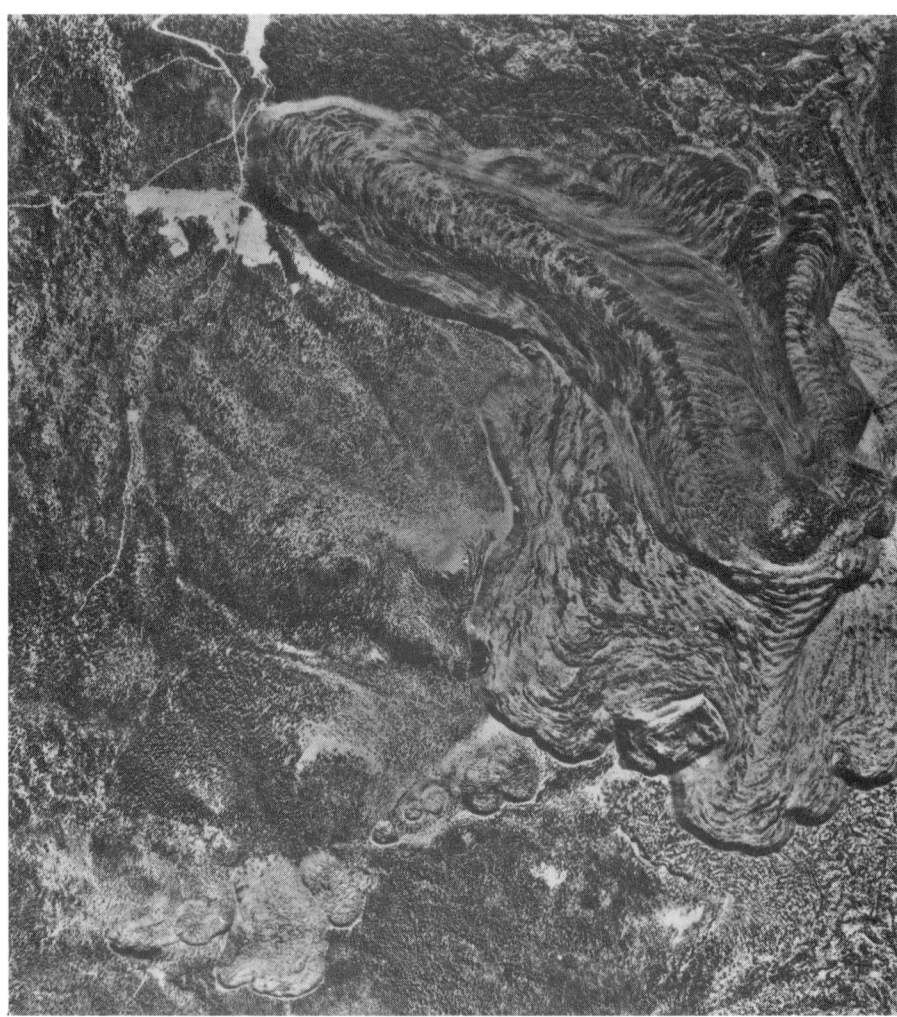

Fig. 5.7: Pressure ridges on the surface of a rhyolite obsidian flow, Big Glass Mountain, California. Note the lobate flow fronts and generally rugged relief of the flow surface. US Forest Service Photograph DDC-3P-27, July 1954.

terrestrial lavas by Theilig and Greeley (1986) and also to selected Martian lavas with surface festoon structures revealed an unexpectedly high figure for estimated interior viscosity (10^8 Pa s). They suggest that relatively elevated viscosities are necessary for the development of festoons and flow ridges, even in basic lavas, and that their development is characteristic of flood-type eruptions involving large, single-unit basalt-like flows.

Baloga and Pieri (1986) recently have emphasized the time-dependency of flow emplacement, investigating the combined effects of time-dependent effusion rate and spatially varying viscosity on the thickness profile of moving flows. In applying these results to Martian sheet flows at Alba Patera, they conclude that unexpectedly high and constant apparent viscosities are implied, rather similar to those characteristic of terrestrial basaltic andesites. Baloga (1987) also draws attention to the possible interpretation of flow profiles by kinematic wave theory and shows how flow behaviour at source propagates faster downstream than a flow advances.

5.5   MAGMA DISCHARGE RATES

The rate at which magma is erupted is largely dependent upon the rheology (in particular, the viscosity) and the size of the eruptive conduit. In nearly all eruptions, discharge rates are highest during the early stages of activity. At Hekla, for instance, in March 1947, during the first half hour of the eruption magma was erupted explosively at a rate of 7.5×10^4 m^3 s^{-1} (Thorarinsson, 1950) which is equivalent to 1.7×10^4 m^3 s^{-1} of dense lava. During the ensuing half hour the eruption rate fell to an equivalent of 5×10^3 m^3 s^{-1}, while throughout the next 24-hour period lava was extruded at a rate of just over 1×10^3 m^3 sec^{-1} per kilometre length of fissure. During the next day it fell by one order of magnitude. The volume of material blown out during the first hour of activity represented about one tenth of the total eruptive products of the year-long phase of activity at this famous Icelandic volcano. Hawaiian rates of discharge have been found to be comparable to those of Iceland. Pulsatory activity is, however, characteristic, on the scale of a few minutes to a few hours.

Beginning in 1943, (the year this volcano was born) the Mexican volcano, Paricutin, erupted lava more or less continuously for nine years, gradually decreasing its discharge rate while changing in composition from olivine basalt to basaltic andesite. During the five months commencing October, 1945, the eruption rate was between 2 and 6 m^3 sec^{-1}. The composition of lavas discharged during the 1914 eruption of Sakurajima, Japan, was also andesitic and during the early phase of this eruption the effusion rate was 1.6×10^3 m^3 s^{-1}. Andesites erupted in 1973 at Langila, in the Bismarck volcanic arc, were erupted at a steady rate of about 0.2 m^3 s^{-1} for the first 20 days, but then the rate increased somewhat to near 1×10 m^3 sec^{-1} for the next ten days before declining again (Cooke *et al.*, 1976). More silicic lavas also discharge at much lower rates than basaltic ones; for example, the dacitic dome which developed within the caldera of Santorini during 1966 grew at a rate of only 0.2 m^3 s^{-1} for the first few days, before the growth rate fell to half this.

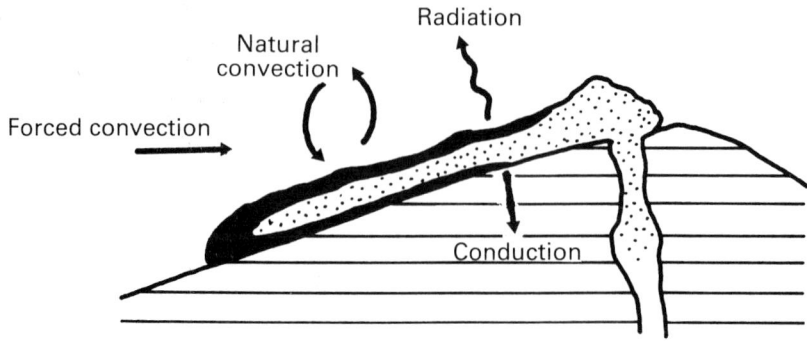

Fig. 5.8: Schematic diagram illustrating the principal heat loss mechanisms affecting a lava flow on a planet clothed by an atmosphere.

5.6   THE COOLING OF LAVA FLOWS

Silicate flows are characterized by low thermal conductivity and high heat capacity, therefore even if of only moderate thickness they are well insulated and give up their thermal energy slowly. Terrestrial flows are known which have retained high internal temperatures for up to six years. Fig. 5.8 illustrates the principal heat loss mechanisms affecting silicate flows.

At the base of a flow heat loss is principally by conduction. It has been found that the temperature at the base of a moving lava is approximately halfway between that of the hot interior and the cool ground below. Such a relationship favours the promotion of crust formation next to the ground surface and this may crack and warp as the flow continues to progress, with the formation of a basal flow breccia. Conductive heat loss also occurs at the top of a flow, and this process may be enhanced if atmospheric convection occurs, thereby increasing the heat transfer rate.

The principal heat loss mechanism is, however, radiation, which can be represented by the Stefan-Boltzmann equation for black-body radiation

$$Q = \sigma T^4$$

where Q is the energy radiated in cm^2 sec^{-1}, T is the absolute temperature and σ is the Stefan-Boltzmann constant (5.67×10^{-12} J s^{-1} cm^2 $°C^{-4}$). A cursory glance reveals that implicit in the fourth-power temperature relationship is the fact that a small degree of cooling greatly reduces the thermal energy loss. Furthermore it has been shown that such thermal losses control the eventual plan area of the flow (Baloga and Pieri, 1986). Since a flow's surface is at a temperature very much higher than the surroundings there will be a rapid temperature drop at the interface, so that the lava acquires a skin whose temperature is approximately that of the surroundings. Within this thermal boundary layer the viscosity and yield strength increase rapidly and in time it will grow inward from all surfaces until the flow becomes rigid and stops moving. Sometimes the flow exterior becomes rigid long before the interior of the lava ceases to move, whereupon motion continues through a thermally insulated tube. Such a tube-flow mechanism is much more conducive to the growth of long flows than that of open channel flow, since the insulating properties of the tube walls minimize the thermal energy loss. Field measurements indicate that flowing lavas experience little cooling through most of their length, expecially if initial temperatures exceed 1100°C. To cite a specific example, at Paricutin, where the vent temperature was 1070°C, 5 km downstream the temperature was only a trifle less (Krauskopf, 1948). Similar data have been collected for the 1859 Mauna Loa eruption.

An idea of the quantities involved in the cooling of typical flows can be obtained from the equation

$$Gz = (E/\kappa x)(d/w)$$

where E is the effusion rate of the flow, κ if the thermal diffusivity of the lava, x the distance from the vent, d and w the depth and width of the flow respectively. Gz is the dimensionless *Gratz Number* which describes the distribution of temperature across the flow; its value is proportional to the distance from the vent. Near to the vent Gz is large and the temperature of the magma is more-or-less uniform across the width of the flow and also approximates to the eruption temperature. At greater distances from the source region, Gz becomes smaller and the cooler thermal boundary layer at the flow margins gets thicker. As the cooled regions move inwards so the central temperature drops. By the time Gz = about 30, it will have fallen by about 1% of the difference between the eruption and external temperatures. The distance required for this to happen may be calculated from the above equation if certain properties of the flow are known. There is both theoretical (Hulme and Fielder, 1977) and observational (Pinkerton and Sparks, 1978; Walker, 1973) evidence that for terrestrial flows, by the time Gz has fallen from an initially high value to about 300, flow will have ceased. At this point the maximum length (X) of the flow should be

$$X \approx Ed_c/300\kappa w_c$$

where E is eruption rate, d_c is the maximum thickness of levees, k is the thermal diffusivity (probably 3×10^{-7} m^2 s^{-1}) and w_c is the width of the central channel, assuming the flow has developed levees.

The critical value of Gz for cessation of flow motion is unlikely to be the same for all planets. In particular, the high ambient temperatures at the surface of Venus (about 520°C) should lead to much lower cooling rates. Calculations by Head and Wilson (1986) reveal that during the first hour after leaving source, surface temperatures of Cytherean flows would be less than those of the Earth, largely because of the much more effective convective thermal energy loss into the very dense atmosphere of Venus. At longer flow life times, however, temperature of Venusian Cytherean flows would be higher than those on Earth, as the surface temperature would approach asymptotically the higher ambient temperature of Venus. The thickness of rigid flow crusts immediately after the flows leave their vents are also predicted to be greater on Venus than the Earth; however, after a period of about 30 min, by which time crusts with thicknesses of about 60 mm should have formed on Cytherean lavas, crusts would be thicker on terrestrial flows. Short-lived flows (less than 1 hour) would be systematically shorter on Venus than Earth, but longer-lived flows (more than 1 hour) would be correspondingly longer. Their calculations indicate that cessation of flow motion on Venus would cease when $Gz = 270$.

The final length of a flow, while in part a function of viscosity, is determined by various other factors. Walker (1973) concluded that effusion rate was the most important single factor governing the length, but Malin (1980), challenging this, argued that flow volume and cross-sectional area must play an equally important role. Wood (1984) notes that both Walker and Malin should be correct, since high effusion rates would be generated by high-pressure eruptions yielded by large magma chambers and large flow volumes would emerge since such chambers erupt high magma volumes. Support for the view that rapid discharge of large magmatic volumes produces long flows is provided by Stephenson and Griffin (1976), who note that olivine-basalts erupted in North Queensland at liquidus temperatures of between 1470 and 1520 K advanced up to 160 km from their vent over slopes of less than 0.5°. Favouring emplacement times of about one week for these flows, they see large flow volumes, coupled with high mass effusion rates, as being the principal factors in generating these lengthy flows.

5.7   LATENT HEAT OF CRYSTALLIZATION EFFECTS

The effects of energy released by the latent heat of crystallization have been studied by various workers. Of particular interest is olivine which has the largest latent heat of crystallization per gram of all the common rock-forming minerals; as a result its crystallization significantly offsets thermal losses produced by other mechanisms. A particularly interesting study of the 1965 Makaopuhi lava lake was made by Settle (1979b), who noted that, once a crust had formed, the net effect of the thermal energy released during olivine crystallization was to severely retard the overall cooling rate. Thus if a 10 m-thick column of magma was to cool by 1°C, the release into the flow of approximately 4×10^3 J would be achieved by the latent heat of crystallization, principally of olivine. Settle surmises that magmas in which olivine is the sole liquidus phase probably could be stored in near-surface chambers over relatively lengthy periods under isothermal conditions.

This may be of particular significance for basalt-like flows extruded on Mars, since it has been proposed (McGetchin and Smyth, 1978) that they may have been particularly enriched in MgO and FeO due to the partial melting of high-density (possibly Fe-rich) mantle material. Experimental data for mafic silicate melts with model Martian conditions, shows that olivine is the sole liquidus phase and will continue to separate until approximately 50% of crystallization is completed. If this is so, it provides at least one mechanism for the generation of very long (i.e. 300-500 km) lava flows on that planet, since Martian lavas erupted over a comparatively wide range of sub-liquidus temperatures would be thermally buffered against the normal cooling mechanisms through a major part of their cooling history.

5.8 RHEOLOGICAL ANALYSIS OF VOLCANIC FLOWS

On Earth, the rheology of flowing lava often may be measured in the field, as may the eruption temperature, gas content and mass flow rate; analysis of samples subsequently provides data on chemical composition and the crystallization history of the magma. This information may then be utilized in assessing how each factor or combination of factors affects the final planimetric form of individual flows. It is also possible to time the duration of flow emplacement and monitor local details such as crust development, the generation of lava tubes and channels, levee growth and any branching of the flow front, should it occur. Some of the most detailed studies have been accomplished in Hawaii and at Mount Etna. The recently published US Geological Survey's Professional Paper 1350 is a two-volume compendium entitled *Volcanism in Hawaii* and runs to over 1600 pages. It contains a wealth of information on all aspects of Hawaiian activity, including basalt flow development and rheology. Numerous studies exist of Mount Etna; the recently published book *Mount Etna* (Chester *et al.*, 1985) crystallizes much of the volcanological research which has focussed upon this active centre in recent years. Even where *in situ* observation and measurement are possible, however, it is often difficult to identify which morphological characteristics of a flow are a direct manifestation of which rheological properties, although it does seem clear that the initial morphology is determined largely by the yield strength and the plastic viscosity. The uncertainties which remain are a direct reflection of the extreme complexity of the flow emplacement process.

It is entirely predictable, therefore, that the planetary volcanologist has to struggle in an effort to characterize flow rheology. Since he or she can study only remotely sensed data, which reveals not the active process of flow emplacement but its consolidated products, to constrain any or all of the lava's rheological properties or arrive at reasonable estimates of the mass eruption rate and flow emplacement time, the worker must needs base interpretation solely on the analysis of flow parameters which can be derived from spacecraft images. These then must be interpreted by one or other of the quantitative models developed in recent years which relate measurable parameters of flow geometry to eruption conditions.

Some of the basic premises upon which quantitative rheological models are dependent have already been discussed; they have also been reviewed recently by Steve Baloga (1987). Each suffers, to some degree, from inadequacies, either within the model itself or in the assumptions which have to be made before it can be applied. These arise in response to the necessity for simplification in modelling – it simply is not possible to deal with every potential variable in any single model – and in having to make assumptions about physical properties of materials that cannot be studied at first hand and which then have to be inserted into model equations. Furthermore there is not yet complete agreement within the geological community regarding which particular physical factors are of the most importance in constraining the eventual form of a solidified volcanic flow. Despite such daunting problems, several groups and individuals have attempted to derive data for volcanic flows on the Moon, Mars and Io, using models developed from basic rheological theory, checked and calibrated from closely studied terrestrial locations. The results of some of this work is summarized in the ensuing section.

5.8.1   Quantitative models

Studies of the effects of thermal losses on flow behaviour were first undertaken more than a decade ago (Shaw and Swanson, 1970; Harrison and Rooth, 1976). These early projects endeavoured to derive a relationship between radiative losses and the dimensions of terrestrial lava flows, the models being based on the assumption that a flow is isothermal in every vertical cross-section. For this to be valid, however, a flow must be thermally mixed across each vertical cross-section normal to the direction of flow. While turbulent flow might achieve such a condition, it is unlikely that it can apply in the majority of cases and, indeed, was observed not to be the case in Hawaii during the 1984 Mauna Loa eruption (Lipman and Banks, 1987). It is not surpising, therefore, that Pieri and Baloga (1986) have shown that "fully mixed" models produce too great a heat loss to satisfactorily explain what appears to be a sympathetic relationship between eruption rate and plan area observed for some Hawaiian lavas. Nevertheless, in the absence of suitable alternatives, single-component *Gratz Number correlation* and *yield stress* models have been widely used.

As a result of such inadequacies some workers have been exploring *two-component models*, whereby a flow surface is divided into one fraction which radiates at a temperature characteristic of an inner core that is only partly exposed, and another fraction, a crust, which radiates at a much cooler temperature and which may thicken with time, as cooling occurs (Crisp and Baloga, 1989). The relevance of such models is evidenced by numerous field observations, such as the appearance of bright cracks on the surface of mobile lava, the development of a crusted central zone to an active channel which is distinguished from the stationary levees by bright (hotter) shear margins, the carrying and shedding of cooler rafts or boats of lava above the hotter core regions, and the frequent breakup of pieces of lava crust when the flow encounters sudden changes in slope. By dividing a flow vertically into an inner core and an overriding crustal layer it is possible to investigate the relative radiation losses for each fraction, given different assumed eruption temperatures and emplacement times. Because such investigations are in their preliminary stages it is not yet possible to ascertain how relevant they will be for future planetary studies; however, temperature changes predicted by this "unmixed" model for the crust and core of a flow, with distance from the vent, are comparable with the very limited terrestrial observations that are available. Preliminary analysis suggests that lava flow dimensions may be influenced most strongly by thermal dynamics in the flow core if (i) there is a lengthy emplacement time, (ii) the fraction of exposed core is maintained at a high level, or (iii) the initial eruption temperature is low.

Complexities are also introduced by the time-dependency of flowing lava. In one of the earliest studies of basaltic flows, Walker (1973) concluded that eruption rate is the dominant influence on lava flow morphology, particularly the flow length. Although later studies of Hawaiian flows questioned the validity of this conclusion and proposed several refinements (Malin, 1980; Pieri and Baloga, 1986), one of the suggestions made by Walker in seeking to explain the dispersion seen in a plot of eruption rate versus flow length, was that it could be attributed to time variations in the effusion rate. He thus anticipated the model of Wadge (1981) which described the observed change in effusion rate of some terrestrial flows by a relatively rapid rise during the early stages of an eruption, followed by a more gentle, exponential-like decay. Baloga and Pieri (1986) in developing this idea, built "waxing" and "waning" phases into a model which investigated the combined effects of a time-dependent effusion rate and a spatially-varying viscosity on the thickness profile of moving lavas. Their investigations suggest that time dependence in the depth of a flow at its source and the form of the viscosity variation are significant influences on the dimensions and morphology of lava flows. Such time-dependent source conditions may, therefore, explain the disparities noted between certain field and laboratory estimates of lava viscosity.

Subsequently Baloga (1987b) reinterpreted this work, producing a *kinematic wave model* which described the time-dependent emplacement of lava flows based on local and global conservation of lava volume. Kinematic wave propagation can best be likened to the unidirectional flow of traffic along a highway, where the variable of interest is the number of vehicles per unit length of road. Starting with the initial conditions of a smooth-flowing traffic column, which may be represented by a kinematic wave of constant amplitude (number of cars per unit length), should one of the vehicles brake suddenly, the flow rate immediately behind it would diminish very rapidly. As a result the density of vehicles increases sharply, and a kinematic wave of slowing propagates backwards through the column of traffic. Those vehicles originally responsible for the slowdown may be many kilometres away before the backward-propagating wave effects have subsided. Thus the shape of the wave changes from an initially constant value as a result of the slowdown, and invariably changes again as each driver sees more open highway ahead.

The implications of kinematic waves for volcanic flow development are obvious. Application of the theory to lavas reveals that, in general, source behaviour propagates downstream more rapidly than a flow advances. In this way only a fraction of the source time dependence is reflected in the form of the kinematic wave. Consequently the residual of the time-dependent source effects are deposited at the flow front, the lateral margins or the solidifying crust, or may be dissipated by other unmodelled processes.

5.8.2   Yield strength models

Where a flow has developed levees it is possible to model it as a Bingham plastic of finite yield strength according to the model of Hulme (1974). This and other yield stress models have been applied by several workers to terrestrial, lunar and Martian flows (Hulme, 1974; Hulme and Fielder 1977; Moore *et al.*, 1978; Schonfeld, 1979; Zimbelman, 1985; Cattermole, 1987). While it is not entirely clear that the Bingham model is germane to even most flows, the formation and presence of initial levees (Sparks *et al.*, 1976) does find a satisfactory explanation in this model which also accounts for various other features of flow movement often observed in the field, as for instance in the Makaopuhi lava lake. As has been observed by Moore (1987) at Mauna Loa, a flow may leave its vent with Newtonian rheology, behave as a Bingham fluid for much of its course and become a pseudoplastic fluid in more distal regions.

One important consequence of the Bingham model is that the yield strength of the lava may be derived from some or all of the measured parameters: flow thickness, flow width, levee width and the gradient of the ground over which the flow has moved. Three equations for yield strength (S_y) can be utilized, the selection of a specific one being dependent on the nature of the flow and quality of the available data:

$$S_y = 2w_b g\rho\sin\alpha^2$$
$$S_y = \rho g\sin\alpha H$$
$$S_y = \rho g H^2/w_f$$

where H is flow thickness, w_f is flow width, w_b is levee width, g is acceleration due to gravity, ρ is lava density, and α is slope. It is also necessary to input the planetary gravity constant and the density of the lava into this equation. Since Booth and Self (1973), in directly measuring a flowing terrestrial basalt, found a density of 2.6×10^3 kg m^{-3}, this value is usually used in most calculations.

Sufficient high-resolution terrestrial and lunar stereoscopic aerial photographs exist for the three-dimensional morphology of individual flows to be mapped and contoured. This may be achieved by transferring flow edges and levee crests visually to topographic charts or may be achieved by the use of a stereoplotter. For Mars the situation is less satisfactory; there is little stereoscopic coverage within the Viking data set and what there is was obtained from very high altitude. While limited stereo measurements can be made, the errors inherent in determining flow thickness are usually greater than the measured thicknesses and it is usually advisable, indeed necessary, to appeal either to sophisticated photoclinometry or image processing techniques (see Moore et al., 1978; Cattermole, 1987). In theory, flow thickness can be measured by accurately determining the width of shadows cast by flow edges, inputting the data into a suitable shadow method equation (van Diggelen, 1951; Fielder, 1965); knowing the spacecraft attitude and illumination geometry it is then possible to arrive at an approximate relative height difference, i.e. thickness of scarp edge or levee bank. In reality the "shadows" seen on pre-processed Viking Orbiter hardcopy may be spurious due to enhancement techniques applied to them before printing. Thus the determination of true shadows necessitates using raw data tapes. In this technique, after a DN histogram for each image is obtained, several obvious impact craters or subsidence pits are "zoomed" onto and the DN (digital number) range of such true shadows inspected via manual pixel counting. Having thus established the DN range of true shadows, flow margins are similarly inspected, using a zoom facility, counting those pixels as true shadows only if their DN values fall within the DN range established above. The shadow lengths thus obtained are then subjected to the shadow method treatment and the relative height differences calculated.

A similar problem arises with slope: gradients are well constrained for both the Earth and Moon but this is not so for Mars. Existing topographic maps of Mars are well below their lunar and terrestrial counterparts with respect to the accuracy of both horizontal and vertical contouring. In consequence significantly larger errors occur in the data for Mars.

5.8.3   The derivation of effusion rate
The amount of lava erupted per second can be related to flow geometry (Hulme, 1976) according to the equation

$$F = E\eta(g\rho)^3(\alpha/S_y)^4$$

where E is the effusion rate, g is the acceleration due to gravity, ρ the density of the lava, a the slope, η the viscosity and S_y the yield strength. The parameter F is intimately related to the total flow width (w) and the combined width of its marginal levees ($2w_b$). Thus for a flow which is wide compared with its thickness

$$F = 2W^{5/2}/15 - W^2/4 + W/6 - 1/20, \text{ where } W = w/2w_b$$

It will be seen that it is necessary to know the apparent viscosity of the lava. Hulme (1976) based his studies on the assumption that lunar and Martian lavas would have similar viscosities to terrestrial lavas of comparable yield strength. He therefore devised a graph relating silica content to plastic viscosity, using data derived from actual measurements made on lavas at Mauna Loa (1.7×10^5 N m^{-2}) and Paricutin (3.6×10^6 N m^{-2}) as a calibration, and input the appropriate values into the above formula.

The effusion rate can also be estimated from the dimensionless Gratz number that is related to the cooling of a hot fluid moving through a cool pipe (Hulme and Fielder, 1977; Wilson and Head, 1983)

$$F = Gz\kappa wL/H$$

where Gz is the dimensionless Gratz number, κ is the thermal diffusivity of the fluid (taken to be 7×10^{-7} m^2 s^{-1} for mafic lavas (Pinkerton and Sparks, 1976) and L the total length of the flow. There is considerable field evidence that flow ceases in terrestrial lava channels when Gz reaches a value of 300. This may also apply to the Moon and Mars but is likely to be smaller on Venus, due to the high ambient temperatures, which would favour the development of longer flows.

Where volcanic flows have not developed levees, a *radiative heat loss model* may be utilized (Pieri and Baloga, 1984). While such a model does not permit calculation of either yield stress or viscosity, it does allow fair estimates of a flow's average effusion rate and emplacement time. Based on the premise that for most flows the principal heat loss mechanism is through radiation, which is represented by the Stefan-Boltzmann equation for black-body radiation, it also presumes that such thermal losses determine the planimetric form of the emplaced flow. The predicted correlation takes the general form:

$$<Q> = (\text{constant}) A$$

where (constant) is the Stefan-Boltzmann radiation coefficient and A is flow plan area. This model has been used in studies of both Hawaiian and Martian basalt-like flows (Pieri and Baloga, 1986; Baloga and Pieri, 1985; Cattermole, 1987) and provides the capability of modelling for flows where the outer crust was continually renewed and therefore thermally homogeneous (i.e. "thermally mixed") or where the molten lava was enclosed in a solid skin that radiated at the solidus temperature as a maximum and consequently was thermally inhomogeneous ("thermally unmixed"). The principal shortcoming of this model is that it requires knowledge of the total length of a flow (in order to calculate the area), but this is often difficult to establish, particularly on Mars, since the upstream sections of many flows associated with shields and paterae are buried by later flows. If flows show a strong radial distribution around a shield, therefore, it is usual to assume that a flow originates at or very near the summit. Fortunately, since the upper reaches of such flows are usually rather narrow (in contrast to the often very broad distal sections), the errors introduced by such uncertainty are often not too large.

Like the yield stress models described above, calculations require the input of lava density and also realistic values for both emissivity and specific heat (say, 1 and 8.4×10^6 ergs gm^{-1} K, respectively). Assuming that the outer crust of the mobile lava was continually renewed and therefore thermally homogeneous, the eruption rate (Q_m) is represented by

$$Q_m = 3[T^{-3}_{solidus} - T^{-3}_{original}]^{-1} [\varepsilon\sigma/\rho C_p]A$$

while if the molten lava was enclosed in a solid crust that radiated at the solidus temperature as a maximum, and therefore was thermally inhomogeneous, the eruption rate (Q_u) is represented by

$$Q_u = T^4_{solidus}[T_{original} - T_{solidus}]^{-1}[\varepsilon\sigma/\rho C_p]A$$

where Q is the eruption rate, A the plan area of the flow, ε the emissivity, σ the Stefan-Boltzmann radiation constant (5.67×10^{-8} W $m^{-2}K^{-4}$), ρ the density and C_p the specific heat.

5.9   ESTIMATES OF VISCOSITY

The viscosity of a flowing lava may be determined if it is possible to measure the velocity of the flowing lava at the centre of a channel (V_o), the channel width and make accurate estimates of flow depth during eruption. These data may then be inserted in an equation for laminar flow in semi-elliptical channels derived by Johnson (1970), thus

$$\eta = \rho \sin\alpha/V_o\{B^2/2[(B/A)^2+1]+[(B/A)^2+1][\tau_y/\rho g \sin\alpha]^2/2 - \tau_y B/\rho g \sin\alpha\}$$

where B is the depth at the channel centre and A the half-width of the channel or flow.

Naturally such a method cannot be used for the "fossilized" flows imaged on the Earth and on other planetary surfaces. However if the Gratz number method is used to determine effusion rate and the yield strength is known, it is possible to calculate the plastic viscosity (η). Thus

$$\eta = w^3Y \sin^2\alpha/24S_y \quad \text{for } r < 1, \text{ or}$$

$$\eta = w^{11/14}Y^{5/4} \sin^{3/2} \alpha/24S_y \, g^{1/4}\rho^{1/4} \quad \text{for } r \geqslant 1$$

where r is the ratio of channel width to total levee width. This method has been used by Zimbelman (1985) for flows surrounding the summit of the Martian shield, Ascraeus Mons, and the results show close similarity to field measured viscosities for basaltic andesite lavas at Arenal (Cigolini et al., 1984).

5.10   RHEOLOGICAL DATA

In the following section a summary of rheological measurements made by various workers is presented. For reasons of clarity and space, error information is not given but can be consulted in the original publications. Yield strength and viscosity determinations are shown in Table 5.1.

5.10.1   Yield strength and viscosity

The yield strength values for lunar flows are in general accord with those for terrestrial basaltic lavas, although Imbrium lavas tend to be lower in yield strength than all but a few Hawaiian flows. Lunar impact melt flows tend to have higher yield strengths than the basin-filling lavas, while data for Martian flows indicate yield strength and viscosity values appropriate to a mafic composition. Most workers have concluded that Tharsis lavas are somewhere between basalt and basaltic andesite in composition, but in view of the inability of rheological data firmly to establish chemical composition (particularly silica content – see Moore *et al.*, 1978, Baloga, 1989), "basalt-like" is probably a better term.

5.10.2   Effusion rate and flow duration

The problems associated with measuring effusion rate have already been outlined and have been discussed by various workers. Walker (1973) notes that during the first few hours of many Hawaiian and Icelandic eruptions the eruption rate is very much higher than the average rate. For example, during the first 8 h of the 1961 Askja eruption, the average effusion rate was estimated to be 800 m^3 s^{-1} but then fell rapidly to one-tenth this value, giving an average rate of about 33 m^{3}s^{-1} (Thorarinsson and Sigvaldason, 1962). More recently, during the three-week-long emplacement of a 27-km-long aa flow at Mauna Loa in 1984, where lava was erupted at a rate of 8.06×10^2 m^3 s^{-1} for the first 6 hours, this quickly fell to 1.2×10^2 m^3 s^{-1} during the ensuing few days, then dropped to below half this for the next two weeks before declining to much lower rates as the eruption reached its dying stages (Lipman and Banks, 1987). In Table 5.2, therefore, average rates of effusion are listed for a variety of terrestrial, lunar and Martian locations. For terrestrial flows "average" implies the volume of material divided by the duration of activity at the source vent.

Data on the duration of flow activity for non-terrestrial flows is very limited. Hulme and Fielder (1977) estimated that lunar impact-generated lavas had emplacement times of 0.2 days for Aristarchus flows, 13 days for Tycho flows and 15 days for flows inside Copernicus. Hulme (1976) arrived at a figure of 85 days for emplacement of one Olympus Mons flow, while Zimbelman (1985) gives 100 days for Ascraeus Mons flows. On Alba Patera, average eruption rates derived by the radiative heat loss model predict that the narrow summit flows were emplaced in between 1 and 6 days and the more extensive sheet and tube-fed flows in between 2 and 35 days. Average eruption rates calculated by a Hulme-type method for leveed flows around the summit calderas give durations of between two and three times those quoted above. If this relationship is extended to the larger non-leveed flows, flow durations for sheet and tube-fed flows rise to between 50 and 150 days.

Table 5.1: Yield strength and viscosity estimates for lava flows on the Earth, Moon and Mars.

Location	Yield strength (Pa)	Viscosity (Pa s)	Lava type	Author
Earth:				
Mono craters, Calif	$1.2\text{-}3.3 \times 10^5$		rhyolite	4
Mt. St. Helens	1.5×10^5		andesite	4
Teide, Tenerife		4.4×10^7	phonolite	2
Paricutin		3.6×10^6	andesite	2
Arenal, Costa Rica		10^7	bas.andesite	14
Mount Etna	9.4×10^3	9.4×10^3	basalt	5
Mount Etna	$10^3\text{-}5 \times 10^4$		basalt	11
Meru, Tanzania	6.6×10^4		phonolite	12
Oldonyo Lengai		10-100	carbonatite	13
Columbia River	$<7 \times 10^3$		basalt	6
Columbia River		$5.0\text{-}4 \times 10^3$	basalt	16
Mauna Loa, Hawaii	0.4×10^4		basalt	4
Mauna Loa, Hawaii		1.7×10^5	basalt	2
Mauna Loa, Hawaii	$3.5 \times 10^2\text{-}7.2 \times 10^3$	$1.4 \times 10^2\text{-}5.6 \times 10^6$	basalt	10
Makaopuhi, Hawaii	$70\text{-}8 \times 10^3$	$7 \times 10^2\text{-}4.5 \times 10^3$	basalt	14
Kilauea, Hawaii	$1.5 \times 10^3\text{-}5 \times 10^4$		basalt	8
Hawaii	$0.23\text{-}1.1 \times 10^5$		trachyte	4
Mt.St.Helens 1980	$0.4\text{-}1.4 \times 10^3$	20-320	mudflow	17
Moon:				
Mare Imbrium	4.2×10^2			3
Mare Imbrium	1.5×10^2			1
Aristarchus	1.3×10^4			3
Aristarchus	1.94×10^4			4
Copernicus	1.8×10^4			3
King	2.41×10^4			4
Necho	2.25×10^4			4
Tycho	$1.2\text{-}2.0 \times 10^4$			3
Tycho	$4 \times 10^2\text{-}1.5 \times 10^3$			3
Mars:				
Alba Patera	$1.9 \times 10^3\text{-}2.8 \times 10^4$	$1.7 \times 10^5\text{-}1.9 \times 10^6$		9
Arsia Mons	$0.39\text{-}3.1 \times 10^3$			4
Ascraeus Mons	$3.3 \times 10^3\text{-}8.3 \times 10^4$	$6.5 \times 10^5\text{-}2.1 \times 10^8$		7
Olympus Mons	$1.8\text{-}5.3 \times 10^4$			4
Olympus Mons	$8.8 \times 10^3\text{-}4.5 \times 10^4$	$2.3 \times 10^5\text{-}6.9 \times 10^6$		2

Author key: 1 Moore and Schaber, 1975; 2 Hulme, 1976; 3 Hulme and Fielder 1977; 4 Moore *et al.*, 1978; 5 Pinkerton and Sparks, 1978; 6 Murase, 1981; 7 Zimbelman, 1985; 8 Fink and Zimbelman, 1986; 9 Cattermole, 1987; 10 Moore, 1987; 11 Kilburn, 1985; 12 Cattermole, (unpubl.); 13 Dawson, (unpubl.); 14 Cigolini *et al.*, 1984; 15 Shaw *et al.*, 1968; 16 Murase and McBirney, 1973; Fink *et al.*, 1981.

Table 5.2: Estimated average effusion rates for terrestrial, lunar and Martian flows. Qm = thermally mixed model; Qu = thermally unmixed model. $T° = 1500$ K.

Volcano	Effusion rate ($m^3 s^{-1}$)	Reference
Earth:		
Mauna Loa, 1984	8.6×10^2	Moore (1987)
Mauna Loa, 1851	3.0×10^2	Walker (1973)
Mauna Loa (overall)	10-1000	Malin (1980)
Kilauea (overall)	2-600	Malin (1960)
Kilauea, 1955	16	Walker (1973)
North Queensland	12×10^3	Stephenson and Griffin (1976)
Laki, Iceland	6.0×10^2	Walker (1973)
Askja, Iceland, 1961	33	Thorarinsson and Sigvaldason (1962)
Etna (overall)	$0.1 - 75$	Lopes and Guest (1982)
Etna, 1669	70	Walker (1973)
Columbia River (Rosa)	11.6×10^3 (/km vent)	Swanson *et al.* (1975)
Paricutin, 1945	0.5	Walker (1973)
Cerro Negra, 1968	1.8	Walker (1973)
Karkar, 1974	0.4-15	McKee *et al.* (1976)
Ulawan, 1973	~60	Cooke *et al.* (1976)
Langila, 1973	0.5	Cooke *et al.* (1976)
Moon:		
Mare Imbrium	8.2×10^4	Hulme and Fielder (1977)
Mare Imbrium	2.5×10^5	Hulme and Fielder (1977)
Mare Imbrium	2.1×10^5	Hulme and Fielder (1977)
Sinuous rille lavas	8.0×10^5	Hulme and Fielder (1977)
Copernicus	260	Hulme and Fielder (1977)
Tycho	1060	Hulme and Fielder (1977)
Aristarchus	300	Hulme and Fielder (1977)
Mars:		
Olympus Mons	470	Hulme (1976)
Ascraeus Mons	18-60	Zimbelman (1985)
Alba Patera (Qu)	$27-151 \times 10^3$	Baloga and Pieri (1985)
Alba Patera (Qm)	$42-233 \times 10^3$	Baloga and Pieri (1985)
Alba Patera (leveed)	$155-5846 \times 10^3$	Cattermole (1987)
Alba Patera (sheet, Qu)	$8-877 \times 10^3$	Cattermole (1987)
Alba Patera (sheet, Qm)	$13-1348 \times 10^3$	Cattermole (1987)
Alba Patera (tube, Qu)	$134-1047 \times 10^3$	Cattermole (1987)
Alba Patera (tube, Qm)	$206-1610 \times 10^3$	Cattermole (1987)

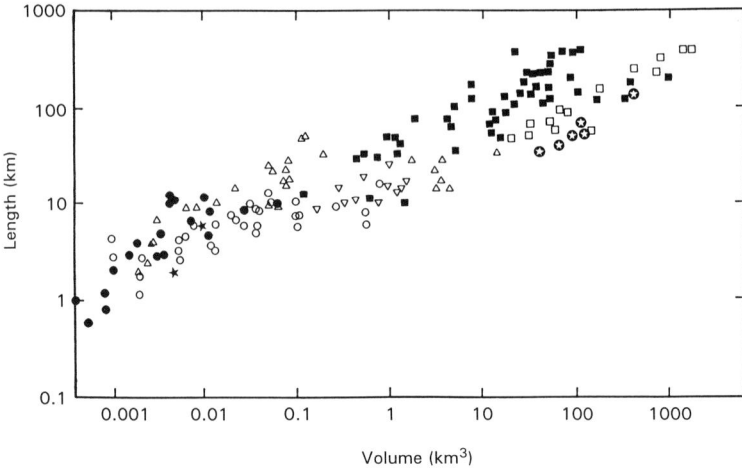

Fig. 5.9: Plot of flow length versus subaerial volume for a variety of terrestrial and Martian flows. Key: solid circles – Kilauea (after Malin), open triangles – Mauna Loa (after Malin), black stars – Ulawan and Langila andesites (Papua New Guinea, open circles – Etna (after Lopes), inverted open triangles – Olympus Mons (after Lopes), solid boxes – Alba Patera leveed flows, open boxes – Alba Patera sheet flows, white stars in black circles – Alba Patera tube-fed flows. All Alba data after the author.

5.10.3   Analysis of quantitative data

In an earlier section it was shown how the rate of effusion will be controlled by the lava rheology, the conduit shape and the magma supply rate. Thus fissure vents will tend to be associated with higher effusion rate activity than circular conduits (Wilson and Head, 1981). The morphological parameters of individual flows, in particular their length and width, has been the subject of much study, beginning with the work of Walker on Etnean lavas (1973). He concluded that effusion rate was the single most important factor controlling the final distance a flow progressed from its source and also noted that the longer lavas on Etna originated in low-altitude vents (Walker, 1974). He extended the study to Hawaiian and other volcanoes which, apparently, confirmed his initial conclusion. Subsequently, however, Malin (1980) questioned the validity of this conclusion and pointed to the importance of being able to distinguish the "actual effusion rate", i.e. the volume of a flow divided by the length of time it was fed, and "average effusion rate", i.e. volume of flow divided by total duration of an eruptive episode (during which time a flow may only by fed and active for a fraction of the total time). Malin also pointed out that while there was a good correlation between flow length and average effusion rate for many flows on Mount Etna, that the correlation was less good for Hawaii, where, for similar eruption rates, flows tended to be significantly longer. He concluded that both volume and cross-sectional area (probably dependent on the viscosity, cooling rate and slope) played equally important roles as effusion rate in determining the final length of flows, a conclusion which is not entirely unexpected in view of the fact that to increase the length of a flow the supplied magma volume also must increase. In a sense, as Wood (1984) has observed,

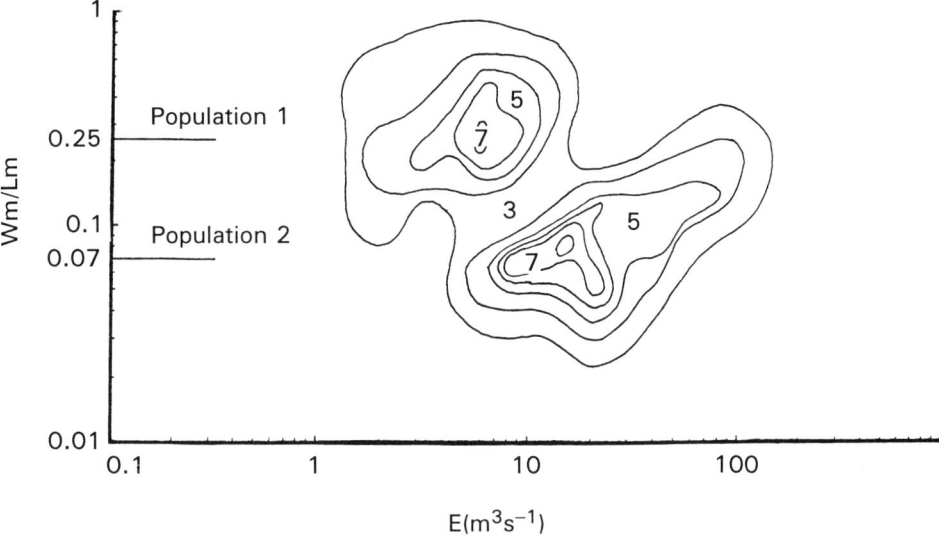

Fig. 5.10: Logarithmic plot showing variation of W_m/L_m with average effusion rate for Etnean flows and identifying the two main populations (after Lopes and Kilburn, 1987.)

both Mike Malin and George Walker should be correct! A length versus volume plot for a large number of terrestrial and lunar flows illustrates the not-unexpected overall broad correlation between these two parameters

Other factors must, to varying degree, also have important effects. Premier amongst these are whether flows are channel- or tube-fed, the state of the magma as it emerges from the vent and the local topography. With respect to the first of these, Swanson (1973) indicated that magma enclosed within closed tubes flows virtually isothermally, cooling at a rate of ~1°C km^{-1}; thus for an eruption temperature of 1160°C, such low cooling rates could generate very lengthy flows (about 12 km) at very low effusion rates (approximately 4 m^3 s^{-1}) in Hawaii. Tube formation also seems to be enhanced by sporadic eruption, where multiple flow units are formed interspersed with periods of quiescence; this style of activity characterized a six-year effusive episode at the Hawaiian centre, Mauna Ulu (Greeley, 1971). The common development of flow tubes in the mainly pahoehoe lavas of Hawaii is almost certainly instrumental in governing the much greater lengths to which these flows extend, for the same effusion rate or lava volume, than those on Mount Etna.

Recent studies of aa flows on Etna have shown that different flow populations are attributable to systematic differences in the duration, effusion rate and flow decay rate of the lavas (Lopes and Guest, 1982; Lopes and Kilburn, 1987). Thus statistical analysis of length, width, area and duration for Etnean lavas shows that two distinct flow populations occur (Fig. 5.10). Population 1 flow-fields have $W_m/L_m = 0.07$, effusion rates greater than 10 m^3 s^{-1} and durations

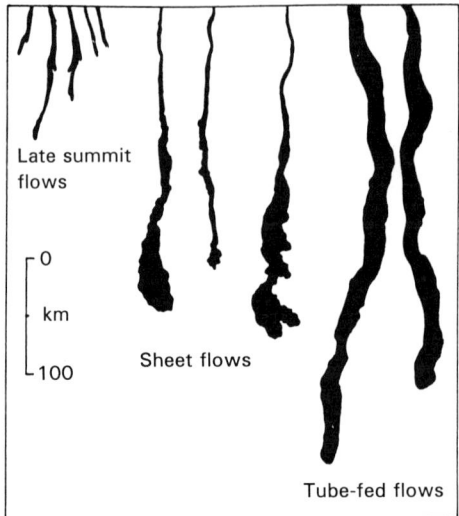

Fig. 5.11: Principal types of flow exposed on the Martian volcano, Alba Patera.

of less than 15 days. In contrast, population 2 flows have $W_m/L_m = 0.25$, effusion rates lower than 10 m³ s⁻¹ and durations of more than 15 days. Despite such clear differences, both populations show similar degrees of spreading about their longest extension downstream. From this it may be inferred that it is the duration of an eruption which has the most direct influence on the population to which a particular flow-field belongs, while the average effusion rate either plays a secondary role or exerts a much less sympathetic influence. This suggests that a time-related process governs the axial dimensions of many aa flow-fields, since the chances of any width-increasing factor coming into play must increase with time. Thus once cooling has significantly reduced or actually halted progress of the longest lava stream, subsequent eruption from the vent area must then preferentially contribute only to spreading and thickening. Such a conclusion confirms the conclusions of Baloga and Pieri (1986).

Recently a study has been made of flow-field development associated with the Martian volcano Alba Patera (Cattermole, 1989). Here visual inspection allows the differentiation of three main flow types on the volcano's flanks: (i) relatively short, narrow and often leveed flows which cluster around the summit region, (ii) longer sheet-style flows which often show terminal broadening on the lower flanks, and (iii) long tube-fed flows (Fig. 5.11). When a more rigorous quantitative analysis is made, the distinctiveness of these flow types is highlighted and interesting facts concerning flow-field development in both space and time emerge. A linear plot of maximum length (L_m) versus area for all types of flow is shown in Fig. 5.12 and distinguishes three principal groups: Group 1 lavas, which include most summit sheet flows that show a relatively minor increase in area per unit length, Group 2 flows- the sheet flows of the flanks, which exhibit distal broadening and a wider scatter of points; and Group 3 flows, which include tube- and channel-fed lavas

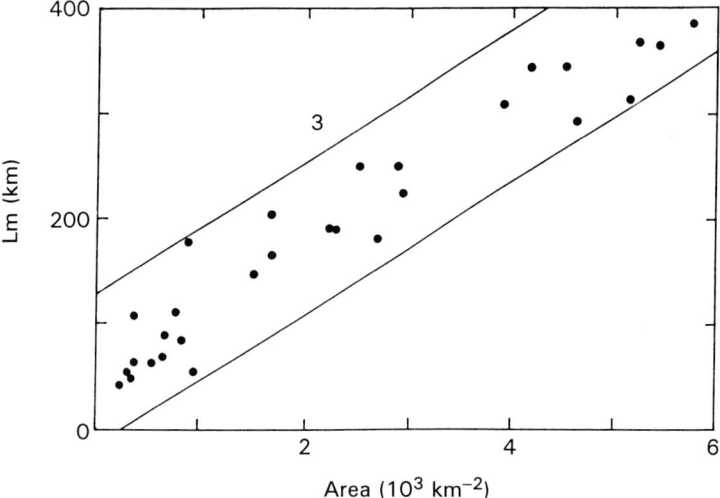

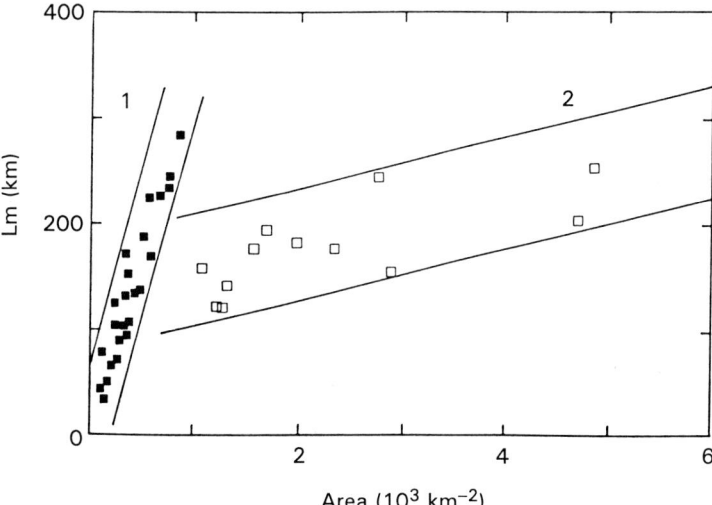

Fig. 5.12: Linear plot of maximum length (L_m) against area (km²) for flows on Alba Patera: (above) tube-fed flows (Group 3); (below) leveed and sheet flows (Groups 1 and 2 respectively). The three groups of flow are quite distinct.

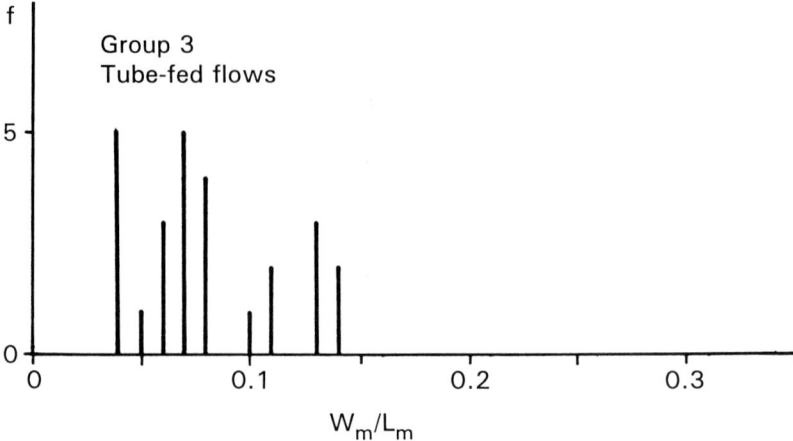

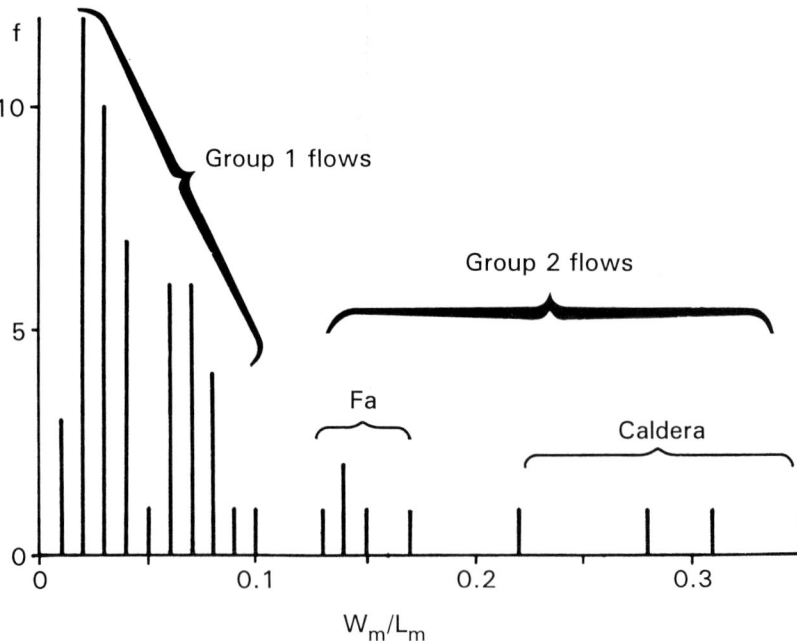

Fig. 5.13: W_m/L_m values for flows on Alba Patera. The greater degree of terminal broadening developed in some flows within some large flank flow-fields (Fa) is manifested in some higher Group 2 peaks (peak at 1.4).

which, like Group 2, show a relatively large increase in area per unit length. The distinct geomorphological imprint afforded to the circum-caldera region by the concentration of group 1 flows, is largely a function of the changing flow-field style which accompanied the volcano's evolution and also reflects a degree of topographical control on the lateral spreading of flows in the lower flanks.

Since the ratio of maximum width to maximum length (W_m/L_m) was used successfully by Lopes and Kilburn (1987) to segregate Etnean aa flow-fields into two populations, a similar analysis of Alba flows was made. The shorter circum-summit flows showed the lowest values for this ratio, peaking at 0.02, while the longer flows of this group peak at 0.07. The Group 3 population show a relatively large range in values, but is distinct from both of the other two groups (Fig. 5.13). The longer Group 1 lavas of Alba exhibit a comparable range of W_m/L_m values to the Etnean population 1 flows. A logarithmic plot of L_m versus area reinforces the distinctiveness of the Group 1 lavas (Figure 5.14). Group 3 flows also cluster into a well-defined field, of similar slope to Group 1 but with a different y-axis intercept; the implication here being that tube-fed lavas increase in area much more rapidly per unit length than the narrower summit flows. The Group 2 population plots close to the Group 3 field but shows an even greater tendency to increase in area per unit length – a function of much distal broadening within the group. When volume is plotted against length for all the flows (Fig. 5.14), the distinctiveness of Group 1 is again very clear, while Groups 2 and 3 exhibit a mutually similar trend, with a relatively large volume increase per unit length, as would be expected in the case of lavas of very high volume. The overall shift from Group 2 and 3 flow styles towards the late-stage Group 1 flows must indicate a gradual change in the plumbing of the volcano with time.

If the basic premises of Lopes and Kilburn (1987) regarding the attribution of different Etna flow populations to systematic differences in the duration, effusion rate and flow decay rate are accepted, then Alba flow-fields with small W_m/L_m values (Group 1 flows) should represent eruptions of relatively short duration that were generated at higher average effusion rates than the broader sheet and tube-fed flows of the lower flanks. Thus the latter would be expected to widen and thicken downslope as eruptions took place over more protracted periods, the steadier eruption conditions being more conducive to the generation of fresh lava streams while earlier ones remained active, giving rise to increasingly broad flows. Interestingly, should Hawaiian effusion rates be more realistic for these very large Martian flows than those estimated by the models described above, the emplacement times for the larger flows could rise to as high as 500 years.

5.10.4 Rheology and lava composition

For terrestrial and lunar lavas it is possible to relate magma chemistry, in particular silica content, to rheology; currently this is not possible for either Mars or Venus. Rheological calculations for lunar basalts confirms that they had lower yield strength (and probably viscosity) than most terrestrial basalts. This probably accounts for the general lack of discernible flow margins and other positive relief features over very large areas. Furthermore the mare-filling lavas must have been erupted at very high rates indeed (around 1×10^6 m^3 s^{-1}). While significant errors are known to attach to Hulme-type methods of deriving yield strength, largely because they assume a uniform lava density for each flow

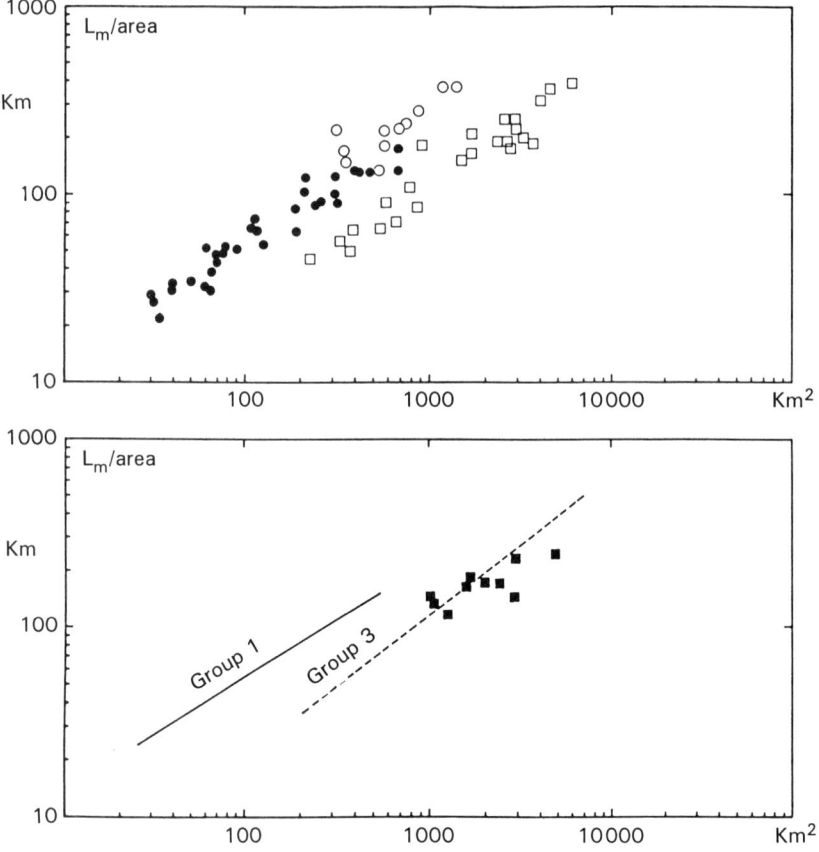

Fig. 5.14: Logarithmic plot of L_m against area for flows on Alba Patera: (above) Group 1 flows (solid circles) and projected lengths of these flows if originating in summit region (open circles); Group 3 flows (open boxes); (below) Group 2 flows (solid squares) with trends of Groups 1 and 3 flows superposed.

(a condition seldom if ever achieved in nature), even if the current data set for Mars were subject to a 50% error, the rheological properties remain those of low yield strength and low viscosity, implying a basic or ultrabasic composition. This is consistent with revised estimates of Martian mantle composition by Goettel (1980) and with the refined spectroscopic data of Singer (1980) which places Martian rocks in the 5-10% modal olivine range. Melting of a Martian mantle with two to three times the iron content of Earth's would yield iron-enhanced basic or ultrabasic lavas comparable in their rheology to the observed Martian flows (Kaula *et al.*, 1981).

As has been noted by Fink and Zimbelman (1988) inherent limitations in the Viking data sets render elaborate modelling of Martian lava composition, at best, very approximate.

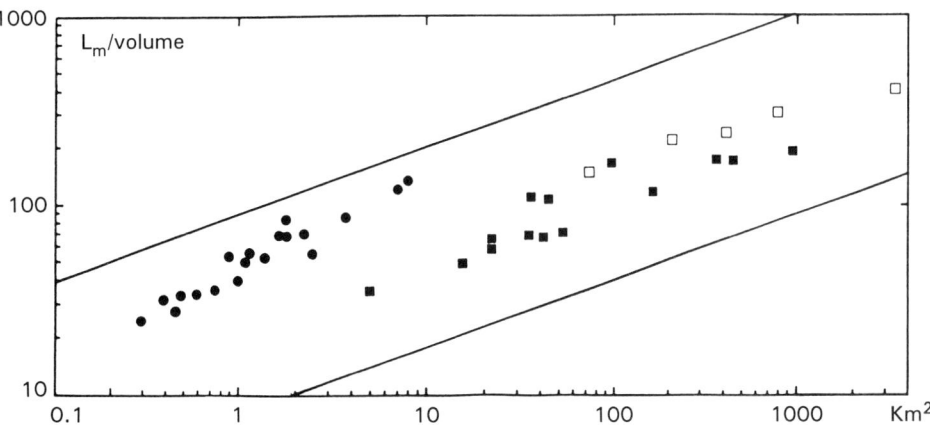

Fig. 5.15: Plot of L_m against volume for all flows measured on Alba Patera together with extrapolated limit lines from the work of Lopes for Mount Etna flows (symbols as above).

In an evaluation of simple models and their applicability to Hawaiian flows, i.e. yield strengths calculated from flow margin heights, viscosity calculated from observed flow velocity, they confirmed that yield strengths for flows ranging in length between 5.3 and 8.4 km, was consistent with a basaltic composition. After "degrading" their data to make it comparable with Viking imagery, results for a 11-km-long Ascraeus Mons flow indicated basaltic yield strengths, but for a slope substantially less than that in Hawaii.

Rheological estimates for Venusian flows must await detailed analysis of Magellan imagery. However, the extensive volcanic plains, the dimensions and morphology of flow-fields, widespread development of low shields and shield fields, all are indicative of effusive activity of fluid lavas of low yield strength, i.e. basalt-like rheology. The very long channellized flows observed on some Magellan images suggests that many such lavas flowed beneath insulating crusts, i.e. were tube-fed (Gregg and Greeley, 1993) Such geochemical data as exists confirms basalt-like chemistry. The occurrence of steeper-sided volcanic structures such as pancake domes, however, is indicative of more viscous, evolved lavas, suggesting perhaps that a greater degree of magmatic evolution has taken place on Venus than on Mars.

5.11 PYROCLASTIC FLOWS

Pyroclastic flows generally are believed to form by the gravitational collapse of a rising eruption column (Sparks *et al.*, 1978). Such a conclusion is based on analysis of their deposits and, to a lesser extent, on observation of small-scale Plinian eruptions. The principal controlling factors on the height at which collapse of the eruption column takes place are vent radius (R), gas content (N) and initial gas velocity (W). By modelling within the range $R = 50 – 600$ m, $N = 0 – 5\%$, and $W = 200 – 600$ m sec^{-1}, Sparks *et al.* were able to demonstrate that column collapse occurred at height between 0.6 and 4 km above a

Fig. 5.16: Blocky texture of massive mudflow associated with caldera collapse of Mount Meru, N. Tanzania. Angular blocks of a wide range of sizes are suspended in a matrix of smaller clasts and mud-sized material. Photo: Peter Cattermole.

vent. The initial velocities of flows ranged between 60 and 310 m sec^{-1}. They also noted that entrainment of atmospheric gases during collapse could lower the temperature of the cloud of collapsing material by as much as 350°C. Collapse is favoured by diminishing gas velocity, decreasing water content and increasing vent radius. These parameters lower the minimum velocity of the convecting cloud such that it may reach zero, whereupon, the density of the column being greater than the surrounding atmosphere, it collapses.

On the basis of their modelling, thick incoherent ignimbrite sheets could be shown to have been generated either by magmas with high water content or low initial temperature. For high gas content magmas to undergo column collapse during eruption, wide vents are required. Densely-welded, thin, ignimbrite sheets are produced from magmas with either low gas content or or high intitial temperature.

Ignimbrites are frequently preceded by a reversely-graded air-fall pumice deposit which is assumed to form during the initial stages of a Plinian eruption, when the vent radius and gas content are suitable for production of a Plinian eruption column. This may well occur when the initial eruption emanates from a fissure. As the vent widens and localizes, (typically accompanied by ejection of lithic clasts within the deposits) so the frictional effects of the vent walls on the escaping gas diminish, and the gas exit velocity will increase. Reverse grading is one manifestation of such a process. Furthermore, widening of the vent pushes the eruption towards conditions under which column collapse will occur. Once collapse does take place, it would be extremely difficult for convective recovery to take place unless there was a hiatus in the eruptive cycle.

Within a few kilometres of the vent, pyroclastic flows are deduced to segregate into a high-concentration basal zone within which larger clasts settle downwards. Inside this high-density zones fines are generated by intense crushing and are fluidized by exsolving gases, giving rise to a pyroclastic flow with a high concentration of fluidized particulates. It is the fluidized fines that carry along the larger clasts. The upper (dilute) portion of the flow and the fine ash washed out by the action of the gases, contribute to the formation of the widely dispersed ash-fall deposits which often are as voluminous as the pyroclastic flow generated by the same eruption. The motion of the lower part (the pyroclastic flow) disassociates itself from the upper part (turbulent cloud of gas and ash), the latter gradually mixing with the atmosphere to generate a convective plume.

5.12   LAHARS (MUDFLOWS)

Lahars or mudflows are commonly observed in association with stratovolcanoes but are rarely associated with basaltic shields. They were observed forming during the 1980 eruption of Mount St. Helens, much of their material having derived from loose debris mixed with ice melted during the rise of heat beneath the huge dome on the volcano's northern flank. Their flow effectively silted up local streams and rivers over a wide area.

The characterization of rheological properties is rendered difficult because they leave different types of deposit at different points in their path. For example, lahars typically concentrate boulders ahead of themselves and at their margins, leaving coarse debris rather like glacial moraines. Yield strengths of such material will, therefore, be very different from mid-channel deposits which, in turn, may differ from the fine outwash that frequently exudes from their toes (Fig. 5.16).

Having said this, by sampling at different points along a flow, average properties may be calculated which may enable one flow to be distinguished from another (Fink et al., 1981).

Fink et al. (1981), using models developed for determining properties of debris flows based on the geometry of their deposits (Johnson, 1979) and Johnson and Hampton, 1969), calculated yield strengths for three individual flows formed during the 1980 eruption of Mount St. Helens. Calculated yield strengths were respectively 1100, 1000 and 400 Pa, with maximum flow velocities of $10 - 31$ m sec^{-1}, volumetric flow rates of between 300 and 3400 m^3 sec^{-1}, and plastic viscosities of between 20 and 320 Pa sec^{-1}. They noted that Bingham-type models explained most of the observed structures of such low- to medium-volume lahars but that extrapolation to large volume flows might require modification of rheological models.

5.13   SULPHUR FLOWS

Sulphur flows may be expected to behave rather differently from silicate lavas. This is a reflection of the rather exceptional behaviour of liquid sulphur (expanded upon in Chapter 11). Under surface conditions, molten sulphur remains molten down to 120°C – nearly 1000°C lower than typical basaltic magmas – but the viscosity depends very strongly upon the temperature. In contrast to silicate melts, pure sulphur exhibits a drop in viscosity of a thousand-fold as the temperature falls between 167°C and 157°C. Thus, while it has the

the consistency of stiff treacle at high temperature, it is closer to hot engine oil when it cools. Detailed studies of sulphur rheology are unavailable at the time of going to press.

6

Volcanic landforms

The association of active volcanism with lithospheric plate boundaries is a characteristic of the Earth. At divergent plate boundaries, rising, mantle-derived melts give rise to mafic volcanism both within and along the flanks of the median rifts which develop along the crests of mid-oceanic ridges. These active spreading zones are typified by basaltic rocks of relatively low viscosity and restricted chemical composition, which are effusively erupted. Where plate convergence takes place, the more complex interplay of forces associated with subduction zones gives rise to magmas of a wider compositional spectrum. Basaltic melts do occur but there is a much higher proportion of andesitic, dacitic and rhyolitic magmas and a greater incidence of explosive volcanism, giving rise to ash and tuff deposits as well as lavas.

Volcanism may also occur within plates. In these *intra-plate* settings, magmatism is believed to be associated with rising plumes of hot mantle material. This manifests itself within oceanic crust at oceanic island locations such as Hawaii, where fluid basaltic magmas are generated in large amounts. In contrast, continental intra-plate volcanism is represented by the more explosive style of eruptions seen in parts of Africa, like Tibesti and Jebel Marra, where, amongst other things, ignimbrites are common. Volcanic activity is also concentrated along continental rifts, like the East African Rift, where a wide variety of magma types is found. These give rise to basaltic, phonolitic, nephelinitic, highly potassic, carbonatitic and kimberlitic rocks. One of the Earth's strangest volcanoes, Oldonyo Lengai, is situated in the Tanzanian sector of the Rift and currently erupts not silicate, but sodium carbonate lavas!

6.1 THE DIVERSITY OF VOLCANIC LANDFORMS

Not surprisingly, because of the wide range of magma types found on the Earth, the resultant volcanic landforms are diverse. The size and shape of these landforms are largely a function of the viscosity and quantity of the erupted materials. Relatively large dimensions but low profiles characterize volcanoes and associated landforms built by mafic magmas. On the other hand, volcanoes associated with more silicic magmas are generally steeper, less extensive and accompanied by landscape features produced by explosively generated

Fig. 6.1: Sequence of flood basalts with vertical dykes exposed in the cliffs of Isla La Gomera, Canary Islands. Photo: Peter Cattermole.

materials. When extra-terrestrial volcanism is discussed, a further complexity is introduced – the effects of different atmospheric pressure. This can have a significant effect on the shape and extent of volcanic landforms, particularly those built by pyroclastic materials. The behaviour of air-fall pyroclastic material will, for instance, be very different on the airless Moon from on the Earth. The lack of a lunar atmosphere will mean that ballistically ejected materials can extend for very great distances before settling, whereas on Earth atmospheric drag will bring them back towards the surface very much more speedily. The effects of planetary gravity differences generally are less important.

Wide diversity in landform morphology appears to be a feature of the Earth which, with the exception of Io, has remained the most dynamic of all the planets and moons with solid surfaces. The less active worlds, in consequence of their more limited internal activity, present a less diverse array of volcanic landforms but this is counterbalanced to some extent by their having developed structures quite un-earthlike in either morphology or size. The Martian patera volcanoes could be cited in this context, and so also could the lunar sinuous rilles. There are numerous studies of terrestrial volcanic landforms, among the classic texts being those of Cotton (1952), Rittman (1962) and MacDonald (1972), while Greeley (1994) has done an exemplary job with planetary landforms in a wider context. Because of the very widespread occurrence of landforms believed to have been generated by mafic magmas on the terrestrial planets, there has been a much greater concentration of research upon basaltic volcanism within the planetary community than on the more silicic style. One of the more major projects completed in recent years was

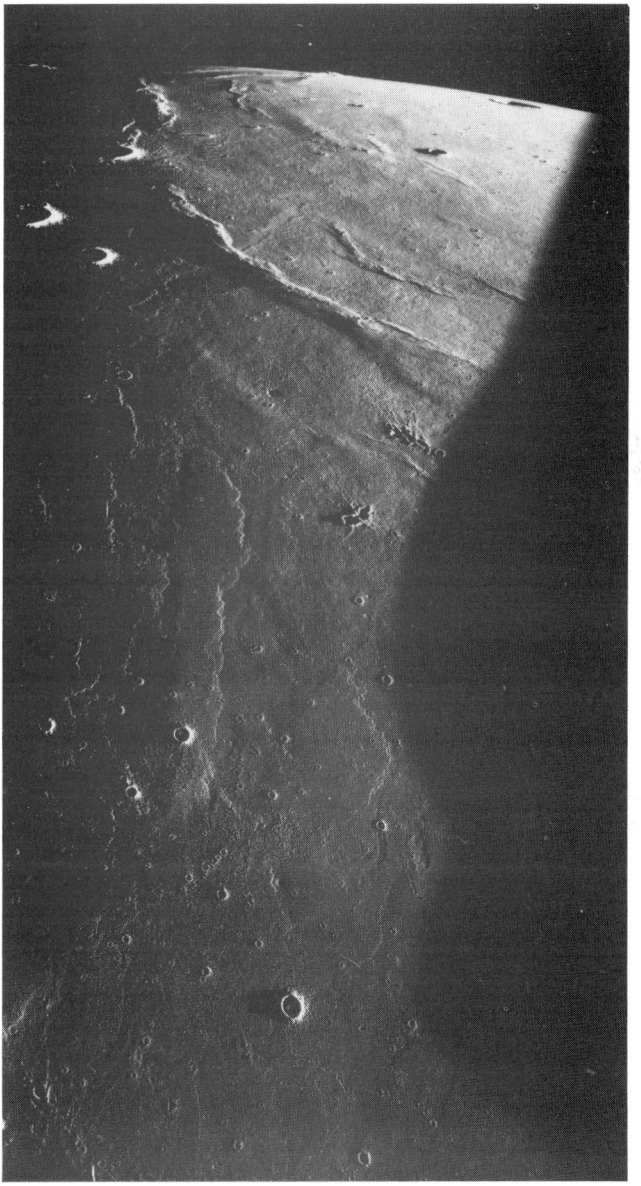

Fig. 6.2: Flood lavas on the western side of Mare Imbrium. These high-volume lavas typically have broad lobate flow fronts, individual units may be up to 30 km across. Several prominent wrinkle ridges lie north of the flow terminations. Apollo 15 mapping camera frame 1559.

published in 1981 as *Basaltic volcanism on the terrestrial planets*, a compendium of the activities of various groups, sponsored by the Lunar and Planetary Institute. This includes a survey of volcanic landforms associated with basaltic rocks. This is referred to subsequently by the abbreviation BVSP (Basaltic volcanism study project). Other surveys which may be mentioned are those of Green and Short (1971), Fielder and Wilson (1975), Wood (1979), Carr and Greeley (1980), Head *et al.* (1981) and Whitford-Stark (1982).

6.2   FLOOD LAVAS

Volcanic plains cover large areas on all of the terrestrial planets and, while totally unexciting from an aesthetic viewpoint, they are among the most important manifestations of basaltic volcanicity. *Flood basalts* are erupted through fissures, emerge in large volumes and spread out as extensive flows which pile up one on top of the other (Fig. 6.1). When the flood-basalts of the Columbia River Plateau were emplaced during the Miocene period, nearly 160 000 km^3 of lava was erupted from linear vents and fissures. These flows outcrop over an area of 200 000 km^2 and buried the pre-existing topography to a depth of 2 km. Similar flows underlie the Deccan Plain of northwestern India. These are even more widespread and the existing 250 000 km^2 of basalt is believed to be but an erosional remnant of an even greater accumulation of lavas. The Parana Basalts of Brazil and Uruguay are even more extensive, having a thickness of 3 km and a volume of 800 000 km^3.

Most terrestrial flood basalt provinces are associated with the early stages of continental divergence. Such an activity does not appear to have characterized any of the other terrestrial planets, with the possible exception of Venus; however, plateau-like lava sequences of large areal extent outcrop on both the Moon and Mars, and are suspected to occur on Venus. These appear to have been erupted from fissures generated during periods of strong crustal extension. The identification of such lavas on planets where erosion has been minimal or even absent is achieved by mapping scarp-like flow margins and diagnostic surface textures such as flow tubes and low domes and using spectral reflectance data which reveals extensive plains composed of basalt-like rocks (Fig. 6.2).

In common with their terrestrial equivalents, source vents of flood lavas on both the Moon and Mars are hidden, having been covered by their own products and by flows from other sources.

Individual flow units within the Columbia River sequence may be very large; for instance, the Roza Member of the Yakima Basalt has an areal extent of 40 000 km^2 (Swanson *et al.*, 1975). Flood lavas erupted from the Icelandic Laki fissure are less voluminous, but still cover an area of 565 km^2. Very large dimensions characterize flood lavas on both the Moon (10^5 km^2) and Mars.

Somewhat less extensive than true flood basalts are flows allied to a style of activity termed *plains volcanism*. This term was introduced by Greeley (1976), after exhaustive study of what originally was believed to be a simple extension of the Columbia River basalts into the Snake River Plain of Idaho. In this region, basaltic lavas are associated with four types of structures: low-profile shields, fissure flows, tube-fed flows and canyon flows, the relationships between which are illustrated in Fig. 6.3. Much of the region is composed of coalescent *low-shields* that have an accumulated thickness of up to 1500 m and which

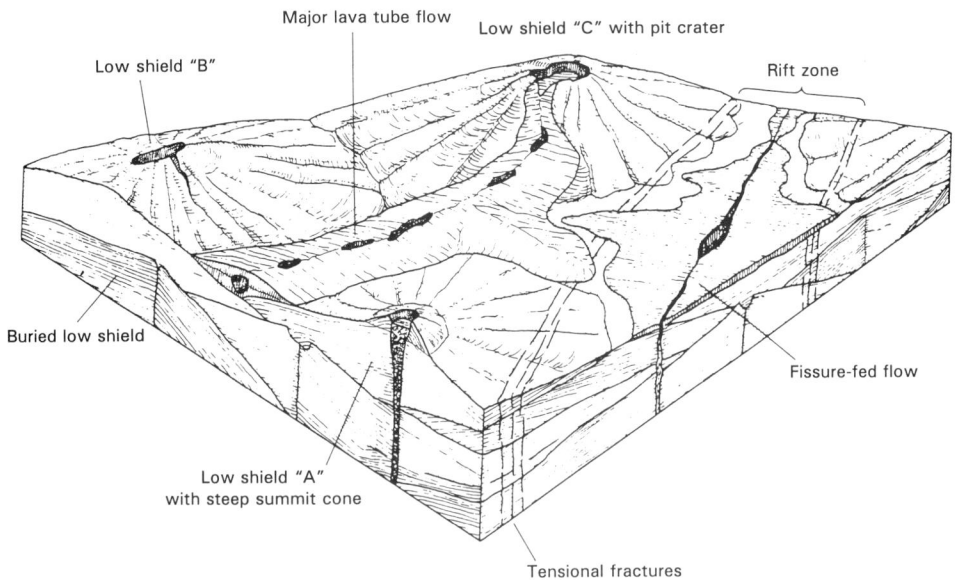

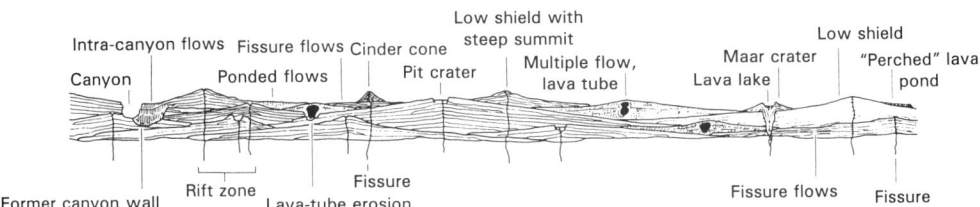

Fig. 6.3: Schematic block diagram and cross section of plains-type volcanic province, illustrating the relationships between low shields, major tube-fed lavas and fissure flows. (After Greeley and King, 1977 and Greeley, 1982.)

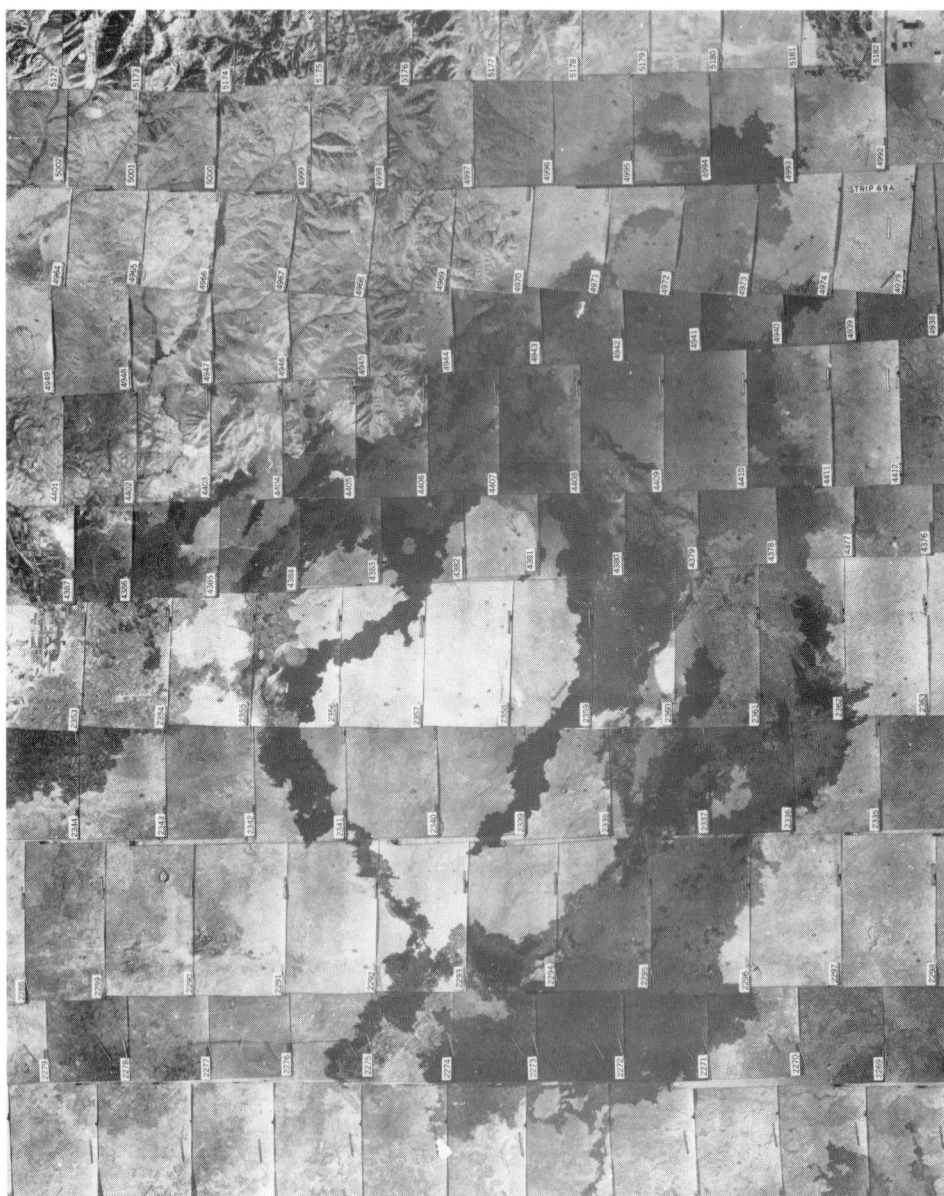

Fig. 6.4: Synoptic photomosaic of the King's Bowl lava field, East Snake River Plain. This is a compound, fissure-fed flow that was erupted both as sheets from a major fissure and also as localized eruptions from point sources along it. U.S. High-altitude Photography Program, Project 121.

Fig. 6.5: The interior of a large collapsed lava tube in the Snake River Plains, Idaho. Note the jointed basalt in the tube walls and the subsequent skim of smoother material left after tube evacuation. Photo: Ron Greeley.

frequently have little topographic expression. The 300 km² Wapi lava field is associated with one of these and is a typical compound pahoehoe flow about 35 m thick in which collapse depressions, pressure ridges and flow ridges are common. On the distal flanks of the dome, slopes are only about 0.5° but steepen to 5° near the crest where short aa flows piled up late in the cycle.

Much of the eruptive activity in Idaho is associated with the Idaho Rift System (Prinz, 1970) which is a series of N/NW-trending discontinuous fractures and aligned vents. Some of the fractures served as flow sources, rather like those of the Columbia River, the two youngest fissure-fed flow-fields being those of King's Bowl and the Craters of the Moon, which comprise compound or multiple flow units. Those of the Craters of the Moon field consist of several pahoehoe and aa flows, generally less than 1.5 m thick which were erupted from fissures along the Great Rift (Fig. 6.4). They cover an area of nearly 1500 km² which, although less than typical Columbia River units, still represents a substantial volume of lava. However, unlike the flows of the Columbia River Plateau, where nearly all the lavas were erupted from fissures or linear vents, in the Snake River Plain the fractures generally controlled the development of active volcanic centres rather than being flow sources, the locations of these foci repeatedly changing with time. The exposed Snake River succession therefore represents a complex of overlapping extrusions from a number of different centres, a feature they share with several Martian volcanic fields and also of certain regions of the Moon.

Fig. 6.6: Region of tube- and channel-fed flows on the west flank of Alba Patera, Mars. Several prominent apical tubes and channels can be seen, some tubes being almost completely collapsed (centre and bottom right of photo) and others only partly so (top right-hand corner). The image is approximately 60 km across. NASA Viking Orbiter frame 007B20.

Widespread also in the Snake River Plains are *tube- and channel-fed lavas*. Small lava tubes (less than 5 m across) are found in many low-shield and fissure-related flows; however, more major tube-fed lavas were emplaced by tubes commonly more than 10 m wide (Fig. 6.5). These produced compound flows with long, narrow and somewhat anastomosing profiles whose courses were very much predetermined by existing topography. Such flows generally have a roughly triangular cross-profile which contrasts markedly with typically flat-topped flood lavas. The successive eruption of similar flows often led to the partial burial of older flows by younger ones, while later lava of various types often infilled the declivities between adjacent tube-fed flows, giving rise to "canyon flows" which often show the effects of ponding.

Studies of Mauna Loa (Greeley *et al.*, 1976) show than over 80% of exposed lavas involved motion partly via lava tubes or channels and field studies indicate the tube flow mechanism to be a very common one. Flow within an enclosed tube is an extremely efficient mechanism for transporting molten rock over great distances, since the carapace of chilled lava act as a thermal insulation. The contribution of the tube mechanism can be gauged by noting that one prehistoric tube-fed hawaiite flow in North Queensland flowed for over 100 km over slopes as low as 0.5° (Stephenson and Griffin, 1976). Many tube-fed lavas on the flanks of the Martian volcano, Alba Patera, are known to have flowed for twice this distance (Cattermole, 1987, 1989) and to have been emplaced by tubes up to 100 m wide (Fig. 6.6). Some lunar basalts are also known to have flowed via the tube mechanism and the enhanced thermal transfer produced by turbulent lava motion is believed to have accounted for the thermal down-cutting which accompanied lunar sinuous rille formation (Hulme, 1973).

Tube formation can occur by the roofing over of an active lava channel for periods of several hours or more (Greeley, 1971) or, as was observed during the 1969-71 activity at Mauna Ulu, Hawaii, where a master tube carried lava almost continuously for a period of 10 months, for very much longer periods (Peterson and Swanson 1974). Subsequently, parts of the tube roof may collapse or flow may occur via an open channel, carrying slabs of broken crust along with it. Another mode of tube formation involves the upswelling of toes of pahoehoe basalt which then burst, producing fresh toes as a result. The repetition of this activity ensures the formation of a tube system which continuously feeds the front of the lava. Activity of the latter kind appears to have been important in some Martian flow-fields.

6.3   SHIELD VOLCANOES

The plains style of volcanism described above is intermediate in character between flood volcanism and the activity associated with shield volcanoes. True shield volcanism is characterized by activity from a low-profile convex structure with gently sloping flanks, known as a shield, which is built from at least 90% lava. The largest terrestrial shields are those of Hawaii, which rise up to 10 km above the ocean floor and may be over 100 km in diameter; despite their great size they do not make dramatic landscape features (Plate 1). The subaerial parts of Hawaiian shields have slopes of less than 5°, but they steepen to around 10° beneath the sea. Icelandic shields are very much smaller but have similar flank

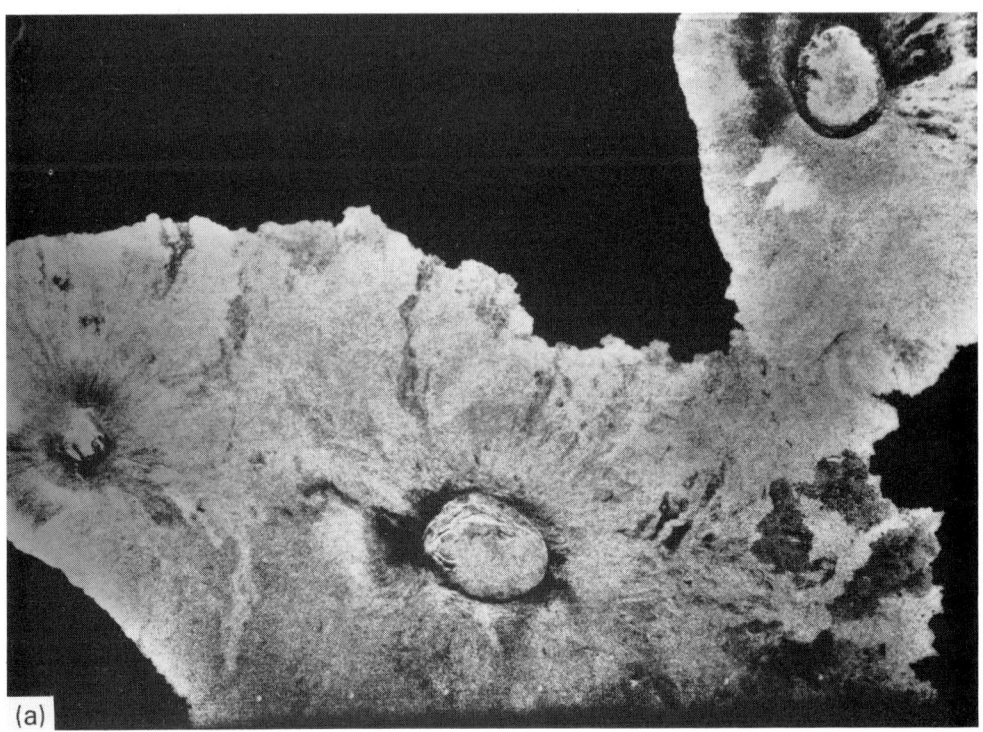

(a)

0   Km.   5

(b)

Fig. 6.7: (a) Shuttle radar image of the volcanic calderas atop three approximately 1000-m-high shield volcanoes of the western Galápagos Islands, Ecuador. The image, taken in 1981, shows a portion of Isabella Island close to the active spreading centre of the Galápagos Rift. The image is 80 km wide. (b) Cross-section through the shield of Isla Fernandina, situated 60 km northwest of the above, but typical of Galápagos shields in general.

slopes, while those of the Galápagos are intermediate in size but unusual in having slopes which may approach 35°. The smallest terrestrial shields are just a few kilometres across at their base and have very low slopes (about 1°), like the low-shields found in the Snake River Plain. While many are approximately circular in plan, others are elliptical due to lateral effusive activity from subvolcanic rifts.

The construction of very large shield volcanoes is achieved by successive eruptions of large volumes of fluid mafic magma which may take place from the summit, lower on the flanks or along rift zones that transect the volcanic edifice. If a particular episode involves relatively small volumes of magma, steep-sided, lava and spatter cones may developed on the summits of the main shield. Both Hawaiian and Galápagos shields may have at their summits large pits or *a caldera*. Such depressions form in response to the depletion of magma from beneath the volcano's summit, either from summit effusions or through flank eruptions. Calderas are often sites of extended lava lake activity (Fig. 6.7).

Compared with terrestrial shields, those of Mars are immense. Concentrated largely in the regions of Tharsis and Elysium, several volcanoes have diameters in excess of 500 km and heights of over 20 km. In accord with their large size, the summit calderas of Martian shields are complex and may be 100 km or more in diameter, whereas those of the larger terrestrial shields seldom exceed a few kilometres. Very long run-out volcanic flows are another feature of Martian shields, these spreading out to radial distances often approaching 500 km or more. Indeed, some Martian volcanic plains actually are built from large-volume flows which emanated from shields and which spread out over the flat ground surrounding them. The distal margins of such flows are often lobate and may coalesce to form terminal scarps hundreds of kilometres long.

Similarly large shields are observed also on the surface of Venus. About 800 such structures within the diameter range 20 – 100 km have been mapped. Fifty have diameters of between 100 and 350 km. Associated with Venusian shields are summit calderas and both summit-related and flank flows, most with lobate margins (Plate 8). A very large number of small shields also occur within the volcanic plains of Venus. Over 30 000 of these (diameters <20 km) have been mapped, the majority occurring in distinct clusters that constitute *shield fields*. On average the shields within such complexes attain diameter of less than 8 km and have small summit depressions <700 m across.

Low shields also occur on lunar maria; however these tend to be rather small, their diameters generally being within the range 5 – 10 km. They often are associated with tube-fed lavas, a feature they share with the low-shields typical of the East Snake River Plain. A particularly well imaged series of such domes is to be found in the Herigonius region.

6.4   PATERAE

This term has been applied to certain very large, low-relief volcanic landforms found on Mars and also on Jupiter's Galilean moon, Io. Perhaps the best Martian example is Alba Patera, situated on the northern flank of the Tharsis bulge. The volcanic flows associated with Alba extend for at least 1000 km from the summit which rises about 6 km above the surrounding plains and is crested by two complex calderas. Alba is unique in having a

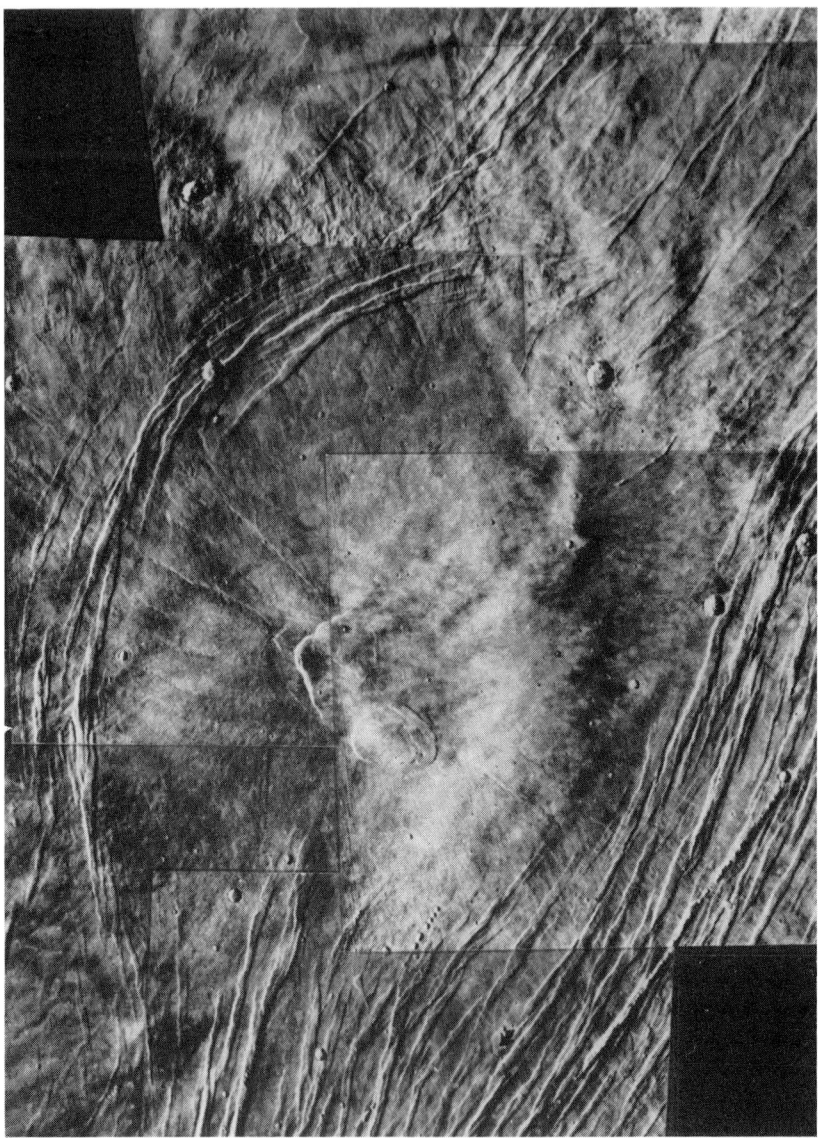

Fig. 6.8: Photomosaic of the Martian patera volcano, Alba Patera, situated in northern Tharsis. The zone of ring fractures has a diameter of 550 km and surrounds the summit region which is crowned by two large complex calderas. Note the prominent volcanic flows towards the northwest corner of the mosaic. Viking Orbiter frames 783A12-17.

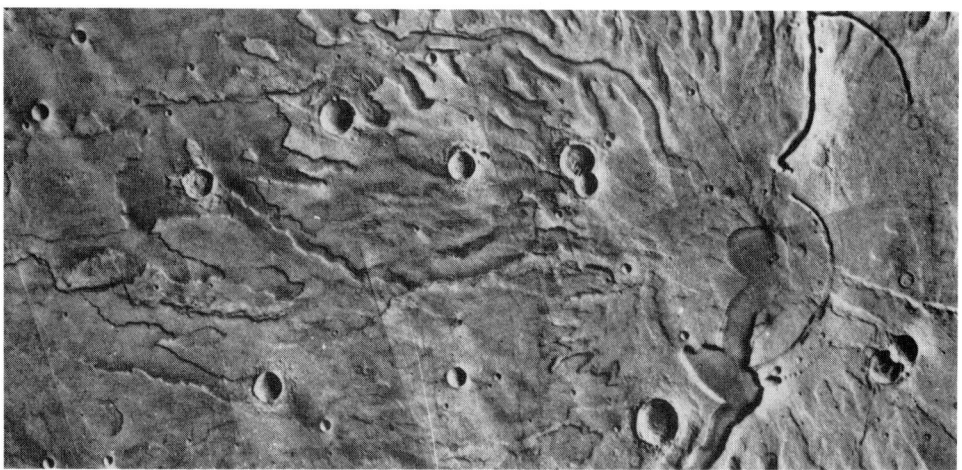

Fig. 6.9: The Martian highland volcano, Tyrrhena Patera, located at 20°S, 252°W. The circular depression at the left of the mosaic is 45 km across and has annular fractures associated with it. Deeply dissected flat-lying sheets may be seen around the summit; these are believed to be eroded ash deposits. Viking mosaic 211-5730.

550-km-diameter annular ring fracture belt which encircles the summit, and although individual volcanic flows are seen to have travelled several hundred kilometres down the flanks of the patera, its slopes are seldom more than 0.1° (Fig. 6.8).

Alba Patera, with its crisply defined volcanic flows, is evidently a very large but very low shield volcano whose profile largely is the result of effusion of large volumes of very fluid lavas. Rather different in morphology are what have become known as *highland paterae* (Plescia and Saunders, 1979). These are deeply eroded and very ancient structures which are located in the cratered southern hemisphere of Mars. They have little vertical relief but have what appear to be central caldera complexes, radiating flow-like features and/or channels (Fig. 6.9). One interpretation of these enigmatic structures is that they represent collapsed shields built in response to massive ash eruptions which occurred very early in Martian history (Greeley and Spudis, 1981; Reimers and Komar, 1979, Greeley and Crown, 1989).

The paterae of Io are also very low-profile structures with calderas that average about 40 km in diameter. Extensive lobate volcanic flows are associated with these.

6.5   VOLCANIC CONES

Central vent activity of the type which builds shields and paterae is largely effusive. The explosivity index of such volcanoes is low. Where a greater proportion of the eruptive products are expelled explosively, volcano profiles become steeper, the intercalation of lavas and pyroclastic deposits giving rise to *strato-volcanoes*, also known as composite volcanoes. These are typified by the symmetrical andesitic cones of the circum-Pacific belt, the volcanoes of the African Rift and the Mediterranean, where Plinian eruptions are

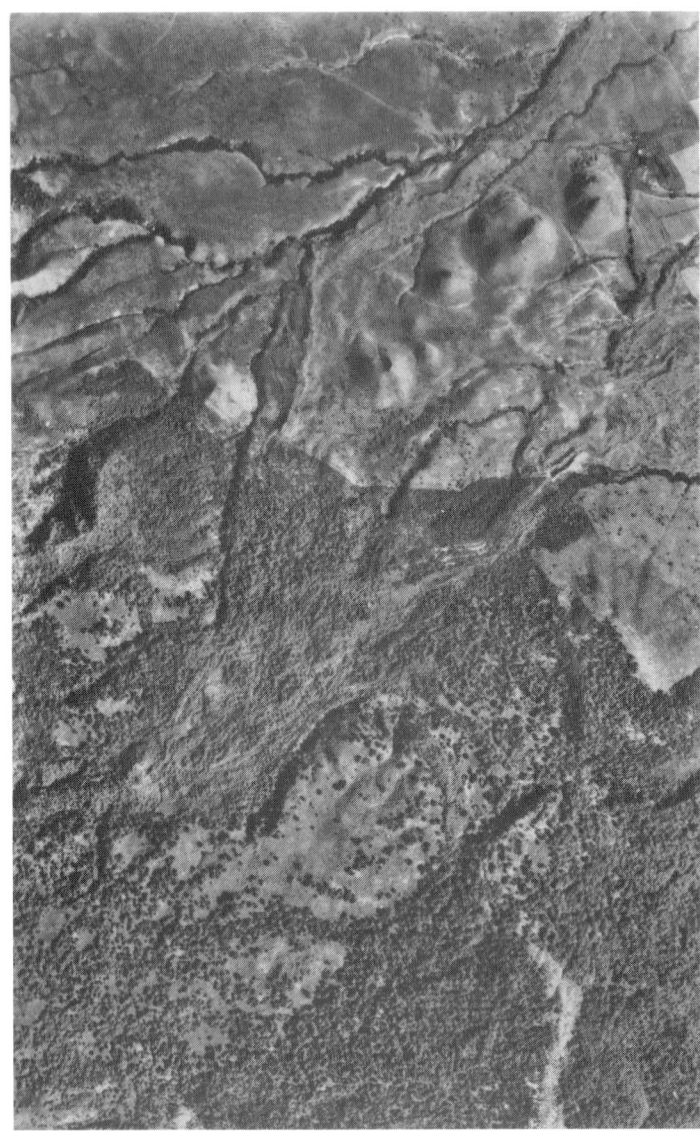

Fig. 6.10: Group of aligned monogenetic cones on the north flank of Meru volcano, Tanzania.
To the south lies a larger intrusive dome. Width of image approximately 6.5 km.

common (Plate 2). Modern strato-volcanoes usually have a small crater at or near the summit and may be intruded by a plethora of dykes, sills and plugs which strengthen the sub-structure. The high explosivity indices of many such volcanoes is reflected in the steepness of their profiles and in a relatively large proportion of air-fall pyroclastics and volcaniclastic deposits like *lahars* (mudflows), which slurry down the flanks with rheological properties akin to many lavas. The term lahar comes from an Indonesian word and it may be noted that the destructive force of these gravity-aided slurries, some hot and others cold, is often very great, causing massive devastation in densely populated regions of the Indonesian archipelago such as Bali and Java.

Because of their complexity it is usual to subdivide the products of mixed volcanism into vent and alluvial facies, or into central, proximal and distal facies (Williams and McBirney, 1979). A prime example in this context is the 4000-m-high Tanzanian strato-volcano Mount Meru, whose central, near-vent facies comprises a bewildering complex of thin lava flows intercalated with volcaniclastic rocks, such as agglomerates and air-fall tuffs, plus intrusive tholoids, domes and dykes. Lavas are very much subordinate to pyroclastic deposits. The proximal facies of the volcano's lower flanks is characterized by somewhat thicker lavas, tuffs and massive poorly-sorted laharic breccia flows which, on the east side of the mountain, extend up to 35 km from the summit. Many of these debris flows have carried along huge blocks which now have been dumped several kilometres from the base of the cone and apparently were able to override small cones situated several kilometres from the volcano's flanks. In distal locations, plains surrounding Meru are composed of an apron of fluvially reworked volcaniclastic deposits, thinner lahars and alluvial deposits, plus a small number of fluid lavas that found their way into pre-existing gullies.

In contrast to the complex structures described above, *monogenetic cones* are single-event central structures, generally of much smaller size than strato-volcanoes, which are formed over periods measurable in days or a few years. Those formed from lava give rise to small shields, while those built largely from pyroclastic materials may form spatter cones, cinder cones or maars (Fig. 6.10). Terrestrial pyroclastic cone volumes seldom exceed 4×10^7 m^3, while lava cones may be larger at 9×10^9 m^3 (Wood, 1979). Most small cones have a summit crater. Studies of suspected lunar and Martian spatter and cinder cones indicate that generally they are only one quarter as large as those on the Earth.

Maars, a special case of small cone, are produced by highly-explosive, phreatic activity where a rising column of hot magma intersects subsurface water. Vaporization of the water generates additional thermal energy, which, in turn, increases the explosivity index. As a result the lava is comminuted into tiny particles which may be spread out by base-surges around the vent, giving rise to a somewhat broader rim than is usual (Peckover *et al.*, 1973). Several classic examples of such craters are located in the Lake Myvatn region of Iceland. Maars are also likely to have been formed on Mars, due to the explosive interaction of magma with subsurface ice.

Fig. 6.11: Large phonolite dome, approximately 3.5 km in diameter, intruded into the lower flanks of Mount Meru, N. Tanzania. Photo: Peter Cattermole.

6.6   VOLCANIC DOMES AND THOLOIDS

Relatively viscous magma may be intruded between more solidified flows or bedded pyroclastic deposits as tholoids which arch up the beds, or may be squeezed into a volcano's substructure, giving rise to steep-sided domes.

The growth of the dacitic dome which preceded the catastrophic 1980 eruption of Mount St.Helens was well documented (Foxworthy and Hill, 1982) and it has been established without question that it was the gradual build up of volatile-laden silicic magma beneath this viscous capping which eventually blew out the whole of one side of the mountain. On Earth, most steep-sided domes are associated with relatively silicic magma. Domes are found also on Venus, where a distinctive variety, or *pancake dome*, is typical. Venusian domes typically have diameters between 20 and 30 km and may be 1 km high. Many have either a central summit pit or one placed asymmetrically. In form generally they are circular and have steep margins, consistent with formation by extrusion of relatively viscous magmas (Fig. 6.11).

6.7   PYROCLASTIC FLOWS

One of the major products of highly-explosive (sometimes called *Peléan*) terrestrial eruptions are pyroclastic flows. It was such a flow, termed a *nuée ardente*, which rushed down the slopes of Mont Pelée in 1902, devastating the town of St. Pierre and killing 28 thousand people. The deposits of such flows have become known as *ignimbrites*, and consist of a poorly sorted amalgam of pumice, lithic fragments and a fine-grained matrix of pyroclastic materials (Plate 3). Because they are erupted exceptionally quickly much of the heat is retained within the ejected pumice fragents which, in the middle third of a flow, may become welded together. This welding produces a characteristic eutaxitic texture, developed because the originally plastic pumice shards have become flattened and drawn

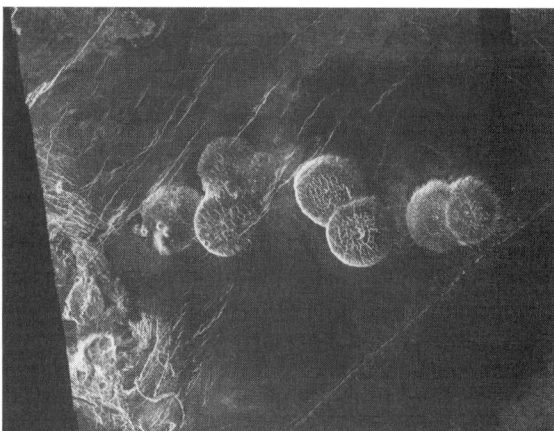

Fig. 6.12: Circular steep-sided volcanic domes in Apha Regio, Venus. Each is approximately 25 km across, has a heavily-fractured carapace, steep margins and a small collapse pit. Magellan image P-37125.

Fig. 6.13: Aerial view of large mudflow on eastern side of Meru volcano, showing distinctive hummocky topography superimposed on broad lobate flow. Radiocarbon dating of bottom sediments obtained from the Momella Lakes (seen towards bottom right-hand corner) dates the episode of lahar generation at 7000 BP, which is presumed also to be the date of caldera collapse.

out parallel to the direction of flow.

During Peléan eruptions, material leaves the vent at a somewhat lower rate than is typical of Plinian activity. For this reason, the cloud of pyroclasts and gas may only just spill over the lip of the crater, moving down the volcano's flanks as a fast-moving density current. Alternatively the particle-laden cloud may simply collapse upon itself, forming a dense flow that swoops down the flanks of the cone. Whichever is the case, such flows appear to have similar rheological characteristics to basaltic lavas and, like them, may pond up in hollows. The topographic expression of such flow deposits is, therefore, in many ways rather similar to basaltic pahoehoe flows, as is their areal extent.

The rheological behaviour of pyroclastic flows generated by nuées ardentes is a reflection of the make-up of the eruption cloud. Within it each particle becomes mantled in an envelope of gas, separating it from its neighbours and allowing the whole to behave as a low viscosity flow. The lower part of a flow is usually devoid of coarse particles and it may be cushioned at ground level by a layer of air which acts as a kind of lubricant. The entrainment of gas and also air within the cloud, allows these incandescent debris flows to travel great distances (more than 100 km) over very low slopes (less than 2°). Often their momentum allows them to override substantial topographic obstacles, a quality they share with certain types of hot mudflow.

Mudflows or *lahars* may be treated here also. In the field, non-welded ignimbrites are often indistinguishable from lahars; often the two are closely associated. Lahars are typically badly sorted and, like ignimbrites, are finer-grained towards the base. The larger fragments tend to move upwards through them, while very large blocks may be carried along their upper surfaces, a phenomenon which was observed during the Mount St. Helens eruption and which frequently accompanies laharic activity in Indonesia. At Mount Meru, blocks the size of a two-storey house commonly may be seen sitting on the consolidated surface of lahars (Plate 4). Some flows are hot, being generated by Plinian eruption through a lake or ice cover, while others are cold, having been produced simply by, say, the breaching of the dam enclosing a crater lake. Either way, where the rate of flow is high, the top surfaces of lahars are usually characterized by sub-parallel ridges and troughs aligned in the direction of flow, while hummocky topography typifies the more distal regions of flows where a loss of momentum was experienced on reaching flatter ground (Fig. 6.13).

One of the two main Meru lahars extends 50 km from its point of origin, is between 15 and 25 m thick and covers an area of more than 1000 km^2; its volume is certainly in excess of 15 km^3. Both it and a similar flow can be shown to have overridden volcanic cones more than 60 m high and one of them washed to a similar height up the western flank of Kilimanjaro, 40 km to the east. Simple potential energy/kinetic energy calculations indicate that, at these points, flow rates must have exceeded 35 m s^{-1}. Studies by Fink and his co-workers (1981) of mudflows generated during the 1980 eruptions of Mount St. Helens produced very similar figures for flow velocity (10-31 m s^{-1}), while volumetric flow rates were found to have been between 300 and 3400 m^3 s^{-1}. In the latter respect, the velocity of mudflows is very similar to that of nuées ardentes. Like pyroclastic flows, lahars also are highly destructive volcanic products (Plate 5).

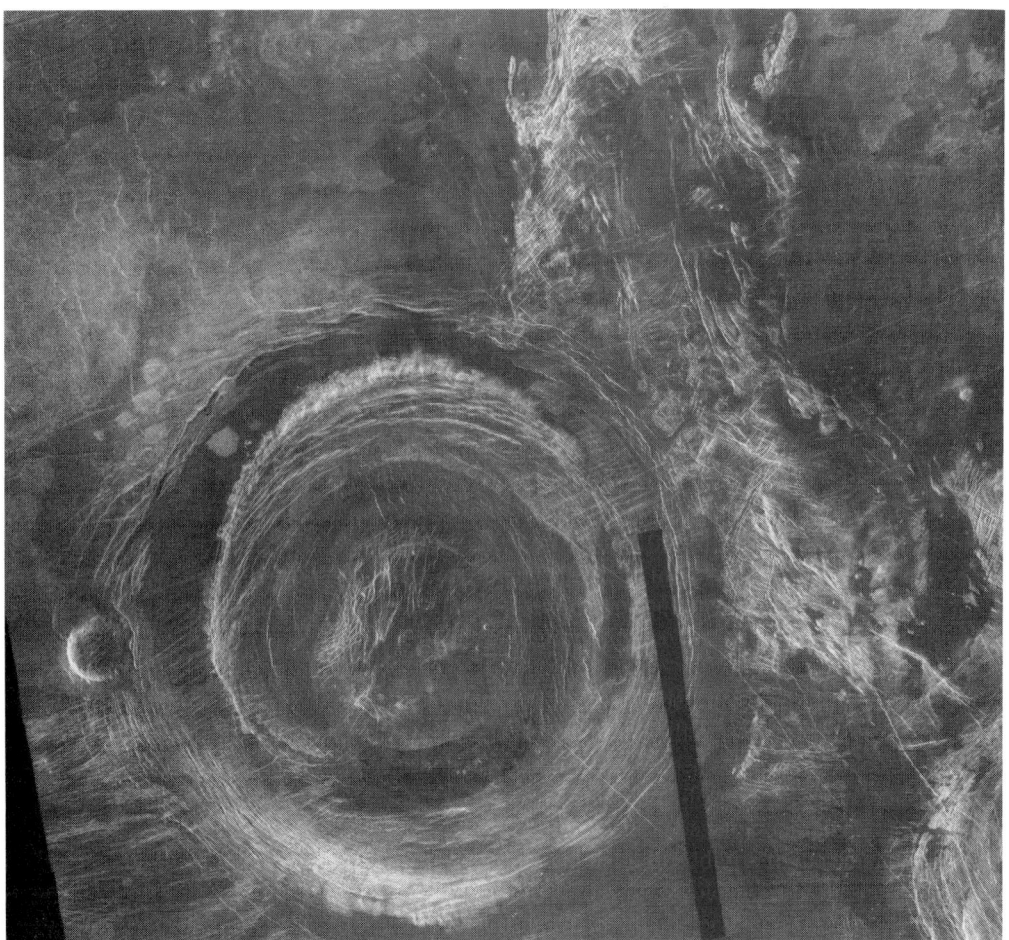

Fig. 6.14: The corona, Aramaiti, located at 18°E, 16.5°S. Note the central plateau, circumferential moat and both radial and annular fractures. Magellan image C1-15S026. Framewidth 690 km. A steep-sided dome lies on the western rim.

6.8 CORONAE, NOVAE AND ARACHNOIDS

Coronae were first discovered on Venus during the Venera 15/16 fly-bys, and are among the most typical of Venusian landforms. No terrestrial equivalents are known. Ranging in diameter between 60 and 2000 km, *coronae*, together with related volcanotectonic structures called *arachnoids* and *novae*, have a range of morphology and it is possible that the three represent an evolutionary sequence (Janes *et al.* 1992; Stofan *et al.* 1992). The general form of this group of landforms is a circular to ellipsoidal shape, a central plateau or dome being surrounded by a moat that has annular fractures associated with it. Radial fracturing is common also, while lobate lava flows and shield fields are common associates. Novae – thought by some workers to be young coronae (sensu lato) – tend to lack annular faults but have a central dome-like edifice and accompanying radial fracturing. Coronae (sensu stricto) are typified by Aramaiti, located on the plains of Aino Planitia, Venus (Fig. 6.14). This is a concentric structure, the opposite rims crests of which are 350 km apart. The rim rises 800 m above the surrounding plains which are separated from it by a 350-m-deep moat with a width of 30 km. This shows up as a darker, smoother, region beyond the radar-bright rim which has a hummocky outer slope. The central structure lies at roughly the same elevation as the exterior plains and is slashed by graben which are about 1 km wide and aligned approximately northeast-southwest. Beyond the moat are both radial and annular extensional faults, some of which transect younger radar-dark lava flows that have partially flooded the moat floor. The annular graben occupy a zone 85 km in width.

Arachnoids are the most fractured of the corona-like landforms and were given a "spider-and-cobwebs" label due to the typical appearance of several such structures in the same area, i.e. strongly-developed concentric and radial fractures. It has been suggested that they represent the middle stages of corona development, when flattening has occurred above a once-active mantle plume (see Fig. 7.16).

6.9 MAARS

Maar craters typically are circular in outline and have virtually no rims. They are generated by single-event, explosive, eruptions, typically where magma interacts with groundwater, converting it almost instantaneously into steam. Fine examples of such landforms can be found in the Puy de Dôme region of the Auvergne, central France. Associated deposits include coarse breccias and ash. Maar-type structures as yet have not been recognized on other planets.

7

Volcanic plains and their development

Flat-lying volcanic deposits cover extensive regions of the Earth's continents and are known to floor the ocean basins. The majority of these plains are built from mafic lavas, but trachytic and phonolitic sheet floods are important locally, particularly in continental rift environments, where there may also be a contribution from silicic ash-flow deposits. Widespread volcanic plains have also been identified on the other terrestrial planets. A combination of direct sampling, orbital chemistry, photogeological analysis and spectroscopic studies strongly suggest these too are built from flows with a basalt-like disposition.

Largely on account of their characteristic geomorphological expression in the Western Isles of Scotland, those subaerial mafic lavas which accumulated in the Brito-Arctic province during Tertiary times were termed *plateau basalts* by Sir Archibald Giekie (1903). Subsequently G. W. Tyrell rightly suggested it would be more appropriate to use the term *flood basalt* since most lavas of this kind accumulated in broad subsiding basins, only to become plateaux at a later time by virtue of uplift. This term is used herein. Viewed in the broader context of the terrestrial planets as a group, mafic and other flood lavas can be said to give rise to *volcanic plains*, which have a rather different geomorphological signature and distribution to the deposits of shields, strato-volcanoes or monogenetic cones. Such plains, whether at their original level or epeirogenically uplifted to form plateaux, are typified by wide areal extent, large volume and eruption from fissures or linear vents associated with major crustal extension.

7.1 BASALTS OF THE EARTH'S OCEANIC PLAINS

On the Earth, extensive basaltic lavas are erupted from rifts that mark zones of lithospheric plate divergence within the ocean floors. *Ocean floor basalts* produced in this way have geochemical characteristics which distinguish them from their continental counterparts and from the mafic lavas of oceanic islands. They are classified as mid-ocean ridge basalts, known commonly by the acronym MORB.

The great majority of these flows originate in a 65 000 km long sub-oceanic ridge system which, if the waters were removed, would stand out as the most spectacular large-scale topographic feature of our planet (Fig. 7.1).

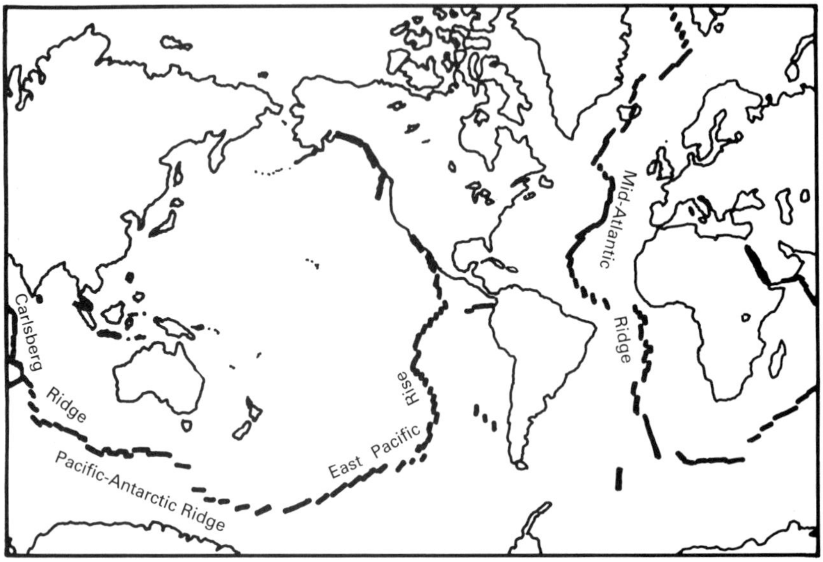

Fig. 7.1: The 65 000-long sub-oceanic ridge system which traverses the Earth's ocean basins. This feature marks sites at which new basaltic crust is generated and of spreading axes from which it diverges laterally.

This great sub-oceanic feature marks the site of spreading axes which are considered to sit above hot mantle plumes rising beneath *en echelon* rifts which incise the ridge crest. Because the lithosphere on either side of ridge axes is inexorably dragged apart, the diverging slabs of newly created ocean floor slowly cool, becoming denser, and subside. In this way successively younger volcanic material is added to the vast abyssal plains at the following margins of oceanic plates, the plains becoming older with increasing distance from the ridge axes. Because of lithosphere re-cycling along subduction zones, oceanic basalt plains are never older than about 200×10^6 years.

Detailed knowledge of the morphology of mid-oceanic ridges is in large part due to the FAMOUS Project, an international study of the mid-Atlantic Ridge at 36°N, near the Azores. This synthesized intensive studies of bathymetry, magnetics, heat flow, seismic refraction and gravity, plus samples collected from selected locations, scientists making extensive use of American and French submersibles. The details of its findings were presented in the April and May 1977 issues of the *Geological Society of America Bulletin*. At this particular site the mid-Atlantic Ridge consists of a broad rise with a central rift valley between 1.5 and 3.0 km wide and between 100 and 400 m deep. The broadly symmetrical rift contains a discontinuous central ridge up to 240 m high and between 800 and 1300 m in width (Fig. 7.2). Occasionally a 200 to 600-m-wide central trough takes its place.

The youngest volcanic rocks were collected from the ridge crest and were found to be remarkably fresh glassy tholeiites. Characteristically these hold a much higher proportion

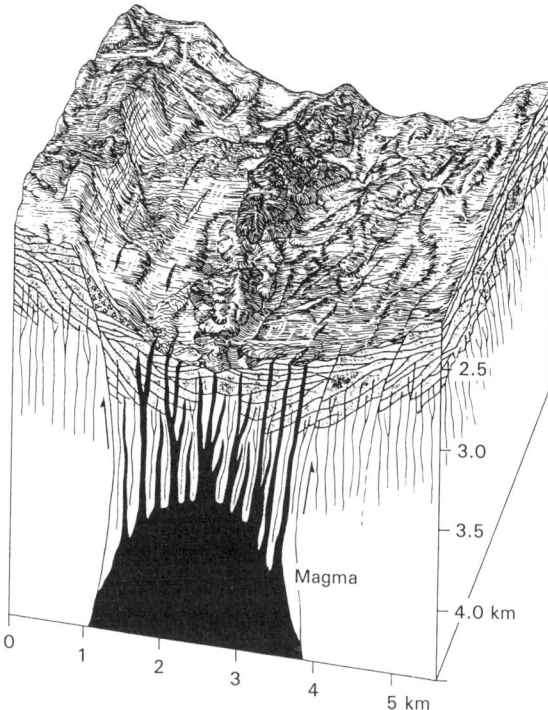

Fig. 7.2: Schematic drawing of a part of a median rift valley floor and its walls. (After R. Hekinian, J. G. Moore and W. B. Bryan (1976) *Contrib. Mineral. and Petrol. 58*, p.107, with permission.)

of olivine phenocrysts than either plagioclase or clinopyroxene and are characterized by very low K_2O and TiO_2, high CaO and exceptionally high K/Rb (see Table 7.2). Basalts of the ridge flanks, on the other hand, are typified by higher amounts of SiO_2, K_2O and H_2O and higher FeO/MgO. This chemical difference appears in large part due to the removal of plagioclase, olivine and clinopyroxene from the flank lavas, which suggests differential tapping of a zoned magma chamber situated beneath the ridge system. Analyses of MORBs from the East Pacific Rise, sampled across the full width of the tropics, are almost identical, save for a slightly lower normative olivine content. Trace element patterns of all MORBs are very similar to those of the chondritic meteorites while $^{87}Sr/^{86}Sr$ initial ratios are low (0.7029 – 0.7035).

Moore *et al.* (1974) estimate that the eruption rates for MORBs are lower than those of Oceanic Island basalts (OIBs). At the FAMOUS site, for instance, they calculate an average eruption rate of 8600 m^3 year[-1], compared with 90 000 m^3 year[-1] in Iceland. However, viewed on the extended time-scale, the budget of material erupted from the oceanic ridges has been orders of magnitude greater than that from any other source of terrestrial volcanism. Estimates of MORB effusion rates range from 5 km^3 year[-1] to 20 km^3 year[-1], compared with 1.5 and 1.7 km^3 year[-1] for hot spot and island arc basalts respectively (Schilling *et al.*, 1978).

Table 7.1: Temporal and dimensional data for terrestrial continental flood basalt provinces.

Locality	Age (10^6 y)	Max. & ave. thickness (m)	Present & original area (km²)	Volume (km³)	Flow thickness (m)
Parana Basin	119-149	1500-1600 ave. 650	1.2×10^6 2.0×10^6	6.5×10^5	ave. 50 max. >100
Karroo Basin	166-206	8000-9000 ave. 1000	1.4×10^5 2.0×10^6	?	ave. 10 sills 1000
Siberian Platform	216-248	3500 ave 1000	1.5×10^6	5.7×10^5 9.1×10^5	ave. 30
Lake Superior Basin	1100-1200	8000-12000 ave. ?5000	1.0×10^5 1.2×10^5	>3.0×10^5	3-30 ave. 25
N. Australia	540-570	1500	3.5×10^4	?	?
Columbia River	6-17	1500	2.0×10^5	2.0×10^5	10-45
Iceland	1-16	3000-12000	1.0×10^5	Post-glacial 40	10
Deccan Traps	50-65	2000	5.0×10^5	1.0×10^6	?
Greenland	55-65	?700	1.6×10^5	?	10-50
Thulean Prov.	56-66	3000	6.5×10^3	?	120-150
Faeroes	52-60	3000	1.4×10^3	?	?

7.2   TERRESTRIAL CONTINENTAL FLOOD BASALTS

Episodes of basaltic lava flooding have punctuated the geological record of the Earth's continents from the Pre-Cambrian to the present and, as many workers have noted, whatever their age, they share a number of common characteristics. Typically they give rise to subaerial basaltic floods of wide areal extent and large volume; individual volcanic units are often of immense size. The lava floods have their origin in fissures or rifts and usually are associated with extensive dyke swarms. Although within individual provinces there is quite a wide range of composition, overall they have a fairly uniform tholeiitic chemistry, while trace element and isotopic data indicates that most of the lavas are considerably evolved when compared to primitive mantle melts. Table 7.1 presents the ages and dimensions of the larger terrestrial continental flood basalt occurrences, while Table 7.2 shows selected basalt analyses.

The more ancient flood basalt provinces, such as those of the Lake Superior Basin and the Siberian Platform, are now located well within continental cratons, but post-Triassic ones can clearly be linked with the initiation of continental rifting and eventual drift. This can be illustrated nicely by a reconstruction of the southern continents prior to the Mesozoic (Fig. 7.3), from which it is clear that there was a pre-Mesozoic connection between, firstly, the flood basalts of the Drakensburg and Antarctica, secondly, those of Southwest Africa and the Parana Basin of Brazil and Uruguay and, thirdly, those of the Deccan Traps and Madagascar.

Because of the relatively evolved chemistry, isotopic imprint and the association of flood basalts with continental crust, considerable debate has centred upon the extent to which their mantle source melts have become contaminated with sialic material. To date

Fig. 7.3: Reconstruction of the southern continents prior to continental drift during the Mesozoic. Black areas show the outcrop of flood basalt provinces.

there is a dichotomy of opinion upon this matter since the goechemical evidence is somewhat ambiguous. Two alternatives appear to present themselves: (i) the source melts were contaminated by relatively radiogenic sialic materials, or (ii) some regions of subcontinental mantle were inhomogeneous and anomalously radiogenic, giving rise to partial melts with more differentiated signatures than unevolved mantle-derived basalts such as MORBs.

7.2.1 Iceland

The active volcanic province of Iceland sits astride the seismically active Reyjkanes-Jan Mayen Ridge, which is an integral part of the mid-oceanic ridge that bisects the Atlantic. It is also part of a transverse topographic rise which joins the ridge to Greenland and the Faeroes, roughly along the Arctic Circle. This, the most recent flood basalt province, began as a continental rift but current activity is associated with the divergence of two oceanic plates. Therefore it may be seen as a link between continental and oceanic volcanic plains formation. During its continental phase, in Tertiary times, extensive volcanic activity occurred and also gave rise to the flood basalts of the Faeroes, East Greenland, northwest Scotland and northeastern Ireland (Fig. 7.5).

Situated above an active divergent plate margin and its associated rising mantle plume system, Iceland is the Earth's largest oceanic and volcanically most productive landmass, with an area of about 10^5 km^2 and a current magma production rate of 0.04 km^3 year^{-1}. The island is dissected by two major rift zones, both active, and along these major volcanic

Table 7.2 Chemical analyses of different basalt types.

	1	2	3	4	5	6	7
SiO_2	49.20	43.15	45.50	43.20	52.05	54.50	49.68
TiO_2	2.03	2.70	3.44	3.93	1.78	1.95	2.13
Al_2O_3	16.09	13.46	15.71	16.30	12.43	13.59	16.99
Fe_2O_3	2.72	4.52	3.61	8.09	5.18	3.28	3.45
FeO	7.77	8.22	8.64	4.69	10.08	8.80	8.99
MnO	0.18	0.11	0.22	0.16	0.24	0.18	0.27
MgO	6.44	10.80	5.37	5.16	3.95	3.84	2.79
CaO	10.46	9.80	9.43	10.74	7.33	7.22	5.46
Na_2O	3.01	3.47	3.47	3.12	2.76	3.05	5.78
K_2O	0.14	1.63	1.38	1.64	2.07	1.45	1.90
P_2O_5	0.23	0.75	0.29	0.88	0.28	0.21	0.48
H_2O^+	0.70	1.21	0.60	2.03	1.90	0.78	1.77
H_2O^-	0.95	0.15	2.49	-	0.36	0.92	0.34

Key: 1. Oceanic tholeiite (MORB), Mid-Atlantic Ridge; 2. Alkali olivine basalt, (MORB) St. Paul's Rocks; 3. Ocean island basalt, St. Helena; 4. Olivine basalt, Isla La Gomera, Canary Islands. 5. Flood basalt, Karroo volcanics; 6. Yakima Basalt, Columbia River; 7. Mugearite, Thulean Province, Skye.

episodes have taken place during the past 4 million years. Post-glacial eruptions have poured out 40 km³ of lava, mainly from central vents or from groups of fissure-aligned craters. In 1783, around 12 km³ of lava emerged from the 25-km-long Laki Fissure in southeast Iceland, covering an area of 565 km² and comprising the Earth's most voluminous single-event historic extrusion (Noe-Nygaard, 1974; Sigvaldason, 1974). Flanking the highly active central volcanic zone on both the west and east is a series of Tertiary age flood basalts, erupted between 16 and 1 m.y. ago, which cover about 40% of the total area of the island. These attain a thickness of 12 km on the east coast. Tholeiites predominate (48%), but there are also olivine basalts (23%), feldspar phyric basalts (12%), andesites (3%) and rhyolites/dacites (8%), while pyroclastics and sediments account for 6%. The widespread tholeiites and olivine basalts largely emanated from fissure vents which opened in response to extensional stresses.

Iceland's most recent volcanic activity, naturally, is associated with the mid-oceanic ridge and therefore may not be viewed as being directly associated with the generation of flood basalts. However, since recent Icelandic activity can be viewed as a continuation of crustal processes which began in the Tertiary, there is some justification for digressing briefly in order that some petrogenetic considerations can be discussed. Fig. 7.4 shows that lava types have a bilateral symmetry about the ridge axis, with alkali olivine basalts outcropping in the Snaefellsnes Fracture Zone and in the more southerly portion of the Eastern Volcanic Zone (Jakobsson, 1972).

Moving northwards along the Eastern Volcanic Zone, the alkali olivine basalts grade first into transitional basalts and then tholeiites. In the extreme northeast of the region, the predominant type is olivine tholeiite. The more evolved Icelandic tholeiites and Fe/Ti-basalts have higher Fe, Ti, K and light REE, and lower Al and Mg than MORBs and thus

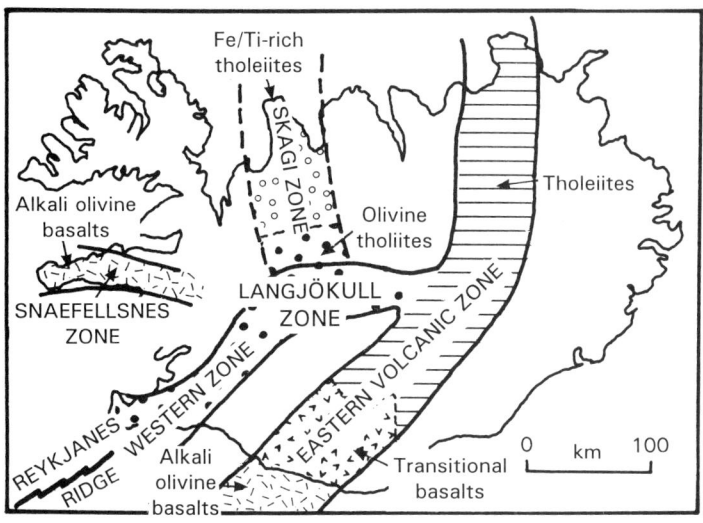

Fig. 7.4: The volcanic zones of Iceland. (After Sigvaldason, 1974.)

are quite distinct from them. However, some basalts from the central zone and the northern part of the Eastern Volcanic Zone are almost indistinguishable from MORBs that outcrop on the Reyjkanes Ridge.

This diversity needs some explanation and has led to a wide range of ideas concerning basalt genesis. For instance, O'Nions and his co-workers (1976), on the basis of $^{87}Sr/^{86}Sr$ and chondrite-normalized Ce/Yb ratios, argue that the chemical features of all Icelandic basalts largely are a reflection of mantle inhomogeneities established at least 100-200 m.y. ago. In contrast, Langmuir et al. (1978) prefer to distinguish two separate trends involving several mantle sources: beneath Iceland itself they see magmas derived by mixing of an enriched alkali basalt source and a depleted mantle source, while in the neighbourhood of the Reykjanes Ridge they envisage a heterogeneous depleted oceanic source and a more homogeneous intermediate mantle source being involved. The solution to this problem must await further geochemical studies.

7.2.2 Greenland and the North Atlantic Province

Iceland is just a part of a much more widespread flood lava province which extends from Greenland to northwestern Britain (Fig. 7.5). In East Greenland, Tertiary basalts cover an area of over 16 000 km[2] to the north of Scoresby Sund, where they are 700 m thick; south of Scoresby Sund they are more extensive, covering an area of 60 000 km[2] and reaching a thickness of 6.5 km. Along the west coast, similar basalt flows cover an area of 40 000 km[2] and attain a thickness of 3 km. How much basalt is covered by the permanent ice cap remains unknown. The very wide extent of the lava pile, which is built from flood lavas

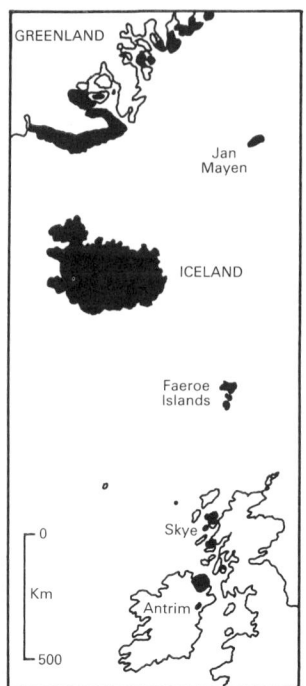

Fig. 7.5: Map showing the distribution of Tertiary flood basalts in the Brito-Arctic (Thulean) Province in the North Atlantic region.

predominantly of quartz tholeiite and olivine tholeiite composition, is typical of fluid mafic magmas erupted at high effusion rates. The fact that some effusions took the form of massive (10-50 m thick) units, while others emerged as thin (less than 10 m) pahoehoe and aa flows, indicates some variability both in the magma eruption rate and probably viscosity during the magmatic event.

Flood basalts also characterize the Tertiary province of northwest Britain, where basalt volcanism commenced 65×10^6 y ago and continued for about 9 m.y. and where Tertiary volcanic rocks cover an area of approximately 5550 km². Associated with the basalts on the Hebridean islands and the Scottish mainland are northwest-trending dyke swarms and spectacular gabbro/quartz-monzonite ring-dyke and cone-sheet complexes representing the eroded substructures of several large central volcanoes. Stocks of more silicic plutonic rocks also occur. The largest expanse of flood basalt occurs in Ireland where, in the Antrim Plateau, 4000 km² of basalt outcrops. Overall, the Hebridean suite shows a trend from basalt-hawaiite-mugearite, but trachytic and rhyolitic types also occur. Four hundred kilometres to the north of Scotland sit the Faeroe Islands. Covering an area of 1400 km², they are built almost entirely from flood basalts which were extruded between 60 and 52 m.y. ago. They have a total thickness of 3 km and a volume in excess of 3000 km³.

Studies of each of these regions indicates that lava flooding originated either in fissure-aligned vents or very low shields, the emplacement of feeder dyke swarms occurring

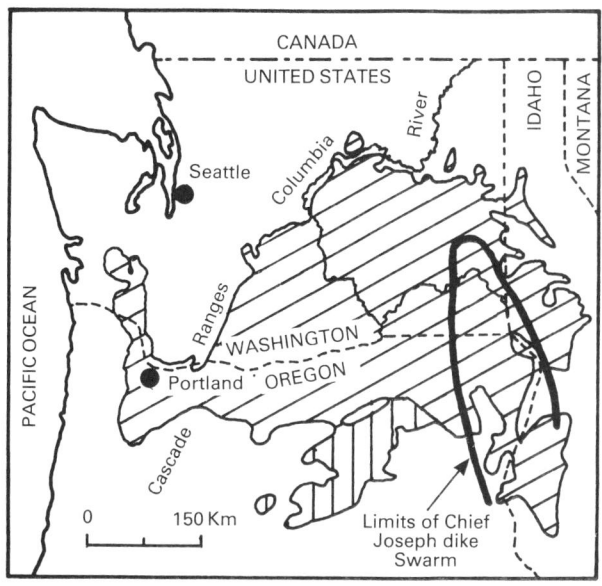

Fig. 7.6: The margins of the Columbia River Plateau. The approximate extent of the Yakima Basalts (diagonal lines), Picture Gorge Basalts (vertical lines) and Chief Joseph Dyke swarm (heavy black line) are shown. (Modified from D. A. Swanson, T. L. Wright and R. T. Helz (1975), *Amer. J. Sci. 275*, p.879, with permission.)

during extended episodes of crustal tension related to continental break-up (see, for instance, Noe-Nygaard, 1974). Some idea of the scale of dyke intrusion may be gained by noting that the northwest/southeast-trending suite of mafic dykes which traverses northwest Britain, and which largely fed the surface floods, extends over an area of 112 000 km².

7.2.3   The Columbia River Plateau

This immense flood basalt province covers an area of 2×10^5 km² and has a volume estimated to approach 2×10^5 km³ (Waters 1962). Produced during the Miocene period (17-6 m.y. ago), approximately 99% of the basalt pile was generated during a 3-m.y. period at the beginning of the episode. Very fluid lavas were introduced into a wide basin situated between the Cascades and the Idaho Batholith, at a distance of between 200 and 600 km from the Pacific coast, and eventually covered a tract which extends through northern Oregon and central and eastern Washington to western Idaho (Fig 7.6). Subsequently they were uplifted to form the present plateau.

The earliest volcanism is represented by the Picture Gorge Basalts, a series of just-undersaturated tholeiites with a volume of 40 000 km³, which were extruded in the mid-Miocene. These were succeeded by the late-Miocene/early-Pliocene Yakima Basalts, with a volume of 120 000 km³ and quartz-normative chemistry. The latter stages of this extended volcanic episode saw the eruption of late Yakima and Ellenburg flows, during the early-

Pliocene, giving rise to a sequence of Fe-rich quartz-normative basalts with a volume of 20 000 km³. Chemically these later lavas are similar to the ferrobasalts of the Galápagos Islands (see Table 7.2). Interestingly, during the same general time period (Pliocene to Recent), equally extensive flood basalts were erupted over the area of eastern Oregon, to the south of the Columbia River region; however these flows were totally different, being high-alumina basalts, accompanied by a small but significant volume of andesitic and rhyolitic magma.

While typical Columbia River flow units have volumes within the range 10 – 30 km³, some members of the flow complex have volumes which are measurable in hundreds of cubic kilometres. This is particularly true of the Yakima Basalt, the most widespread and voluminous formation within the Columbia River region. The Roza Member of the Yakima Basalt has a volume of more than 1.5×10^3 km³, covers an area of 40 000 km² and is on average 30 m thick (Swanson et al., 1975). Both field evidence and theoretical considerations predict that high-volume flows of this size must have been erupted quickly during very vigorous effusive episodes. Shaw and Swanson (1970) suggest that the entire Roza Member could have been emplaced in a few hundred years, while single cooling units may have taken as little as a few days to be emplaced. The implied rapid eruption rates would necessitate a large vent system to accommodate this.

The Yakima flows, which usually consist of three thick units, and the less extensive Ice Harbor Flows, have in fact convincingly been shown to have originated in linear vent systems which parallel the trend of the widespread Chief Joseph dyke swarm. The latter covers a region at least 200 km wide and 450 km long (Fig. 7.6) and many of the 21 000 dykes mapped apparently were flow feeders. Each vent system strikes northwest and is confined to a linear zone measuring several tens of kilometres long and a few kilometres wide. Associated with each are sets of parallel dykes and the remains of spatter and tuff cones and near-vent collapsed pahoehoe. This realization ensued from detailed mapping of the Roza Member of the Yakima Basalt (Swanson et al., 1975) which revealed the existence of 19 vents within a linear zone 120 km long and 15 km wide (Fig. 7.7) and 18 similar vents along a 55-km-long outcrop for the Ice Harbour Member. Each vent area is characterized by the presence of spatter and/or pumice which is very similar in most respects to the spatter generated at Kilauea during fire-fountaining. Dense, apparently de-gassed, flows commonly either are interbedded with the tephra or overlie it. Collapsed pahoehoe, typical of near-vent locations in Hawaii, is also found. The largest cone appears to have had a basal diameter of between 100 and 200 m.

Calculations by the same workers indicate that average eruption rates during the entire Columbia River episode were between 10^{-1} to 10^{-2} km³ day⁻¹ km⁻¹ of vent, but that during the emplacement of the more voluminous Roza flows, average rates rose to 1 km³ day⁻¹ km⁻¹ of vent. By way of comparison, the maximum observed eruption rate along the rift zones of Kilauea is 10^{-2} km³ day⁻¹ km⁻¹ of vent, however, this only occurs very early during an eruption and quickly drops to between 10^{-3} and 10^{-4} km day⁻¹ km⁻¹ of vent. On this basis it can be shown that while the Ice Harbor flows may have been erupted at a rate roughly equal to the sustained Hawaiian rate, the Roza flows must have been erupted at rates 3 to 4 orders higher.

The rate at which magma was produced beneath the Columbia River Plateau also must have been high but nevertheless would have been less than the average eruption rate. The

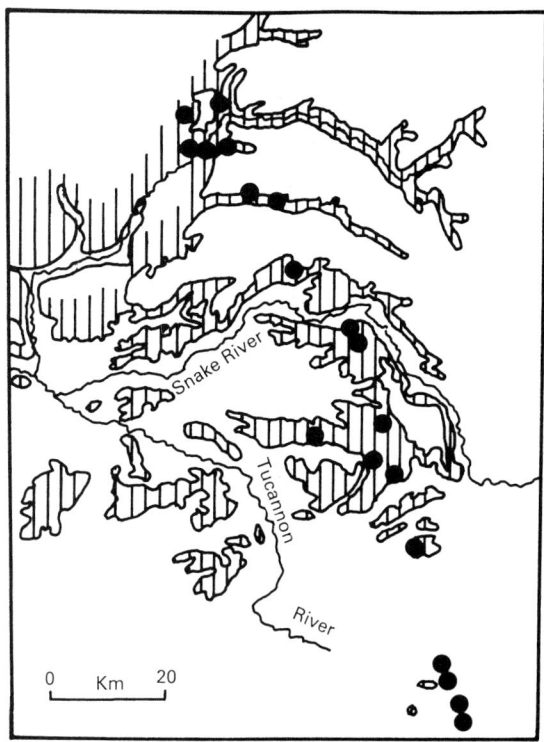

Fig. 7.7: Distribution of the Roza Member of the Yakima Basalt in southeastern Washington state. The vent areas for the two Roza flows are shown by black circles. (After D. A. Swanson, T. L. Wright and R. T. Helz (1975), *Amer. J. Sci. 275*, p.884, with permission.)

difference between the two would have been accounted for by periods of quiescence, during which subcrustal magma reservoirs re-filled. Using the present-day Hawaiian rate of magma production (10^{-1} km^3 year^{-1}), the estimated total volume of the Columbia River Basalt – most of which is accounted for by the Yakima Basalt – could have been produced in 2 m.y. On the basis that the age range of the Yakima Basalts is about 3 million years and the complete span of Yakima magmatism perhaps twice this, it would appear that the average magma production rate was less than that of modern Hawaii.

Bearing in mind the tectonic environment of the Columbia River region during the Tertiary, a more meaningful comparison could, perhaps, be made with the ocean ridge basalts of Iceland, where the Icelandic fissure system, like that of the Columbia River, is about 400 km long. Estimates of the volume of lava erupted in Iceland during the past 10^4 years range from 4×10^{-2} to 4.8×10^{-2} km^3 year^{-1} (Thorarinsson, 1967; Jakobsson, 1972); the comparable figure for the Yakima Basalt is 7×10^{-2} km^3 year^{-1}. This would imply an average production rate of between 1 and 1.2×10^{-4} km^3 year^{-1} per kilometre of vent on Iceland and 1.8×10^{-4} km^3 year^{-1} per kilometre of vent in the Columbia River during

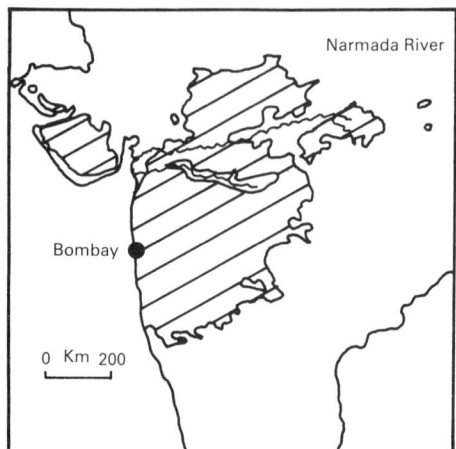

Fig. 7.8: Distribution map of the Deccan Traps in northwest and central India.

Yakima times, suggesting that, to within a factor of two, Yakima rates were similar to those of modern Iceland. Iceland may not, however, be typical of all oceanic ridges and several workers feel that Icelandic production rates may be several times those of oceanic ridges elsewhere (Baksi and Watkins, 1973; Moore *et al.*, 1974) which is why the island exists. Should this be so, the implication is that the Yakima production rates were significantly higher than those typical of mid-oceanic ridges generally; should it not, then there is a close similarity between the two.

7.2.4   Other flood basalt provinces
Numerous other flood basalt provinces exist, all of which share the common characteristics of wide lateral extent, tholeiitic chemistry, and eruption into subsiding basins. Three, in particular, deserve mention here, the first being the Karroo Basin of southern Africa which covers an area of one quarter of a million square kilometres. The infilling of this structure with continental sedimentary rocks of Permo-Triassic age was followed, between 190 and 154×10^6 y ago, by the outpouring of flood lavas (Stormberg Basalts) and the simultaneous intrusion of innumerable sills, inclined sheets and dykes of similar magmatic type. This major magmatic event was an integral part of the break-up of the super-continent of Gondwanaland (Walker and Poldervaart, 1949). The average thickness of the volcanic sequence is believed to be about 1000 m, but the maximum thickness approaches 9000 m. The present area covered by these rocks is 140 000 km^2 but originally the basalts may have covered 2×10^6 km^2. One of the largest remnants of this great lava flood may be seen in Lesotho, where a 2000 m thick series of lavas covers an area of 25 000 km^2. The majority of these predominantly quartz-normative basalt flows are of low-K/Ti type (see Table 7.2). Most flow units are about 10 m thick and are intruded by many sills and inclined sheets that may attain thicknesses of 1000 m, together with vertical dykes between 3 and 10 m wide which, in some instances, can be traced for upwards of 50 km across country. Further to the southeast, so many dykes are found that they may account for one-

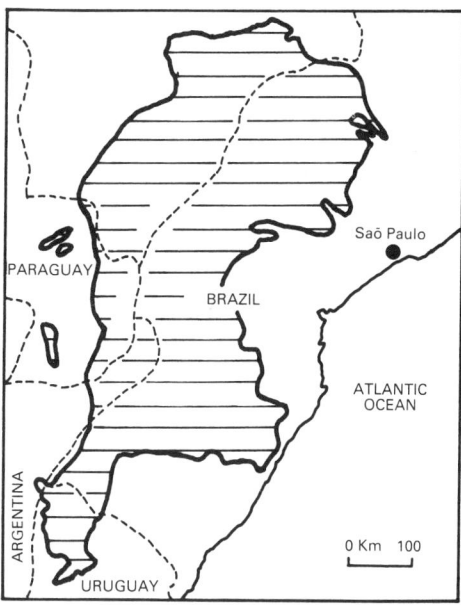

Fig. 7.9: Distribution map for the Cretaceous flood lavas of the Paraná Basin.

fifth of the total outcrop and clearly mark the focus of the volcanic activity. The fissure-related extrusive activity was complemented by the intrusion of massive gabbroic ring complexes in the northeast of the region.

More recent than the Karroo basalts are the flood basalts of the Deccan Plateau, which cover an area of 500 000 km² in northwest and central India (Fig. 7.8). Adjacent to the west coast these are 2000 m thick, but they reach only 100-200 m in the east. Erupted between 65 and 50×10^6 y ago, they comprise two series: a lower which is uniformly quartz-normative and which occupies about 80% of the sequence in the east and central part of the region, and an upper which is more important in the west and of much more varied composition, the rocks ranging from nephelinites through basalts to monzonites (Ghose, 1976). The tholeiitic members of the sequence were erupted from fissures associated with two major rift zones. The more undersaturated lavas, on the other hand, emanated from central vents, some of which may have been active until 35×10^6 y ago. The regional variations in thickness and age of the Deccan lavas generally is considered to be related to the gradual westward migration of the magma source over a period coeval with the development of the Indian Ocean, uplift of the Himalayas and slow northward drift of the Indian Ocean.

Lastly, the flood basalts of the Parana Basin of Brazil, Uruguay and Paraguay, which were poured out onto a floor of post-Triassic sedimentary rocks and which are very similar to the basalts of the Karroo province, are even more extensive (Fig. 7.9). The sequence of 50-100 m thick flows forms a pile which currently extends over an area of 1.2×10^6 km²; however, mapping of dykes peripheral to the main basalt outcrop, suggests the basalts may have extended over almost twice this area. The approximate volume of the lava pile is believed to be close on 650 000 km^3, but the original volume may have approached 2×10^6 km^3.

7.2.5   Characteristics of terrestrial flood lava sequences

The first characteristic is that regional subsidence typically has accompanied or immediately followed major episodes of flood basalt emplacement. Only rarely does it precede it. For this reason most of the older volcanic piles are covered by thick continental sequences. In the Columbia River province, which is not alone in this respect, major subsidence occurred to one side of the principal zone of fracturing. In general, however, the basalt plateaux themselves and their cover prevent us from determining if subsidence was actually centered over the main feeding fissures which supplied magma from the mantle.

Secondly, there is little difference between pre-Tertiary and later flood basalt magmatism in terms of the style of eruption and its extent. Whatever the age, all flood lava plains are characterized by substantial thicknesses of flat-lying lavas which were erupted from widespread centres on a global scale. Effusion rates typically were very high, the basalts very fluid, and lava sheets of wide lateral extent. Among the largest are the Deccan Plateau and Parana Basin basalts, each of which may have covered an area of at least 2×10^6 km^2.

Thirdly, the predominant rocks are quartz-tholeiites and, despite being of large size, individual eruptive units are surprisingly uniform in composition. Most primitive basalts have Mg values lower than other terrestrial tholeiites; therefore, if the flood magmas are primary, they must have been generated from a more iron-rich mantle than either MORBs or OIBs. The wide range in Nd and Sm in the Columbia River Basalts and the Deccan Traps suggests some degree of crustal contamination may have affected the source magmas. However no convincing conclusion can currently be reached regarding whether the magmas are coming from a deep source, with little high-level storage, or are contaminated by crustal material.

7.3   CONTINENTAL RIFTS

Flood lavas of both basaltic and other chemical types are typical also of continental rift valleys. In these rather different structural settings, broadly concurrent crustal uplift, extensional faulting and volcanism has included the copious outpouring of basaltic flood lavas, accompanied by more alkaline rocks like trachytes, phonolites and rhyolites. The nature of typical rift valley rocks generally is taken to indicate a greater depth to the magma source region, due to melt segregation beneath regions of lithosphere which are both thicker and have lower thermal gradients than vigorously spreading or converging plate margins. It should be noted, however, that compared with the average thickness of continental crust beneath cratons, beneath sites of major rifting there is considerable attenuation.

7.3.1   The East African Rift

The classic example is the East African Rift which extends for 3700 km from Mozambique

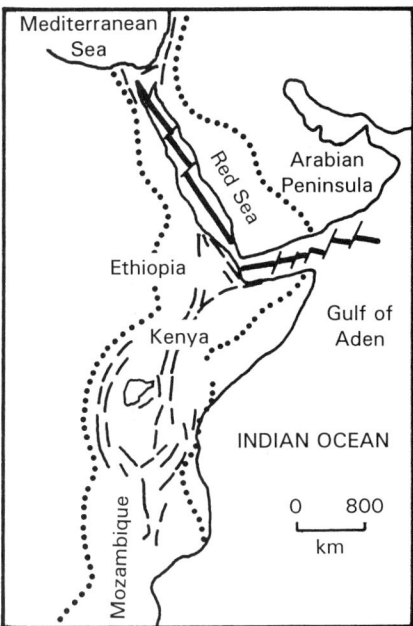

Fig. 7.10: The East African Rift Valley. The spreading axis of the Red Sea – Gulf of Aden is shown by heavy black lines, major rift faults by thin lines and outlines of major domes by heavy dots.

in the south to Ethiopia in the north, where the rift splits. One arm continues through the Red Sea Rift and continues for a further 2400 km into Turkey, while the other arm diverges through the Gulf of Aden and eventually merges into the Carlsberg Ridge of the Indian Ocean floor (Fig. 7.10). Generally the rift is about 80 km wide, but it flares to a width of 480 km in the Afar Triangle.

This vast region evolved during the Cenozoic, by a process of broadly concurrent crustal upwarping, extensional faulting and volcanism which included the copious effusion of mafic flood lavas, together with peralkaline trachytes and rhyolites, and a smaller volume of extremely alkaline rocks. In contrast to continental plateau-like floods, which are associated with basin subsidence, magmatism in continental rifts usually follows crustal upwarp. During the early Tertiary, broad domes were formed in both Ethiopia and the Red Sea regions and associated with these was the outpouring of over 10^6 km^2 of alkali basalts, which reached a thickness of 1 km in places. After a period of relative quiescence, separation of the continental crust gave birth to the Gulf of Aden in Miocene times and the Red Sea in the Pliocene. Both evolved into true oceanic rifts, with typical low-K olivine tholeiites being erupted.

In the mid-Miocene, continental extension (but not separation) began in Ethiopia and this was accompanied by the effusion of basalts transitional in chemistry between alkali and tholeiitic types. Their volume was only a small proportion (0.2%) of the initial basaltic

Table 7.3 Chemical analyses for a variety of rift valley volcanic rocks.

	1	2	3	4	5
SiO_2	45.2	50.5	61.09	42.1	73.46
TiO_2	2.3	1.7	0.91	2.6	0.25
Al_2O_3	16.0	18.0	16.64	12.4	13.91
Fe_2O_3	6.5	3.9	3.37	3.3	0.96
FeO	8.1	5.5	2.40	7.4	0.09
MnO	0.17	0.1	0.18	0.11	Trace
MgO	7.6	2.4	0.59	7.8	0.13
CaO	9.3	6.7	1.52	12.4	0.04
Na_2O	3.2	5.5	6.49	4	5.33
K_2O	0.81	2.8	5.89	2	5.27
P_2O_5	0.39	0.89	0.13	0.80	0.08
H_2O^+	1.0	0.3	0.68	3.5	0.12
H_2O-	2.5	0.25	0.28	1.3	

1. Olivine basalt, E. Uganda; 2. Trachyandesite, E. Uganda; 3. Flood trachyte, Mt. Suswa; 4. Olivine nephelinite, E. Uganda; 5. Rhyolite, New South Wales.

flooding. Effusive basaltic activity continued through the Pliocene and into the Pleistocene but more voluminous trachytes and peralkaline rhyolites were poured out, the latter mainly as ash-flow deposits. These more alkaline lavas had a volume of 3×10^5 km³. Chemical analyses for a variety of rift volcanics are shown in Table 7.3.

Continental extension was delayed in Kenya where, although basaltic eruptions were widespread both in space and time, they were overshadowed by the eruption of nephelinites and phonolites. The mainly Miocene age phonolites mostly were erupted from fissures, while the Pliocene trachyte floods and rhyolitic ash-flows tended to emerge from central vents. In the Western Rift, particularly in Uganda, the most significant development was the eruption of highly potassic volcanics, while carbonatites become important in the southern part of the Eastern Rift in Tanzania and in Malawi.

7.3.2   The Rio Grande Rift
Of similar age to the East African Rift is the Rio Grande Rift which extends through Colorado and New Mexico. This comprises a series of axial rift valleys offset to form a number of *en echelon* basins. At its widest the Rift is about 120 km wide. It developed in response to crustal extension behind a waning magmatic arc (Lipman, 1980). The ensuing uplift, extensional faulting and volcanic activity are believed to reflect the localized upwelling of the asthenosphere and its interaction with the overlying lithosphere. Of particular interest (particularly in the context of Venus) is the existence of numerous transverse lineaments which, in many places, coincide with offsets between discrete segments of the main rift valley. The intersections of the rift with the offsets typically coincide with major volcanic centres, both on the rift floor and on its flanks. Originally it was believed that at least one of these traced the movement of a hot spot (Suppe *et al.*,

1975), but later work involving the dating of basalts along the same section, showed that the transverse zone had been volcanically active over its entire 800 km length during the last 5 million years of its activity, militating against the plume idea.

Along the ridge axis the rocks are dominantly tholeiitic and are considered related to the segregation of partial melts at relatively shallow depth (Golombek and McGill, 1981). However, as is the case with other rifts, away from the ridge axis the lavas become progressively more alkalic and undersaturated, implying a greater depth of origin and perhaps different degree of partial melting. The progressively alkalic character of the flank lavas most likely is a function of lateral differences in crust/mantle structure and its thermal evolution and arises largely from the greater sub-lithospheric heat flow beneath the rift axis, resulting in a more pronounced uprise of the underlying asthenosphere, general uplift and crustal attenuation. The fact that rocks of rather diverse chemistry were erupted more or less simultaneously implies that several sources were being tapped at the same time.

7.3.3   Continental rifts – the wider context
Seen in a broader context, continental rifts may be seen as a manifestation of a failure of the brittle lithosphere arising from tensional stresses. The important question which requires an answer is whether such stresses are induced by magma uprise or whether extension is brought about passively by plate movements, volcanism being an effect, rather than a cause.

Looking wider afield than Africa and the western USA, it can be noted that both the Rhine Graben and the Baikal Rift have clear associations with lithospheric plate movements. The former was a direct result of the stress field caused by continental collision between the Eurasian and African plates, while the latter was related to a broad belt of deformation generated by collision between the Indian and Eurasian plates. The deformation so typically produced in such situations most likely is a response to membrane stresses produced in the elastic lithosphere (Turcotte, 1974). As Dewey (1975) has noted, the geometrical incompatibility of plate boundaries in motion requires that intraplate stresses and deformation must occur.

Returning to the Rio Grande Rift – this developed along the eastern boundary of a broad zone of crustal deformation which extends inland from the Basin and Range Province. Since rifting generally preceded both doming and volcanism, Golombek and McGill (1981) were led to conclude that its formation was due to extensional forces in the lithosphere related to the rest of the North American plate. Turcotte and Oxburgh (1973) concurred with such a view, noting that the extensive volcanic rocks of both the Rio Grande Rift and the Basin and Range Province appear to have accompanied crustal extension rather than being a direct result of much more localized plume activity. The rift, therefore, is viewed as a passive feature. On this basis, the magmatism associated with rifting generally is most likely to be associated with some form of pressure release melting concomitant upon the uprise of hot mantle rocks in response to crustal attenuation. Using the same line of argument, the East African Rift could be seen to have formed in response to ridge push forces associated with the Red Sea and Gulf of Aden, these being responsible for the failure of the African plate and the eventual upward movement of hot mantle materials

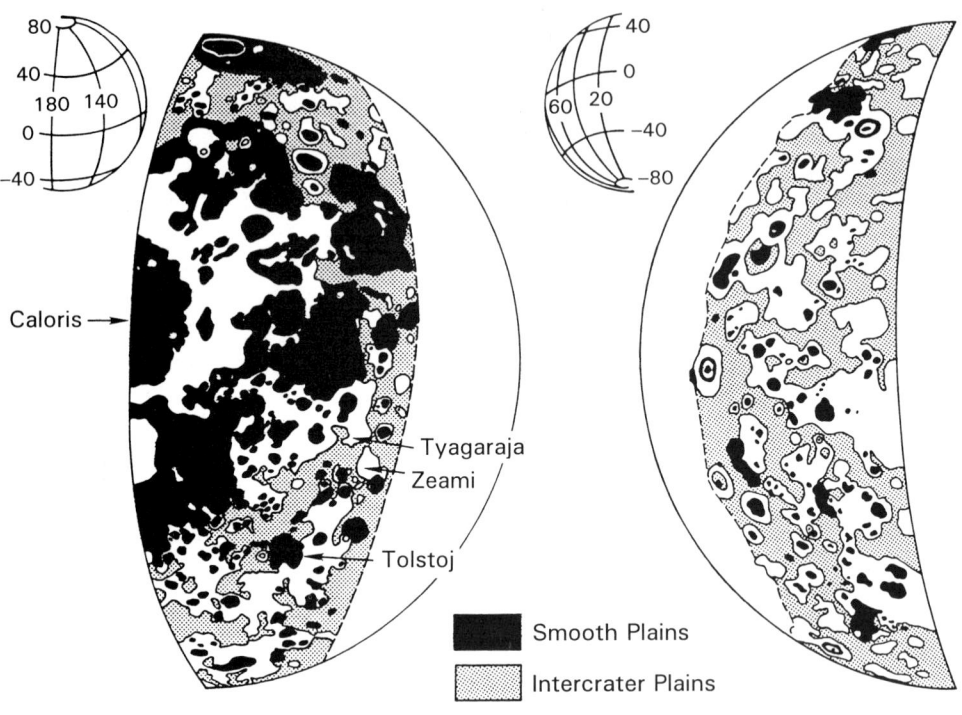

Fig. 7.11: The distribution of principal plains units on Mercury. (After Trask and Guest, 1975).

7.4   MERCURIAN PLAINS VOLCANISM

Although 2700 images were collected by Mariner 10 in 1974, it provided photographic coverage of only 45% of the planet's surface. Mariner imagery has a resolution of between 4 km and 100 m, which is roughly equivalent to that of the Moon achieved by Earth-based telescopes. However due to constraints imposed by the planet's rotation period and the spacecraft's trajectory, much of the coverage was obtained under high solar elevation angles which rendered the discrimination of surface topography rather difficult. Nevertheless the mission elevated the status of Mercury to a level equivalent to the Moon prior to the Apollo missions. An excellent compendium of imagery and mapping can be found in *Atlas of Mercury* by Davies *et al.* (1978).

The overall appearance of Mercury's surface is rather similar to that of the Moon, with at least one large basin (Caloris), a smaller one (Tolstoj), and large numbers of impact craters of a wide range of sizes. The Caloris basin is multi-ringed and has a diameter of 1300 km. Antipodal to it is a region of peculiar hilly and lineated terrain which disrupts older landforms. Both the basins and craters have an impact origin. Inside the Caloris basin, surrounding it and in the northern polar regions are widespread low-lying areas known as *light plains*. These account for about 15% of the area imaged and extend into what was the unilluminated hemisphere during the Mariner 10 fly-bys. Their concentration in the northern hemisphere between longitude 120° and 190° W. led Strom and his co-workers (1975) to draw parallels with the Moon where the distribution of mare basalts is also quite asymmetrical. These plains and more highly-cratered *intercrater plains* have been the subject of much debate, since there is a very strong chance that they are of endogenous (volcanic) origin. Unfortunately, however, the resolution of Mariner imagery is insufficient to reveal many tell-tale volcanic features, such as flow lobes, rilles and low shields on the plains units. The distribution of plains deposits is shown in Fig. 7.11.

Generally of smooth appearance and lightly cratered, the light plains are the youngest of the terrain units on the planet. On the basis of broad morphological characteristics, Strom (1977), Trask and Guest (1975) and Murray *et al.* (1975) argue for a volcanic origin. Wilhelms (1976) and Overbeck *et al.* (1977), on the other hand, contend that they are equivalent to lunar light plains which they conceive of as basin ejecta, equivalent to units like the Cayley Formation and Orientale smooth ejecta facies. However an impact origin for lunar light plains has been questioned by a number of workers for several different reasons. For instance, Haines and co-workers (1978) suggest that thorium abundances of the plains in the vicinity of Mare Smythii, revealed by Apollo gamma-ray spectrometry, are more consistent with volcanic deposits than ejecta, while Bell and Hawke (1986), note that many lunar impact craters sited within light plains regions have dark haloes whose IR spectroscopic signatures are those of mare basalt. On the basis of the distribution of these dark halo craters, they conclude that >15% of lunar basalts are disguised by a mantle of higher-albedo material with an impact signature. Then again, Hawke and Head (1978) and also Spudis (1978) have shown that the light plains of the lunar Apennine Bench Formation are in fact the result of KREEP volcanism. Finally, Kiefer and Murray (1986) present indirect evidence that the Mercurian plains are volcanic in origin. Thus, in Borealis Planitia, stratigraphic relations show there to be two distinct stages in smooth plains formation, the more recent having volcanic characteristics. They also note distinct

Fig. 7.12: Mercurian smooth plains southeast of the Caloris Basin. Note the wrinkle ridges, small conical volcanic constructs, the low albedo and low crater density. Frame width 150 km. (Mariner 10 frame FDS 72.)

colour contrasts in the hilly and lineated terrain and also within the Tolstoj basin between smooth plains and adjacent units. The same workers also identify a rimless collapse pit inside Tolstoj which appears to have an endogenic origin. When all these observations are considered, obvious parallels can be drawn between the lunar light plains and those on Mercury, and the view is taken here that they have a volcanic origin.

The older plains units – the intercrater plains – show a much wider range of crater ages than the less heavily cratered light plains (Watkins and Strom, 1984). Thus some intercrater plains appear to be older than the craters of the heavily cratered terrain (Trask and Strom, 1976), while others are seen to embay the craters (Strom, 1977; Leake, 1981). This appears to argue against an impact origin which would be characterized by a relatively restricted age range indicative of formation during the period of early bombardment. Furthermore the intercrater plains do reveal some positive evidence for a volcanic origin. For example, Dzurizin (1976) has mapped several linear features which he suggests may be fissure-related extrusives, while Malin (1978), in reviewing the evidence for a volcanic origin for the plains units, points to the presence of several likely volcanic constructs, including a low dome along the edge of the prominent Discovery Scarp (50˚S, 35˚W), and also a 70 km-wide elevated area at 64˚S, 152˚W which has a prominent peak at its summit, a large summit pit and an associated linear ridge which could be due to extrusive fissure volcanism (Fig. 7.12). While the intercrater plains are somewhat more enigmatic than the light plains, their wide range of crater ages and general morphological similarity to the lunar maria also seem to lobby for a volcanic origin.

The morphological characteristics of Mercurian plains units suggests they are built of mafic flood lavas. Where plains units have flooded larger craters or basins, the observed form of ghost craters suggests the plains cover is relatively thin and thus entirely appropriate for such volcanic materials. By analogy with terrestrial basalt plains, it has to be assumed that they were extruded from fissures generated in the planet's crust by extensional stresses. The subsequent development of a compressional stress field, manifested in the long compressional scarps unique to Mercury, may have led to the cessation of volcanic activity by shutting off feeder conduits to the surface flows.

Some idea of the composition of the near-surface layers may be gained from the limited spectral and colour difference data which exists. The colour ratio of Mariner 10 images indicate that Mercurian ray systems are bluer than the average surface – quite the opposite of the lunar case, where the rays are redder (Hapke *et al.*, 1975). The most reasonable explanation of this is that the surface rocks here are low in Ti, Fe^{3+} and metallic Fe compared with the Moon. This is confirmed by the planet's lower ultraviolet reflectivity and generally higher albedo than the lunar maria surfaces. Some Mercurian basalts may, therefore, be somewhat different from those of the Moon. It should be noted, however, that spectral reflectivity curves for Mercurian highland and mare plains are very similar to those for the Moon, sharing two features: a positive uniform slope and a shallow absorption band near 0.95 μm (McCord and Adams, 1977). The accepted explanation for this is that the slope is controlled by glass containing titanium and iron in the lunar regolith, and that the absorption at 0.95 μm is largely an expression of pyroxene in the regolith. This mineral may thus be an important constituent of Mercurian rocks.

In summary, opinion is divided with respect to the origin of the plains units; however, there appears to be more evidence in favour of a volcanic origin for both the light and intercratered plains than against it. If they are volcanic, as is assumed herein, it is quite feasible that the emplacement of both the intercrater and light plains occurred during a single, very extended phase of volcanism which terminated when the build-up of compressive stresses shut off feeder conduits from the Mercurian mantle. Since the darkest smooth plains on Mercury have an albedo which is higher than that of average lunar mare material, if they are volcanic they may be equivalent to lunar low-Ti basalt (Hapke *et al.*, 1975).

7.5 VOLCANIC PLAINS ON VENUS

Early insights into the nature of Venusian plains were provided by the Arecibo facility and by the Russian Venerae 15 and 16, which obtained high-resolution (1.5-km) imagery of the northern mid- to high- latitudes – approximately equivalent to high-resolution images obtained from Arecibo. Most recently, the Magellan spacecraft has allowed the details of plains and highland morphology to be elucidated and the wide extent of volcanic plains confirmed.

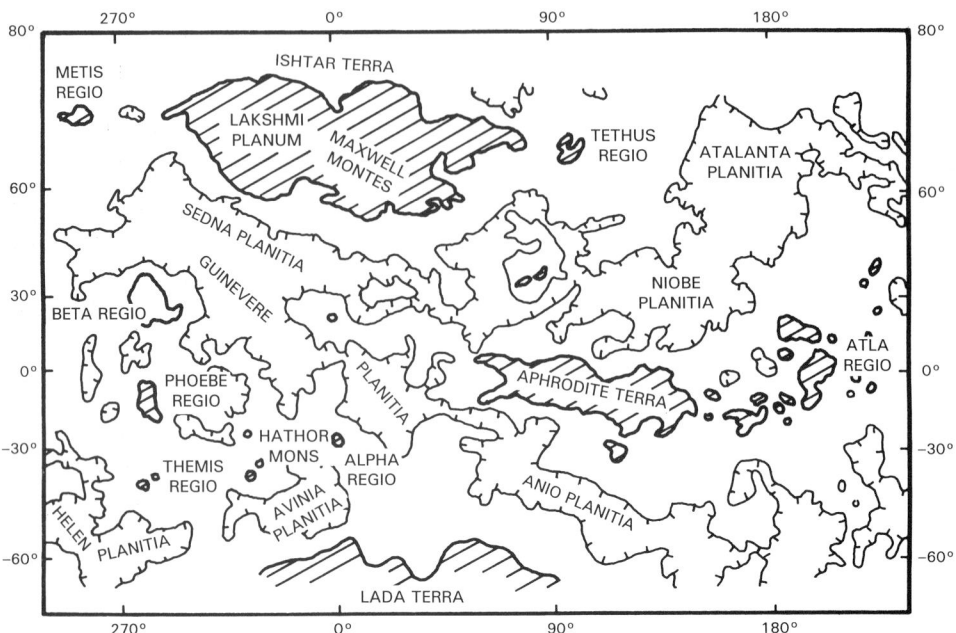

Fig. 7.13: The principal topographic units of Venus. Plains (0-2 km above datum), no pattern; highlands (more than 2 km above datum), oblique lines; lowland plains (below datum), short ticks on inward-facing boundaries. Polar regions are principally plains.(After Masursky *et al.*, 1980.)

7.5.1   Venus: general physiography

Radar mapping completed in the late 1970s and early 1980s, both from Venus orbit and Earth-based, gave the first detailed information concerning that part of the planet between 74°N. and 63°S. It revealed three broad terrain types: *upland rolling plains* which rise to about 2 km above datum and which constitute 65% of the surface then imaged, *lowlands* which lie below 0 km and account for the 27%, and *highlands* which rise above the 2 km level up to 11 km and cover about 8% of the surface. The highlands include two continent-sized regions, Aphrodite and Ishtar Terrae, and two smaller ones, Beta-Phoebe Regiones and Lada Terra, separated by lower-lying regions. A series of major canyons or rifts connects most of the higher ground at low latitudes, the deepest of these being Diana Chasma which is over 3000 km long, up to 100 km wide and descends 4 km below mean datum.

The arrival, during 1983, of the Russian Venera 15 and 16 probes extended the mapping to the northern polar region, which was found to be an extensive plain traversed by ridge-and fracture-belts. The mapping of the polar regions was completed by the U.S. Magellan spacecraft which commenced detailed mapping in September, 1990 and which, during four Venus-year-long mapping cycles, provided high resolution imagery, altimetry and gravity data. The greater resolution achieved by this last mission, has enabled planetary scientists to define a larger number of terrain types which included several different kinds of volcanically-produced plains units which cover 85% of the planet's surface.

The distribution of major terrain features is clearly not random. Phillips and his co-workers (1981) have noted the obvious alignment of the upland terrain, including Aphrodite Terra and Beta Regio, along a great circle inclined at 30° to the equator. This prominent global-scale feature became known as the "Equatorial Highlands" and the suggestion has been made that it could represent an ancient divergence zone. Furthermore, the same workers cite the Pioneer-Venus gravity data as implying that the region may still be tectonically active. Schaber (1982) extended the argument and identified two further global-scale zones of possible tectonic origin. The two longer ones, the Aphrodite-Beta and Themis-Atla zones, extend for 21 000 and 14 000 km respectively. The former describes a northeast-trending great circle that runs from the southern slopes of Ovda and Thetis Regiones at the western end of Aphrodite Terra, to the west flank of Beta Regio. In doing so it passes through Atla Regio, an elevated region in the eastern end of Aphrodite Terra. The latter trends northwest from a point closely east of Themis Regio (south of Aphrodite) and ends on the northwest side of Atla Regio. Magellan mapping reveals Aphrodite Terra to be built from a number of volcanic rises and raised blocks of complex ridged terrain (tesserae), traversed by rift zones and transverse faults.

The shortest of the major tectonic zones runs north-south from Beta Regio to Phoebe Regio, a distance of about 6000 km, and is a discontinuous rift zone, part of which transects a major volcanic rise. McGill et al. (1981) consider Beta Regio itself to be a broad dome produced by crustal uplift, implying continued dynamic activity in the Venusian upper mantle. Campbell and his co-workers (1984), in discussing high-resolution imagery (2 km radar resolution) of Beta Regio, confirmed the presence of a major rift system and associated volcanic features, arguing that magmatic activity may well have been relatively recent. Magellan images have confirmed this view and added much detail to the broad picture. In particular, it has been realized that the more northerly of two suspected shields, Rhea Mons, is actually a rifted block of complex ridged terrain. Several groups of workers have pointed to gross similarities between this structure and the East African Rift Valley, an analogy which presumably can be extended to other major rifts which traverse the Venusian crust.

It is no mere coincidence that Beta and Atla Regiones, broad shield-like uplifts with which are associated radial flow-like features and caldera-like structures, are situated where intersections of the major linear zones occur. These volcanic rises should be characterized by eruptions of mantle-derived, basalt-like magmas. The Soviet Venera probes numbers 9 to 14 landed at various locations on and between Beta and Phoebe Regiones and recorded (i) densities similar to terrestrial basaltic silicate rocks, (ii) major element chemical compositions similar to terrestrial oceanic basalts, and (iii) limited trace element compositions indicative of both terrestrial oceanic and continental igneous rocks (see Barsukov, 1982; Surkov et al., 1977). A consistent picture thus emerges.

With the exception of the uplifted massifs of complex ridged terrain (tectonically-deformed highland areas) the majority of Venus's surface is covered by volcanic products. The extensive plains are host to a plethora of individual volcanic structures. These include 556 shield fields, 274 intermediate-sized volcanoes (average diameter 25 km), 156 large

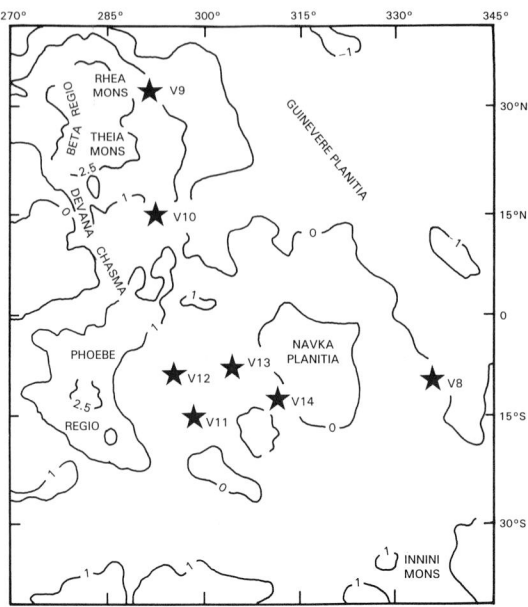

Fig. 7.14: Location map for the Russian Venera landers in the equatorial regions of Venus.

volcanoes (average diameter 400 km), 86 calderas, 176 coronae, 259 arachnoids and 50 novae. In addition over fifty substantial flood lava fields have been identified, with an approximate area of 128 200 km². The latter tend to be concentrated in the regions of Lavinia Planitia and Alpha Regio, on the flanks of Aphrodite Terra and Atla Regio, in both Beta and Phoebe Regiones, and in Sedna Planitia. Mapping of the flow direction within such flow-fields indicates that flood lava eruption has contributed widely to the flooding of the lowest regions on the planet's surface. Caldera-topped shields are not, however, confined to the low-lying plains; for example, the two prominent radar-dark circular depressions Colette and Sacajawea, situated on Ishtar Terra in the vicinity of Maxwell Montes, likely represent volcanic depressions.

7.5.2   Local surface composition and characteristics
The local composition of the Venusian surface was revealed by the series of Venera landers which set down in the general region of Beta-Phoebe Regiones (Fig. 7.14) and by the more recent Vega probes. The Vega 1 and 2 probes landed in Rusalka Planitia and between that location and the eastern edge of Aphrodite Terra respectively.

Chemical analyses from several sites (e.g. V14 and Vega-2), conducted by XRF-spectrometry, are most reasonably interpreted in terms of tholeiitic basalts or gabbros (Surkov et al., 1977; Barsukov et al., 1982; 1986); however, at the V8 and 13 sites, rather more alkaline basalt species are indicated (Table 7.4). The restricted data for trace elements (U, Th and K) obtained by Veneras 8, 9 and 10, although somewhat equivocal, is generally in line with the rocks being akin to terrestrial oceanic basalts. Gamma-ray measurements

Table 7.4: Composition of surface materials at Venera and Vega sites compared with terrestrial oceanic basalt.

Constituent	Venera 13	Venera 14	Vega-2	Oceanic basalt
SiO_2	45 ± 3	49 ± 4	45.6 ± 3.2	51.4
Al_2O_3	16 ± 4	18 ± 4	16.0 ± 1.8	16.5
MgO	10 ± 6	8 ± 4	7.7 ± 3.7	7.56
FeO	9 ± 3	9 ± 2	11.5 ± 3.7	12.24
CaO	7 ± 1.5	10 ± 1.5	7.5 ± 0.7	9.4
K_2O	4 ± 0.8	0.2 ± 0.1	0.1 ± 0.1	1.0
TiO_2	1.5 ± 0.6	1.2 ± 0.4	0.2 ± 0.1	1.5
MnO	0.2 ± 0.1	0.16 ± 0.08	–	0.26

for K, U and Th made at both Vega sites indicated a tholeiitic chemistry. The basaltic nature and general morphological appearance of the lowlands at the V14 site led Masursky and his colleagues (1980) to equate them with the lunar maria, although, of course, the former were not emplaced within ancient impact basins.

The details of the surface itself were also revealed by the lander probes and seen to be characterized by flat-lying blocks and laminated pavement material – interpreted by some workers as a duricrust – between which are darker, fine-grained materials, and upon which is a small amount of soil cover (Florensky et al., 1977; 1983). Studies by Garvin and colleagues (1984) interpreted the pavement materials to be of volcanic origin, and in favour of this interpretation is the close similarity between the ground at the Venera sites with fractured slab pahoehoe surfaces both in Hawaii and the Snake River Plains (Fig. 7.15). Studies by Pieters et al., (1986), who corrected multispectral images for the effects of the orange-coloured incident radiation, indicated that at visible wavelengths the Venusian surface is dark grey and in accord with mafic materials.

7.5.3   Volcanic nature of the Venusian plains
Volcanism has been the predominant process responsible for plains development on Venus, thus, by implication, volcanic processes may have moulded at least 85% of the surface. Compared with the volcanic plains of the Earth's ocean basins, radar studies reveal that Venusian plains are more rugged; furthermore they are characterized by numerous closely spaced circular and linear features which are concentrated in broad linear zones of global extent. The "rolling plains" are typified by somewhat higher slopes (2.2-3.5°) than the smoother "lowlands".

High-resolution images of the southern part of Ishtar Terra and the plains regions of Guinevere and Sedna Planitia to the south, were obtained at Arecibo in 1983. These topographically low regions are characterized by patches of radar bright, dark and intermediate terrain often with lobate and flow-like margins. Magellan confirmed the latter to be extensive flow-fields and that associated with the plains are volcanic shields, shield fields and domes. Circular to oval coronae, nova and arachnoids between 60 and

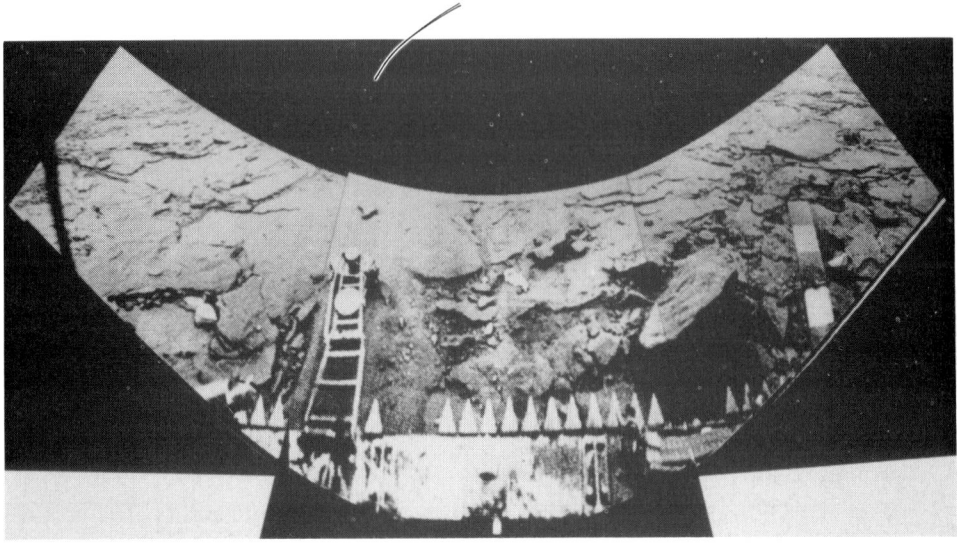

Fig. 7.15: Geometrically-transformed image of Venusian surface at the Venera 14 landing site. (Courtesy, J. W. Head.)

300 km in diameter are abundant, as well as linear fault scarps and ridge- and fracture-belts (Fig. 7.16). The interiors are most corona-lie structures appear to have been flooded by lavas. Aubele *et al.* (1988) mapped 21 000 dome-like features between 2 and 15 km in diameter on these and other plains, of which about 15% are aligned. The overall density of such features is around 28 per 100 km^2.

The plains south of Ishtar were subdivided into a number of sub-units by Stofan *et al.*, (1987), and while all appear to contain evidence for a volcanic history, they are not of the same age. One of the geologically young provinces they recognized, in Sedna Planitia, had earlier been studied by Head *et al.* (1985) and interpreted as volcanic plains. Lying between 0 and 1 km above mean datum, the plain has on it, close to its western margin, a broad depression 140 km across but less than 500 m deep. Flow-like features are very common, having low to high radar backscatter returns and often lobate form. These are seen to embay the tectonically disturbed belt which borders Lakshmi Planum to the north. Some of these flows, now known to be lavas, are associated with vents and fractures.

The generally similar aspect of large tracts of plains suggests that effusive volcanism associated with low shields and domes has been very widespread on Venus. The morphological similarity between such plains and those of the Snake River Plains, leads inevitably to the conclusion that plains volcanism – intermediate in character between the fissure-related style of terrestrial basalt plains formation and major central shield activity, as it expresses itself on Mars and in other regions of Venus – has been an important volcanic style on the planet.

Other parts of both Sedna and Guinevere Planitiae lack recognizable flow features or vent areas. Such regions are generally smooth, radar-dark, and have associated with them

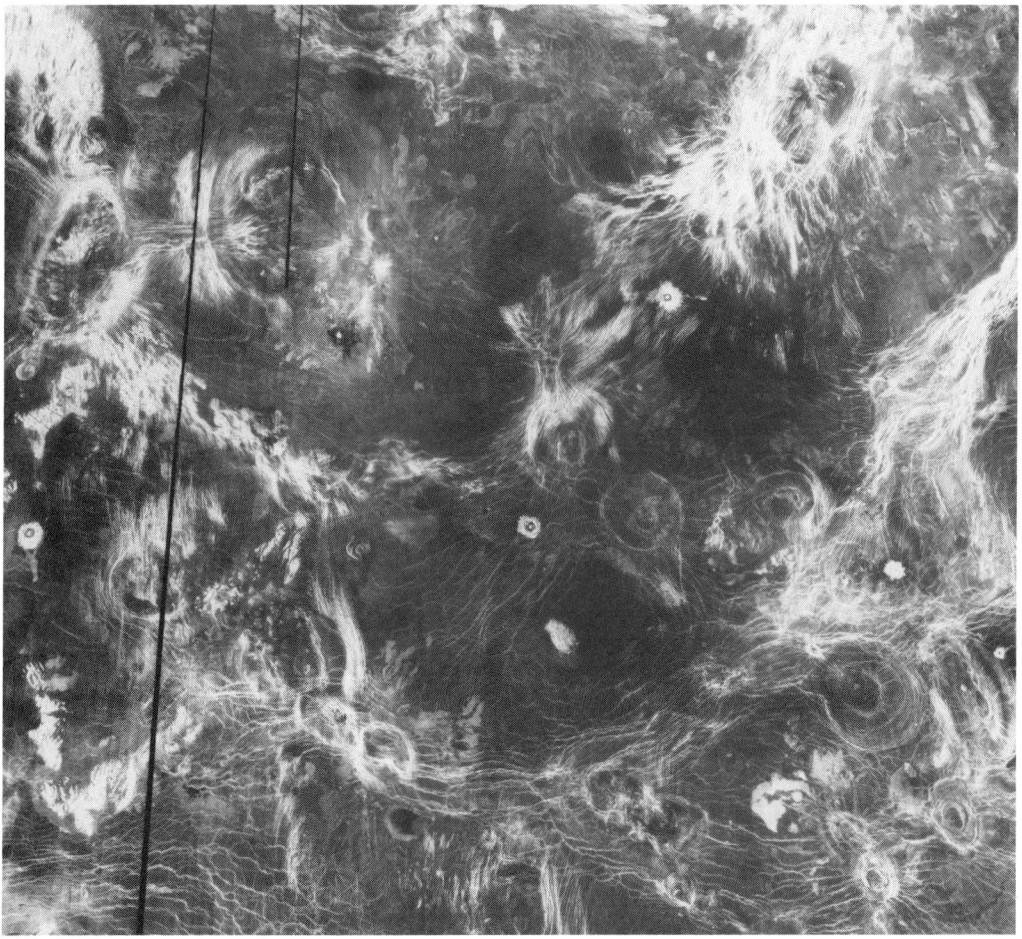

Fig. 7.16: High-resolution Magellan image of the Bereghinya Planitia region of Venus. The mosaic is 1843 km wide and is dominated by volcanic features. The most prominent structures are arachnoids which range in size from 60 to 250 km diameter. Also visible are lobate lava flows, impact craters, coronae, volcanic domes and fracture belts. Magellan image P-37948.

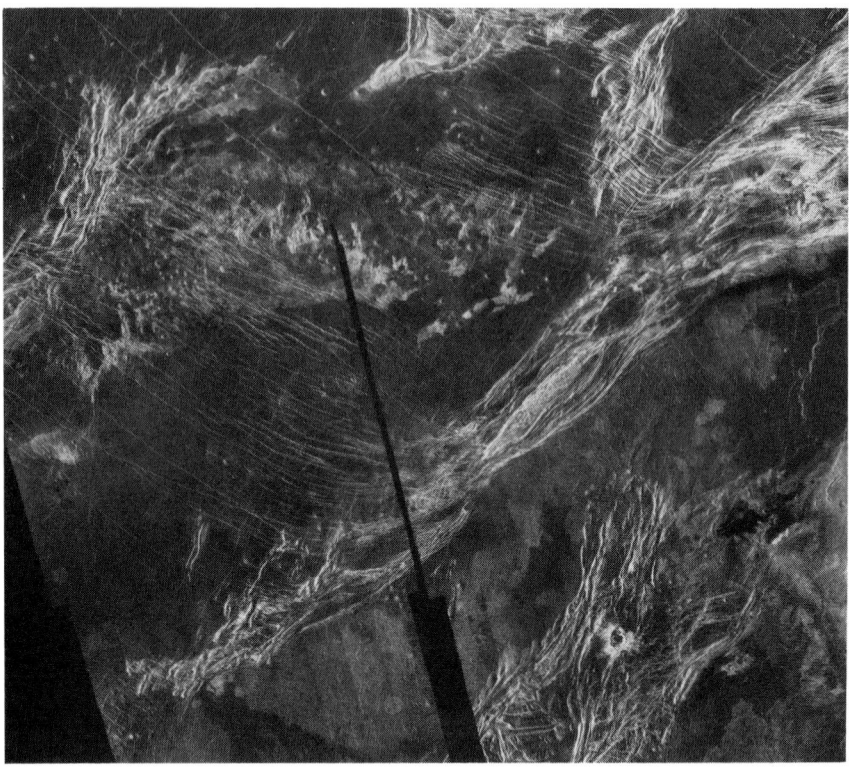

Fig. 7.17: Magellan image of the plains region of Lavinia Planitia. Tectonically-deformed ridge belts cross the plains, which are host to small shield volcanoes, wrinkle ridges and, at bottom right, to extensive volcanic flow-fields. Frame width 1800 km. Magellan image P-38293.

large numbers of radar-bright coronae and groups of lineations, interpreted to be faults. Some of the latter appear to be an integral part of the Beta Regio rift system. Some appear to have become flooded by radar-dark materials, presumably lava flows, since they have radar-dark centres, while in some instances even the peripheral rings have been breached by such flows, indicating more extensive lava flooding. Basilevsky *et al.* (1986) have interpreted coronae to be related to hot spot activity and gravity tectonics. These smoother plains, which occupy the greater part of Guinevere and Sedna Planitiae, are apparently older than the units containing broad domes and vents. They may be closer in their genesis to terrestrial flood basalt provinces or, alternatively, since they appear to be older than the vent-bearing plains, they may simply represent more degraded regions of plains-style activity.

The distribution pattern of the major plains units on Venus also implies a measure of structural control. On Mars, volcanism is entirely associated with extensional tectonics, but the situation is evidently more complex on Venus, where both extensional and compressional deformation apparently has occurred. Where sufficiently convincing evidence is available for Venus, it seems that extensional faulting accompanied active plains volcanism, as would be expected from terrestrial experience, and that compressional tectonics generally played a later modifiying role or was linked to more centralized activity.

Lowlands, with their darker and smoother signatures, have been likened to lunar maria and interpreted as lava-flooded basins (Masursky *et al.*, 1980). Because they also lack the significant numbers of impact craters recognized on the rolling plains and had small gravity lows recorded over them, at this time they were widely accepted as being regions of thinned low-density crust covered in basalt-like lava flows. It also transpired that not only were impact craters present, but also volcanic flows, domes and shields, and extensive systems of ridge-belts (Barsukov *et al.*, 1986). What the improved resolution of the Venera and Magellan data has shown is that the lowlands were much more complex than hitherto had been realized.

Atalanta Planitia – the deepest regional depression on the planet – rather than being simply a featureless lava basin, is an extensive depressed plains area, the greater part of which is traversed by ridge belts which continue across the north polar region (Fig. 7.17). However, the western part, adjacent to Tellus Regio, is rather different and is host to clusters of dome-like hills and a network of radar-bright ridges. Indeed, the ridge belt crossing Atalanta turns out to be the longest and most prominent on Venus. Guinevere Planitia is in many respects similar to Atalanta but also has several large coronae on its surface; additionally, in the south, Guinevere becomes rather hummocky, as though older units are modified or buried by plains materials. Similar developments occur widely within the plains elsewhere. Where this occurs the elevation difference between what clearly are volcanic plains and hummocky plains is generally in the 0.5 to 1 km range. This has led several workers to surmise that where hummocky plains development is found, we are seeing the effects of mantling of older tesserae units by younger plains materials (Nikishin, 1990). This, in turn, raises the interesting question of whether or not the volcanic plains of Venus, in general, rest on an older basement of fragmented tessera material.

Magellan data (Head *et al.*, 1991; Saunders *et al.*, 1991a; Solomon *et al.*, 1991) allowed recognition of the fine detail on the plains' surfaces for the first time, including the highly complex volcanic and tectonic history of coronae and related features. In the initial Magellan mission summary, Saunders *et al.* (1991) recognize four major plains types: (i) *smooth plains* which, as the name implies, are relatively featureless, have no discernible volcanic flow features and also few linear structures or dome-shaped hills. They tend to be radar-dark but range up to moderately radar-bright. These are seen as having an origin in volcanic flooding, presumably by very fluid lavas or by the coalescence of a plethora of low volcanic shields. (ii) *Reticulate plains* are typified by having one or more sets of somewhat sinuous, radar-bright lineaments which often cannot (at the radar resolution of Magellan) be identified positively either as ridges or grooves; whichever these are, however, they are spaced on average less than 5 km apart. Their morphology suggests an origin in volcanic flows or low shields which either have embayed older units, or have been tectonically deformed, or both (Fig. 7.18(a)). *Gridded plains* (iii) are rather distinctive in having intersecting orthogonal sets of radar-bright lineaments, regularly spaced, which extend for hundreds of kilometres. The spacing of these features tends to be closer than those in (ii) and typically is less than 5 km. Complex deformation is a characteristic (Fig. 7.18(b)).

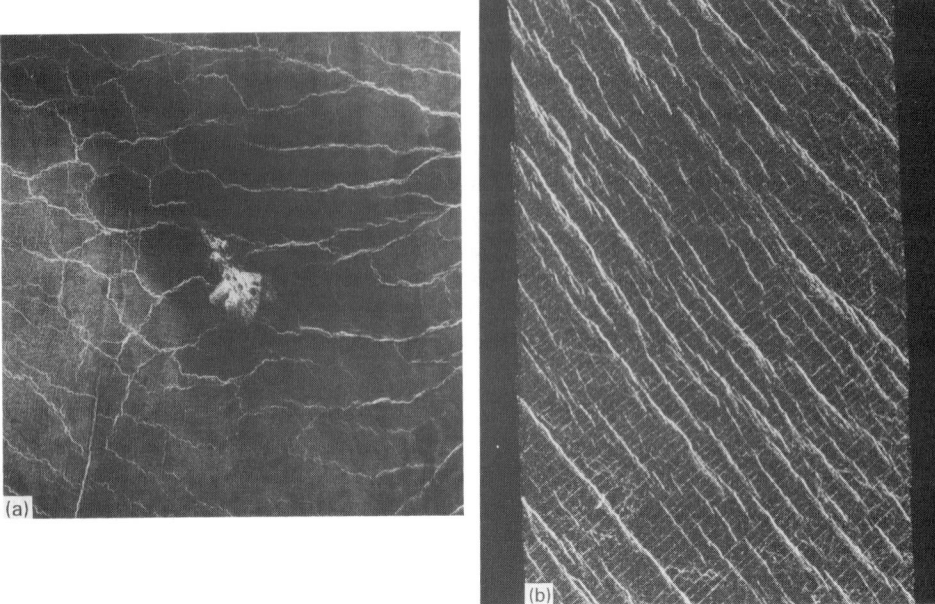

Fig. 7.18: (a) Reticulate plains, with impact crater just above centre and shield field towards bottom right. Magellan image F-MIDRP 25N119:1 (b) Gridded plains with intersecting lineaments. Magellan image P-36699.

Lobate plains (iv) comprise overlapping lobate flow features with variable radar signatures which extend for tens to hundreds of kilometres. Such plains are traversed by few if any linear structures. Fractures, where developed, and local topography, appear to have controlled the emplacement of the lobate plains materials, presumed to be complexes of volcanic flows (Fig. 7.19(a)).

Detailed study of plains in the region of Lavinia Planitia, reveals that most of the different types of plain are traversed by *wrinkle ridges* (Squyres *et al.*, 1992) which show up as radar-bright lines. They generally are <1 km wide, although a few are much wider (several kilometres); in length they range between a few tens of kilometres to in excess of 100 km. As a rule, individual ridges are located between a few kilometres to 20 km apart. As is the case on both the Moon and Mars, where similar structures abound on the plains units, the ridges show quite strong preferred orientations within different regions, as can be seen on Fig. 7.19(b). In the region of Lavinia Planitia, Squyres and his colleagues (1992) observe that the ridges decrease in concentration from older to younger units, while the trend of the ridges appear to be related to location rather than age. This implies that ridge formation – presumably by compressional forces – was, and may still be, an ongoing process connected to local stress fields that operated while plains units were being laid down.

Other common radar-bright lineaments which cross the plains as parallel families tend

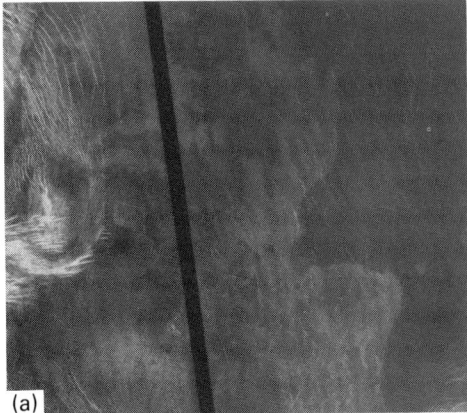

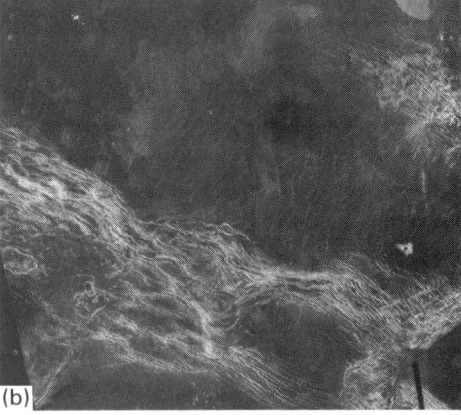

Fig. 7.19: (a) Lobate plains with intermediate radar-backscattering characteristic. Magellan image F-MIDRP 05S177;1. (b) Wrinkle ridges on a part of Lavinia Planitia. The radar-dark plains units are host to one radar-bright impact crater and many small volcanic shields. A ridge belt crosses the bottom part of the image. Frame width 600 km. Magellan image P-37135.

to be longer and less sinuous than wrinkle ridges. Where resolution allows, these can be identified as narrow grooves that may extend for anything between 25 and 200 km across the Venusian surface, with spacings of between 30 to 100 km. In contrast to wrinkle ridges, these are extensional features and many (perhaps most) appear to be graben. As with the ridges, they tend to share a common strike over quite large areas but always lie normal to the ridges, which gives the plains of regions like Lavinia and Guinevere Planitiae a distinctive orthogonal imprint.

With the benefit of all the available imagery, there can be little doubt that volcanism has been responsible for generation of the Venusian plains, that is, over three quarters of the planet's surface. The kinds of primary structures developed – long flowlike features, caldera depressions, domes, low shields – are typical of plains-forming basaltic volcanism (Basilevsky and Head, 1988; Barsukov et al., 1986), and a basaltic composition is also implied by the Venera and Vega lander geochemical data. In addition to the smaller-scale features there are 156 large volcanoes >100 km across, 274 volcanoes in the range 20 – 100 km, 86 caldera-like landforms between 60 and 80 km diameter which are not associated with large shields, 550 clusters of small <20 km volcanoes), called *shield-fields*, 175 annular concentrations of fractures and ridges termed *coronae*, 259 *arachnoids* and 50 *novae* (foci comprising radial fractures forming stellate patterns).

7.5.4   Age of Venusian plains

In the absence of absolute ages for Venusian rocks, estimates of the age of plains surfaces can be derived only by reference to the cratering record. However, there are significant differences between Venus and other terrestrial planets due to the great density of the planet's atmosphere. Recent modelling exercises on the atmospheric filtering process (Phillips *et al.* 1992) using data both for the total crater population and the population of crater clusters imaged by Magellan – the latter representing the cut-off limit for breakup of bolides in the Venusian atmosphere – shows that the smallest crater whose numbers are not affected by atmospheric filtering is circa 32 km diameter. On this basis it is possible to compare the size-frequency distribution curve for craters >32 km diameter with that of the Moon. In this way a notion of the resurfacing history of Venus may be obtained. Thus far an average surface age of 300-500 Ma has been derived, indicating that the latest resurfacing event must be at least as recent as this.

7.5.5   Eruption characteristics on Venus

The very elevated surface temperature (470°C) and pressure (98 bars) experienced at mean planetary radius (MPR) have the potential to influence surface and near-surface volcanic processes on Venus. Theoretical consideration has been given to the rise and eruption of Venusian magma (Head and Wilson, 1986), and to melt generation and crustal melting under Venusian conditions (Hess and Head, 1990). Suffice it to say that the atmospheric pressure in the Venusian lowlands is so extreme (about 9×10^6 Pa) that at least 2 wt% H_2O (or 5 wt% CO_2) would need to be dissolved in any crustal melt before pyroclasts could commence forming. In the Venusian highlands, where the pressure is lower, the appropriate figures would be 1 wt% and 3 wt% respectively. This contrasts sharply with the Earth, where fire fountaining would result from a mere $0.1 - 0.4$ wt% dissolved H_2O. So, pyroclastic deposits would be less likely to form on Venus than on the Moon, Earth or Mars.

The prevailing high temperatures would have different effects. Certainly they would inhibit the radiative cooling of Venusian lava flows; however, the high atmospheric density more than counteracts this, since it means that convective cooling initially will be more effective, with the result that, overall, flows will actually cool faster than they do on the Earth's surface (Head and Wilson, 1986). The consequence of this is that, for the same mass eruption rate, a lava flow on Venus might be expected to be about one fifth longer than its terrestrial counterpart, but to undergo a pahoehoe/aa transition more quickly.

On Venus, where atmospheric pressures are high, explosive disruption of magma would be rare, with the result that lavas extruded on to the surface would retain most of their exsolved volatiles. For this reason their bulk density would remain relatively lower than comparable terrestrial flows. Since there is also a marked pressure gradient with altitude, both volatile exsolution and the density structure of the upper crust are affected; furthermore, it affects whether or not neutral buoyancy zones can form and at what depth (Head and Wilson, 1992a). Near to mean planetary radius most magmas would rise directly to the surface, assuming volatile contents similar to those of Earth; however, at altitudes of 2 km or more above MPR, roughly half of any magma generated (assuming

Table 7.5: Size, distribution and area of volcanic features on Venus.

Feature mapped	Number average	Approximate average diameter (km)	Approximate area covered (km^2)
Shield fields	556	150	17,700
Intermediate volcanoes	274	25	490
Large volcanoes	156	400	125,600
Calderas	86	60	2,900
Coronae	176	250	49,000
Arachnoids	259	115	10,400
Novae	50	190	28,300
Flood lavas	53	350*	128,200

*Length

similar compositions to terrestrial ones) would be erupted directly, while the rest would stall at some depth, in neutral buoyancy zones. The relevance of this to the global distribution of volcanic rocks on Venus is that widespread extrusion of high-volume lava flows would be favoured in low-lying regions, while smaller-scale activity sourced by melts residing in shallow reservoirs/neutral buoyancy zones would characterize higher regions. Theory predicts that the depth of such zones would increase with increasing elevation, a factor of between 2 and 4 being appropriate for a height difference of between 1 km below MPR and 4.4 km above it (Head and Wilson, 1992a).

7.5.6 Flowfields and lava channel systems

Volcanic landforms on Venus share many of the characteristics of terrestrial structures: thus extensive lava flows, lava channels and small shields, together with volcanic domes and large volcanoes have a generally Earth-like aspect, although the large volcanoes may exceed terrestrial dimensions by a factor of 2 or 3. In addition, peculiarly Venusian features also occur on the plains; these include coronae, novae and arachnoids – all fairly major volcanic features – which have no terrestrial counterparts. Table 7.5 summarizes the size, distribution and areas of those volcanic features identified on Magellan imagery.

The radar backscattering properties of Venusian plains vary from low (radar-dark) to high (radar-bright). While a number of factors contribute to this, surface roughness is believed to dominate the radar signature over most of the planet's surface. A recent quantitative analysis of both Arecibo and Magellan backscatter data, and a comparison of it with terrestrial flows, supports that the very extensive radar-dark plains are floored by lava floods with relatively smooth surfaces, akin to terrestrial pahoehoe (Campbell and Campbell, 1992). Confirmation of this comes from the images received from Venera landers that showed slabby surfaces very reminiscent of the Snake River slab pahoehoe. However, not all volcanic plains regions are radar-dark, although it does appear to be

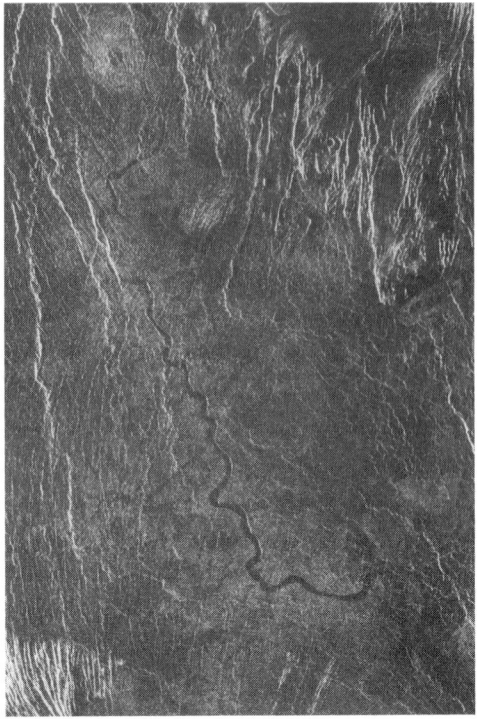

Fig. 7.20: 200-km-long segment of a sinuous Venusian lava channel in Lavinia Planitia. The channel is 2 km wide. These meander across the plains and are presumed to have been incised by low viscosity lavas behaving rather like water. Magellan image P-39226.

the case for the oldest flows. Younger flows may be either radar-dark or radar-bright, the signature being a reflection of surface texture. Some flows show a change in backscattering from proximal to distal locations, others show local changes which may be due to the breaking of slab crust as it either encounters obstacles or flows down a steeper slope; some are homogeneous over very large areas. Strangely, most show the radar signature of terrestrial pahoehoe flows, despite theoretical prediction that there should be a rapid pahoehoe to aa transition on Venus (Head and Wilson, 1986).

Fifty lava channel systems also have been identified; many with greater lengths than 250 km. Interestingly, only in a relatively small number of locations has it been possible to detect central flow channels bounded by raised levees, so commonly associated with many Martian flows. It seems unlikely, bearing in mind the considerable width of Venusian flows, that this simply is a function of resolution, and is more likely to be a manifestation of the rheological properties of Venusian lavas at the very high ambient temperatures found at the surface. The longer channels have been termed "canali" (Baker *et al.*, 1992), the longest attaining 6800 km length and being between 1 and 3 km wide (Fig. 7.20).

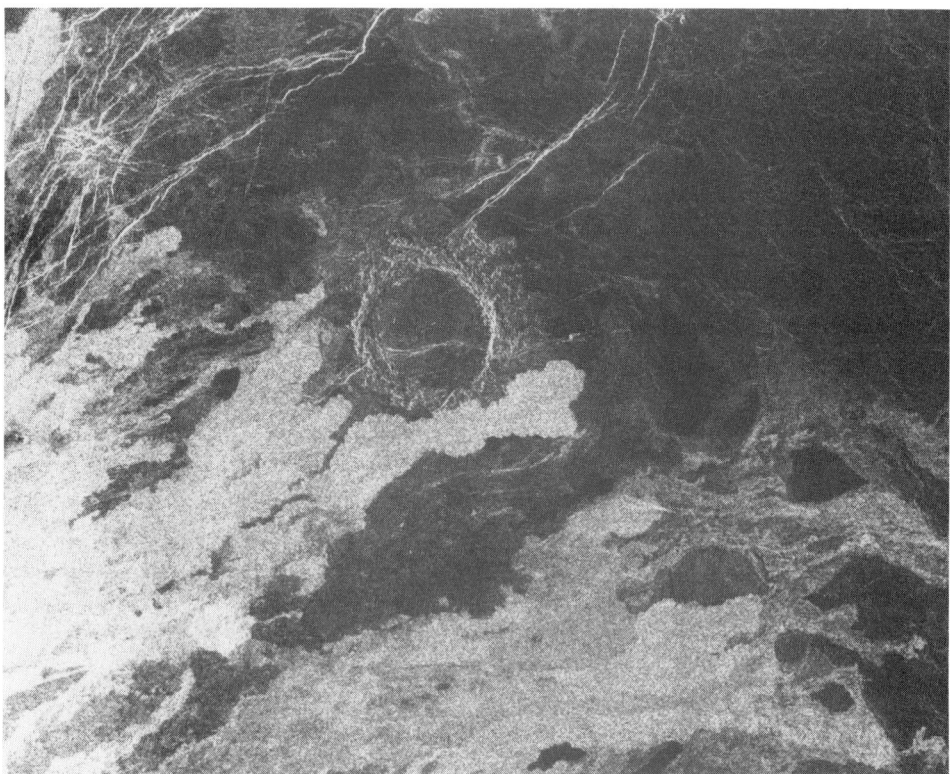

Fig. 7.21 Lava flow fields on the eastern flank of Sapas Mons shield volcano. The bright lobate flows measure between 5 and 20 km in width and between 50 and 100 km in length, and contrast with the radar-dark signature of flows comprising the underlying plains. Magellan image P-38690. Frame width 140 km.

Gregg and Greeley (1993), after analysing Magellan data and applying numerical models, concluded that Venusian lavas rapidly form insulating crusts due to the greater efficiency of total cooling compared with the Earth. They also suggest that Venusian canali-forming lavas are most likely to be within the compositional range: tholeiite-lunar basalt-komatiite, since these are characterized by low viscosities and are all stable under Venusian surface conditions (sulphur flows and carbonatites are not).

By terrestrial standards, many of individual Venusian flow-fields are large; some have volumes as great as 8000 km^3. One particularly prominent radar-bright flow-field outcrops in southern Atla Regio, about 900 km to the south of the volcano, Ozza Mons. This family of lobate-marginned flows has a total length of around 1000 km and covers an area of 180 000 km^2. Superposition/transection relations indicate there were at least four discrete emplacement episodes, each of which saw extensive flows inundating fractured and faulted plains. This particular flow-field appears to have been erupted from a series of graben associated with

Fig. 7.22: The region of Mylitta Fluctus, in northern Lada Terra. This is largely inundated by a massive sequence of radar-bright lobate flows which emanate from a shallow caldera-like structure to the south. Magellan image P-38088.

regional extensional faulting. In one location it is possible to discern a well-defined flow channel with levees, and also to see at a bend in the channel that the flow crust has been broken into huge slabs, exposing radar-dark lava beneath. At another site, one of the flows floods a small 500-m-high shield. By noting that nearby unflooded shields have a basal diameter of around 500 m, it has been estimated that the flow thickness is of the order of 100 metres.

Particularly spectacular are the flows associated with Mylitta Fluctus, in southern Lavinia Planitia, which inundated an area of at least 300 000 km². Like the Atla flow-field and that associated with Sapas Mons, Mylitta Fluctus flows are radar-bright and fairly uniform in texture (Fig. 7.22). The complex actually is composed of six smaller flow-fields, each of which appears to have emanated from a large shield volcano 400 km across which has a single summit caldera 40 km in diameter situated at the southern end of the complex (Roberts *et al.*, 1992). The individual flow-fields have very large areas, the largest covering about 120 000 km^2. While there is little evidence that eruptions were fissure fed, Roberts *et al.* (1992) admit that rift-related effusions may have preceded centralization, as is believed to have happened at certain Martian eruptive foci. The source volcano itself is located along a proposed rift zone that runs along the northern flank of Lada Terra.

Roberts *et al.* (1992) estimate effusion rates to be of the order of $460-4600 \times 10^3$ m^3 s^{-1}. This is of the same order as was reported by Cattermole (1987) for areally extensive sheet and tube-fed lava flows developed on the flanks of the Martian volcano, Alba Patera ($17-2120 \times 10^3$ m^3 s^{-1}), which also are thought to have issued from a centralized source. Using these derived values, Roberts *et al.* then calculated preliminary eruption durations, estimating that the first major outpouring took place over a period of between 10 and 70 years and generated the largest volume of lava (1.7×10^4 km^3). This early event is believed to have flooded the original Mylitta Fluctus rift zone and constructed the asymmetric shield on its northern flank. Subsequent individual eruptions are thought to have lasted only a matter of days and to have gone on over periods of a few months, rather than years.

Since over fifty major flow-fields now have been recognized on Venus, it is clear that the rapid effusion of large volumes of fluid, mafic or ultramafic lavas was characteristic of Venus, at least during certain phases of its geological evolution. Comparable flows are located on Mars where they have been equated (Cattermole, 1987) with fluid terrestrial lavas erupted at high effusion rates, such as those of the Columbia River Plateau, at least in terms of rheology and composition (Shaw and Swanson, 1970). The more voluminous of terrestrial flood lavas can be shown to have erupted from long fissures or from numerous vents sited along them; thus some of the Columbia River flows emerged from a fissure system 175 km in length. The same is probably true for many volcanic plains on Venus, although the source fissures are not discernible, probably since later flows have covered up the evidence, as they have on both the Moon and Mars. However, there is on Venus evidence for subsurface dykes which can be traced for upwards of hundreds of kilometres, and these support the notion of at least some measure of fissure-related activity. When the overall volume of these Venusian flow-field complexes is compared with terrestrial flood basalts, they are between 1 and 2 orders of magnitude smaller. The more modest volumes of flood lava involved on Venus may reasonably be interpreted as having originated in large-scale mantle upwelling probably along pre-existing rift zones, rather than in any form of divergent plate margin situation that is global in scale.

The distribution of major flow-fields is not entirely random; for instance, there is a distinct concentration in the region of eastern Lavinia Planitia and Alpha Regio, on the flanks and sometimes on the surface of Aphrodite Terra and Atla Regio, in both Beta and Phoebe Regiones, and in Sedna Planitia. Where it has been possible to determine the direction of flow, it supports the notion that major outpourings of fluid lava have contributed significantly to the resurfacing of topographic lows such as the lowlands. In several instances it has been possible to trace lava channels for several hundred kilometres, indicating that these – in much the same way as lunar sinuous rilles – provided major pathways for the downslope emplacement of basaltic flows.

The dense Venusian atmosphere also dictates that altitude-dependent factors come into

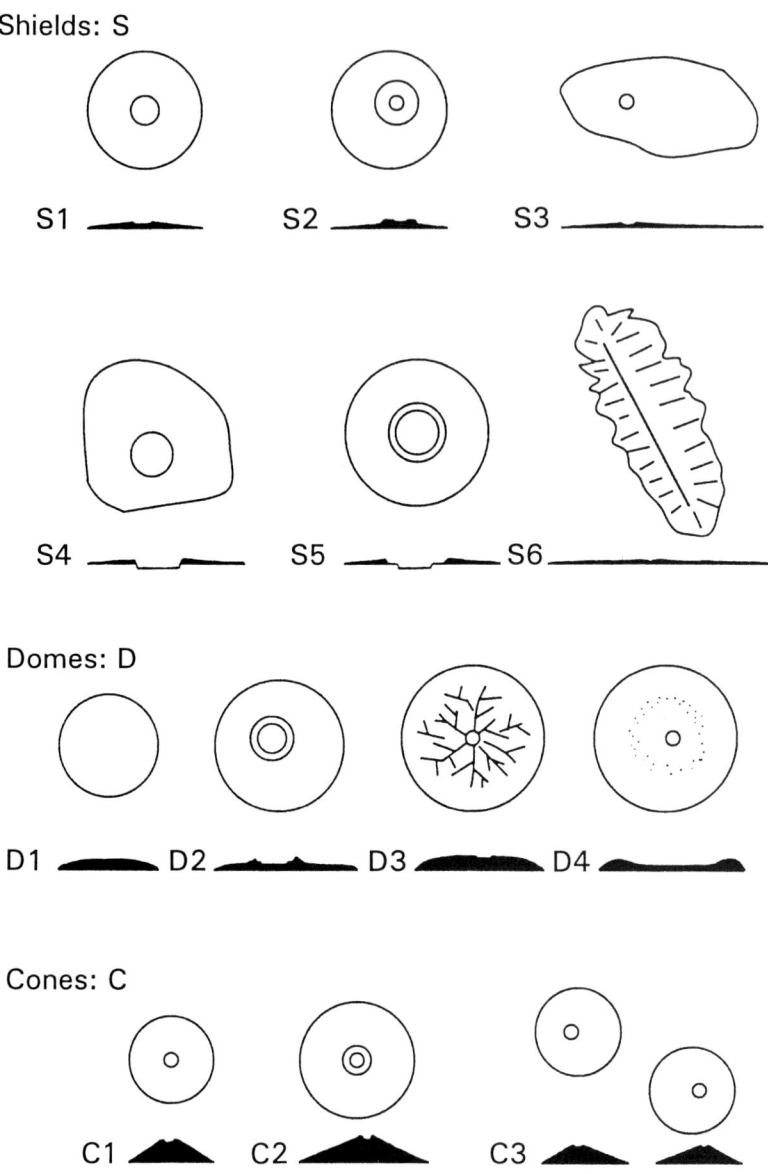

Fig. 7.23: Different styles of volcanic shield on Venus. After Guest *et al.* (1992).

play; thus, at low levels magma will be able to rise directly to the surface but as altitude increases it will of necessity be stalled and collect in neutral buoyancy zones whose depth will increase proportionally with height above MPR (Head *et al.* 1992). This also accounts for the greater proportion of shields, domes and calderas away from the lowland regions.

7.5.7   Distribution and characteristics of small volcanic structures

A huge number of small volcanic structures are located on the plains surfaces. These can be classified into three groups: shields, cones and domes. The most common of the three types are *small shields* – generally less than 20 km in diameter – of which over 22 000 were identified on the quarter of the globe imaged by Venera 15/16 (Slyuta *et al.*, 1988) the Magellan tally is considerably greater than this, and on the basis of calculations by the above and by Aubele and Slyuta (1990), Head *et al.* (1992b) derive concentrations of between 0.2 – 0.3 shields per 10^3 km². The majority are circular in plan and have gently inclined convex slopes; most have a single summit pit whose average size is around 700 m, but many of the smaller features appear either to lack a pit, or host one that is too small to be resolved by Magellan radar. While shields may be as large as 20 km across, most fall within the range 2 – 8 km and are less than 200 m high (Fig. 7.23).

Other shield-like landforms are of similar dimensions but have flat tops, while a further expression sees much broader plateau-like summits with less uniform flank slopes. Some shields appear to have a central plateau-like region with radar-dark signature surrounded by a roughly circular region of higher radar backscatter; these are believed to be shields with a summit lava lake surrounded by a spatter rampart. Guest *et al* (1992) describe six different styles of small shield-like structure, and note that if most are analogous to terrestrial shield volcanoes then the individual lava flows from which they are built must be <75 m across since they apparently are irresolvable in Magellan images. However, they also note that some such landforms may be built from a single flow surrounding a central source vent.

While the majority of such landforms are located on the Venusian plains, others are associated with coronae, arachnoids and other major volcanic edifices. Small shields frequently occur in groups, forming *shield fields*. This is true of large areas within Guinevere Planitia, for instance, where a group of 55 shields, ranging in diameter from 1.3 to 6.5 km, is superimposed on fractured dark plains.

Random sampling on a global basis serves to give an impression of the numbers of such structures and their sizes, the distribution of volcanic associations known as *shield fields* being shown in Fig. 7.24. There is a clear inverse relationship between numbers and size, down to a diameter of 2 km, beyond which the numbers fall off dramatically. The latter is almost certainly a function of resolution rather than a real reversal of the otherwise inverse trend. The most abundant small structures are found to be those whose basal diameters fall within the range 2 to 8 km.

A more specific study of the type and distribution of the three landform groups within the region of Niobe Planitia showed there to be 234 shields, 219 cones, and 163 domes out of a total of 616 small structures present (Guest *et al.*, 1992). In the same area, 393 summit pits were counted. Most of the volcanoes clustered together in large groups associated with fracture belts crossing the region (Fig. 7.25). This seems to be the

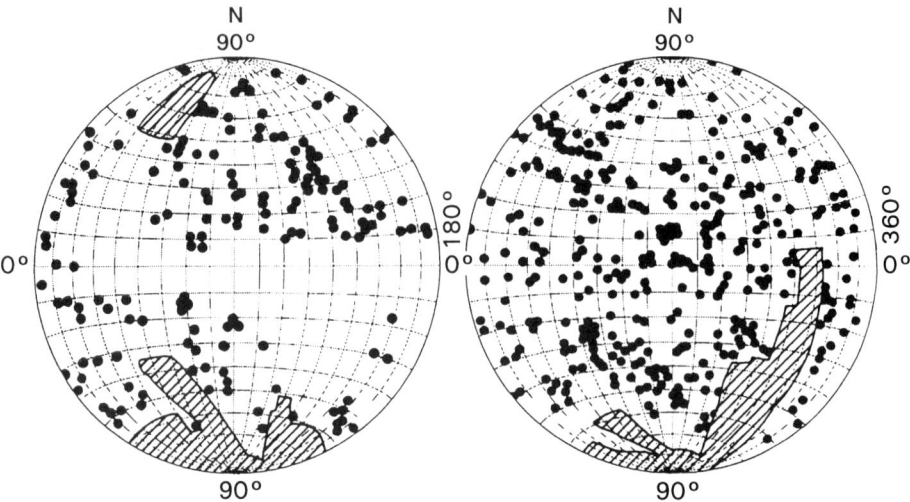

Fig. 7.24: Global distribution of shield fields over the surface of Venus. Areas for which data are absent are shown by hatching. From Head *et al.* (1992).

characteristic of such small structures Venus-wide.

On Earth, similar shields typically have summit depressions that are linked to subsurface magma conduits. Frequently the depression fills with lava, forming a lava lake which may overflow the rim and spread downslope as a flow; when the reverse occurs, drainback into the conduit forms the pit. Such is the pattern, for instance, at Mauna Ulu in Hawaii. Shield-related flows have been identified around some of the larger, intermediate-sized, shields, many of which are somewhat elongated, and related to subsurface structures believed to be dykes. They are also distinctive in having associated well-defined volcanic flows that spread away from the summit region. Such flows typically are radar-dark towards the point of origin, but have a higher radar backscatter distally, as if becoming slabby or rougher towards their terminations (Fig. 7.26).

There is also a strong association of shields with regional structural patterns. Thus many summit pits are located on graben which may also transect the summit area or have elongated lines of depressions running along their axes. These relations suggest that some faults may be the surface manifestation of dykes which supplied magma to the summit region. The most likely source for shield clusters are thermal anomalies in the Venusian mantle, whereas other, solitary, edifices could form anywhere along regionally-propagating dykes. Either way, the shields and the surrounding radar-dark plains, appear to have their origins in the same parent magma.

Most Venusian small volcanic *cones* are less than 15 km across; they tend to have a higher radar-backscatter than shield surfaces and higher flank slopes. They are believed

Fig. 7.25: The association of shield fields and fracture belts on the surface of Niobe Planitia. Magellan image F-MIDR 45N119E;1.

to have built up from cindery lava in much the same way as terrestrial cinder cones. *Domes* are larger, typically 20 – 30 km in diameter and up to 1 km high, that is, larger than comparable terrestrial examples. Many have either a central summit pit, or one placed asymmetrically. They generally are quite circular and have steep margins, characteristics consistent with their having been produced by eruption of magmas more viscous than basalt, i.e. with the effective viscosities of dacite and rhyolite (Fig. 7.27). The characteristically radar-bright returns from fractures and rough surfaces along their margins are consistent with furrowing and/or a degree of collapse of brecciated material at flowfronts. Guest and his colleages (1992) note that if such domes represented extrusion from a single eruptive episode on Earth, it would have given rise to extensive ash flow deposits such as ignimbrites. However, such activity may have been prevented by the dense Venusian atmosphere, with the result that massive domes were formed instead.

Some domes have deeply scalloped margins and an overall aspect that sets them apart from the rest, and from any known terrestrial counterparts. Such landforms – eighty have been identified – may have flat, convex or concave tops, and show all stages from steep margins with a single steeply-backed alcove, to domes whose margins are entirely scalloped (Fig. 7.28). Their morphology is consistent with their having experienced peripheral collapse by slope failure, a view which finds support in the frequent presence of debris

Fig. 7.26: Synoptic view of a part of Atla Regio showing a variety of intermediate-sized volcanic shields, many with summit depressions and associated radar-bright lobate volcanic flows. A number of long north-south graben cross the landscape; these clearly are younger than the volcanic deposits. Magellan image P-38281. Frame width 350 km.

aprons beyond their collapsed sectors. In one case, a breached skin with associated outflow of lava has been identified, presumably representing damage to the carapace by hot lava beneath (Guest *et al.*, 1992). Such collapse events are not uncommon on the Earth and usually are a response either to oversteepening or explosion at dome margins, which may set off relatively small-scale avalanches of rocks and ash. However, the scale of such events on Venus – they may involve as much as 200 km^3 of material – is orders of magnitude greater than terrestrial ones, perhaps because the elevated ambient temperatures experienced on Venus allow only relatively weak carapaces to form.

Volcanic lava domes may occur singly or in groups, like small shields. Frequently they are located within zones of fracturing. Groups of domes are common and there may be some overlapping of individuals. There is little doubt that domes represent the expression of relatively viscous lavas from central conduits that punctuate the plains. The high degree of circularity which typifies them implies that the plains surfaces over which they flowed must have been horizontal. McKenzie *et al.* (1992), after studying the form of such domes, argue that their morphology is consistent with the spreading of magma within the viscosity range 10^{14} and 10^{17} Pa s, which implies magma temperatures of 610 – 700°C in dry rhyolite magma. Such temperatures agree well with laboratory measurements of the solidus

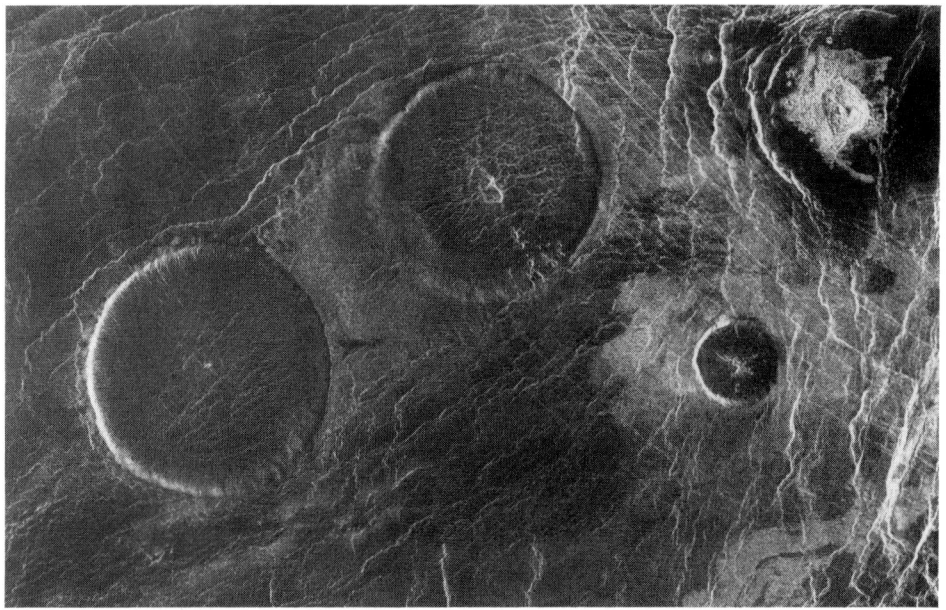

Fig. 7.27: Magellan image of three flat-topped "pancake" domes in Eistla Regio. The largest measures 65 km across and rises almost 1 km above the adjacent lava plains. Note the summit pits and the fracturing of the dome carapaces. The plains in this region are traversed by numerous wrinkle ridges. Magellan image P-38388. Frame width 250 km.

temperatures of wet rhyolite, from which it follows that dome growth would naturally follow surface eruption and degassing of viscous magmas generated by wet melting at depths of >10 km.

7.6   LUNAR VOLCANIC PLAINS

The basalt flooding which accompanied maria formation on the Moon is sufficiently distinctive in both its style and extent to warrant separate treatment. For this reason, the nature, distribution and genesis of lunar mare basalt and light plains units are discussed in Chapter 9.

7.7   VOLCANIC PLAINS ON MARS

The surface of Mars can be divided broadly into two units – densely cratered terrain and sparsely cratered plains – the boundary between which corresponds approximately to a great circle inclined at 28° to the equator (see Fig.7.29). Generally speaking the former is

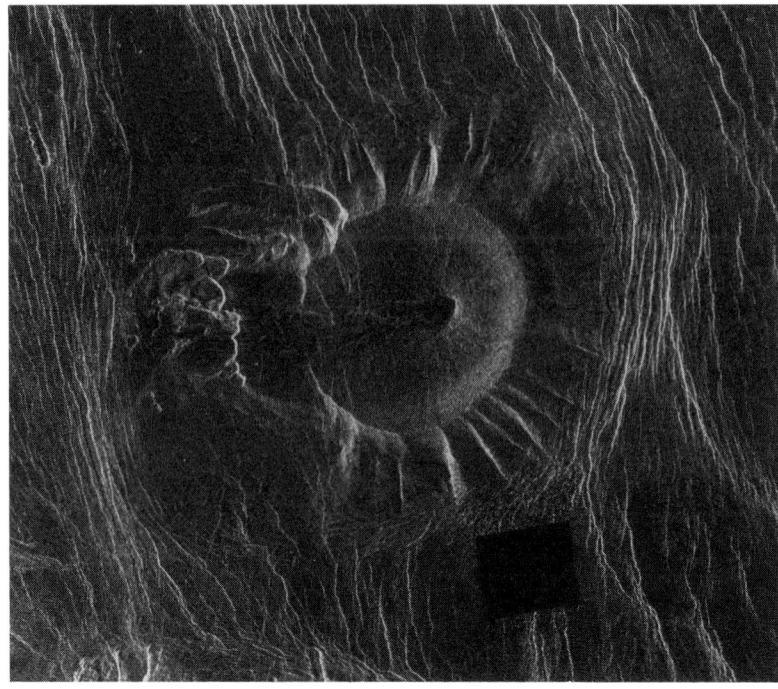

Fig. 7.28: Scalloped dome in Alpha Regio. This landform has a 5-km-diameter summit depression, a concave central area and a narrow rim enclosed by radiating spurs and troughs. Magellan image F-MIDR 20S003.

topographically higher, stratigraphically older and outcrops largely in the southern hemisphere. The latter lie below Mars datum and largely occupy the northern hemisphere, although a large area protrudes southwards between longitudes 80° and 140°W, at the southern edge of the Tharsis bulge. The distinction between the two is not, however, obvious everywhere. Thus in several locations the sparsely cratered plains overlap the densely cratered terrain, while elsewhere intercrater plains within the upland plateau are visually indistinguishable from the older plains of the sparsely cratered northern hemisphere. The more Viking images that one studies, the stronger becomes the realization that the Martian plains are excessively complex, almost defying classification due to their lateral diversity, however it also reveals evidence for widespread volcanic activity, even amongst the oldest terrains.

Approximately 60% of Mars is smooth on the kilometre scale and while positive identification of volcanic features such as flow lobes, sinuous rilles and low domes sometimes can be made, for large areas of plains topography direct evidence for a volcanic origin is lacking. This absence of evidence is partly a function of inadequate image resolution over large regions of the planet, but also is a manifestation of the mode by

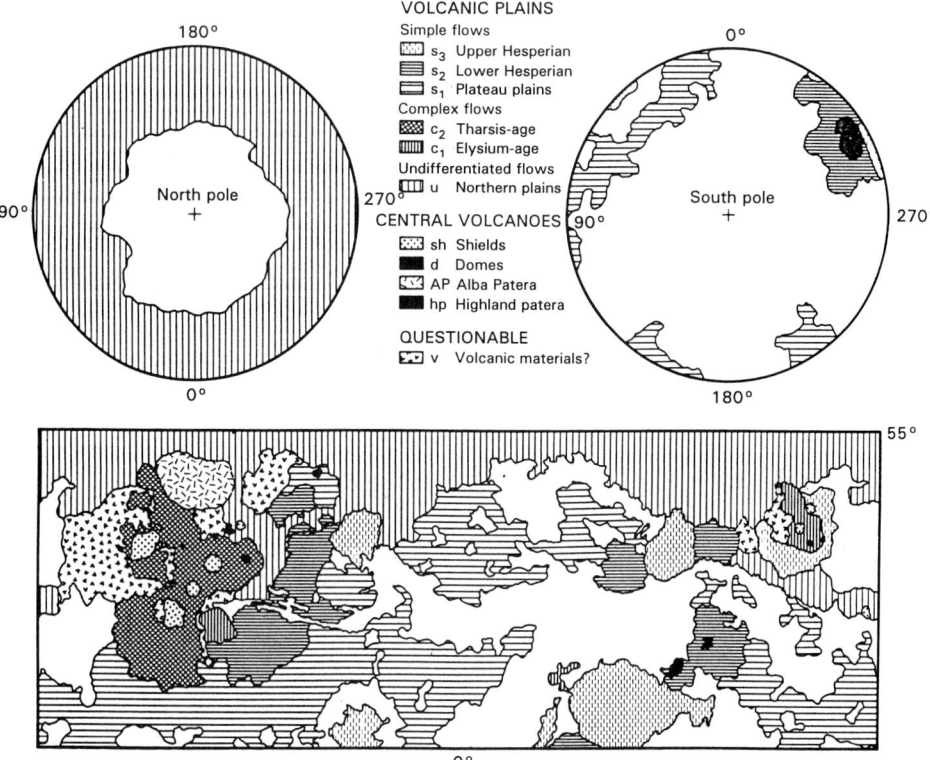

Fig. 7.29: Map showing the principal volcanic regions of Mars according to volcanic type. (After Greeley and Spudis, 1981).

which plains units were emplaced and their subsequent modification and/or burial. It is ironic that the very discovery which generated so much excitement amongst the planetary community – namely, that Mars had a very complex and un-Earthlike geological history – was the very one which renders the interpretation of many terrain units – in particular the plains – equivocal.

7.7.1   Noachian plateau plains – the cratering record

Plains units are widespread between the impact craters of the heavily cratered hemisphere. These *plateau plains* of the cratered plateau are amongst the more equivocal of the Martian plains units, a feature they share with the Mercurian intercrater plains. An important aspect of the cratered plateau is the general paucity of craters smaller than 5 km among the impact crater population produced during the early phase of intense bombardment and the significant modification of larger craters within the range 5-30 km. This has been

Fig. 7.30: Noachian-age plateau plains south of Elysium Planitia. Note the rather poorly-preserved channel networks and prominent ridges crossing the smooth, low-albedo floor of the large impact crater at centre left and its continuation on to the plains to the north. Frame width 315 km. V.O. frame 629A43, centred at 9.42°S, 226.89°W.

noted by various workers (Hartmann, 1973; Jones, 1974; Chapman and Jones, 1977) and has important ramifications when the evidence for volcanic activity during the earlier period of Martian history is assessed. It is almost certainly due to a period of intense crater obliteration which accompanied the decline in impact flux which occurred around 4×10^9 y ago. The population of fresh-looking smaller craters which now is observed on the cratered terrain is considered to have been produced largely subsequent to this obliteration phase. While obliteration undoubtedly was in part achieved by continuing impacts, there is morphological evidence that flood volcanism also occurred about this time and that it may have also played a role in the resurfacing process.

7.7.2   Plateau plains – morphological characteristics

In many places old plains partially submerge the heavily cratered plateau surface, such that only the rims of buried impact craters protrude. Elsewhere the intercrater surfaces may be cut by channel networks or may be ridged; ridges may also be seen traversing the flat floors of large impact craters which appear to have been flooded by lavas (Fig. 7.30). Spudis and Greeley (1978) estimate that the ancient plains have an areal extent of about 2.9×10^7 km², while about 36% of the ancient cratered surface is covered by intercrater plains with ridged surfaces (Greeley and Spudis, 1981). Extensive remnants of Noachian-age ridged plains are located eastwards from Noachis Terra, in Memnonia and the south part of Sirenum Terra (see Fig. 7.30).

Cratering studies suggest that they are among the oldest plains units on Mars, dating back to the Middle Noachian (Scott and Condit, 1977; Scott and Tanaka, 1986). The ridges are extremely similar to lunar wrinkle ridges and their development has been invoked by many workers to support the notion of a volcanic origin. As will be shown in the next section, while wrinkle ridges themselves do not necessarily have a volcanic origin – although many may – from lunar experience they do tend to form only in resilient rocks, like lava flows. At the very least, therefore, there is strong circumstantial evidence for the presence of extensive lava plains of Noachian age, since impact breccias or unconsolidated sediments would be insufficiently resilient to support ridges.

More specific evidence for Noachian volcanism comes from the plateau plains south of Protonilus Mensae between 300°W and 340°W, originally mapped from Mariner 9 images as impact breccias (Scott and Carr, 1978). While large areas of the plains are relatively rugged there are substantial regions which, on Viking imagery, are seen to have a distinctly smoothed appearance. This smoothness may in part be due to mantling by aeolian material but there is compelling evidence that volcanism too has played a significant role in their development. This derives from observations of numerous features that are difficult to interpret other than as volcanic flows, overlapping flow lobes and what appear either to be exhumed dykes or spatter ridges (Fig. 7.32). The same observation applies in lesser degree to other regions of similar units.

The principal evidence for volcanism on the cratered plateau comes from the presence of wrinkle ridges and the scattered occurrence of flow lobes and associated volcanic features. This can be augmented, however, by noting also that there is a close correlation between the occurrence of plateau plains and floor-fractured impact craters, which have been interpreted by Schultz (1977) as being impact craters modified by volcanic processes. The limited amount of information currently available appears therefore to strongly implicate volcanicity in the development of the plateau plains; such an episode would have taken place during middle to late Noachian times, that is, prior to 3.9×10^9 y ago (Greeley and Guest, 1987). This early volcanism provides a mechanism which considerably alleviates the crater extinction problem on the cratered plateau. As has been observed by Carr (1984), there is a strong likelihood that high rates of volcanism would have attended not only the brecciation and heating produced by intense bombardment, but also would be favoured by the enhanced radioactive decay and elevated rates of accretional energy dissipation typical of the period prior to 4×10^9 y ago. Early flood volcanism would have been a most effective resurfacing process and one which must have played a part in modifiying the cratering record in the southern hemisphere.

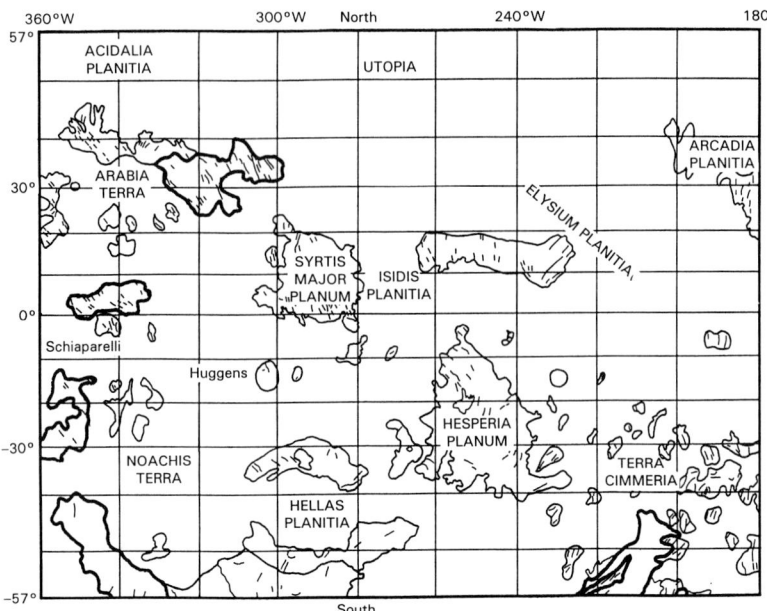

Fig. 7.31: (a) Distribution of ridged plains units in the western hemisphere of Mars.

7.7.3   Hesperian-age ridged plains

Ridged plains with a relatively high incidence of impact craters outcrop in the western hemisphere within a broad zone which attains a width of about 1000 km on the eastern flank of the Tharsis Bulge and covers an area approaching 4×10^6 km^2 (see Fig. 7.31). It extends from Tempe Terra at 30°N, 70°W, passing southward through Lunae Planum, then across Valles Marineris to include the eastern ends of both Sinai and Solis Planae. Similar plains also outcrop in the eastern hemisphere, the largest expanse occurring in Hesperia Planum, northeast of the basin of Hellas. Upland topographic depressions like Syrtis Major Planum and floors of the large impact basins Hellas, Argyre and Isidis, and the western part of Amazonis also show a development of such plains. The ridged plains of Hesperia Planum, Lunae Planum and Syrtis Major Planum are all of Lower Hesperian age, the greatest crater age derived being that of 3.6×10^9 y for Syrtis Major Planitia (Hartmann *et al.*, 1981).

Ridge development evidently was related to both regional and local tectonic regimes, since the trends of ridge segments are seldom random but typically aligned over wide areas, such as on Syrtis Major Planum where they trend predominantly northwest-southeast. The ridges themselves have the typical asymmetric cross-profile characteristic of lunar wrinkle ridges and are spaced between 20 and 70 km apart (Fig. 7.33). In addition to the ubiquitous ridges, flow fronts and lobes can be discerned widely in both Solis and Hesperia Planae, while on the western side of the latter province, smoother ridged and lobed plains

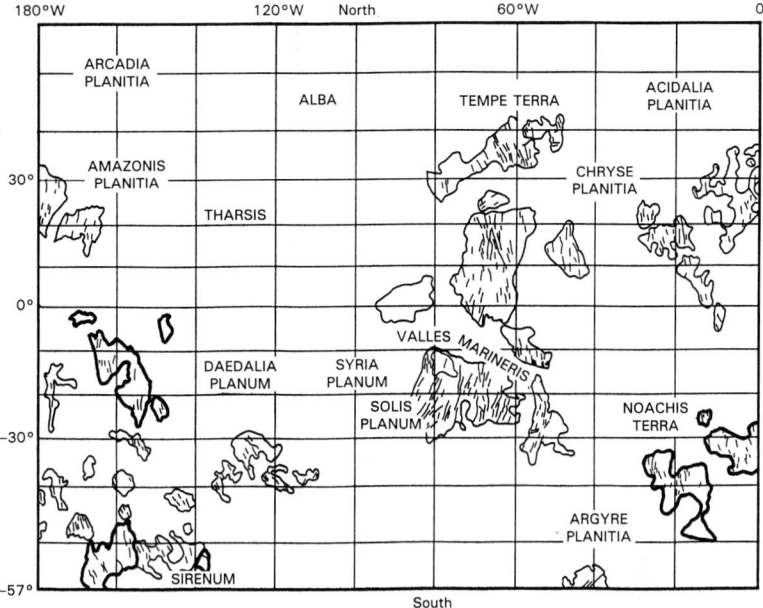

Fig. 7.31: (b) Distribution map of ridged plains units in the eastern hemisphere of Mars.

deposits are associated with the highland volcanoes, Hadriaca and Tyrrhena Paterae. A 400-km-broad concentric swath of ridged plains also occurs around the margins of Hellas.

Narrow flows, small cratered domes and what appear to be either exhumed dykes or spatter ridges are clearly discernible on the borders of Isidis (Fig. 7.34), where volcanic activity, which extends westwards into Syrtis Major Planum, post-dated the impact which produced the Isidis Basin. Within Syrtis Major Planum itself there is convincing morphological evidence for the existence of two very low volcanic shields (Schaber, 1982) whose products are superimposed on the ridged plains unit (Figure 7.35). Crater statistics suggest the flows may have an age of around 2.6×10^9 y (see Meyer and Grolier, 1977) while spectral reflectance data for the region are in accord with a composition of oxidized basalt (Adams and McCord, 1969).

Less heavily cratered (Upper Hesperian) ridged plains outcrop in Chryse Planitia and south of the Elysium bulge. The landing of Viking 2 in the former locality indicated the presence of what appear to be vesicular basaltic blocks on the surface of the plain and confirmed there were basalt weathering products in the surface soils. While this does not prove a volcanic origin for the plain, it is strong circumstantial evidence in support of it and of its surfacing by mafic lavas.

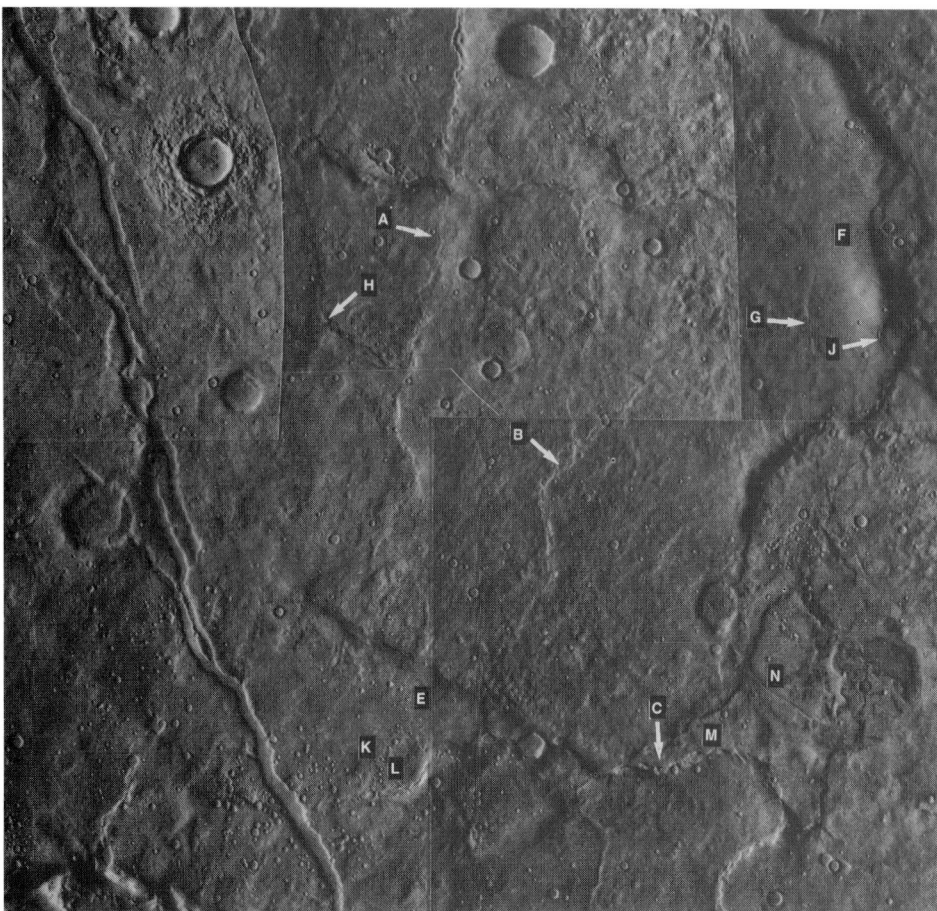

Fig. 7.32: Noachian-age smooth facies ridged plains northeast of Arabia (centred at 34°N, 311°W). Note the prominent flow scarps (A, B and C) and wrinkle ridges at E and F. What appear to be narrow volcanic flows outcrop at G and H, while a lava channel and associated small flow sits on the ridge crest at J. Linear ridge segments at K, L, M and N may represent spatter ridges or exhumed dikes. Frame width 100 km. V.O. frames 641A02-07.

7.7.4   Significance of plains ridges

While the ridges have the general appearance of lunar wrinkle ridges, their development does not necessarily prove a volcanic origin for the plains, proof relying upon the presence of other diagnostic landforms. There is a substantial volume of literature which supports origin of lunar wrinkle ridges by both volcanic and tectonic processes (Howard and Muehlberger, 1973; Luchitta, 1977; Sharpton and Head, 1982; Solomon and Head, 1979). Lunar ridges usually have a preferred orientation both concentric and radial to the basins the mare lavas fill. A concentric arrangement also is clearly evident for the ridges on the eastern flank of the Tharsis Bulge. This leads to the inevitable conclusion that ridge

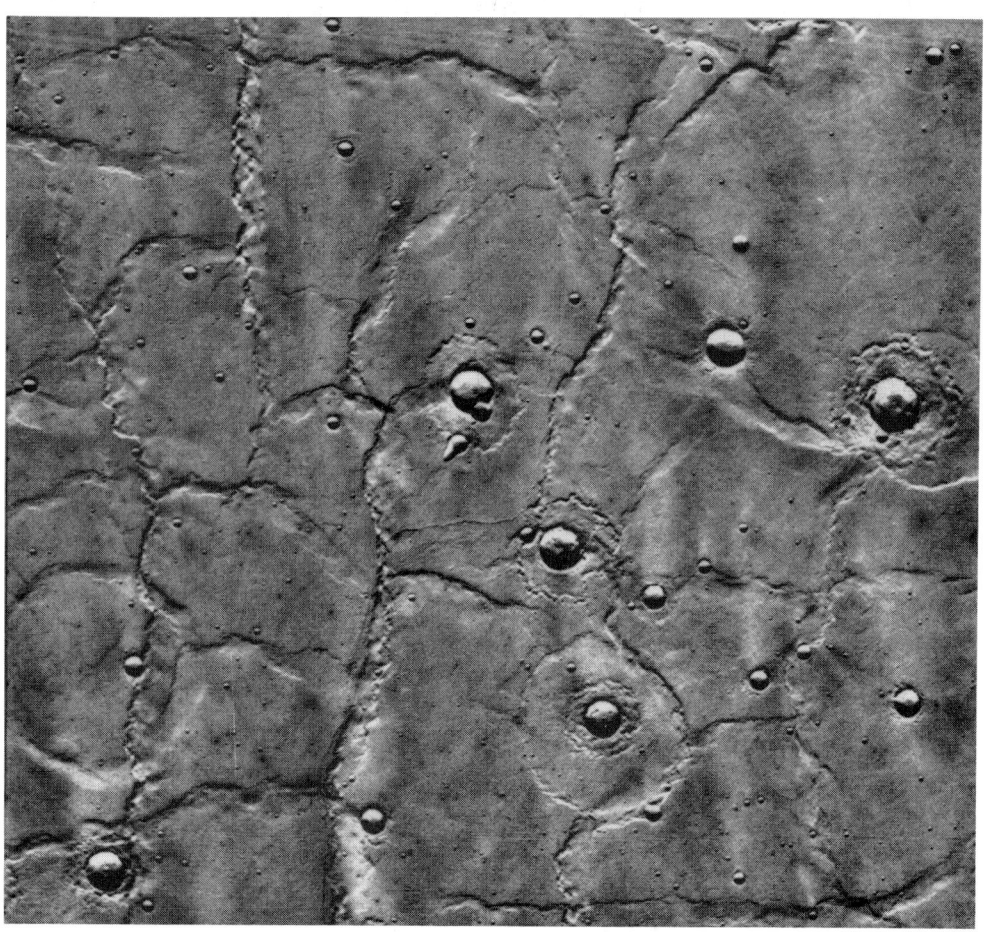

Fig. 7.33: Wrinkle ridges crossing the plains near Hesperia Planum. Note the asymmetric profile of the ridges and their rather sinuous courses. Frame width 185 km. V.O. frame 417S05.

Fig. 7.34: Plains units on the borders of Isidis, showing possible volcanic flows (A and B) and spatter ridges or exhumed *en echelon* dykes (C and D). Frame width 250 km, centred at 4.97°N, 268.12°W. V.O. frame 067B63.

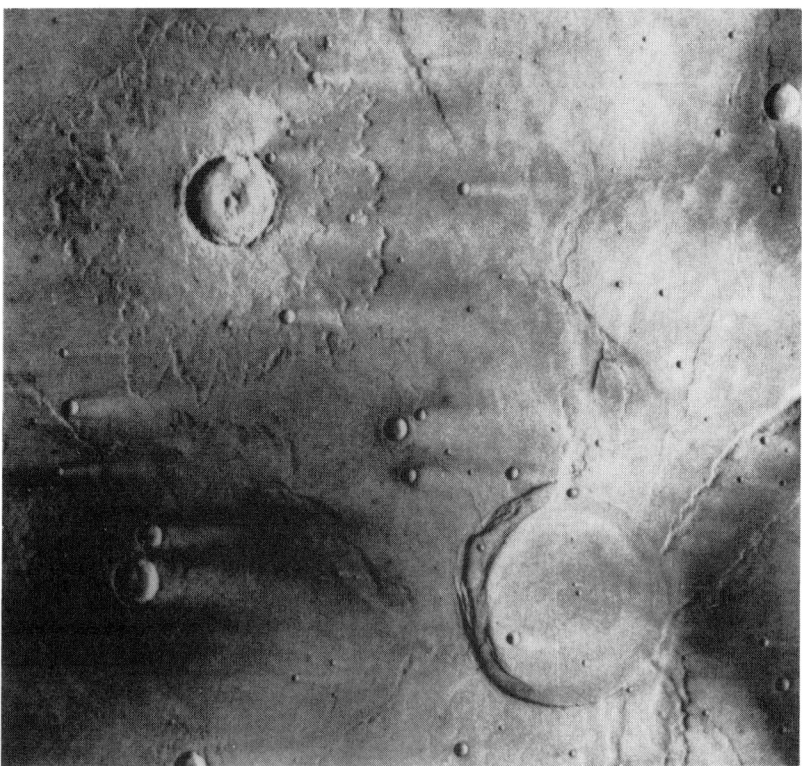

Fig. 7.35: 50-km-diameter caldera structure on the ridged plains of Syrtis major Planitia. It has been suggested that this lies at the summit of an extremely low volcanic shield. Frame width 200 km. V.O. frame 372S32.

formation here must, in some measure, have been a function of regional deformation. The contentious issue, however, is whether the ridges were generated by post-emplacement deformation or whether they grew contemporaneously with lava emplacement, as a result of the upwelling of lava onto the surface, or some kind of thin-skin tectonics on cooling lake lavas.

Ridge development on the plains east of the Tharsis Bulge can be shown to have been consequent upon the presence of the Bulge and not coeval with its growth. This fact emerged after geophysical modelling by Phillips and Ivins (1979) and Phillips *et al.* (1981), who studied the stress distribution inside the planet resultant upon the Tharsis loading. Their plots of stress distribution show a clear-cut orthogonal relationship between the direction of principal stress and the alignment of wrinkle ridges. The implication is, therefore, that these ridges have a tectonic origin. Notwithstanding this, there is one significant characteristic of lunar wrinkle ridges, shared with their Martian counterparts, that throws some light on their significance. On lunar maria the ridges have a two-

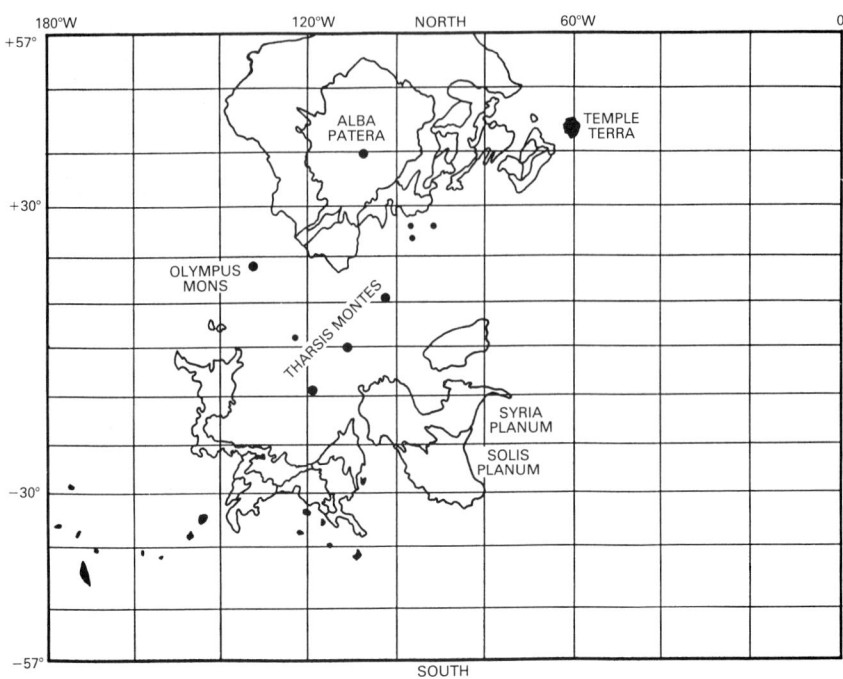

Fig. 7.36: (a) Outcrop of Hesperian-age flow plains in the western hemisphere of Mars.

component morphology, comprising a broad arch up to 10 km across and a superimposed narrower summit ridge, usually offset into en echelon segments. When they pass onto highland terrain, however, they continue as simple scarps and it appears that wrinkle ridge development on the mare surfaces is a reflection of the relative competency of the mare lavas compared with the brecciated rocks of the highlands. By analogy with the Moon, while the Martian ridges may have been produced largely by tectonic deformation, the plains units they corrugate were evidently composed of materials more competent than impact breccias. Almost by default, therefore, one is inclined towards the view that they are composed of competent lava flows and, furthermore, probably high-volume flood lavas which left few discernible traces of either vents or flow features over large regions. Such a conclusion is borne out by such photogeological evidence as is forthcoming, indicating a continuation of the flood volcanism which began earlier and which left its imprint over the more highly cratered plateau.

7.7.5   Hesperian-age flow plains

The principal occurrence of Hesperian-age *flow plains* is peripheral to the major volcanic provinces of Tharsis and on the flanks of the Elysium shield volcanoes, but there is also a major outcrop of somewhat equivocal volcanic deposits in Malea Planum which may be

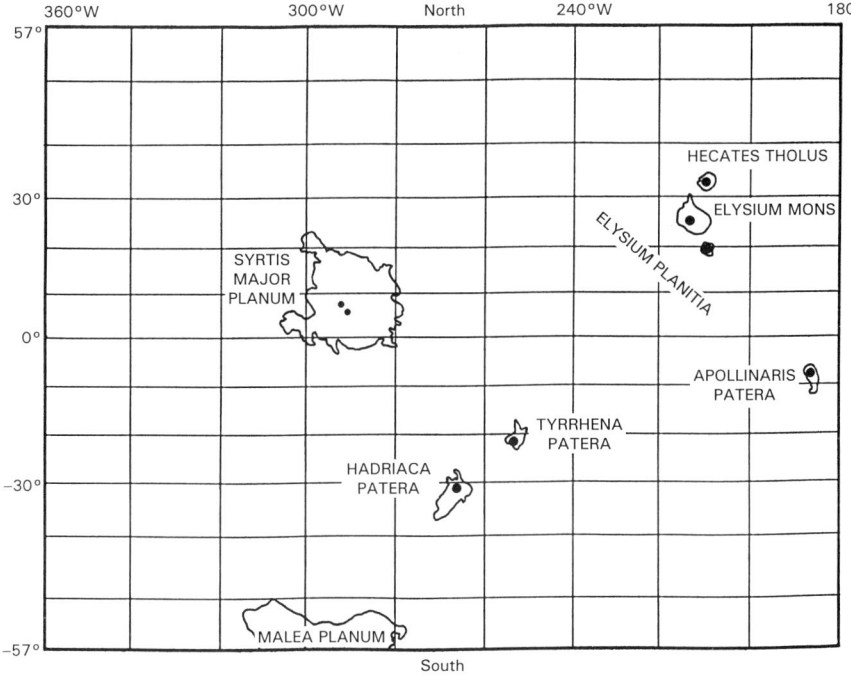

Fig. 7.36: (b) Outcrop of flow plains in the eastern hemisphere of Mars.

fluvially modified volcanic flows (Fig. 7.36). The oldest of these lavas outcrop around Tempe Terra, Memnonia and Ceraunius Fossae. Their eruption marked the first of several major volcanic episodes which resurfaced huge areas of the northern lowlands. Extensive flow plains also were erupted from near the crest of the Syria Rise and now cover large areas of Syria and Solis Planae at the western end of Valles Marineris. Particularly extensive lobate flows of mid- to late-Hesperian age are found east and northeast of the base of Olympus Mons shield and also on the western side of Tempe Terra.

The most extensive series of flows emanates from beneath the low pile of Alba Patera, where a sequence of broad, flat-topped sheet flows extends at least 1500 km from its eruptive focus near the summit; these often coalesce to form long terminal scarps (Cattermole, 1989). Other plains composed of similar flows appear to have originated from foci now situated beneath younger centres like Arsia Mons and Uranius Patera (Fig. 7.37), where occasionally aligned rimless depressions and discontinuous spatter-type ridges indicate the alignment of what are assumed to be linear source vents or fissures, most of which have been buried by the flows themselves or younger ones. Similar flood-type lavas of late-Hesperian to early-Amazonian age occur around Ceraunius Fossae, where they flood fractures incised into older highland terrain.

The flows themselves typically are between 60 and 120 m thick, often composite, and have relatively featureless flow surfaces. Many flows can be traced for hundreds of

Fig. 7.37: Hesperian-age sheet flows northeast of Uranius Patera. The prominent flow (A) has a width of 20 km and is about 70 m thick. More degraded flows outcrop to the east. Note the north/south-trending ridge elements (r) which are similar to supposed spatter ridges seen elsewhere in northern Tharsis and probably are the sites of feeder fissures for some flows. Frame centred at 27.59°N, 88.71°W. Frame width 58 km. V.O. frame 626A66.

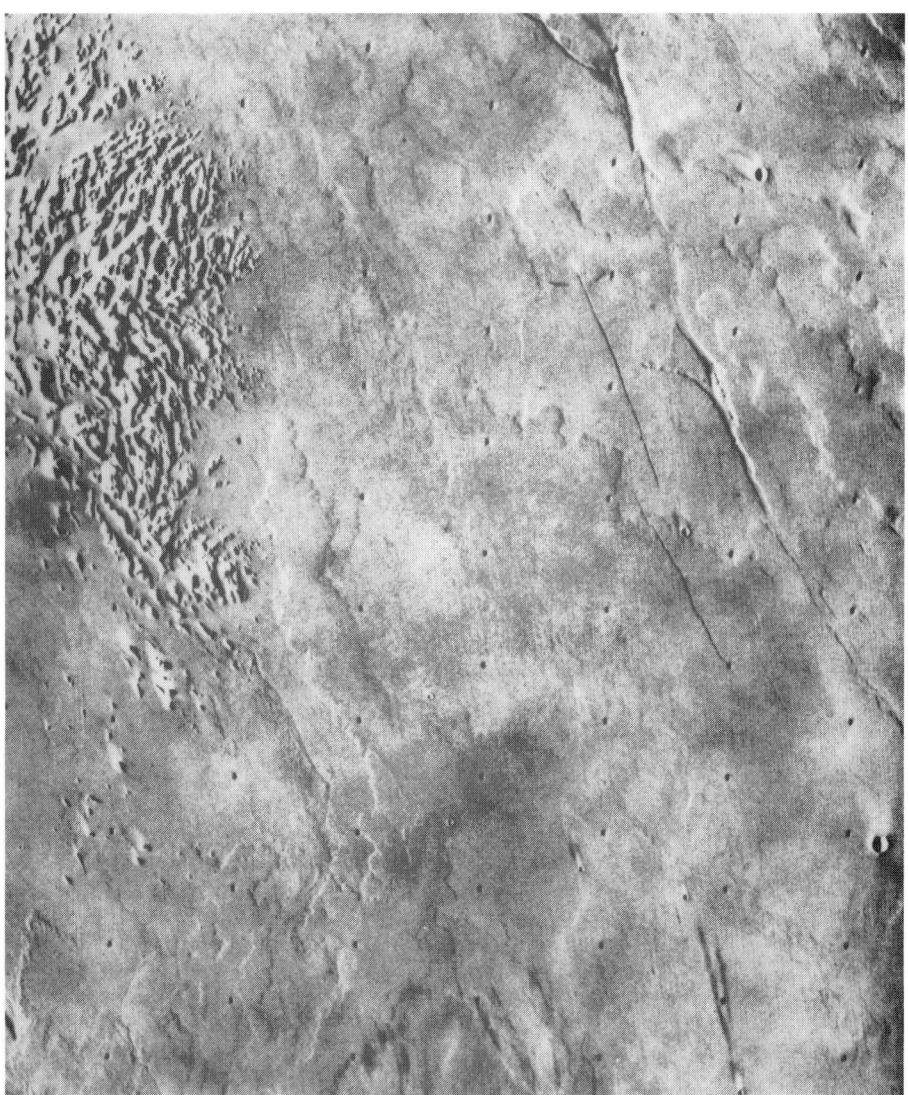

Fig. 7.38: Extensive flood lavas on the Hesperian-age plains on the southern flanks of Tharsis. Frame width 230 km. Centred at 31.53°N, 130.73°W. V.O. frame 056A14.

kilometres and coalesce to form broad, overlapping sheets (Fig. 7.38). Individual flows have volumes in excess of 400 km³. In several places flow channels can be discerned and there are numerous low domes and small depressions, presumably the sources of some flows. By and large, however, the sources of such lavas are obscured by younger flows and it can only be assumed that they issued from fractures or linear vents which were buried by their own products.

7.7.6 Tempe Terra volcanic province

While there is a general lack of specific evidence regarding the sources of the very extensive Hesperian flow-plains lavas, this is not the case with those in the Tempe Volcanic Province. This interesting volcanic complex developed on the uplands which form a northwestward continuation of Lunae Planum, on the opposite side of Kasei Vallis. The volcanic sequence occupies an area of 3.4×10^6 km² (see Fig. 7.36) and comprises three distinct kinds of terrain: (i) rugged hilly terrain, (ii) faulted terrain, and (iii) smoother uplands which are an extension of the Lunae Planum plateau (Scott, 1982).

On its western side Tempe Terra is embayed and partially overlapped by sheet flows associated with Alba Patera and the Tharsis Montes. On the surface of the plateau itself, however, is a suprising variety of smaller volcanic constructs, several major volcano-tectonic structures with little or no relief, and steeper-sided volcanic mountains, which may be dissected shield volcanoes (see Chapter 10). Details of several of these structures have been given by Underwood and Trask (1978), Scott and Carr (1978), Wise (1979), Plescia (1981) and Scott (1982). The major ring structures are somewhat similar to Alba Patera, though with even less vertical expression; the largest is 250 km across. At the focus of one such structure is a 20-km-diameter dome with which are associated several long flows.

Of greatest interest to the discussion of volcanic plains, however, are the smaller volcanic features, in particular the substantial number of low shields which have been built on the resurfaced parts of the fractured plateau. These bear a striking resemblance to those which developed during the Plains-type volcanism of the Snake River Plains and, like their terrestrial counterparts, have well-defined summit pits, often aligned in the direction of the southwest-northeast fracturing, elongate vents and thin sheet flows, some of which appear to have originated in fissures (Fig. 7.39). Plescia (1981) estimates that low shields account for about 75% of the constructional landforms present. Their obvious development along rift faults, as well as the presence of fissure-fed sheet flows and elongate depressions, reveals, as it does in the Snake River Plain, a close genetic relationship between volcanism and crustal extension. What appears to be absent from the Tempe plains are tube-fed flows, so common a feature in the Snake River province; however, it is possible that such flows were emplaced but that their roofs have not suffered collapse.

In addition to the low shields, there are several steeper-sided constructs with diameters in the range 5-10 km. These are probably composite cones and have no discernible lava flows associated with them (Fig. 7.39). Such features are not found in the Snake River Plains and their occurrence in Tempe may imply a more extended range of volcanic style, involving more viscous, silicic magmas than are common in the Snake River province. Supporting evidence for such a hypothesis is provided by a large dissected patera structure

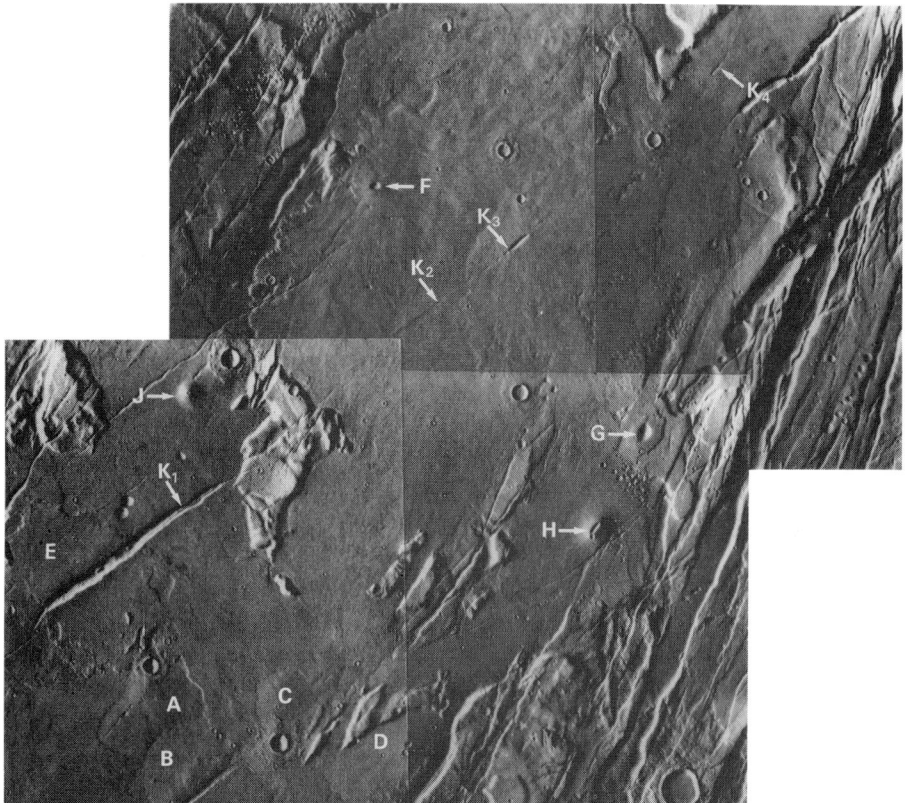

Fig. 7.39: Plains-style volcanic features in the Tempe Terra volcanic province. Several low shields (A, B, C and D) may be seen at lower left. One of these (A) has a row of aligned depressions along its crest, while D has a single ovoid vent. A prominent sheet flow (E) crosses the area and evidently predates the southwest-northeast grabens. Two further elongate depressions sit atop low domes at F and G, while two steeper-sided cones, both with summit pits, occur at H and J. Running diagonally across the mosaic, parallel to the southwest-northeast faults is a prominent volcano-tectonic feature which comprises a rift fault (K1), fissure vent with short radiating flows (K2) and two low shields with elongate pits (K3 and K4). Frame width 270 km. V.O. frames 627A26, 27, 29 and 41.

located in the northeast of the plateau which, by analogy with highland paterae elsewhere on Mars, may have developed during a phase of explosive volcanism that produced ash-flow and air-fall pyroclastic deposits (see Chapter 10).

7.7.7   Volcanic plains of Amazonis, Memnonia and Aeolis

Rather different Plains-style volcanic deposits are encountered in the triangular region which has the central constructs Apollinaris Patera, Biblis Patera and Olympus Mons at its apices (Fig. 7.40(a)). They take the form of a series of discontinuous, flat-lying sheets,

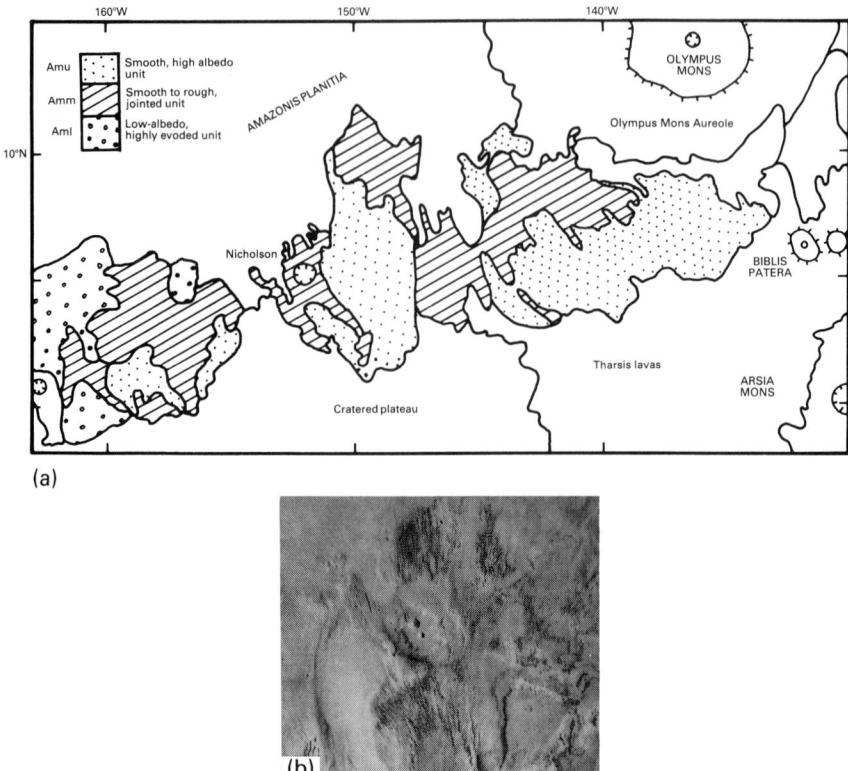

(a)

(b)

Fig. 7.40: (a) Distribution map for volcanic deposits in the Amazonis-Aeolis region. (Modified from D. H. Scott and K. H. Tanaka (1986) *USGS Map I-1802A*.) (b) Diverse morphology of three volcanic units near 5°S, 175°W. The yardangs developed are believed to have been etched in welded tuffs, while the non-welded units have a smoother surface. Frame width 320 km. V.O. frame 635A84.

individually 100 m or so thick, which have smooth, gently undulating or etched surfaces. Some of them show a development of yardangs. They outcrop in a broad but discontinuous zone that strikes east-west along the lowlands-highlands boundary. The morphological signature of these sheets was first recognized as bearing a striking resemblance to terrestrial ignimbrites by Malin (1979), who estimated their aggregate thickness to be around 3 km. Subsequently, Scott and Tanaka (1982) mapped the region in greater detail.

The detailed study of Scott and Tanaka revealed the deposits to cover an area of around 2.2×10^6 km^2 and to have an estimated minimum volume of 3.85×10^6 km^3. Mapping allowed recognition of seven units ranging in age from Lower to Upper Amazonian. The substantial thickening of the sheets at four specific locations suggests that these may have been eruptive foci. Based on work among silicic volcanics in the Pancake Range of central Nevada (Scott, 1969), they note the following features which find analogues in terrestrial outcrops: (i) rounded patches of smooth, high-albedo (non-welded) materials that overlie low-albedo, jointed (welded) flows; (ii) local complementary joint sets in some (welded) materials; and (iii) thick flow sheets of great lateral extent that pursue but subdue the underlying topography (Fig. 7.40(b)).

Naturally nothing is known of the grain size or internal fabric of the Martian deposits; thus Scott and Tanaka's hypothesis must remain just that. However the circumstantial evidence for a pyroclastic origin is considerable. One interesting feature of the Amazonis region is the general absence of volcanic domes, which are common in terrestrial silicic igneous provinces. Theoretical considerations indicate that there should be wider dispersion of pyroclasts on Mars for the same mass eruption rate (Head and Wilson, 1981); therefore the absence of domes may be a function of the lower expected relief and an inability of Viking images to resolve very low features.

In the Basin and Range region of the western USA ignimbrites are closely associated with block faulting where highly volatile silicic magmas were available for energetic eruption. The high concentration of northwest/southeast faults and graben within the Amazonis-Aeolis region, for instance Medusae Fossae, shows that there was indeed extension in the Martian crust and it is possible that such faulting could have produced roof failure in a large magma chamber located beneath the deposits, producing extensive ash-flow deposits. Calculations by King and Riehle (1974) suggest that if Martian ash flows were generated, they would remain in a fluidized condition for between 3 and 9 times the period typical of the Earth, giving scope for very extensive flat-lying sheets which could travel great distances from their source.

7.7.8   Amazonian volcanic plains

Most of the younger, Amazonian, flood lavas appear to be associated with the massive shield volcanoes of both Tharsis and Elysium; therefore their consideration will be left until this topic is discussed in Chapter 10. However it must be noted here that while they are closely related to central volcanoes, they extend outwards for distances quite unheard-of in the terrestrial environment. For instance, Olympus Mons flows extend for more than 400 km from the base of the shield, while Arsia Mons lavas can be traced back more than 600 km to their source.

7.7.9   The northern plains of Mars

The character of plains units in higher latitudes (further than 40°N) is very different from those in lower latitudes. These plains are poorly understood and appear to be blanketed in aeolian debris which softens the topography over vast areas. Little or no direct evidence is found for primary volcanic deposits, although such may well be present beneath the sedimentary cover. All that can be noted is that in several northern locations – for instance northwest of Alba Patera and north of Olympus Mons – flows originating in these shields can be discerned becoming more and more obscure as they extend onto the Arcadian plains north of latitude 45°N. Some of these could be pyroclastic in origin.

7.7.10   Summary

Volcanic plains cover extensive regions of all of the terrestrial planets. Those of Mars, in particular, show analogous features to those observed in terrestrial locations and are presumed to have been emplaced under similar structural conditions. Recent radar imaging of Venus reveals characteristics over wide regions of that planet that are broadly similar to terrestrial and Martian flow plains, including their association with tectonic features; however Venusian plains are also strongly deformed in ridge-and-fracture belts, and punctuated by unique volcano-tectonic features called coronae. Mercurian plains remain somewhat equivocal. All planetary flood lava provinces so far identified are closely associated with crustal extension. Such geochemical data as exists indicates a basaltic composition for volcanic plains provinces over very large areas of all the terrstrial planets. On Earth there has been considerable chemical fractionation, with more silicic lavas contributing to plains units, for instance, in rift valley regions. The same may be true of Venus, but does not appear to have affected Mars.

8

Lunar volcanism

Prior to the landing of Apollo probes on the Moon's surface, the lunar highlands, or *terrae*, were believed to be largely volcanic; however, less than 25% of the returned highland samples show textures reminiscent of basaltic lavas (Trask and McCauley, 1972). Highland terrain is, in fact, dominated by impact breccias, most of which were ejected from a number of very large impact basins but also, of course, from the thousands of impact craters which are a feature of the lunar landscape. In contrast, the maria – which account for about 16% of the total area (30% of the nearside and 2% of the farside) – clearly are volcanic and are believed to have risen as partial melts from within the Moon's mantle. Most highland rocks which do have igneous textures are considered to be impact melt rocks, but several specimens, particularly from the Apollo 15 and 17 collections, do appear to be fragments of endogenic volcanics and, as such, are representative of lavas which were erupted prior to the main period of mare flooding. These rocks thus provide clues about a very significant period in the Moon's volcanic history but a part about which our knowledge is very fragmentary, like the samples themselves.

Radiometric dating of returned Apollo samples gives an upper age of over 4.5×10^9 years for the Moon's ancient highland crust. It also indicates quite clearly that the basins excavated from this material are of different ages ($3.92\text{-}3.85 \times 10^9$ years), and also that a hiatus separated the production of the highland basins and the volcanic extrusions that produced the voluminous basaltic flood lavas which constitute the *maria*. Then, again, the mare lavas themselves are not all of the same age ($3.84\text{-}3.17 \times 10^9$ years), their emplacement spanning a period of time greater than that of the Earth's Ordovician.

The sequence of major ejecta sheets associated with each large impact basin has been used as the basis for a lunar stratigraphic column which has been constrained by a number of absolute ages. This lunar time-scale is reproduced as Table 8.1. in order that the ensuing discussions may be followed.

8.1 HIGHLAND VOLCANISM

The broad geomorphological features of the highlands were produced prior to about 3.85×10^9 years ago during a phase of basin and crater formation which saw the

Table 8.1: The lunar stratigraphic column. Time-stratigraphic units contain all those rock-stratigraphic units that developed on the Moon during a given time period. (After Wilhelms, 1984.)

Time-stratigraphic unit	Rock-stratigraphic units	Estimated duration (years)	Absolute ages (years)
Copernican System	Crater materials and minor mare materials	1.0×10^9	$\leqslant 0.05 \times 10^9$ 0.81×10^9
Eratosthenian System	Mare materials and crater materials	2.2×10^9	3.17×10^9
Upper Imbrian Series	Mare and crater materials	0.6×10^9	$3.26\text{-}3.79 \times 10^9$
Lower Imbrian Series	Orientale basin materials Schrodinger basin materials Imbrium basin materials	0.05×10^9	$3.84\text{-}3.85 \times 10^9$
Nectarian System	Crater and basin materials Nectaris basin materials	0.07×10^9	$3.84\text{-}3.92 \times 10^9$
Pre-Nectarian System	Basin and crater materials (represented by clasts)	0.63×10^9	$4.17\text{-}4.54 \times 10^9$

accompanying widespread ejection of basin-related breccia blankets. Once Apollo samples were returned, it quickly became clear that the major components of the lunar highlands were materials composed of three igneous rock types: anorthosite, norite and troctolite which resulted in its being known by the acronym, ANT. Of the three components, anorthosite is by far the most abundant, on account of which the predominant highland rocks have a characteristic Al-rich geochemical signature. The plutonic rocks which constitute the terra materials generally are considered to have evolved from an early magma ocean by a process of plagioclase flotation (Walker and Hays, 1977). Only a few samples of pristine highland volcanic rocks were found (Norman and Ryder, 1979; Warren and Wasson, 1978); typically potential highland volcanics occur as clasts in impact breccias and are highly shocked or recrystallized.

8.1.1 KREEP

While ANT dominated the returned sample collections, another type of material was found to be widespread among the highland samples. First found as a few small glassy fragments amongst the Apollo 11 breccias at Tranquillity Base, then, more abundantly, at the Apollo 12 site, this acquired the tag, KREEP, an acronym standing for potassium (K), rare earth elements (REE) and phosphorus (P), in which it was significantly enriched. The major element composition of this material has never been well-defined, although it can be said to be intermediate in composition between ANT and mare basalt with respect to Al, and the term really applies to an assemblage of minor and trace elements present in characteristic proportions (Table 8.2).

KREEP was subsequently collected in substantial amounts where Apollo 14 landed

Table 8.2: Major and selected trace element composition of terra anorthosite, KREEP and mare basalt. (Major elements in wt %; trace elements in p.p.m.)

Element	anorthosite	KREEP low-high K	mare basalt
SiO_2	44.1	47-53	37-49
Al_2O_3	35.5	23-16	7-14
CaO	19.7	12-10	8-12
FeO	0.2	10	18-23
MgO	0.1	11- 6	6-17
K	–	0.12-0.54	0.02-0.29
Th	–	5.3-12	0.05-3.4
U	–	1.37-3.2	0.02-0.30
Zr	0.5	480-930	22-560
Total REE	2.9	247-574	20-500

near Fra Mauro, and at both the Apollo 15 and 17 sites. Following the wider sampling, Low- Intermediate- and High-K varieties of this material were identified. The trace elements abundantly present in KREEP do not easily enter sites within those silicate minerals typical of lunar rocks, therefore they would tend either to be concentrated in the first partial melts or in the last residual liquids of fractionating magmatic systems (Taylor, 1975; Irving, 1977; Meyer, 1977). KREEP therefore represents the most abundant highly differentiated terra material that has been returned to Earth.

In addition to those elements referred to above, the large ion lithophile (LIL) elements Ba, Rb, Th, U and Zr are present in enhanced proportions compared to the other terra rocks. Elevated concentrations of the trace elements U, Th and ^{40}K make KREEP far more radioactive than ANT, so much so that it was detected from orbit by the Apollo 15 gamma-ray spectrometer as it passed over Mare Imbrium and Oceanus Procellarum. Elevated levels of radioactivity have also be recorded in the extensive ejecta blankets of large craters such as Copernicus,

Aristarchus and Archimedes which are located within the Procellarum-Imbrium basins, as well as in the Fra Mauro Formation, now known to be a massive ejecta sheet emplaced when the Imbrium basin was excavated. The widespread distribution of radioactive KREEP materials within this region has been used to confirm the existence of a larger, pre-Imbrium basin (the Procellarum Basin) in which were emplaced KREEP-rich volcanic materials prior to the Imbrium excavation (Cadogan, 1981; Whitaker, 1981).

8.1.2   Fra Mauro Basalts

One of the main purposes of the Apollo 14 mission was to sample the Fra Mauro Formation, known to be a massive ejecta sheet associated with the excavation of Mare Imbrium (Fig. 8.1). By doing so not only would it be possible to constrain the age of the Imbrium event but also the composition of the ejected material, presumed representative of the Moon's deep subcrust. After analysis of the returned samples an age of 3.85×10^9 years was established for the Imbrium impact. The Apollo 14 sample collection was also found to be

Fig. 8.1: Oblique view of Fra Mauro (centre of frame), with Bonpland (bottom left) and Parry (bottom right). The strongly textured Fra Mauro Formation can be seen extending in a north-south direction to the west of Fra Mauro itself (arrows) and to the east of Mare Cognitum. Frame width 140 km. Apollo 16 mapping camera frame 1419.

extremely enriched in KREEP materials, more so than any other previously sampled formation. Several KREEP-enriched samples acquired the name "Fra Mauro Basalts", the implication being that they represented truly volcanic rocks. Such a conclusion was not universally accepted, however, and the abundant occurrence of KREEP in the Imbrium ejecta was variously interpreted to imply either that a KREEP-rich layer deep in the lunar crust was excavated by the impact (Ryder and Wood, 1977), or that volcanic KREEP lavas were re-ejected by the Imbrium event (Wood, 1972; Cadogan, 1981), or that the distribution of KREEP is local and laterally variable (Wetherill, 1981).

Distinguishing between impact-melt rocks and endogenous volcanic rocks is not an easy task, because highland samples usually result from the mixing of impact-induced melt with relatively cold shock-generated clasts. Studies of these rocks suggest they may have experienced a two-stage cooling process (Simonds, 1975). Thus during an impact event, impact melt and clastic material is mixed together and emplaced as a mixed clast/ melt sheet. Initially there is very rapid heat loss via thermal transfer from the superheated impact melt to the enclosed, relatively cold, clastic fragments. Subsequently, however, cooling gets much slower and thermal loss is via conduction and radiation, as in a typical volcanic flow. The first stage is over extremely quickly, but the second stage proceeds more slowly, and the rate of cooling is dependent on the position of material within the melt sheet. The material of thin sheets or that near the margins of thicker ones would be expected to retain impact-shocked clasts which are diagnostic of exogenic origin. However, material within thick sheets may develop quite coarse crystals and exhibit intersertal, ophitic or poikilitic textures such as are found in typical basaltic volcanic rocks. Furthermore the longer cooling period typical of these "interior" rocks will allow for more complete digestion of impact-shocked clasts and thus remove important textural evidence for their impact origins. Such rocks are virtually impossible to distinguish from endogenic volcanics.

The KREEP-rich Fra Mauro Formation sampled at the Apollo 14 site is believed to be enriched in impact melt rocks that were melted in the Imbrium target zone and then transported to their present sites among the ejecta (Wilhelms, 1984). These melt rocks give a cluster of younger ages between 3.79 and 3.85×10^9 years, based on both the ^{40}Ar-^{39}Ar method (Turner, 1977) and the Rb-Sr method (Papanastassiou and Wasserburg, 1971) However, it also contains pristine pre-Imbrium volcanics which are among the best dated of the older samples from the Apollo 14 mission. There are several fragments of mare-type basalt and these have ^{40}Ar-^{39}Ar and Rb-Sr ages of between 3.85 and 3.98×10^9 years (Ridley, 1975; Ryder and Spudis, 1980). Not only do the latter group differ in age but they also differ in composition from the former. They are less aluminous than the KREEP-rich impact melt rocks (though more so than the mare basalts) and have lower initial Sr ratios (Papanastassiou and Wasserburg, 1971). The obvious interpretation is that these fragments represent clasts of pre-Imbrium mare-type basalts which subsequently became incorporated in the bedrock breccias before the clasts were redistributed at the Apollo 14 (Cone Crater) site. There are thus at least two distinct groups of pre-mare basalt volcanics represented in the basin ejecta: *aluminous mare basalts* (11-14% Al_2O_3) and *KREEP-rich basalts*.

8.1.3   The KREEP-rich Apollo 15 rocks

Apollo 15, which put down in Palus Putredinis to the southeast of Mare Imbrium, was intended to sample massif material from the Imbrium basin. Despite landing near to the base of the Montes Apenninus, little of their material was collected; rather samples of regolith or debris from colluvial aprons of talus-type material were recovered, whose exact sources were not always clear. However, among the sample collection were the first truly unadulterated, crystallized KREEP samples. Collected as small clasts, these samples showed laths of calcic plagioclase and more tabular orthopyroxenes set in interstitial brown glass (Fig. 8.2). Chemical analyses revealed that these rocks also were free from meteoritic siderophiles, an important feature which, together with the unequivocal igneous textures, led many investigators to conclude they were truly endogenic rocks (See BVSP, p. 274-278).

Some of these crystallized KREEP clasts are thought to have originated in the Apennine Bench Formation (Fig. 8.3), a light-coloured interior plains deposit to the west of the

Fig. 8.2: Photomicrograph of crystallized KREEP, showing laths of calcic plagioclase and tabular orthopyroxene set in a brown glass. (NASA photograph.)

Apollo 15 landing site which, from orbital geochemistry, was revealed to be a region of elevated Th content consistent with a major KREEP component. This unit is clearly older than the mare basalts and the ejecta from Archimedes, but is younger than parts of the Apennine Front, known to be representative of Imbrium massif material. Thus, if KREEP here is endogenic, it must be the product of a phase of volcanism that post-dated the Imbrium event but preceded the extrusion of the mare-filling basalts (Spudis, 1978; Hawke and Head, 1978). Radiometric dating gives it an age of 3.85×10^9 years (Dowty *et al.*, 1976). This would make it younger than the "older basalts" discussed above. The older group of ages imply a further series of pre-KREEP volcanics of Nectarian age about which little is known.

8.1.4   Pre-Imbrium KREEP-rich flood volcanism

The general consensus appears to be that most lunar terra plains have an impact origin, but the origin of the KREEP materials in the Imbrium-Procellarum region remains contentious. While it is possible that KREEP is simply an impact melt which has become fractionated by the removal of siderophiles, the notion that it was produced during an early volcanic episode is a simpler and, to my mind, a more attractive possibility. Certainly there were no lack of suitable heat sources during the earlier phases of lunar history; therefore, extensive flood volcanism is almost to be expected. Furthermore, during an extended and extensive episode of this type, there would undoubtedly be chemical variations, either due to inhomogeneities in the KREEP source regions or to subsequent fractionation, which would explain the variations in potassium content of the sampled clasts.

Cadogan (1981) ventured the opinion that the whole of the region affected by Mare Imbrium was once the site of flooding by KREEP lavas and suggests that the major impact event which was responsible for the pre-Imbrium Procellarum Basin, identified and described by Whitaker (1981), could have been responsible for stripping the Moon's

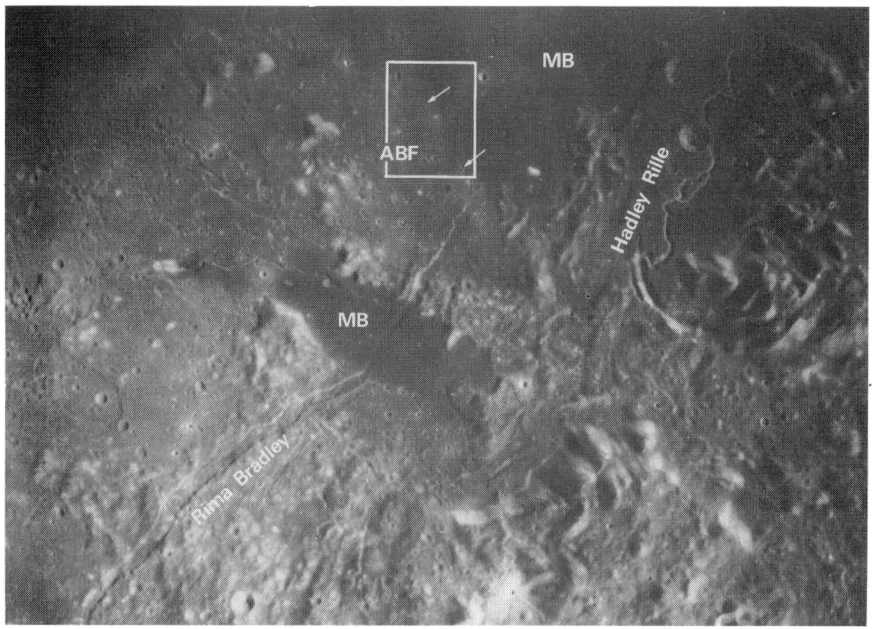

Fig. 8.3: (a) View of the lunar surface around the Apollo 15 landing site. Hadley Rille and the position of the lunar module are shown at the right of the frame. The dark mare basalts of Palus Putredinis (MB) and the underlying material of the Apennine Bench Formation (ABF) are seen here. Frame width 135 km. Apollo 15 mapping camera frame 1820.

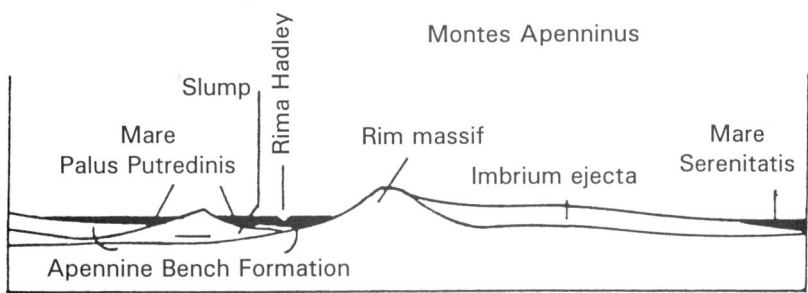

Fig. 8.3: (b) Hypothetical cross-section west-east across the above. (After Wilhelms, 1984.)

Fig. 8.3 (c): Detailed view of boundary between dark basalts of Palus Putredinis (right) and more heavily cratered, higher albedo Apennine Bench Formation (left). Note the lava-eroded bench of older material at bottom right. Frame width 25 km. Part of Apollo 15 panoramic camera frame 9433.

anorthositic crust prior to this volcanic episode, either to receive KREEP-rich rocks or to expose KREEP-rich zones beneath (Fig. 8.4). In support of this, where the normal lava stratigraphy has been disrupted, as at the sites of major impact structures, such as Copernicus, Aristarchus and Archimedes, enhanced radioactivity levels have been measured, in accord with the excavation of subsurface KREEP-rich materials. Then again, gamma-ray traverses show a marked radioactivity hot spot over the Fra Mauro Formation.

8.1.5 Age and origin of KREEP

Some pre-Imbrium volcanic rocks give absolute ages of around 3.86×10^9 years (Dowty *et al.*, 1976; Carlson and Lugmair, 1979); others apparently are somewhat older. However, interpreting the age data is not without its difficulties. This is highlighted by the $0.85 \times$

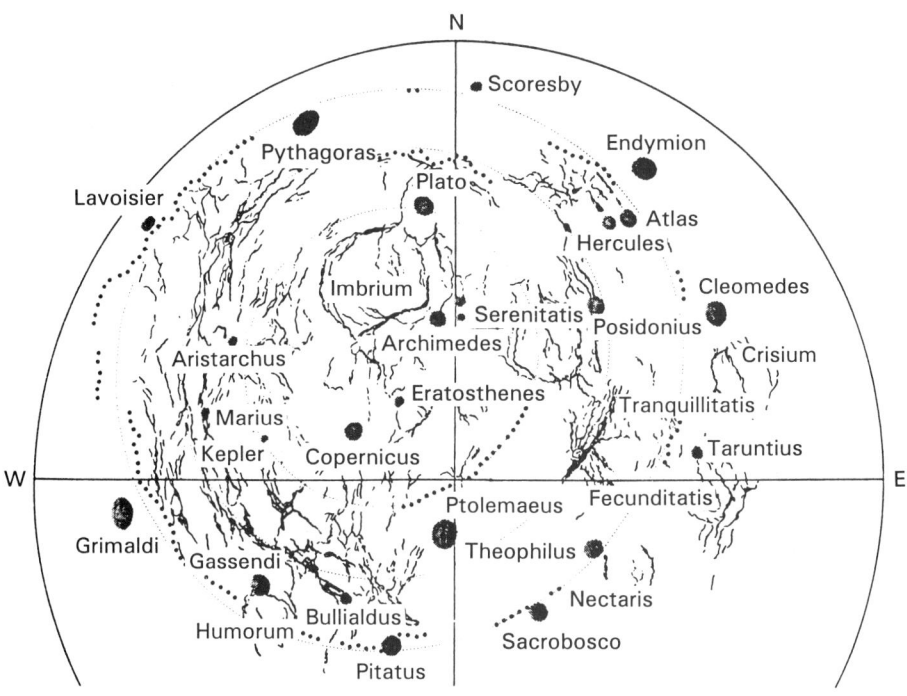

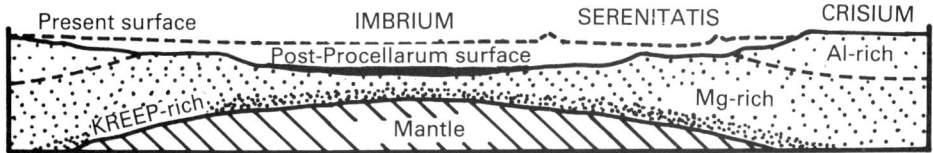

Fig. 8.4: (upper) The Procellarum basin as conceived by Whitaker (1981). Note how well-defined is the basin's outermost ring along its western perimeter and how the next inner ring follows the border of Mare Frigoris. Basin centred at point X. (lower) Hypothetical cross-section through the Procellarum basin from west to east, showing position of KREEP-rich basalt beneath the later mare basalt cover.

10^9 year age obtained for KREEP glass ejected by the Copernicus event, and collected at the Apollo 12 site. If the radiometric clock can be reset by what, by basin standards, is a relatively minor event, then the chances of absolute ages derived from, say, the Imbrium-related Fra Mauro Formation being crystallization ages are small. Whole rock isochrons for KREEP yield ages of 4.3×10^9 years and presumably represent the time of magma separation within the lunar mantle. However, it is not impossible that they could also represent the time of extrusion of the oldest KREEP-enriched lavas. All that can be said with confidence about the age of KREEP is that some lavas enriched in its components were emplaced after the formation of the Procellarum Basin, but prior to the infilling of Imbrium by mare basalts.

Whatever the true ages of the lavas, it can be assumed that they originated in a source melt which will have evolved by mantle processes and, as such, must be considered volcanic. This disregards whether the melting of the Moon was achieved by internal heat sources, the infall of planetesimals or remanent accretional energy. There is a wide range of support for the notion that a lunar magma ocean developed early on in the Moon's history, as the outer regions of the planet melted to a depth of several hundreds of kilometres. Petrological evidence indicates an origin for highland and mare source regions by fractionation from a single parent material, while geophysical and geochemical data indicates that the products of this planetwide differentiation are located at the appropriate depths for those source regions (Wood *et al.*, 1970; Ringwood and Kesson, 1976.). As far as the Apollo 15 KREEP-rich (intermediate-K Fra Mauro) basalts are concerned, Irving (1977) concludes that they were produced by fractional crystallization of a parental melt from a troctolite source deep in the lunar crust or upper mantle. This appears to be a reasonable hypothesis.

Low nickel and low Ni/Co ratios, together with lack of metamorphic contamination, appear to confirm that at least some KREEP lavas are pristine (Wänke *et al.*, 1976; ,Warren and Wasson, 1978). There is also strong evidence that there was three-phase low-pressure cotectic control of compositions parental to KREEP fractionation series (Irving, 1977); however it is still not known whether the rocks are the products of partial melting within the lower crust, or are residual samples of the primaeval magma ocean which separated 4.3-4.4×10^9 years ago. The Rb/Sr initial isochron of around $3.94 + 0.04$, and initial $^{87}Sr/$ ^{86}Sr values of 0.7004 and 0.7007 (Nyquist *et al* 1974, 1975) certainly are compatible with an origin by partial melting at around 3.96×10^9 years, of sources in which the Sr isotopic system was last set at around 4.25×10^9 years. This means the volcanic episode was at least as old as this, but may be older.

8.2   HIGHLAND VOLCANIC FEATURES

A variety of lunar landforms have been attributed to highland volcanism, usually of a chemical type less basic than the maria. These include widespread flat-lying upland plains with higher albedos than the maria and termed *light plains*, domes, irregular mounds, ridges and elongate crater chains.

Fig.8.5: The Descartes region of the Moon, showing the lineated and furrowed deposits and prominent domes of the Descartes Formation and the smoother *light plains* of the Cayley Formation which, prior to the Apollo 16 landing, were suspected to be of volcanic origin. Apollo 16 landing site indicated near the eastern rampart of the degraded ring, Descartes. Frame width 105 km. Apollo 16 mapping camera frame 0440.

8.2.1 The Cayley and Descartes Formations

Wilhelms and McCauley (1971) drew attention to the similarity between certain broad domes, elongate furrowed cone structures and linear ridges exposed in the region of Descartes – eventually chosen as the Apollo 16 landing site – and terrestrial landforms associated with less basic extrusive volcanism. Thus, prior to the Apollo 16 mission, the Descartes region was considered by many scientists to be representative of a highland volcanic province and, since the light plains of the Cayley Formation covered several per cent of the lunar nearside, that light plains elsewhere could be of endogenic origin (Fig. 8.5). The Descartes Formation itself is unusually bright; this, together with its hummocky landforms had led many scientists to propose that it represented relatively silicic volcanoes and associated volcanic rocks. The smoother light plains of the Cayley Formation appeared

Fig. 8.6: Groups of smooth-textured domes typify the Cayley plains to the northeast of the crater, Ritchey (west of Descartes). The prominent 1.5 km diameter dome at X has a small summit pit, while the clusters of similar domes (Y) and (Z) are very reminiscent of many mare domes clearly related to the flood volcanism of those plains. Frame width 56 km. Apollo 16 mapping camera frame 979.

to have flooded the floors of the highland craters in the area (and indeed of a relatively large area of the lunar nearside) and were believed to be highland volcanic lava flows (Milton, 1972). However, exploration of the site by the Apollo 16 astronauts failed to find any evidence for the suspected volcanism, instead they encountered a wide variety of impact breccias and reworked melt rocks largely related to the excavation of the Nectaris Basin and modified in Imbrian times. As a consequence of these findings, lunar *light plains*, formerly thought to be highland volcanic flows, are now generally considered to be nothing more than modified impact breccias (Muehlberger *et al.*, 1972; Oberbeck *et al.*, 1973).

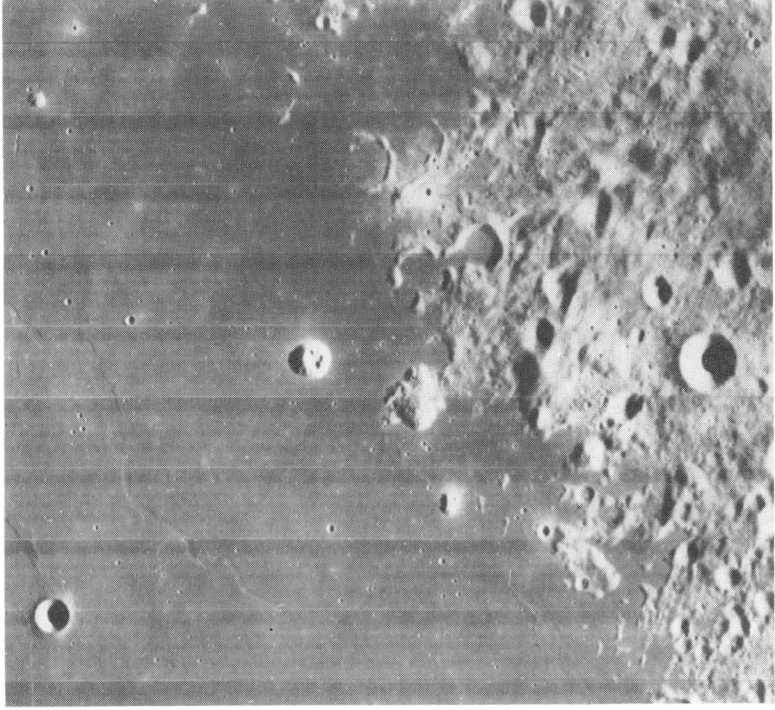

Fig. 8.7: (a) Prominent pre-mare domes on the eastern shore of Sinus Roris. Mairan T, the circular feature near the centre of the photograph, is 6.7 km across and has several summit pits. Similar landforms may be seen to the northwest, north and southeast. In all, over a dozen such features constitute the group. Frame width 130 km. Orbiter 4 frame 159H$_2$.

Despite the general reluctance of the planetary community to believe in extensive volcanism among these light plains, inspection of Apollo imagery leads to a very strong suspicion that, although emplacement of the widespread Cayley plains may have been achieved by ejecta flow, numerous landforms do have a morphology very reminiscent of volcanic domes (Fig. 8.6). This is particularly clear among the plains west of Descartes, where several clusters of small domes, many having rimless summit pits, are difficult to explain other than by endogenic activity.

It should be borne in mind, therefore, that the Apollo 16 team inspected and sampled a very local area, well east of most of the dome-like landforms and there is no objective reason to deny such domes an endogenic origin. There remains, therefore, much geomorphological evidence for volcanic dome formation amongst these highland light plains, presumably in response to the extrusion of relatively viscous lavas, possibly as tholoidal structures which arched up the overlying crust. Similar comments apply also to regions of Fra Mauro Formation and other light plains units.

Fig.8.7: (b) High oblique view of the same region as 8.7(a), showing also the long sinuous rille, Rima Bradley. Apollo 15 Hasselblad frame 93-012730.

Interestingly, among the clasts brought back from the Apollo 16 site were numerous volcanic fragments which appear to represent basaltic flows emplaced prior to the Nectaris event and which add to the inventory of pre-Imbrium volcanics which must have existed on the Moon before the later mare flooding.

8.2.2   Other highland volcanic structures

Among the most likely candidates for highland volcanic structures are a number of groups of broad light-coloured *domes*, of which two, those near Mairan and Gruithuisen, deserve particular mention. The former group is located approximately 225 km northwest of the latter, on Sinus Roris. Mairan T is about 6.7 km in diameter and about 920 m in height; it is surrounded and slightly embayed by (and therefore older than) the mare. It has several rimless summit pits about 3 km across which argue for a volcanic origin. Several similar structures occur nearby (Fig. 8.7).

The Gruithuisen group of domes is located on the northwestern edge of Mare Imbrium, just south of Sinus Iridum. Gruithuisen γ is about 15 km across and 1.7 km high; Gruithuisen δ is somewhat larger (Fig. 8.8). They have a rather pitted, rough texture which marks them out from the somewhat smoother highlands to the west and also are considered to be volcanic in origin. These domes clearly post-date the Iridum impact, giving them a greater age than the last multi-ringed basin. On the basis of their morphology and relationship with the surrounding, darker mare units, they are believed to be either endogenous domes or possibly tholoids which have arched up the overlying layers. The peninsula extending southwards from γ may be a viscous flow associated with the lateral breaching of a lava dome.

The light-coloured Gruithuisen domes and a similar dome near the crater Billy on Oceanus Procellarum, share certain spectral characteristics: they are prominent in colour difference images obtained at 0.40/0.56 μm but not at 0.95/0.56 μm (Head and McCord, 1978; Pieters, 1978). This redness is unusual, is due to strong UV absorption and appears to indicate that the domes are mantled in mature (agglutinate-rich) regolith with a lower Fe and Ti content than typical mare regolith but the precise chemical composition which manifests itself in the spectral characteristics is a mystery. Malin (1974) has suggested that KREEP-rich materials may be responsible and this certainly would not be inconsistent with the view that these domes represent vestiges of pre-mare volcanism which have escaped both impact degradation and total burial by mare lavas.

Similar high-albedo, rounded, landforms are scattered over the lunar surface. Many are found, for instance, in the Fra Mauro region; others near Lansberg. The modern consensus is that these are degraded terra landforms which have been smoothed by impact and mass wasting mechanisms and embayed either by mare lavas or basin ejecta flows. There are, however, numerous rather equivocal features which may represent vestiges of highland volcanism. Several of these have been discussed by Schultz (1976).

One particularly interesting discovery was made several years ago when the author was studying high-resolution Apollo stereo frames of the Aristarchus – Harbinger Montes region. The Harbinger Montes lie along a projection of the most prominent raised ring of the Imbrium basin and their highest peaks are highland massifs which have been equated with the Apennines and Carpathian Montes (Wilhelms and McCauley, 1971) (Fig. 8.9). Volcanism in the Harbinger Montes area is a well-established fact and has been dated at around 3.7×10^9 years (Zisk et al., 1977). Much of the region has been mantled by dark pyroclastic material with a distinct red spectral signature; on the steeper slopes of the Harbinger Montes this has been shed, to lie as talus at the feet of the high-albedo massifs. Study of these massifs on high-resolution stereo pairs clearly reveals the tree-bark pattern on the steep slopes, the rather rounded forms and the subdued nature of the massif surfaces (Fig. 8.10). The more northerly of the two massifs shown on Fig. 8.10 has a height of about 1900 m and a base area of 110 km². Situated on a flat col between its twin summits is what appears to be a perfectly preserved small cone with a rimless summit pit, while extending away from its western flank are what appear to be two short lobate flows. The cone has a height of 145 m and a diameter of 0.8 km, giving it a volume of around 0.025 km³, very similar to many small terrestrial cones. There seems no sensible alternative to

Fig. 8.8: Oblique view of Gruithuisen γ and δ, on the edge of Mare Imbrium. Both are embayed by Imbrium lavas. Apollo 15 Hasselblad camera frame 93-12711.

this being a small cinder or lava cone, in which case it opens up discussion of the real origin of these massifs and certainly provokes the question of whether pre-mare volcanism has played a significant role in their development.

There are also several landforms among the rugged highlands of both the near and far sides of the Moon which appear to have a volcanic origin. One such is located on the floor of the large farside ring Mendeleev (Fig. 8.11). Here, amidst the light plains of the basin's floor, is a prominent 6-km-diameter dome or low cone with a summit pit. It is difficult to envisage any but an endogenic origin for this feature.

Naturally, caution must be exercised in interpreting such features as volcanic on the basis of morphology alone, but in the absence of other evidence it seems reasonable to speculate in this way, if only to indicate that highland volcanism may have been more widespread than currently it is fashionable to admit.

Finally, volcanic structures are also found on the floors of large impact craters, like Copernicus and Tycho. These include leveed flows, ponded lavas and steep-sided domes, often with summit pits, and have been described by several workers (see, for instance, Schultz, 1976; Greeley, 1985). These in a sense must be included within the broad context of highland volcanism, although it is clear that the melts which give rise to these landforms were, in most cases, the product of impact processes. While the origins of crater floor 'volcanics' may be somewhat different, their ultimate morphology remains very similar (Fig. 8.12).

Fig. 8.9: The region northeast of Prinz, showing the upland massifs of the Harbinger Montes, surrounded and embayed by Imbrium lavas. (The boxed area is shown enlarged in Fig. 8.10.) Frame width 100 km. Apollo 15 mapping frame 2193.

8.3 MARE VOLCANISM

Lunar mare deposits are exposed over an area of about 6.3×10^6 km², and have a fundamental asymmetry, with the great majority of post-Imbrium mare materials outcropping on the nearside (see Chapter 7). Such materials are found either as fillings to (i) the central regions of multi-ringed impact basins (e.g. Mare Serenitatis) or peripheral troughs (Lacus Veris), (ii) less regular major depressions (e.g. Oceanus Procellarum) or, (iii) the floors of large impact craters (e.g. Plato) or low-lying areas within the highlands. As such they outcrop within regionally low areas.

Mare volcanism took place over an extended period, from 3.8 to about 2.5×10^9 years ago and, since the lavas infilled what currently are low-lying areas, apparently did so in a hydrostatic mode. Some ancient mare deposits have apparently been admixed with highland breccias, giving rise to patches of what have been termed *cryptomare* within the boundaries of some ancient impact basins. This is particularly true of the huge South Pole – Aitken basin but such mare rocks also outcrop within ringed basins such as Schickard and Riccioli. The fact that the farside crust is significantly thicker than the nearside (farside averages 74 km, nearside 48 km – see Kaula *et al.* (1974) and the lavas were emplaced hydrostatically, may explain the scarcity of farside mare lavas, even in the deeper basins. Many lava sources are associated with crustal fractures; this fact, when taken together with geomorphological observations, geochemical data and estimates of lava eruption rates, suggests that the lavas originated at sub-crustal depths, finding their way to the surface

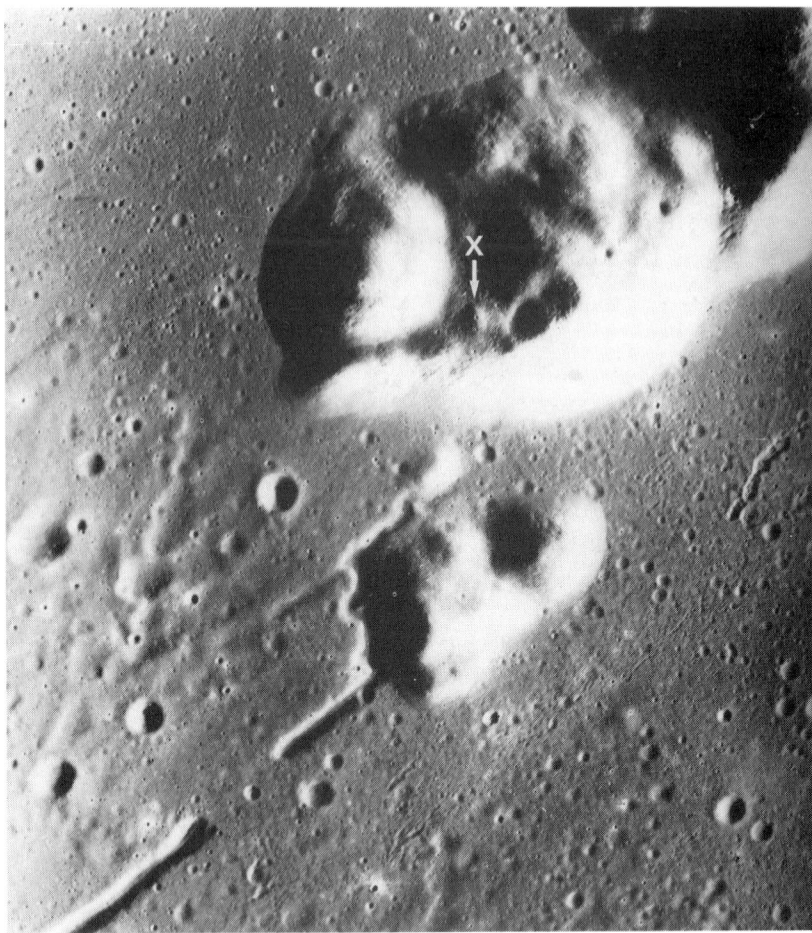

Fig. 8.10: Enlarged view of two Harbinger massifs, showing the typical rounded topography, tree-bark texture and basal talus aprons. A prominent small volcanic cone with summit pit is situated on the northern mountain, at X. Two short lobate flows extend from the west side of this structure. Frame width 25 km. Apollo 15 panoramic frame 315.

via the fractured, passive, lunar crust. The infilling of individual maria must have been determined largely by the pre-fill geometry, isostatic state and degree of degradation of the basins they eventually occupied. Spudis (1982) has presented a scenario which relates differences in basin/mare morphology to a lithosphere which gradually thickened with time. At one extreme, early basins such as Crisium, because of being impacted into relatively thin lithosphere, quickly readjusted isostatically; at the other extreme, the Orientale basin – generated in very thick lithosphere – underwent much slower rebound and there was minimal lava flooding. The geomorphological features of the maria surfaces indicate that mare volcanism was characterized by flood and plains-style activity.

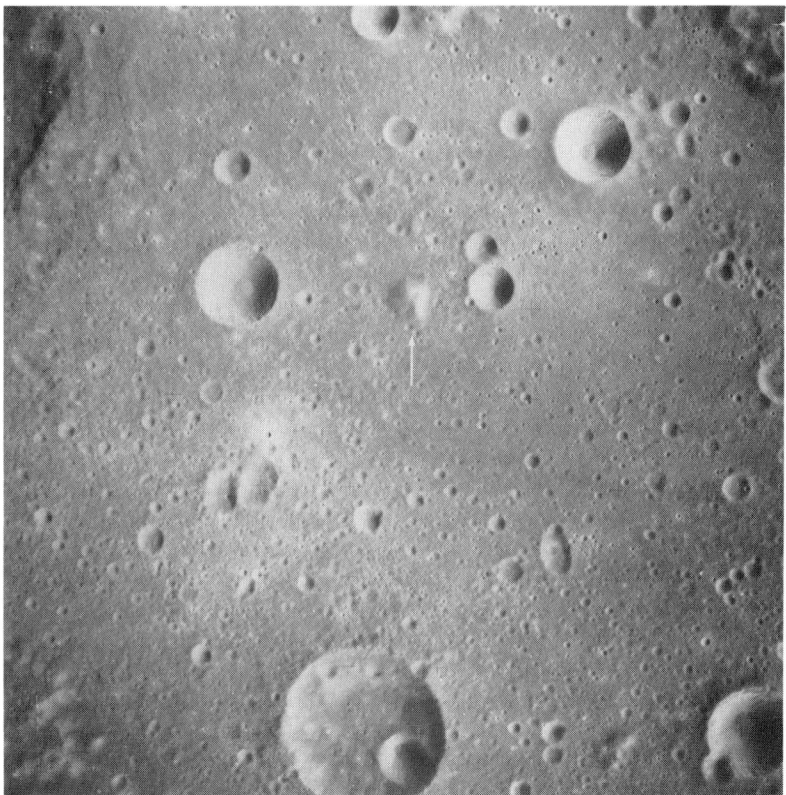

Fig. 8.11: Prominent volcanic dome or low cone within the light plains of the floor of the farside ring, Mendeleev. The cone is about 6 km across and has a summit pit. Frame width 110 km. Apollo 16 mapping camera frame 0343.

8.3.1   Thickness of mare deposits

An idea of the average thickness of mare materials may be obtained by studying the depth of burial of impact craters, the original crater depths being estimated from their diameters, assuming depth/diameter ratios characteristic of fresh craters. One of the earliest attempts to constrain mare thickness was that of Marshall (1961), who determined profiles over Oceanus Procellarum. This method was subsequently refined by DeHon (1974, 1979) and by DeHon and Waskom (1976) who found that the materials infilling the eastern maria averaged between 0.2 and 0.4 km, and the western, nearer to 0.4 km. These workers note, however, that thicker lenses occur, these being up to 1.5 km thick; furthermore, the diminished density of craters in the central regions of maria implies that the fill must be thicker there than over the peripheral shelves. Assuming an average fill of thickness 0.4 km gives a total volume of mare deposits over the whole Moon of 2.4×10^6 km^3.

Fig. 8.12: Dome-like landforms within the ponded units on the floor of the farside ring, Waterman. The smooth profile and generally morphology suggests these are volcanic domes; the most northerly has a summit pit. Close inspection of the original image with a lens indicates the presence of what appears to be leveed lava flow at the base of the massif west of the lava pond. Frame width 45 km. Apollo 15 Hasselblad frame 94-012815.

Head (1981) suggests that thickness measurements based on crater burial techniques may underestimate the actual thickness of mare fill. Using a modelling technique providing volume estimates of mare fill by artificially flooding basins of known topography, he concludes that the central regions of Orientale may be flooded to a depth of as much as 6 km, while the marginal lava deposits attain depths of the order of 2 km. By the same technique, Head derives a thickness for the lava fill in the Apennine region of Mare Imbrium of about 2 km, substantially greater than the figure of DeHon and his co-workers.

Thicker deposits are also demonstrated by the positive gravity anomalies (mascons) detected from orbit and shown to exist over several of the regular maria (Muller and Sjogren, 1968; Solomon and Head, 1980). The largest mascons were modelled over Maria Imbrium, Serenitatis, Crisium, Humorum, Nectaris, Smythii and Orientale, and the centre of the smaller Grimaldi ring, all of which are regular in outline. Mare fill is estimated from the gravity data to be between 2 and 4 km within the central regions of those maria having mascons. There is a general tendency for flooding to commence at a basin's lowest point and to extend

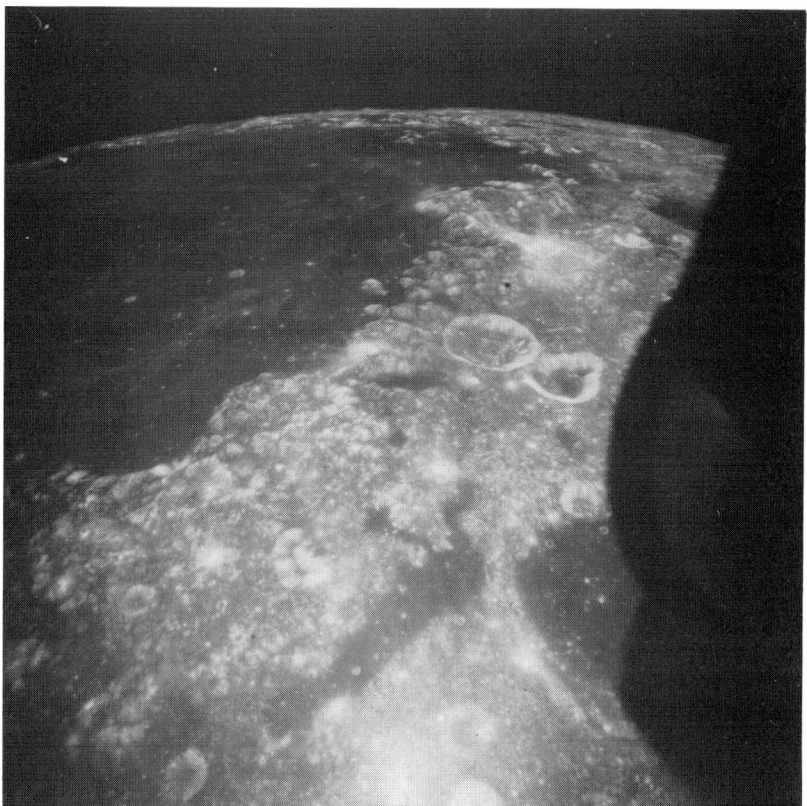

Fig. 8.13: Oblique view looking north across Mare Crisium, showing the smooth, dark mare lavas and the rugged, higher-albedo highland massifs marking the rim of the Crisium basin. The prominent crater is Picard. Apollo 15 frame 15-1495.

upwards and outwards. Stratigraphic studies of Maria Serenitatis (Howard *et al.*, 1973) and Crisium (Head *et al.*, 1978) show that the oldest lava plains outcrop around the borders of the maria and are covered by younger, thinner lavas towards basin centres. The relatively thick accumulation of earlier mare lavas must have caused loading of the lunar lithosphere and, presumably, a degree of deformation of the newly-emplaced deposits. On the basis of modelled thicknesses, such downwarping could exceed 2 km towards basin centres.

There is also a broad correlation on the lunar nearside between the depth of mare fill and the age of the infilled basins (Head, 1976). Thus the relatively older basins have the lesser thickness of mare lavas, while the relatively younger ones show the thicker successions. The most likely explanation for this phenomenon is a structural one: the older basins presumably decreased in depth with time, due to isostatic readjustments, these having largely been accomplished prior to the main phase of mare volcanism. Additional contributions to this may have come from earlier, pre-mare basalt, volcanism and plains formation whose remains are obscured by younger materials.

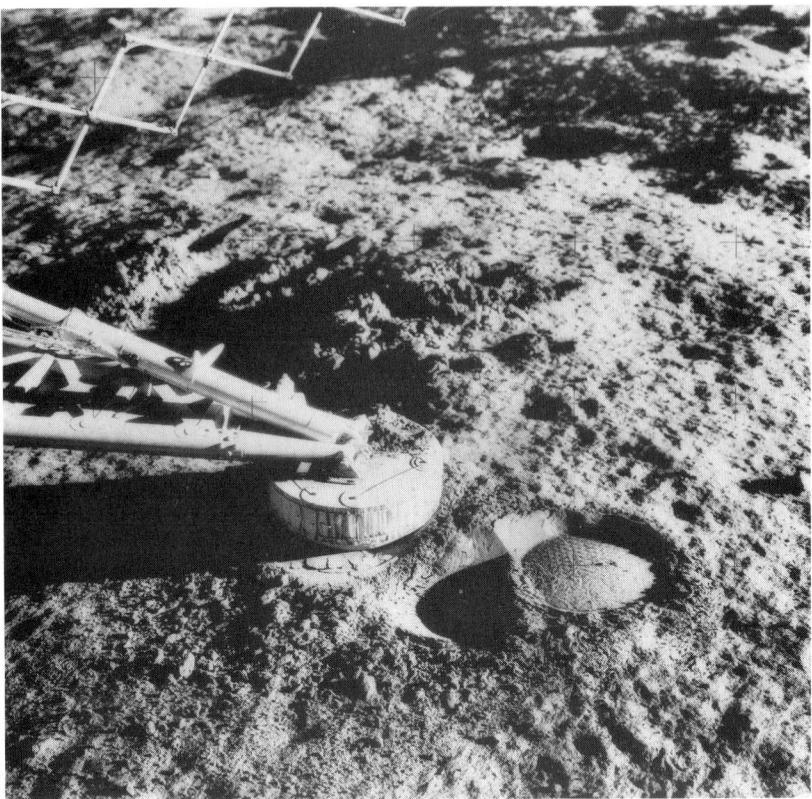

Fig. 8.14: Imprint of Surveyor III footpad into the lunar regolith, as photographed by the Apollo 12 LM crew. This fine-grained fragmental layer, together with larger blocks, has been accumulating on the surface of Oceanus Procellarum since the mare basalts were emplaced, over 3.5×10^9 years ago. Apollo 12 photograph 4994-7110.

8.3.2   Mapping and subdivision of mare units

Unlike the highlands, whose origin was an enigma prior to the Apollo landings, the maria had long been assumed to be volcanic plains (Fig. 8.13). Their low albedo, generally flat appearance and morphological simplicity had led many observers to correctly conclude that they were floods of basalt-like lavas that had solidified in large depressions within the lunar crust (Kuiper, 1959; Fielder, 1961; Shoemaker and Hackman, 1962; Baldwin, 1963). During the 1960s and 1970s, samples either analysed in situ or returned to Earth from the Surveyor, Luna and Apollo missions confirmed they were indeed basaltic but that they were very much older than people had suspected. Furthermore they were significantly different from terrestrial basalts, implying that the evolution of the Moon had proceeded somewhat differently from that of the Earth.

The basaltic nature of the maria was first substantiated by the Surveyor 5 and 6 analyses, obtained in 1967 (Gault *et al.*, 1968). Since then basalt of one kind or another has been returned from every mare site visited, i.e. Apollo 11 (Mare Tranquillitatis), Apollo 12 (Oceanus Procellarum, near to Surveyor 3 lander), Apollo 15 (Hadley Rille) and Apollo 17 (Taurus-Littrow). Detailed subdivision of mare flow units subsequently has been achieved by remote-sensing techniques for studying colour and spectral characteristics and geochemical properties. When this information is added to photogeological and returned sample data, we arrive at the obvious conclusion: that the mare lavas are among the best understood rock units outside of the Earth.

The most obvious difference between the terrae and the maria is the lower albedo of the latter. However, there are significant differences between one mare and another and between units within a single mare. It was discovered during the Apollo missions that these differences were due to the effects of the regolith which has been accumulating on the lunar surface for several aeons and which is largely generated by micrometeorite impact (Fig. 8.14).

Within the regolith are large numbers of agglutinates: small complex glass-bonded particles produced when micrometeorite impacts melt and weld together rock and mineral fragments. As a rule of thumb, such agglutinates accrue in proportion to the time the regolith is exposed to impacts. It has been found that albedo is controlled largely by the composition of these particles (which is related to the chemistry of the underlying rocks) and by the length of time they have been exposed at the lunar surface. If sufficient time has elapsed, the proportion of agglutinates reaches a steady state, whereby new impacts destroy the same amount of agglutinates as they generate. Where a regolith has attained this steady state, it is termed mature. The chemical composition of mature regoliths (which most mare regoliths are) strongly affects their albedo, such that regoliths enriched in Fe and Ti are the darkest (Adams and McCord, 1973; Pieters, 1978). Since Ti varies to a greater degree than Fe among lunar lavas, differences in the former element probably contribute most to albedo variations. Where regolith has been disturbed, as in a major impact, the fresher material ejected by such impact will be of higher albedo than the undisturbed regolith but will gradually darken with the passage of time. Thus it is that bright rays may be seen radiating out from relatively recent impact craters, such as Tycho and Copernicus.

Since Fe and Ti absorb strongly in the near infrared (IR), spectral reflectance measurements made at wavelengths of between 0.3 and 1.1 µm have become a very powerful tool in mapping planetary surfaces. Peaks in the absorption spectra obtained from Earth and from orbiting spacecraft, such as those shown by elements like Fe and Ti, will characterize the chemical state of those elements at individual sites on the Moon. It has therefore been possible to draw up geochemical maps of the lunar surface showing the distribution of known rock types, such as Ti-rich basalt and, indeed, certain materials which have not yet appeared in the returned sample collections. Fig. 8.15 shows the major basalt types for the nearside of the Moon, based on spectral reflectance studies, which are now briefly summarized. These are shown in colour in Plate 9.

HDWA represents relatively old, dark, high-Ti basalts located at the Apollo 11 site on Mare Tranquillitatis and at Taurus-Littrow (Apollo 17). Most of the low-albedo ring surrounding Mare Serenitatis is also of this type. Some workers suspect the same kind of material underlies certain low-Ti basalts on the central nearside maria. hDWA represents

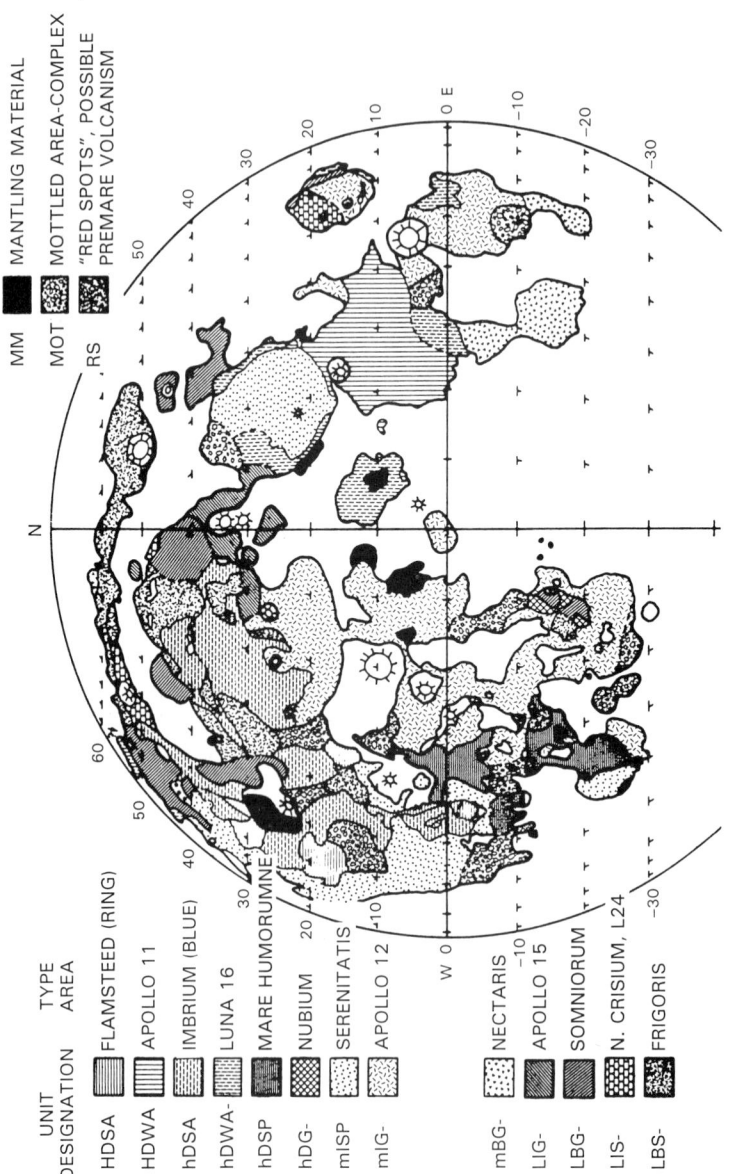

Fig. 8.15: Major basalt types for the nearside of the Moon, based on spectral reflectance data. (After BVSP, Fig.2.2.9.)

medium-Ti basalts sampled by Luna 16 on Mare Foecunditatis; these are surrounded there by more extensive low-Ti types. Similar rocks are believed to underlie low-Ti basalts in Mare Crisium. The low-Ti basalt type, mIG, is widespread, covering large areas in Maria Foecunditatis, Crisium, Nubium, Humorum, Cognitum, and southern Imbrium, as well as in Sinus Aestuum and in peripheral regions of Mare Serenitatis. It was sampled at the Apollo 12 landing site. The low-Ti basalt type, LIG, sampled at the Apollo 15 site near Hadley Rille, outcrops only in the northern hemisphere, occurring in eastern Mare Imbrium and the nearby Mare Vaporum, and occupying a small peripheral area on western Mare Serenitatis. It also floors the large ring, Plato. The very low-Ti basalt (LISP) is found only in northern Mare Crisium and was sampled by Luna 24. These rocks have some of the lowest recorded Ti values of any lunar basalts, and also have relatively high Fe content.

All of the above basalt types were sampled. Others distinguished by the spectral work are not among the sample collections. HDSA and hDSA cover extensive regions in Mare Imbrium and Oceanus Procellarum; both are Ti-rich and therefore similar to some HDWA Apollo 11 basalts, but they differ by virtue of being significantly younger than almost any other group and more highly radioactive. hDSP basalts are also young, and have been found in northeast Mare Humorum and parts of several other southwestern maria. Two further low-Ti basalt types (classified together as mISP) occupy the central part of Mare Serenitatis and parts of western Oceanus Procellarum. Finally, the reddest basalts of all (types LBSP and LBG) are found near the perimeters of basins, for instance in Mare Frigoris and Lacus Somniorum, Sinus Roris and Sinus Iridum, as well as in Mare Nubium and the eastern edge of Crisium.

The distribution and origin of KREEP-rich materials has already been discussed; its mapping relies largely upon gamma-ray traverses. These indicate that KREEP-rich basalts are concentrated in the western nearside quadrant, and were probably extruded throughout the western hemisphere prior to mare filling (Metzger *et al.*, 1973).

The lunar rocks are also coloured; that is, they absorb not only in the IR, but also in the visible spectrum. By obtaining photographs of the same region with IR and UV filters, Whitaker (1972) was able to distinguish individual basaltic flows, as well as characterizing, for instance, the distinctly bluer young interior of Mare Serenitatis and the older, more heavily cratered and redder periphery (Fig. 8.16).

8.3.3   Galileo multispectral data (Clementine mission)

The Galileo spacecraft imaged parts of the western limb of the Moon and the far side during December 1990. This mission provided the first new imaging data for our satellite in 15 years and included observations of the western maria and farside using modern multispectral instrumentation. One of the key elements of this mission focussed on observations of the lunar maria and of related deposits, using the solid state imaging (SSI) experiment (Belton *et al.*, 1992). Clearly this has potential for providing new insights into lunar volcanism and magma evolution. The wavelength range (0.4 – 1.0 μm) is particularly sensitive to the composition of lunar basalts and the abundance of ferrous minerals. The general interpretation of the new data sees confirmation that the limb and farside materials are similar to previous characterizations, i.e. the majority of basalts are of intermediate-TiO_2 composition and most of the highland crust is feldspathic with local variations in

Fig. 8.16: (Left) Unfiltered photograph of part of the lunar nearside, showing Mare Crisium towards the right-hand side of the image, with Mare Foecunditatis to its south, then Maria Tranquillitatis and Serenitatis further west. All maria appear dark. Apollo 10 frame 27-3947. (Right) Colour difference photograph of the lunar nearside obtained by Ewen Whitaker. Note the redder (lighter) interior flows of Mare Serenitatis, as well as the details revealed within both Mare Imbrium and Oceanus Procellarum.

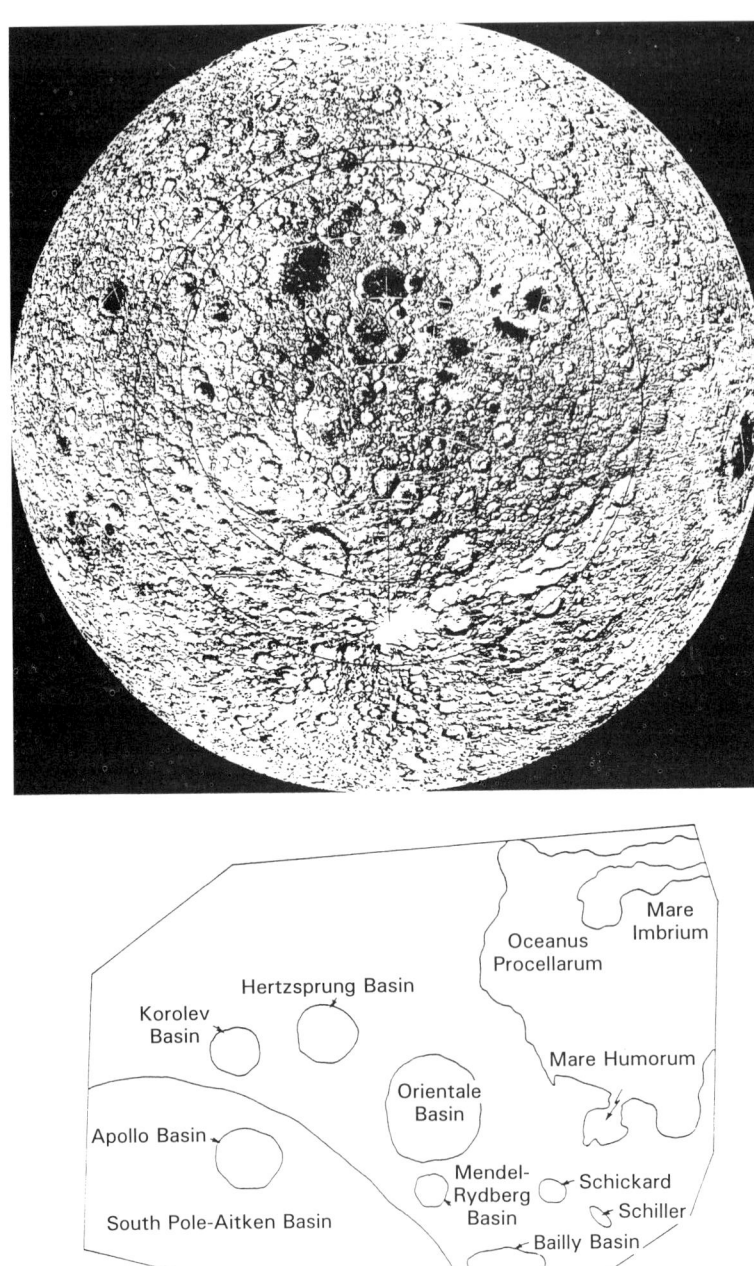

Fig. 8.17: Location of the South Pole – Aitken basin on a view of the Moon from a point above the basin's centre. Mare Orientale is seen close to the right limb.

Fig. 8.18: Cryptomare on the floors of the elongate structure, Schiller. Frame width 130 km. Lunar Orbiter frame IV 155-H$_1$.

mafic content found at large impact craters. Not surprisingly, regions of "cryptomare" – characterized for the first time with modern instrumentation – have spectral properties midway between mare and highland regolith, as might be expected. Surprising, however, is the revelation that limb and farside basalts exhibit a very weak 1 μm ferrous absorption band. This may be taken to mean that in these regions we have a type of basalt characterized by a more Mg-rich composition. Compositional information gleaned for the ejecta from the Orientale impact, on the other hand, indicated that it was excavated solely from the crust and that it lies on pre-Orientale terrain of highland materials and widespread ancient maria units.

Some of the most notable compositional anomalies are associated with the vast South Pole – Aitken impact basin which has a diameter of 2500 km and probably is of pre-Nectarian age (Fig. 8.17). This had earlier been noted by Wilhelms (1987) to exhibit small patches of low albedo material over a wide area. The SSI data confirm this has a far lower albedo than the surrounding highland crust, the darkest, inner, part exhibiting optical properties indistinguishable from low-Ti basalts (Plate 10). The spectral signature widely distributed through the basin can be equated with high-Fe, low-Al soils, meaning that the basin is not composed of "normal" highland crust, but is more mafic. Head *et al.* (1993) consider that this represents *cryptomare* remnants of an early and very extensive volcanic field composed of mafic basaltic rocks. Subsequently this became mixed with non-mare materials during later impact events. This is similar to the situation now know to apply to ancient basins such as Mare Smythii and Australe. Units to the south of the structure show

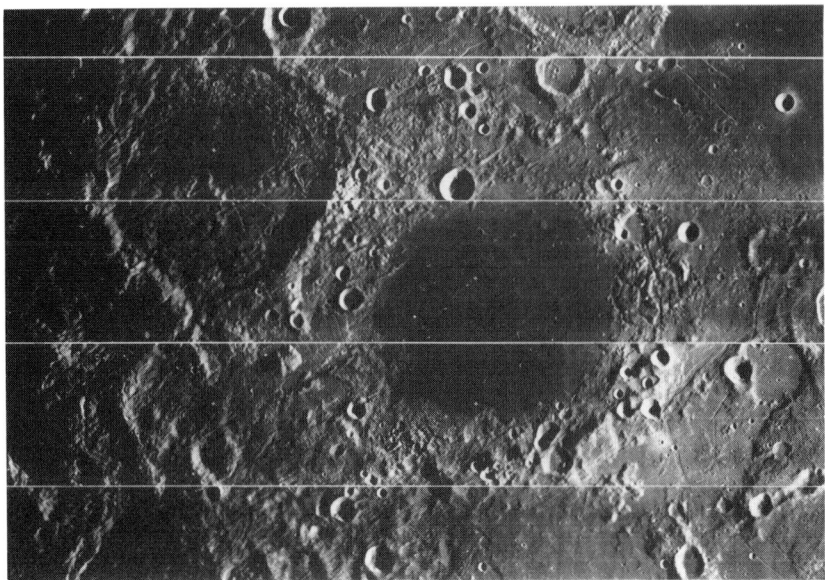

Fig. 8.19: The Grimaldi-Riccioli region of the Moon, showing the extensive low albedo
mare units within Grimaldi and the less extensive ones inside Grimaldi. Lunar Orbiter image
IV 161M. Frame width 400 km.

a strong and broad 1 µm ferrous absorption band, most consistent with abundant olivine
in the rocks. The unusual spectral signature associated with this region suggests strongly
that the impact which created this massive basin excavated mafic-rich lower crust and
perhaps even upper mantle materials (Pieters *et al.*, 1993).

The large smooth-floored basins, Schiller and Schickard, for some time have been
suspected to contain ancient (pre-Orientale) basalts which subsequently have been mixed
with impact-generated highland rocks, i.e. are cryptomare (Hawke and Bell, 1981; Schultz
and Spudis, 1983; Bell and Hawke, 1984 SSI spectra suggest low- to medium- Ti-basalt
soils cover these areas, with patches of slightly enhanced 0.76/0.99 µm ratios that are
more indicative of nearside basalts than those of the western limb and farside. Lighter
plains within and surrounding Schickard are believed to be Orientale ejecta as they have
similar SSI signatures (Fig. 8.18).

The >100-km-diameter dark-floored basins, Grimaldi and Riccioli show SSI signatures
equivalent to medium- to high- Ti-basalts ($<4 - 7\%$ TiO_2). These are believed to be within
the age range: $3.25 - 3.48$ Ga (Greeley *et al,*, 1993). Grimaldi was observed to have one
of the largest positive gravity anomalies on the Moon (mascon), suggestive of there being
thick mare rocks beneath its surface. The mare units within Riccioli are less extensive, the
floor being partially covered by lighter highland rocks (Fig. 8.19).

8.3.4   Dark mantling deposits

Finally, special mention must be made of *mantling materials*, shown in black on the spectral map. Such deposits outcrop on both mare and terra surfaces, but usually near a mare border. Usually they are very dark. The very darkest materials, termed *black spots*, have spectral characteristics dominated by fragments such as the black glass beads collected from Shorty Crater during the Apollo 17 EVAs. These beads, together with the orange and green glass spheres collected during the Apollo 15 mission and from samples amongst the other Apollo collections, all have unique spectral characteristics which imply an origin in fire-fountaining. This kind of explosive activity mantles the terrain adjacent to the lava sources, giving both mare and terra surfaces an unusually red characteristic, as occurs in the Aristarchus-Harbinger region of Mare Imbrium.

The pyroclastic nature of these deposits was confirmed by the Galileo SSI data, the spectra being consistent with the devitrified Ti-rich glass beads found at the Apollo 17 site. However the moderate albedo and the 0.76/0.99 ratio of the dark mantling deposits at Mare Orientale suggest that the local pyroclastic deposits are possibly contaminated with highland ejectamenta.

8.3.5   Sequence of lava emplacement

If KREEP-rich lavas are considered to be volcanic, then these could be the earliest basaltic deposits. Should these turn out to be simply impact melts, then the earliest sampled volcanic rocks are the aluminous basalts, of which clasts were recovered from the Apollo 15 site and for which absolute ages of between 3.85 and 3.98×10^9 years have been established. Studies of lava stratigraphy, petrology and geochemistry, allowed Soderblom (1970) and Soderblom and Lebovsky (1972) to outline the subsequent volcanic history thus. Following on from this early volcanic episode, there was a phase during which early Ti-rich "blue" basalts flooded large regions of the eastern hemisphere during the early Imbrian period (3.8-3.5×10^9 years ago). This activity is well displayed in Mare Tranquillitatis, although similar lavas outcrop along the south-eastern border of Mare Serenitatis. Dark mantling deposits peripheral to Mare Vaporum and Sinus Aestuum show similar spectral signatures, but are overlain by younger lavas. This suggests that the early Ti-rich volcanism in the past may have extended over a much wider area than is now exposed, perhaps covering an area as great as 5×10^5 km^2.

Following on from this early "blue" phase, predominantly "red" basalts with lower Ti content flooded the Serenitatis and Foecunditatis basins, together with extensive areas of the central and western maria in the middle to late Imbrian ($3.5 - 3.0 \times 10^9$ years ago). Such lavas dominate the central regions of regular maria like Imbrium and Serenitatis and the periphery of Humorum (Fig. 8.21). The final episode saw eruption of "blue" Ti-rich basalts in Mare Imbrium and the western hemisphere from about 3.0×10^9 years ago until mare volcanism effectively ceased. These flows give rise to the prominent and much-described flow fronts seen in Apollo imagery.

Ron Greeley and Paul Spudis (1978), after an exhaustive analysis of volcanism in the Herigonius region of Mare Humorum, were able to illustrate how numerous vents were able to supply mare-filling lavas both northward to Oceanus Procellarum and southward to Mare Humorum, many by flow through sinuous rilles (considered to be lava tubes and

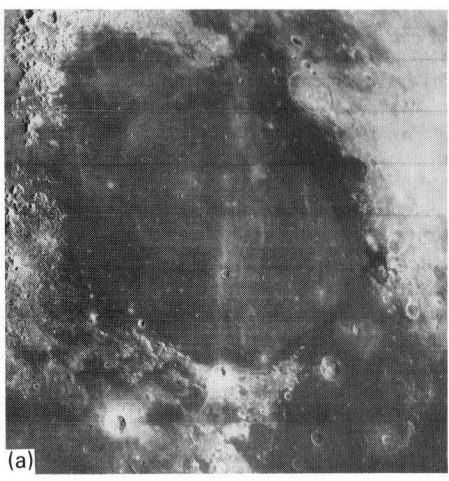
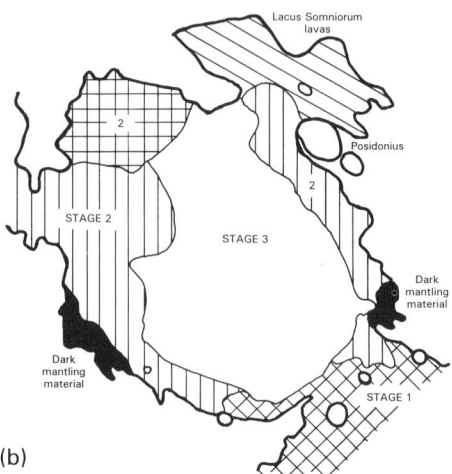

Fig. 8.20: (a) Lunar Orbiter photograph of Mare Serenitatis. (Part of Lunar Orbiter frame IV 109-M.) (b) Distribution of different lava types in Mare Serenitatis. (After Solomon and Head, 1979). The present distribution of lava units has been mapped by Thompson *et al.* (1973). Howard *et al.* (1973), and Solomon and Head (1979). The earliest basalts (Stage 1) are exposed around the southern rim and are continuous with those of Mare Tranquillitatis to the south and of the Apollo 17 landing site. These have low albedo, are "blue" and of spectral class HDWA. Ages range between 3.65-3.85×10^9 years, contemporaneous with the dark-mantling deposits on the mare margins. After their emplacement these flows were downwarped toward the centre of the basin and a series of circumferential graben were generated in them near the margins and on the surrounding highlands (Fig. 8.21). The next stage (Stage 2) in flooding of the basin was the emplacement of medium-Ti flows (spectral class hDWA) which currently are exposed along the eastern and southwestern shores. They are of Imbrian-Eratosthenian age, have also been downwarped and may fill grabens formed in the older group. Finally the central part of the Serenitatis basin was flooded by low-Ti basalts of spectral class mISP, which have a "redder" signature and show a strong development of wrinkle ridges, indicative of substantial contemporaneous deformation (Stage 3).

channels). The earliest phase of this activity saw extrusion of low-Ti basalts and these were related to tubes and channels which subsequently have been partly destroyed or flooded by later flows. The morphology of these flows suggests lower rates of effusion than some later episodes, but prolonged, if sporadic, eruption. Following this early phase was an episode of major flood eruption which produced the most widespread basalt unit, the high-Ti basalts. High rates of effusion are indicated but source vents, as is customary with such eruptions, are not seen. Following on from the flooding episode, a third stage saw the gradual filling of topographically low regions of Oceanus Procellarum, while topographically high areas of Mare Humorum were eventually filled to the point where the third ring of the Humorum basin was breached, lavas then spilling over the divide,

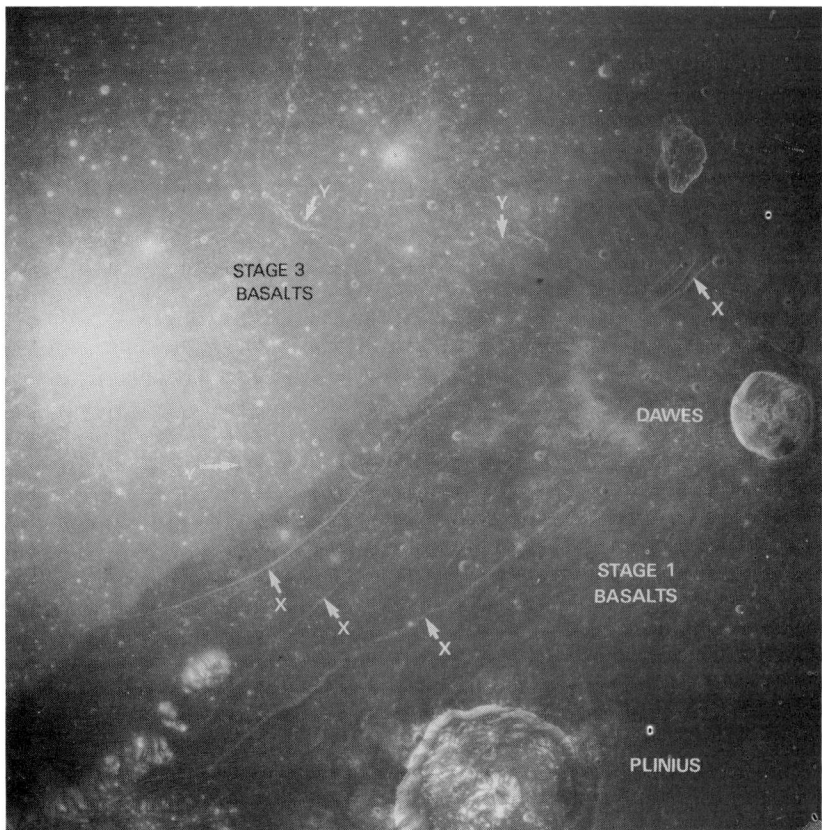

Fig. 8.21: Distribution of Stage 1 and 3 basalts on the southern border of Mare Serenitatis. Note the development of long circumferential graben faults (X) in the former, which are not seen in the overlying stage 3 flows. Prominent wrinkle ridges (Y) on the latter flows indicate a substantial degree of penecontemporaneous deformation. Apollo 15 mapping camera frame 1801. Frame width 220 km.

southwards into Mare Humorum. Some ponding of flows occurred during this stage. Finally, the youngest, intermediate-Ti lavas were erupted from vents located near to Herigonius. These are considered to be more evolved magmas which were extruded in smaller volume during the waning stages of mare evolution.

8.3.6   Petrology and geochemistry of mare basalts
Several overviews of lunar basalt composition, mineralogy and texture have been published; the first four are recommended to those requiring an in-depth analysis: Taylor (1975), chapter 4; Taylor (1982); BVSP (1981), section 1.2.9; and Papike *et al.* (1976), while for a simpler discussion, chapter 3 in Cadogan (1981) is very informative. The following

section presents a brief résumé of petrological and geochemical information gleaned from the returned and surface-analysed lunar basalt samples.

Lunar basalts, like their terrestrial counterparts, are melanocratic, rich in mafics, and characterized by two principal silicate species: Ca-plagioclase and clinopyroxene. Their textures range from intersertal, via intergranular and ophitic to mildly porphyritic, while many are vesicular. Their textures indicate they cooled relatively rapidly, and their mineralogy and chemistry suggest they crystallized from almost dry melts. Samples of quenched basalt, in the form of glass spheres, provide petrologists with the least equivocal information regarding melts which have reached the surface in an unmodified state.

While some lunar basalts are broadly similar to some very young (less than 180×10^6 years old) subalkaline tholeiites dredged from terrestrial mid-oceanic ridges, generally speaking they show striking dissimilarities, many of which are a manifestation of their having crystallized in the absence of volatiles and therefore under conditions of very low oxygen fugacity. The following can be noted:

Lunar basalts contain:

(i)   no detectable H_2O;

(ii)   essentially no Fe^{3+}, but have native Fe metal;

(iii) lower amounts of alkalies, especially Na_2O;

(iv)   higher amounts of TiO_2 (Ti-poor lunar basalts contain approximately similar amounts to average terrestrial basalts);

(v)   lower amounts of SiO_2 and Al_2O_3;

(vi)   higher amounts of FeO and (generally) MgO.

Compared with highland impact melt rocks, they are higher in FeO and MgO and lower in Al_2O_3 and CaO (Taylor, 1975; 1982). Furthermore, while highland rocks show a positive europium anomaly, the majority of mare basalts show a distinctly negative anomaly. The effect of (i) above is that lunar basalts lack alteration, while the implication of (ii) is that they crystallized under highly reducing conditions.

It has been traditional, on the basis of sampled sites, to subdivide mare basalts into low-Ti and high-Ti groups. However, while as a broad working classification this may be useful, spectral mapping has revealed unsampled basalt types which suggest that there may well be a continuum between these two end-members; furthermore, sampling of high-alumina and very low Ti (VLT) types has extended the number of distinctive groups. Since considerable heterogeneity exists at all of the Apollo sampling stations, a more realistic classification would need to include the sample site, as well as mineralogical and chemical data. One of the more widely used classifications is given in Table 8.3.

At all of the Apollo sites, but particularly at the Apollo 17 site (Taurus-Littrow), pyroclastic materials in the form of glass spheres were found. At the latter site the glass droplets were orange and black in colour and have been identified on the basis of their spectral characteristics as being similar to a dark mantling deposit observed telescopically (Pieters et al., 1974; Adams et al., 1974). In the Apollo 17 regolith samples, the droplets, derived from an originally 1.5 km thick pyroclastic sheet, are admixed with lava and highland clasts. They have a high Ti content (9.3% TiO_2) and chemically are similar to both Apollo 11 and 17 basalts. Green glass spheres collected at the Apollo 15 site are, in contrast, low in titanium and highly mafic.

Table 8.3: Classification of sampled basalt types, showing major chemical and/or mineralogical characteristics that distinguish each group. (After BVSP, section 1.2.9.)

Name of group	wt% Al_2O_3	wt% TiO_2	wt% K_2O	wt% MgO
Apollo 11 high-K, high-Ti			>0.3	
Apollo 11 low-K, high-Ti		9-14		
Apollo 17 low-K, high-Ti				
Apollo 12, ilmenite		5-9		7-10
Apollo 12 low-Ti, pigeonite	8-10		~0.1	
Apollo 15 low-Ti, pigeonite		1.5-5		
Apollo 12 low-Ti, olivine				
Apollo 15 low-Ti, olivine				10-18
Apollo 17 and Luna 24 very low-Ti (VLT)		<1.5	<0.04	10-11
Apollo 12 Al, feldspathic				
Apollo 14 Al, feldspathic	10-15	3-5	0.1-0.2	7-9
Luna 16 Al, feldspathic				

As has been described in section 8.3.3 above, two basalt types were retrieved from the Apollo 11 site in Mare Tranquillitatis. An older group, giving ages of 3.55-3.92×10^9 years, are low-K high-Ti basalts and are believed to have come from four flows (Beaty and Albee, 1980) which were emplaced soon after the Imbrium impact. A younger group give an average age of 3.57×10^9 years and are characterized by high-K, low-Ti chemistry. The age, texture and bulk chemistry of the younger samples are consistent with derivation from a single basalt flow (Beaty and Albee, 1978). Different initial Sr ratios indicate the two groups originated from different reservoirs.

The youngest samples of all were collected from the Apollo 12 site in Mare Insularum. This was a mixed bag of basaltic material, in that it sampled material from several overlapping ejecta blankets. Rhodes *et al.* (1977) have suggested that the basalts fall into three major groups and one minor one. Two of the former, a low-Ti olivine (FeMg-rich) one and a low-Ti pigeonite (Si-rich) one are almost certainly from different levels within the same flow unit, and illustrate a degree of pre-emplacement crystal fractionation (Fig. 8.22). Certain inferences about the ejecta deposits suggest that an overlying flow was also sampled, yielding a low- or intermediate-Ti ilmenite basalt group. These samples yielded an average age of 3.17×10^9 years. A single sample of feldspathic basalt also was collected. Its relationship with the other rocks is not known.

The Apollo 14 collections returned two main classes of basalt from Palus Putredinis, both being reddish low-Ti types. The older one is the more dominant and contains modal pigeonite; it is also quartz normative. It appears to have accounted for most of the 60 m face of Rima Hadley (Lofgren *et al.*, 1975). The other type is a low-Ti olivine basalt which is olivine normative. This overlay the former flow and the two clearly had a different parentage. The absolute age of about 3.3×10^9 years yielded by these rocks is the same as the age of the green glass.

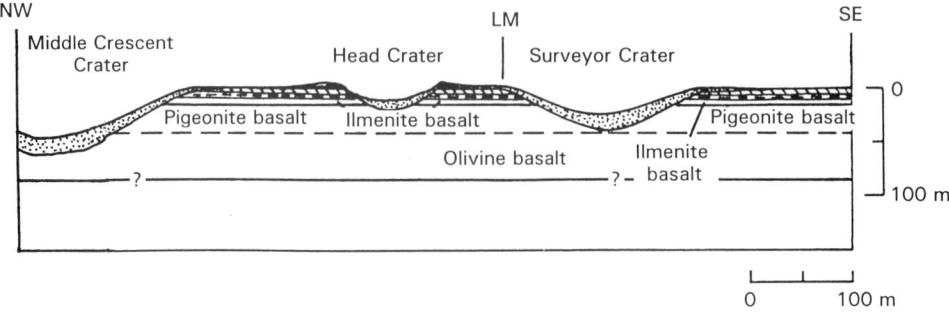

Fig. 8.22: Hypothetical cross-section through basalt sequence at the Apollo 12 site. (After Rhodes *et al.*, 1977.)

The large sample collection returned by Apollo 17 from Taurus-Littrow, a marginal embayment in Mare Serenitatis, yielded a suite of Ti-rich (10-14% TiO_2) rocks containing slightly more Ti than the Apollo 11 samples, but otherwise generally resembling them in chemistry, petrology and age (Rhodes *et al.*, 1976). The somewhat wide variation in absolute ages clustered around 3.68×10^9 years. The glass samples retrieved here gave an age of 3.64×10^9 years.

8.3.7   Petrogenesis of mare basalts

The development of the plutonic anorthositic highland crust and the subsequent extrusion of flood lavas – the mare basalts – arguably is the most significant petrologic development in lunar history. The fact that the terra rocks show a marked positive europium anomaly (Eu enters plagioclase), while the mare basalts have an equally marked negative one, indicates that a primary differentiation must have occurred prior to basalt extrusion at the surface, at about 4.4×10^9 years ago. This is confirmed by a variety of other isotopic and trace element data. Since the lunar mantle is generally considered to be composed predominantly of olivine and pyroxene, any model seeking to account for both terra and mare rocks must satisfy a need to explain their derivation from a single ol-pyx source.

With a view to explaining how these materials evolved, a number of petrogenetic models have been proposed (see BVSP, p.263). The "cumulate model" of Taylor and Jakes (1974) appears to satisfy most of the chemical constraints and is considered the most acceptable here. They, like many others, consider it likely that the primitive Moon would have experienced melting and differentiation down to a depth of at least 500 km (Taylor, 1978). During this early stage, fractionation and sinking, first of Mg-olivine and subsequently orthopyroxene, would have led to an increasing concentration of refractory elements, particularly Ca and Al, which became trapped between the chilled outer skin of the Moon and the denser Ol-Opx layer beneath. When the concentration of Al reached a level of between 12-17% Al_2O_3, Ca-rich plagioclase began to separate and the plagioclase crystals remained suspended or even rose by flotation, eventually giving rise to the 60 km thick, Al-rich anorthositic highland crust. Because of the atomic size of Ca-plagioclase, the only trace elements able to enter the lattice in significant amounts were Sr^{2+} and Eu^{2+}. In this way the terra rocks became enriched in Eu, giving the recorded positive europium anomaly.

Meanwhile Mg-Fe phases continued to accumulate at greater depth, providing the source regions for the mare basalts. Most petrologists believe the mare materials were derived by small degrees of partial melting of the lunar mantle at depths of less than 400 km. The low-Ti basalts probably originated at depths of between 200 and 400 km, the thermal energy required for their production being supplied by the decay of radioactive elements such as U, Th and K (Taylor, 1982). These earlier cumulate sources were relatively enriched in olivine and orthopyroxene and it was not until later that clinopyroxene and ilmenite accrued, to give the younger, Ti-rich basalts. The high-Ti mare lavas contain the greatest proportion of the radioactive elements, explaining why these were the last types to be erupted in abundance.

It has also been proposed that the different mare basalt types were derived from different levels within the lunar mantle. There is general agreement that this was so. Thus the aluminous types came from the highest levels, the high-Ti types from the next highest and the low-Ti types from lower again; deepest of all would have been the green glass (Taylor, 1982). Furthermore there almost certainly were fine-scale inhomogeneities in the mantle layer which contributed to the diversity of basalt types shown to have been erupted more-or-less simultaneously at the Apollo 12, 15 and 17 sites. Experimental work by Delano (1980) on the depth of melting of lunar rocks confirms the above hypothesis, suggesting that the basalts originated in a zone of complex cumulates at a depth of around 400 km. This is consistent with sinking and perhaps convective overturn of the cumulate layers during crystallization within the primitive magma ocean (Delano and Taylor, 1980).

8.4   EMPLACEMENT OF MARE LAVAS: MORPHOLOGICAL EVIDENCE

The evidence presented above clearly establishes that the lunar maria are composed of basaltic lavas; however, the enormous volume of data that has accrued regarding the age, composition and morphology of the mare plains, shows that they are by no means homogeneous in any of these respects. Photogeology reveals landforms characteristic of extensive floods of fluid basalt, as well as of more local activity that gave rise to both effusive and pyroclastic deposits (Greeley, 1976; Guest and Murray, 1976; Head, 1976). Geomorphological studies reveal little or no evidence for Hawaiian-style shield development but show lunar volcanism to have been dominated by flood lava extrusion and plains-style activity. The former effusions are related to fissures of which no really convincing evidence remains; the latter have sources which often can be identified. These usually build volcanic complexes that comprise sinuous rilles, linear vents, mare domes, volcanic depressions and mare ridges. All of these, in broad terms, are genetically related to one another, but for convenience of description they are first treated separately. At the

Fig. 8.23: Eratosthenian-age lobate flows on the southwestern surface of Mare Imbrium. These flows originated in a region of braided lava channels located west of the crater Euler (bottom left) and flowed between northwest and northeast across the centre of the mare plain. The prominent mountain at top right is Mons La Hire. Frame width 165 km. Apollo 15 mapping camera frame 1700.

end of the section, a brief case study of Mare Orientale is presented which highlights the interrelationship between the various mare features.

8.4.1   The morphology of lunar flows

Over very large areas of the maria there are few signs of flow fronts. This has been attributed variously to impact degradation of flow scarps, lack of suitable near-terminator imagery, and properties of the lavas themselves and their mode of emplacement. Of these, the last is almost certainly the most important, the evidence suggesting that, certainly during the earlier phases of mare flooding, fissure-fed flood lavas were emplaced continuously over protracted periods, not only covering their own source vents but also coalescing to generate widespread and uniformly flat deposits. However, during the latter stages of effusive activity (late Imbrian – Eratosthenian) in some maria, a rather different mode of emplacement can be discerned. The most striking evidence for this is provided by flows in the central and southwestern parts of Mare Imbrium, where images obtained under low lighting conditions reveal long, lobate scarps. Similar scarps are occasionally visible in

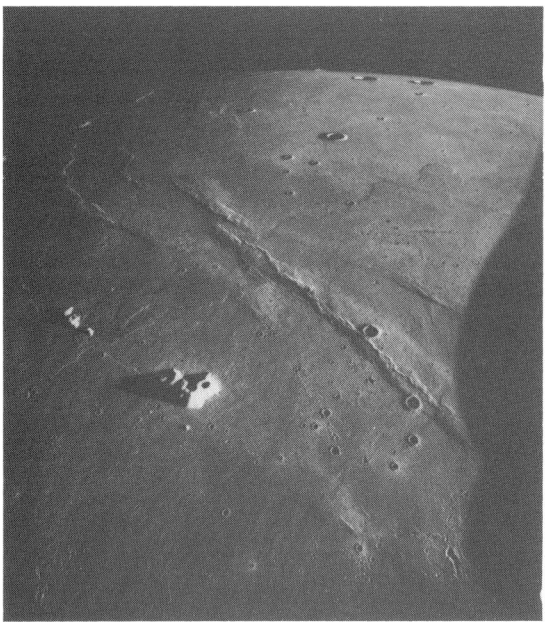

Fig. 8.24: Northeastward continuation of Eratosthenian-age lobate flows depicted in Fig. 8.23. These pass west of Mons La Hire, are disrupted by major WNW/ESE wrinkle ridges and extend towards the crater Carlini. Note the prominent sinuous rille east of Mons La Hire; this undoubtedly represents a collapsed major lava tube, a source depression being visible at its southern termination. The flows are about 15 – 20 km in width. Frame width = 150 km. Apollo 15 mapping camera frame 1555.

other maria, the highest frequency of these being in Imbrian age or younger flows (Gifford and El-Baz, 1978).

The spectacular flows of Mare Imbrium (Figs 8.23, 8.24) were first described in detail by Schaber (1973) but had been earlier noted to correspond to red-blue colour differences by Strom (1965) and by Whitaker (1972). These Eratosthenian-age lavas cover an area of about 200 000 km^2 and have a total volume of the order of $10^3 – 20^3$ km^3. The longest of the series of most recent flood lavas terminates at 36.5°N, 37°W (near Carlini), it and a number of other northeast/northwestward-flowing lavas being traceable back over a distance of 370 km to a source in a zone of braided lava channels bounded by 18°N – 23°N and 28° – 32°W (west of Euler).

Mapping by Gerald Schaber (1973) indicates that earlier Imbrium basin-filling lavas were even more voluminous than these flows, some individual units extending 1200 km from their source vents across the central part of the mare towards Sinus Iridum. Measurements indicate that the younger flows average about 30 m in thickness, although the complete range is 10 to 63 m, and that many flows have axial feeder channels. Morphometric data indicate that these lavas must have been of low yield strength (2×10^2 N m^{-2} (Booth and Self, 1973)) and were erupted at very high rates ($10^8 – 10^9$ kg s^{-1} (Hulme and Fielder, 1977)), the latter being indicated

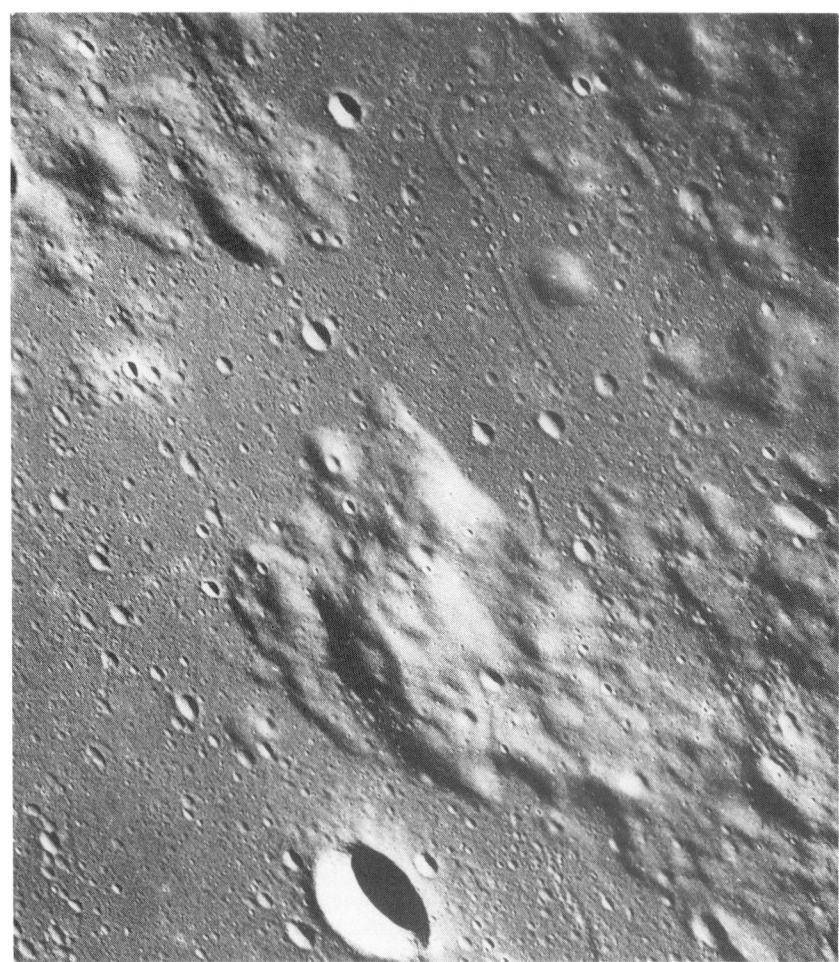

Fig. 8.25: Sinuous leveed flow east of Lansberg. This appears to have originated in a 2.5-km-diameter elongate depression, which itself is associated with a broad linear rille. Frame width 20 km. Apollo 14 frame 10122.

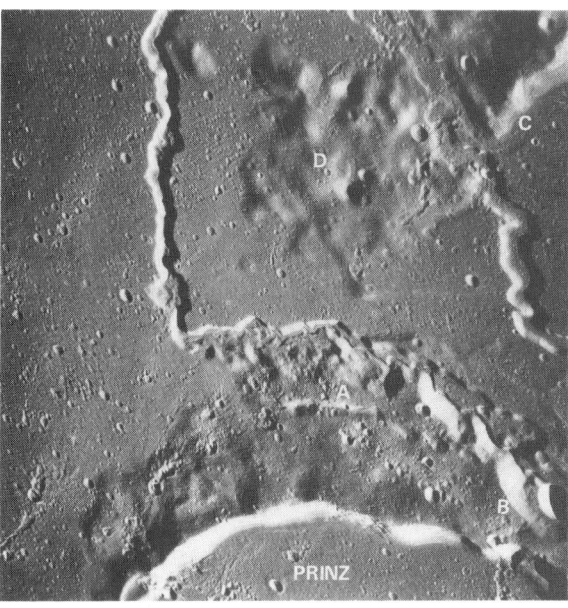

Fig. 8.26: Sinuous rilles north of the flooded crater Prinz. Note the somewhat rectilinear morphology of certain sections of the larger rille, the incised inner rille (A) which hugs one of its walls, and the cobra-head source depression (B). The more easterly rille cuts through the subdued massifs (D) and has several crateriform sections (C), suggesting that collapse has occurred. Frame width 60 km. Apollo 15 Hasselblad frame 96-13044.

by the great length of not only lobate flows but also of sinuous rilles known to have been feeders to some such flows. It appears that the very high driving pressures generated at their vents, was the prime factor in influencing the great length of these flows. To account for such high mass eruption rates it is necessary to invoke feeder fissures significantly wider (10 m) than those typical of terrestrial flood eruptions (0.2-4 m). Wilson and Head (1983) suggest that these may have been favoured either by the production of major impact basins, or an extensional state of stress in the lunar lithosphere, or the latter's early thickening.

By combining data on the kinetics of some appropriate chemical reactions with data on the thermal conductivity and specific heat of returned lunar samples, Brett (1975) derived estimates of about 10 m thickness for lavas from the Apollo 11, 12 and 15 sites. This is in accord with observations and the notion that most lunar flows were thinner than the late Imbrium lavas, perhaps resembling quite closely the relatively thin cooling units observed in the Eastern Snake River Plains.

Only in very few instances are lateral levees preserved, again suggesting that an unconfined style of flow was dominant. Nevertheless one or two examples are recorded on Apollo imagery, perhaps the best being the leveed sinuous flow that crosses the eastern side of Oceanus Procellarum, just east of the prominent crater, Lansberg (Fig. 8.25).

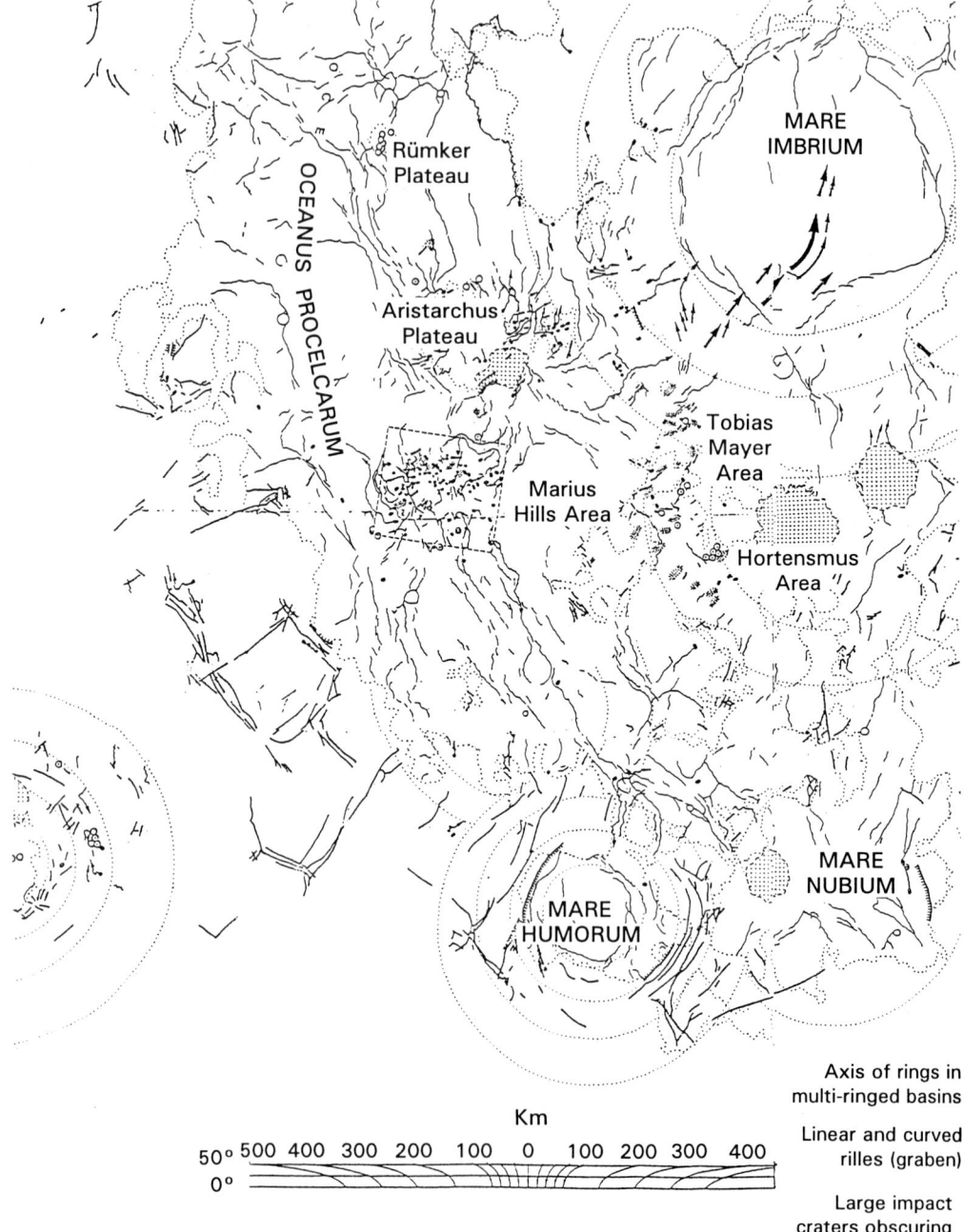

Fig. 8.27: Volcanic and tectonic map of the nearside of the Moon, based on study of Lunar orbiter IV imagery. (After Guest and Murray, 1976.)

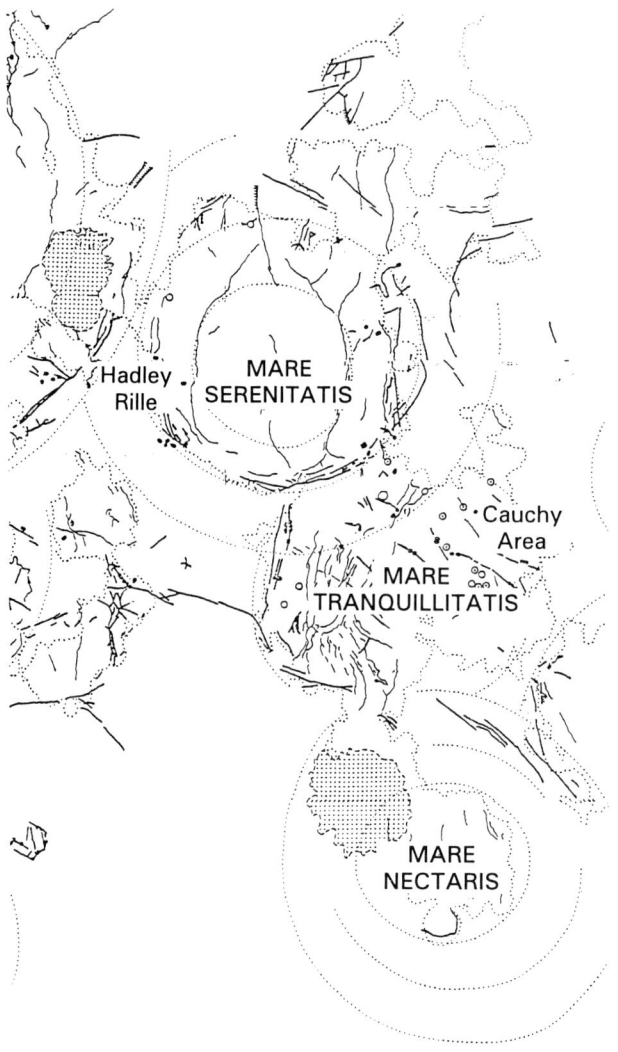

Approximate outline of
maria

Axis of mare ridge, or
ridge complex

Fault scarp (ticks on
downthrown side)

Sinuous rille, showing flow direction

Direction of flow of lava flows

Collapse pit (caldera)

Low dome

Low dome with summit pit

Cone

Chain of fissure cones

Fig. 8.28: Sinuous rilles and other volcanic features in the vicinity of Prinz and Krieger. Several rilles have their origin in prominent depressions (A, B, C, D). One large depression is located on a broad dome (Q), while Prinz (P) itself has a similar location. The general direction of lava flowage is shown by the black arrow. Frame width 300 km. Apollo mapping camera frame 2196.

8.4.2   Sinuous rilles

Amongst the most striking landforms on the mare surfaces are sinuous rilles (Fig. 8.26), which consist of meandering channels that extend up to hundreds of kilometres across the basalt plains. These may range from a few tens of metres to up to 30 km in width, while the longest is over 300 km in length (Schubert *et al.*, 1970). Some wide and rather rectilinear rilles, like Schröter's Valley and several others near Aristarchus, may have narrower sinuous ones on their floors. Most of these distinctive lunar landforms are located near mare borders, some being incised only on the mare surface, others extending onto it from adjacent highlands.

Because of their concentration near to mare borders they tend to be associated with circumferential and radial graben and linear rilles, that is, with structures whose origins are tied to the extension experienced by the downloading of the lunar crust during mare emplacement. While some rilles are relatively isolated, most cluster together in groups, usually as part of volcanic complexes that comprise other endogenous features, such as vents, depressions and domes. The distribution of such features was investigated by Guest and Murray (1976), and a modified version of their volcanic and tectonic map of the nearside is reproduced as Fig. 8.27.

In addition to the rilles near Prinz depicted in Fig. 8.27, there are numerous other meandering rilles in this region (Fig. 8.28). These lie on the flanks of a broad dome and

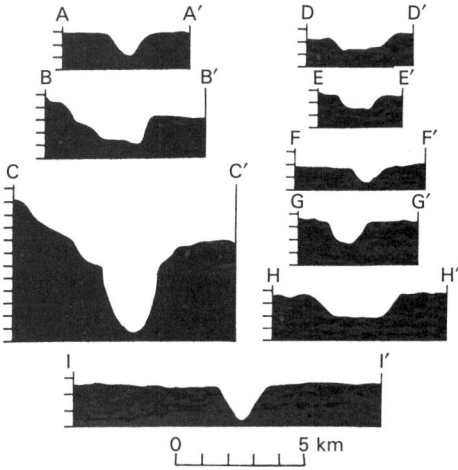

Fig. 8.29: Transverse profiles for three rilles. Vertical exaggeration factor 5. Vertical scale marked at 100-m intervals. A-A´ through C-C´ cross Rima Prinz. D-D´ through H-H´ cross Rima Handel. I-I´ crosses Rima Beethoven. (After Strain and El-Baz, 1976.)

many clearly originate from prominent crateriform depressions. They range in length from 12 to 80 km, have widths of between 0.8 and 4.8 km and depths of up to 300 m (Strain and El-Baz. 1977). Most are flat-bottomed but some approach a V-shaped cross-section for part of their course (Fig. 8.29). The longitudinal profiles of the rilles near Prinz indicate they flowed northwards across the mare surface, over slopes with gradients of less than one degree. These are similar to the slopes traversed by terrestrial lava tubes and channels, which prompted Greeley (1971) to draw an analogy between the two after studying similar landforms in the Marius Hills. Strain and El-Baz favour a similar origin for the rilles.

The idea that such rilles represent major lava tubes or channels is now widely supported (Strom, 1965: Greeley, 1971; Murray, 1971; Guest and Murray, 1976). Many rilles traverse the plains as continuous open valleys for part of their course but then break into small segments or lines of elongate pits (Fig. 8.30). This behaviour finds direct analogy with terrestrial tube-channel systems, such as those of the Snake River Plain, where flow may take place partly through closed tubes and then partly by open channels. As has been shown, many rilles commence in circular or elongate depressions believed to be source vents. Several such depressions lie atop broad volcanic domes, a further occurrence which finds analogy on the Earth. Guest and Murray (1976), having plotted the position and course of 256 prominent sinuous rilles from Orbiter IV imagery, showed that the flows which fed them originated either on the perimeter of the mare surfaces or close by in the highlands, and flowed consistently towards the inner parts of maria.

A mean depth of 100 m was derived for 24 rilles measured by Guest and Murray (1976). This is deeper than most terrestrial examples. Lunar sinuous rilles also are significantly wider than their terrestrial counterparts, a fact which has been used as an

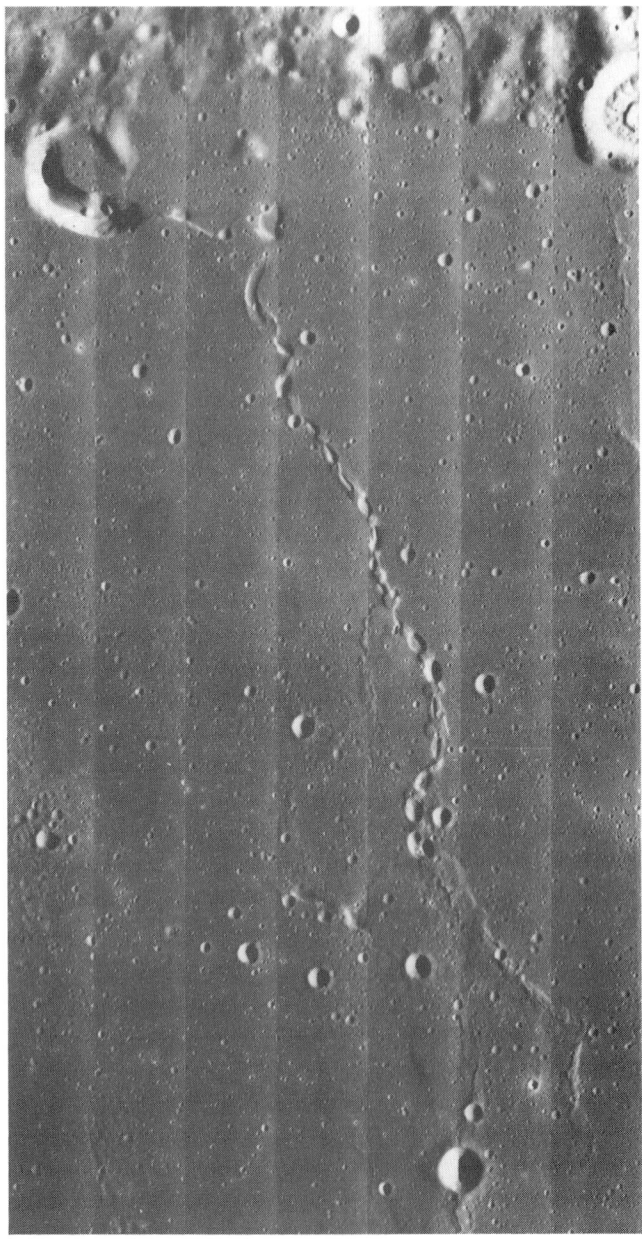

Fig. 8.30: Series of aligned depressions near Gruithuisen, Mare Imbrium. This bears a close similarity to a partially unroofed lava tube and has a source depression near to its northern termination. Frame width 40 km. Lunar Orbiter frame V.182M.

argument against the tube/channel origin. However, Oberbeck *et al.*(1969) have shown that under conditions of reduced lunar gravity, lava bridges could be much wider than those on Earth, thus removing such an objection. In accord with the greater rille dimensions is the large size of rille source depressions. This suggests they, and indeed the rilles themselves, may be partly erosional in origin (Murray, 1971; Carr, 1974), turbulent flow of hot lava having been significant in downcutting through the basalt surface of the maria, sometimes down through pre-existing tube floors, as happens on Earth (Hulme, 1973). Because turbulent flow transfers thermal energy to the underlying substrate more effectively than laminar flow, subflow melting and erosion could have been quite significant on the Moon. Turbulent flow could certainly lead to deepening of pre-existing channels during the later parts of the same eruption or during re-usage of an older channel by later flows. In this context, Carr (1974) in examining the possibility of lava erosion by thermal incision, found lunar erosion rates to have been similar to terrestrial ones. It may also be argued that because thermal incision would have been more effective in the relatively incoherent highland regolith, this may explain why many channels commence from sources within terra massifs, before traversing the maria surfaces.

Geomorphological evidence therefore supports the idea that sinuous rilles are major lava channel/tube systems, and the validity of this hypothesis was finally put beyond question by the geological field work undertaken at Hadley Rille which established the genetic relationship between such rilles and basalt extrusion. With this in mind, clearly the rilles can provide important insights into the mode of emplacement of at least some, if not all, mare-filling lavas. At least two of their geomorphological characteristics imply their emplacement was a manifestation of high mass eruption rates: (i) they are longer on average than terrestrial lava channel and tube systems, and (ii) their source depressions are very large. Studies of the dimensions of the source depressions at the heads of many rilles have led Wilson and Head (1980) and Head and Wilson (1981) to conclude they must represent evidence for very high effusion rates (greater than 0.3×10^7 kg s^{-1}, although the same workers note that extreme care must be taken in estimating eruption rates, since the accuracy of rille measurements made from available imagery may lead to errors of up to 50% (Wilson *et al.*, 1985) and this applies also to the volume of erupted material. The relatively large width and depth of the rilles is most realistically explained by enhanced downcutting and erosion achieved by turbulent flow (Hulme, 1973), while the fact that most sinuous rilles have their sources at mare margins, either in depressions, concentric or radial fractures, or broad domes, strongly suggests that the large volumes of fluid magma that produced the rilles and contributed to the infilling of the basins arose via pre-existing fractures in the lunar crust.

8.4.3   Domes, low shields and other volcanic complexes

The existence of endogenous lunar domes has been alluded to in the previous section. Such features were detected during telescopic study of the Moon and an early catalogue of such domes was published by Moore and Cattermole (1957, 1958, 1959, 1960). Subsequently these have been studied from spacecraft imagery by several workers, including Greeley (1971) and Head and Gifford (1980). Such structures typically occur in groups, being a part of volcanic complexes that usually include sinuous rilles and rimless

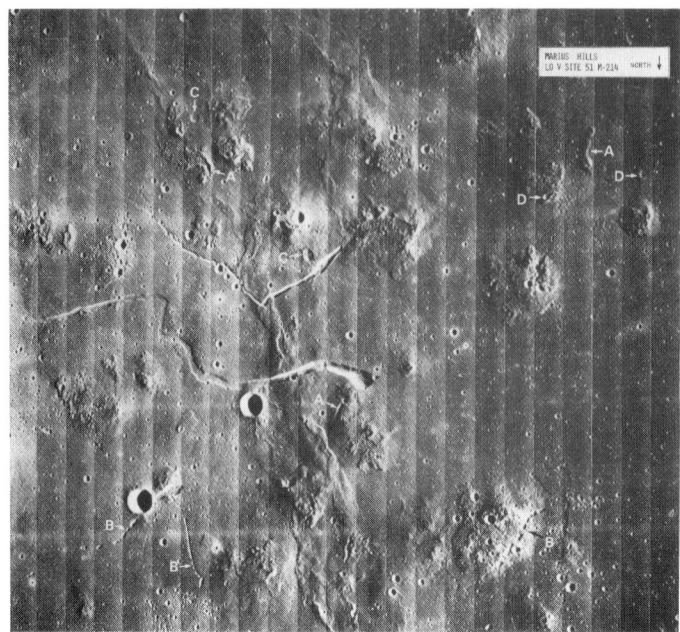

Fig. 8.31: Rough-textured domes, cones, sinuous and linear rilles in the Marius Hills region. Note specifically the linear vents (A), partially unroofed lava tubes (B) and rimless volcanic depressions (C). D are small volcanic cones. Frame width 95 km. Lunar Orbiter V frame 214-M.

depressions. One of the most prominent of such groups occurs in the Marius Hills (Fig. 8.31). Most of the domes here have a rather more rugged aspect than is usual for other complexes on the mare surfaces. This characteristic appears to have evolved in response to superposed volcanic cones and steep-sided flows.

Head and Gifford classified lunar domes into various classes after identification of over 200 individual structures, but noted that while numerous classes could be recognized, there appeared to be two principal modes of occurrence: (i) low, fairly flat-topped and generally circular structures, with flank slopes of less than 5° and with summit pits; and (ii) rather irregular structures with more rugged relief and usually found adjacent to highland terrain.

Because of the variety of inferred volcanic landforms developed, the Marius Hills were one of the stronger candidates for an Apollo landing site, although ultimately it did not become one. For this reason in-depth studies were made and were presented in USGS Open-file Reports (Elston and Willingham, 1969; Karlstrom et al., 1969). The complex includes rough-textured and smoother domes which McCauley (1965) attributed to more- and less-viscous lava respectively. It also has a number of classic sinuous and linear rilles, numerous elongate volcanic vents, partially unroofed lava tubes and prominent volcanic cones, together with a number of stubby flows (see, for instance, Greeley, 1971). Due to

Fig. 8.32 (a): Prominent 9-km-wide domes near Milichius. Apollo 12 panoramic frame 4994-7549.

the large size of imagery which illustrates the finer details of this region, it cannot usefully be reproduced here; thus the interested reader is recommended to study Lunar Orbiter V frame 157-H$_2$.

The diameters of the Marius Hills domes range from 4 to 17 km, while the limited amount of data available suggests they are between 100 and 250 m high. Many of the more rugged domes have the presumed original source craters obliterated by later, steeper-sided domes. Such summit pits as do occur are between 1 and 6 km across, averaging 2.5 km. The relatively high diameter : height ratio suggests that relatively viscous lavas were responsible for their construction, and this is confirmed by the outcrop of several short, steep-sided and stubby flows adjacent to some domes. Guest and Murray (1976) give the length of the longest of these flows as 20 km, most being less, and the average volume as about 4 km^3. Such volumes are similar to viscous flows on the Earth. There is also the distinct possibility that pyroclastic rocks may be involved.

While relatively steep and rugged domes predominate in the Marius Hills, there are also several much smoother, lower-profile domes which, as Greeley (1971) rightly points out, are more closely analogous with terrestrial low shields. Several of these have small summit pits or larger collapse depressions. Such shield-like landforms are located widely in locations close to mare borders and are typical of the complexes located in the Hortensius-Milichius area of Mare Imbrium (Fig. 8.32(a)), in Mare Nubium (Fig. 8.32 (b)), near Herigonius in Mare Humorum, in the Prinz-Aristarchus region of Oceanus Procellarum and on the flooded floor of Mare Orientale.

These smoother-surfaced domes are evidently low shields, similar to those developed in the Snake River Plains. Their morphology and association with collapse depressions and sinuous rilles argues for such an origin. Of 74 domes measured, Guest and Murray derived an average diameter of 10 km (Guest and Murray, 1976), but those seen on the margins of basaltic flood plains in Mare Orientale show a wider variation in size (2.5 – 24 km). The profiles of many of these shields are so low that extremely low solar illumination is required for them to be seen at all and there are numerous instances where dome-like landforms appear to be associated with mare arches and wrinkle ridges. In some instances, it is difficult to ascertain whether dome formation is due to deformation associated with the ridge arches or volcanic construction; in others, the growth of domes evidently was volcanic

Fig. 8.32 (b): Smooth 8 × 4 km dome with summit depression on Mare Nubium. Apollo 16 Hasselblad frame 120-019222.

and clearly preceded the development of mare arches. The Apollo 16 mapping camera series obtained during revolution 60 is particularly revealing, as it imaged the region when it was very close to the terminator, and clearly shows how the mare surface is composed of a plethora of very low shields with diameters ranging from about 10 to 70 km (Fig. 8.33). It seems likely that suitable imagery might reveal similar occurrences in other mare locations.

The very low profile of the Herigonius shields implies there was probably little difference between the viscosity of the lavas which formed them and the more voluminous flood lavas which preceded them, while the similarity in albedo indicates a similarity of composition. The same argument applies to shields in the vicinity of Prinz and the Harbinger Montes. The major difference appears, therefore, to be in the mode of eruption, with the young (mainly Eratosthenian-age) flows being produced in smaller volume from a limited number of central vents, rather than from widespread fissures. Thus it seems that eruption rate is the dominant controlling factor in the production of these lunar shields and low domes. Whitford-Stark and Head (1977) note that low-profile lunar domes probably were produced from lavas erupted at lower rates than those which build terrestrial shields (no more than 10 m^3 s^{-1}), while the steeper ones at even lower rates (no more than 0.1 m^3 s^{-1}). In the Marius Hills, therefore, it is envisaged that an early phase of rapid eruption – implied by the numerous large sinuous rilles located there – was followed by falling eruption rates which promoted the growth of low shields.

Because the great majority of dome-rille complexes are to be found close to the mare borders, it is clear that they preferentially developed where the depth of subjacent mare-filling lavas was relatively thin, these peripheral mare regions being the sites of the final stage of mare development. This was also noted to be the case in Lacus Veris, within Mare Orientale, where low shields are associated with volcanic vents where the depth of

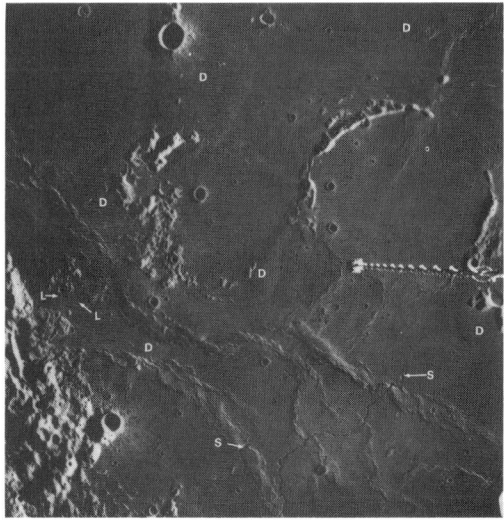

Fig. 8.33: Low shields (D) near Wichmann, Herigonius region, Oceanus Procellarum. Note the prominent lava-inundated highland rings, sinuous rilles with source depressions (S), linear rilles (L) and numerous mare ridges comprising broad arches with superposed steeper segments, all characteristic of lunar volcanic complexes. Frame width 160 km. Apollo 16 mapping camera frame 2838.

mare-fill is not great (Greeley, 1976). These are also the sites of peripheral graben, linear rilles and other tectonic lineaments with which evidently they have some genetic link. This would most likely be accounted for by the propensity for magma to rise up through pre-existing fractures and analysis of tectonic patterns in the regions of Mare Humorum and Oceanus Procellarum suggest that the distribution of Eratosthenian lavas coincides with zones along which fault reactivation has occurred (Raitala, 1977). Some of this might be accounted for by volumetric expansion accompanying the final stages of magma uprise in Oceanus Procellarum. The steeper-profile domes which also concentrate in such regions may represent more silicic magma fractions which had evolved by the time that mare volcanism had reached its waning stages, but there is no actual geochemical data to confirm this supposition at present.

Pyroclastic cones appear to be represented by a number of generally low albedo, conical mounds in the diameter range 2-4 km. These have been studied by McGetchin and Head (1973), McCauley (1969), Mattingley *et al.* (1972). Many are found in the Marius Hills, but they are found elsewhere, for instance in Maria Nubium, Serenitatis and Cognitum (Fig. 8.34). Several of these have elongate shapes and are considered by Wood (1979) to be representative of lunar Strombolian-style activity

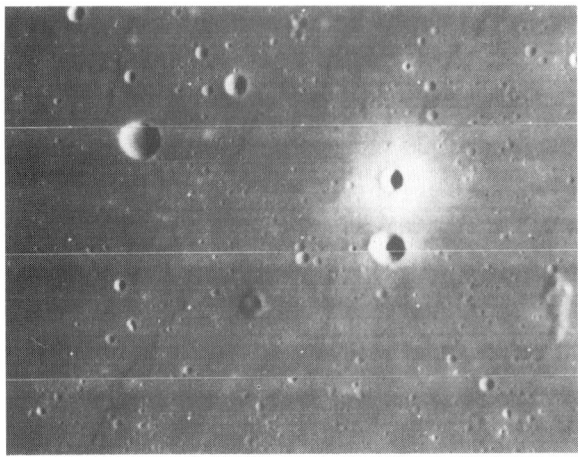

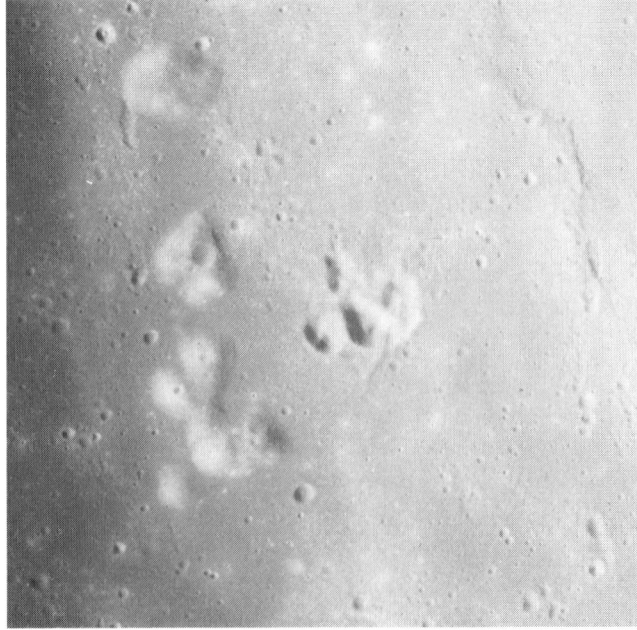

Fig. 8.34: (Top) Small cinder cone in Mare Nubium, west of Lassell. Frame width 55 km. L.O.IV frame 113-H₂. (Bottom) group of cones east of Lansberg in Mare Cognitum. Frame width 40 km. Apollo 12 frame 54-8089.

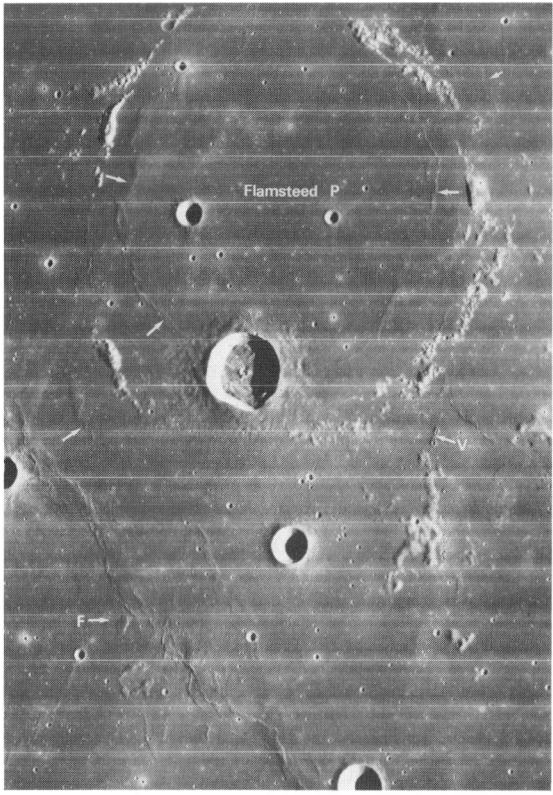

Fig. 8.35: Mare ridges mimicking the ramparts of the buried ring structure, Flamsteed P. To the south a more linear ridge system sits astride a low mare arch and has one prominent volcanic flow emerging from it (F). Closely southeast of Flamsteed P note a prominent linear vent system (V) and its associated linear rille. Frame width 146 km. Lunar Orbiter IV frame 143-H$_3$.

8.4.4.   Mare arches and ridges (wrinkle ridges)

Prominent *en echelon* ridges and complexes of arches and ridges which may be several hundreds of kilometres long are prominent on all maria. Typically they have a concentric distribution with respect to the perimeters of the regular maria and occasionally extend into the bordering highlands; however, in places their trend appears to have little relationship with gross structure. That the ridges are in some way related to the underlying topography is revealed by the frequent occurrence of concentric ridges with buried rings on the mare plains and the large, circular patterns such ridges make with respect to buried basin rings (Fig. 8.35). Individual ridges are up to several kilometres wide and attain axial heights of several tens of metres; where such ridges site atop broad mare arches, they may rise several hundred metres above the general plains level. Various explanations for these features have been given, including igneous intrusion or extrusion from fissures (Fielder, 1965; Strom, 1971) and buckling of lavas or faulting over subsurface strata (Quaide, 1965; Howard and Muehlberger, 1973).

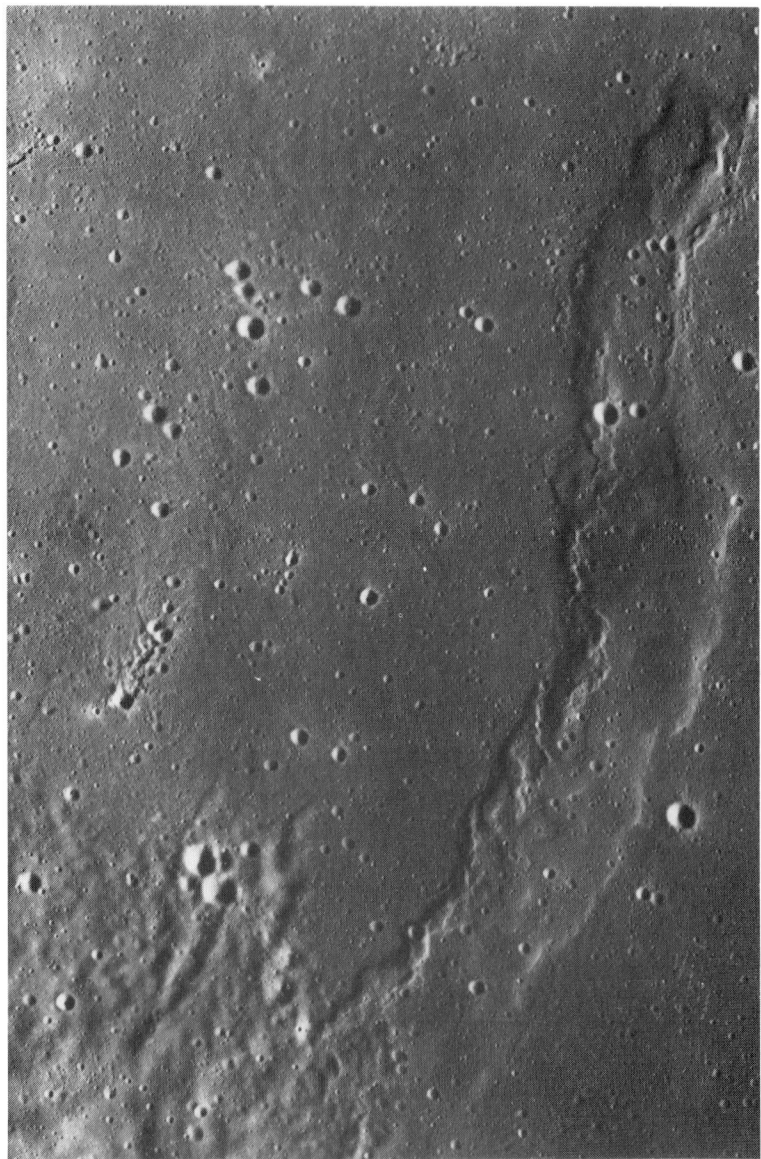

Fig. 8.36: The tortuous, narrow spine of a wrinkle ridge lies on the western edge of abroad
mare arch in the southern two-thirds of this image, but crosses to the opposite side towards
the north. Note that the spines have an *en echelon* outcrop. Frame width 27 km. Apollo 15
panoramic frame 0325 (part of).

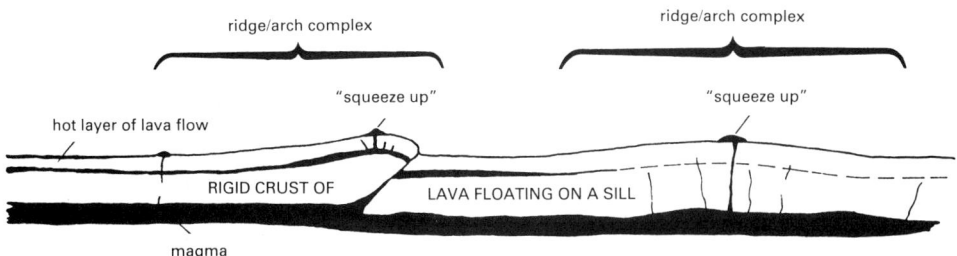

Fig. 8.37: Hypothetical cross-section across a mare arch/ridge complex, showing two different types of ridges. On the right is an arch generated primarily by tectonic processes, while on the left is an asymmetric ridge produced by igneous activity. (After Guest and Greeley, 1977).

Ridges are seen to exist as well-defined spines up to 100 m high and about 200 m wide which often have a tortuous outcrop and which surmount broad mare arches. Lucchitta (1976) notes that arch/ridge complexes frequently have one flank that is offset by a steeper monoclinal flexure or scarp, possibly a fault; the position of the steeper edge may change from one side of an arch to another at different points along its course (Fig. 8.36). Where low-angle illumination provides suitable imagery, broad arches are seen to be much more widespread than was once realized. Arches may be up to about 15 km wide and commonly around 100 m in height, but some arches rise to heights of over 350 m (Muehlberger, 1974). In places the arches comprise broad, overlapping domes, some of which are associated with sinuous rilles and collapse depressions.

During early work on these features, Strom (1971) concluded that broad arches represented either sills or laccolithic intrusions that emerged along fracture systems and there is every reason to suppose that molten magma must have played an important role in their formation. In several locations it can be demonstrated that arches and ridges were formed during the same general volcanic episode. There is thus a strong argument in favour of some ridges being squeeze-ups generated during settling and/or arching of the mare lava crust (Hodges, 1973; Greeley and Spudis, 1978). Elsewhere it has been shown that mare ridges evolved either during eruptions (Greeley and Spudis, 1978) or between successive flow episodes, as in Mare Imbrium (Schaber, 1973; Bryan, 1973) where they were able to divert younger lava flows, while some ridges clearly continued to grow after extrusion has ceased. Bryan (1973), after a study of ridges in Maria Imbrium and Serenitatis, concludes that they were formed in response to localized compression of the relatively thin lava crust which essentially remained uncoupled from the underlying structure. Guest and Murray (1976) make the suggestion that the subcrust material was molten lava, mainly in the form of extensive sills at shallow depth, and that the differential movement of the mare crust above the molten layer below, as the latter cooled and contracted, would have

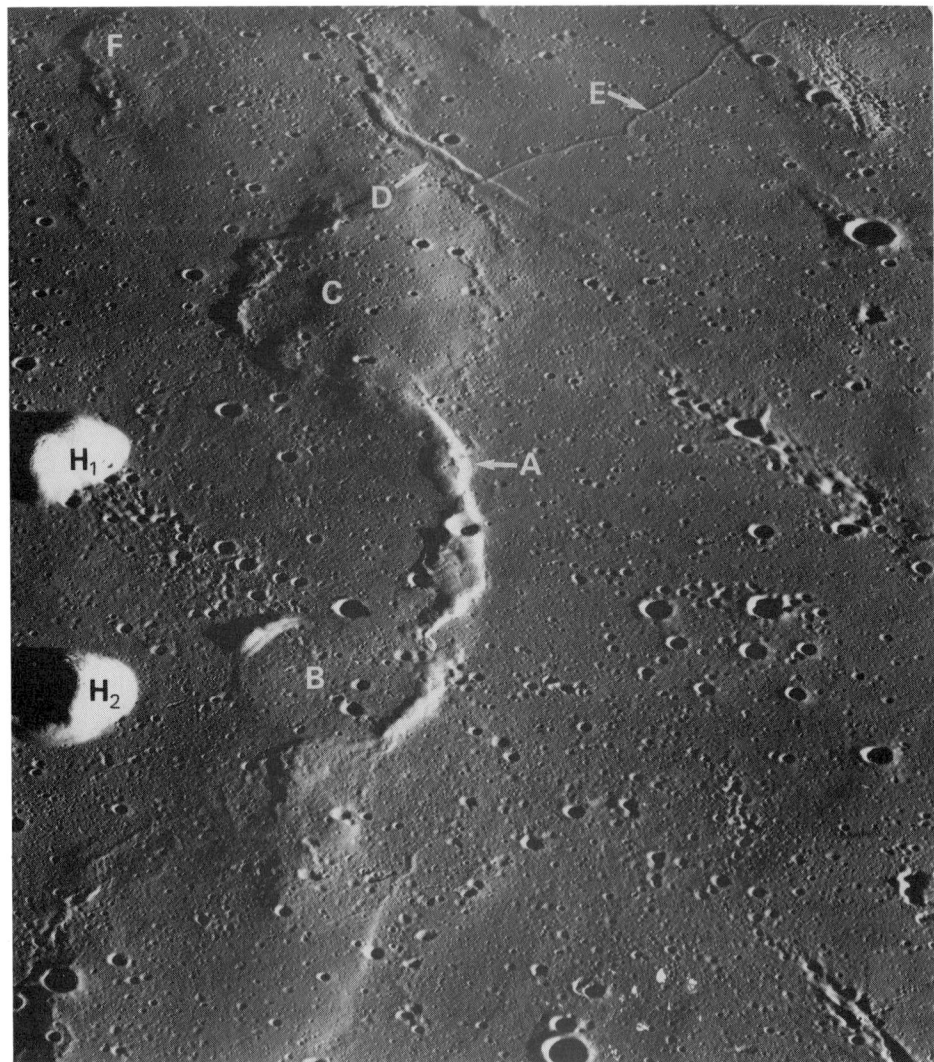

Fig. 8.38: Features of mare arch/ridge system near Harbinger Montes, Mare Imbrium. East of the two prominent Harbinger massifs (H_1, H_2) the arch is narrow and the ridge broad. At A it looks like a viscous dome or tholoidal mass. To the south the system forms part of a buried impact crater (B) beyond which it splits into two narrow sections. North of A a broad dome-like arch can be seen, surmounted by a very narrow ridge that hugs its western margin (C). North of the dome the ridge spine is sharply offset. Obliquely crossing the northeast section of the image is a long narrow positive relief feature that broadens over the dome, and appears to be a volcanic flow associated with a fissure vent. Note at the top left-hand corner a mare ridge formed by lava draping a buried crater ring. Frame width 27 km. Apollo 17 panoramic frame 3135 (part of).

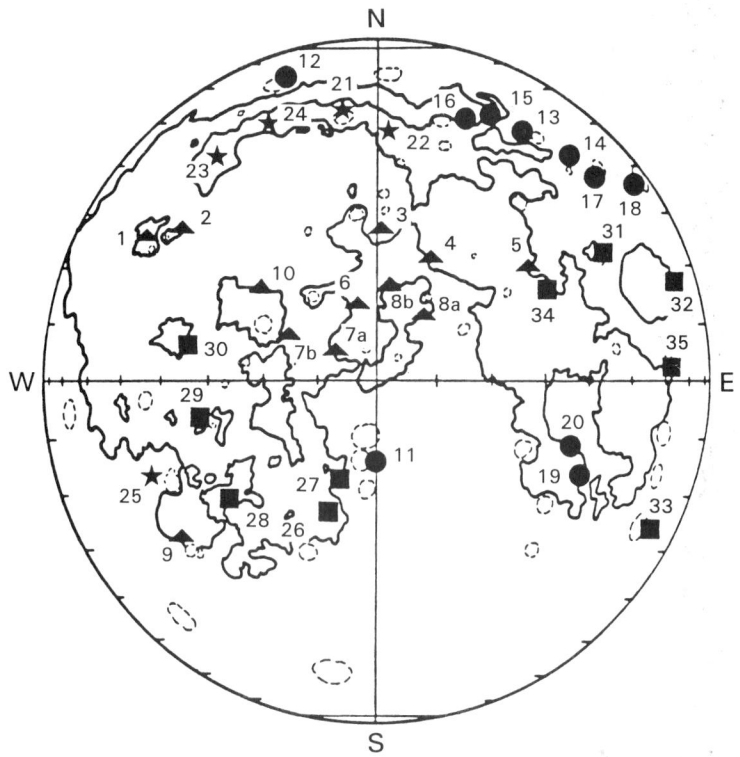

Fig. 8.39: Location of dark-mantling deposits on the nearside of the Moon. (After Gaddis and Pieters, 1985).

caused the solid surface to break into plates. The latter would frequently have regrouped and crumpling would have affected the boundaries, giving rise to the broad mare arches, some of which could have extended beyond the confines of the maria. The steeper-sided spines (ridges) probably represent later extrusions of more viscous lavas that found their way to the surface via fissures (Fig. 8.37).

Colton *et al.* (1972) cite several areas where mare ridges appear to be a manifestation of thrusting of mare units, while Howard and Muehlberger (1973) suggest that compressional forces created discontinuities akin to low angle thrusts along intra-mare gliding planes. Such compression could be supplied where thick lava sequences thin, for instance near mare borders, or where lava drapes buried crater ramparts (Fig. 8.38). Alternatively, there has been the suggestion that the ridges lie along fault systems that include both normal and reversed sections, these together producing compressional stresses at the surface (Lucchitta, 1976, 1977). The complex morphology of individual ridge systems and the variation they show from one location to another implies that their origin is a function of numerous factors and suggestions such as those of Lucchitta are completely consistent with scenarios like those of Guest and Greeley.

The ultimate form of ridge/arch systems evidently derives from the interplay of volcanic intrusion and extrusion with tectonic movements, such as thrusting and normal faulting, while their location is related in various ways to underlying structure and regional tectonic lineaments. The association of volcanic flows and low shields with many ridges certainly suggests that both extensional and compressional tectonic regimes pertained while they grew.

8.4.5   Dark mantling materials

Quite extensive deposits of low-albedo materials can be seen mantling some terra regions adjacent to mare borders, and clearly are related to mare volcanism. These have received considerable attention and generally are considered to represent pyroclastic deposits (Head, 1974). Remote sensing indicates that most such deposits have a spectral reflectance signature that is unique, and which indicates the presence of Fe-rich volcanic glass (Gaddis and Pieters, 1985). Most such low albedo deposits also exhibit low de-polarized 3.8-cm radar returns, while there are a small number of higher-albedo deposits which give similar returns and may also be pyroclastic in origin, but compositionally different. The distribution of dark mantling deposits is shown in Fig. 8.39.

The pyroclastic interpretation was confirmed on the ground at the Apollo 17 site, an area of dark mantling material studied from Earth by several groups (Heiken *et al.*, 1974; Pieters *et al.*, 1974). The regional deposits of this type all occur at or near mare borders, have low albedo and a fine-grained, smoothed appearance in telescopic or low-resolution imagery. However, they have a more rugged relief than the typical mare deposits (Fig. 8.40). In a discussion of the dark mantling deposits near the bright crater Aristarchus, Zisk *et al.* (1977) concluded that this photogeological signature represented the mantling of pre-existing (mare-bordering) topography by explosively generated materials. The occurrence of numerous sinuous rilles and shallow endogenic craters in the same region gave support to this conclusion. A similar origin has been proposed for the other well-documented deposits, i.e. those at Taurus-Littrow (Lucchitta, 1973), Sulpicius Gallus (Lucchitta and Schmitt, 1974), Mare Humorum (Pieters *et al.* 1975), Palus Putredinis (Hawke *et al.*, 1979) and Rima Bode (Wilhelms, 1968; (Gaddis *et al*, 1981). Calculations by Wilson and Head (1981) showed that pyroclastic deposits of regional dimensions could be generated by a continuous eruption of volatile-containing lunar magma; fire-fountaining would be the mechanism responsible for the scattering of these veneers over the surrounding terrain. The sinuous rilles and their source depressions are considered to be the source of these magmas.

The black and orange glasses collected at the Apollo 17 site give the typical low-albedo and corresponding spectral signatures of these deposits, but Gaddis and Pieters (1985) note that there are at least five further areas of higher-albedo mantling materials that give similar radar responses and which may be identifiable, not with orange and black, but with green and yellow glasses. The continued extension of our knowledge of the distribution of such volatile-bearing magmatic deposits by remote-sensing methods

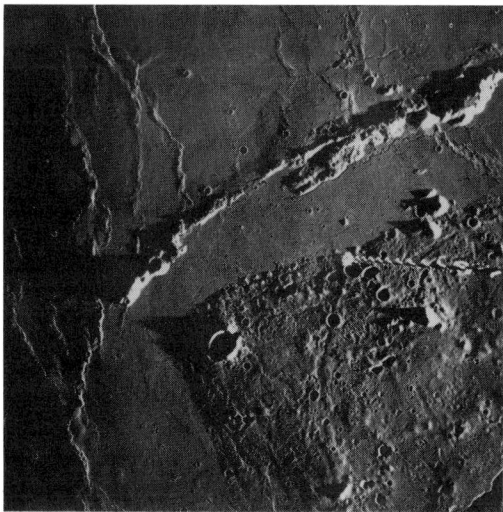

Fig. 8.40: Dark mantling deposits here are seen covering the Aristarchus Plateau, occupying the bottom right-hand part of this image. Note the smoothed, subdued relief but higher impact crater incidence than the adjacent mare. The Aristarchus Plateau gives a distinctly "red" spectral signature. Frame width 148 km. Apollo 14 mapping frame 2033.

will be an important step forward in understanding the compositions of the lunar magma source regions. What is clear is that the explosively generated mantling deposits, with their volatile components, are in direct contrast to the majority of lunar mare materials.

Far more restricted in occurrence are dark-halo craters, perhaps the best-known of these being those which occur on the floor of Ptolemaeus. Such craters are quite widespread but are far more localized than regional dark-mantling deposits. These are considered to be the products of Vulcanian-style eruptions, where the accumulation of volatiles in a blocked volcanic conduit promotes explosive decompression, giving rise to localized expulsion of pyroclastics (Head and Wilson, 1979).

8.4.6   Mare filling as exemplified by the Orientale Basin
In the context of mare emplacement, the Orientale Basin is very informative, and has been studied by Greeley (1976). Geological mapping distinguishes an inner, partially flooded basin centre (Mare Orientale), a less extensively flooded region between the centre and Montes Rook (Lacus Veris), Lacus Autumnii (between the Montes Rook and Cordillera) and a high-albedo plains unit that outcrops in peripheral locations of Mare Orientale (Fig. 8.40). The latter appears to represent early impact melt.

The most extensively flooded part of the Basin, Mare Orientale, is the youngest unit present. It has low albedo, a low incidence of impact craters and upon its surface are numerous mare ridges which have subjacent arches as well as a small number of very low shields (Fig. 8.41). Where older massifs stick through the mare fill they show sharp, steep borders, as though lava flooding effected some erosion of their edges.

Fig. 8.41: Synoptic view of the Orientale Basin showing the distribution of mare plains units and areas shown in following figures. Frame width 600 km. Lunar Orbiter frame IV 187-M.

Fig. 8.42: Northern part of Mare Orientale, showing prominent mare ridge (R) formed near border with submerged higher albedo unit (Maunder Formation), low shields (S) and submerged massif with thermally eroded scarp-like margin (M). Note also several dark-halo craters (H) silhouetted against the ejecta from the large impact crater. Frame width 250 km. Lunar Orbiter frame IV. 195-H$_2$.

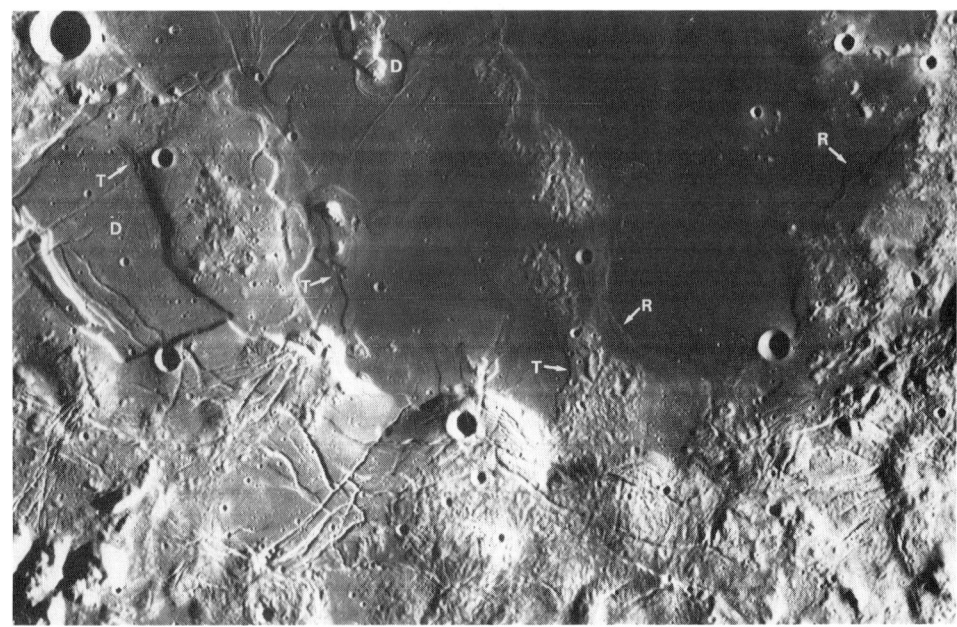

Fig. 8.43: Southern part of Mare Orientale, showing prominent wrinkle ridges (R) mimicking mare border. There is a general lack of sinuous rilles and lava flow fronts. Lava "tide marks" appear in several places (T), while two large depressions (D) suggest that there has been considerable subsidence within the lava sequence during and possibly after its emplacement. Frame width 250 km. Lunar Orbiter frame IV. 195-H$_1$.

Fig. 8.44: Coalescing low shields on the floor of Lacus Veris. Note the linear vent system and several rimless depressions. Frame width 75 km. Lunar Orbiter frame IV. 187-H$_2$.

In the southern part the flooded area, mare ridges parallel the contact between the lavas and the underlying Maunder Formation, indicating there to be a relatively thin cover of lavas there. Perhaps the most obvious feature of the mare lavas, apart from their smooth appearance, is that they have undergone quite extensive subsidence, this being revealed by sharply defined basins with scarps as high as 1 km (Fig. 8.42). Linear arcuate rilles parallel the mare/Maunder Formation contact.

Some of the collapse depressions have scarps that are 1 km high, suggesting that the lava fill must have been at least this thick. The lack of visible source vents and the absence of sinuous rilles suggests that the inner part of the Orientale Basin was filled by flood lavas.

In contrast to this, the mare units of Lacus Veris are strikingly different. Nearly all of the basalt units here have associated sinuous rilles, Greeley (1976) mapping ten of these in Lacus Veris and three more in the mare deposits of Lacus Autumnii. On the surface of Lacus Veris are numerous low domes (Fig. 8.43) which appear to represent coalescing shields. The largest of these is about 12 km across and several have elongate vents on their crests. The basalt fill here, therefore, appears to have been emplaced by plains-style activity, presumably at somewhat lower effusion rates than the central flood lavas. This gave rise to a greater number of identifiable source regions. Since many of the rilles commence on the margins of the Rook Montes massifs, it appears that they carried magma from here onto the low-lying basin floor.

The general sequence of basalt emplacement witnessed in Mare Orientale, the youngest large lunar basin, may provide a blueprint for the processes which operated in most, if not all, the nearside basins. There is certainly a considerable body of evidence that extensive flood lavas produced major thicknesses of early flows and that these were followed by successions emplaced from identifiable source regions, as in the borders of Mare Imbrium, or in regions of major low-shield development. The development of complex ridge/arch systems within the maria indicate a substantial degree of post-emplacement deformation, believed to have occurred in response to movements of slabs of mare crust above a still mobile layer of magma beneath. Pyroclastic volcanism appears to have been confined to the perimeters of mare plains.

9

Shield volcanoes and terrestrial examples

9.1 INTRODUCTION

Shield volcanoes are broad, gently sloping lava cones which usually have a shallow caldera at their summit. Associated smaller pits and spatter cones are typical, these often being concentrated along rift zones. Terrestrial shields show a range of sizes and profiles which range from small low-profile shields such as those of Iceland, through the steeper shields of the Galápagos Islands to the very large volcanoes of the Hawaiian-Emperor Chain. Such groups of shields comprise some of the most volcanically active regions on Earth and exemplify the style and scale of volcanicity associated with mantle hot spots. Smaller, very low shield volcanoes, exemplified by those of the East Snake River Plain, essentially are intermediate between Icelandic volcanoes and flood basalts and have been discussed elsewhere (Chapter 6). The shield volcanoes of Mars and Venus are of immense size and several are orders of magnitude larger than their terrestrial counterparts. Summit depressions and related lobate summit and flank flows are characteristics of shields on both planets.

Patera volcanoes may be of even larger areal dimensions than terrestrial shields and were first identified on Mars. They more closely resemble upturned saucers than shallow soup bowls – the analogy used for shields – and are characterized by lower profiles than shields, flank slopes being as gentle as 0.25°. Most Martian paterae have complex summit calderas and often have well-defined rift zones. While some are believed to be wholly constructed from fluid basaltic lavas, others may have been largely due to ash eruptions. Paterae also exist on Venus; these too are of large size. The paterae of Io have virtually no vertical relief and are associated with a unique style of volcanism which is treated separately in Chapter 11.

Volcanic domes have a range of sizes on the Earth; most are generated by lavas considerably more viscous than basalt. Small domes may be a mere hundred metres or so across; large ones a few kilometres. The carapaces of viscous domes may be breached by short, stubby volcanic flows. Domes also occur on Mars, where they tend to be associated with "plains" style volcanic fields, and on Venus. Those of Venus are particularly interesting, some being quite large (10 – 50 km) and having flat tops; these are termed pancake domes.

Fig. 9.1: Oblique view across South Pit (foreground) and Mokuaweoweo on the plateau-like summit of the Hawaiian shield, Mauna Loa. The larger caldera measures approximately 4 × 5.6 km across and on its floor, just beyond South Pit, may be seen the 1940 cinder cone. (U.S. Air Force photograph, 1966.)

Others have scalloped margins ("ticks") and steep sides. Low relief domes also are located on the lunar maria, usually near to their margins.

Uniquely Venusian are a group of related volcano-tectonic landforms that include coronae, novae and arachnoids. These are characterized by having a circular to ovoid outline, associated radial or concentric fractures, often an encircling topographic "moat", and related smaller volcanic features such as shield fields, domes and volcanic flows. They range in size from 70 to over 1000 km diameter.

9.2   GENERAL CHARACTERISTICS OF TERRESTRIAL SHIELDS

Despite their immense dimensions, even the largest terrestrial shield volcanoes seldom fire the public imagination in the same way as do strato-volcanoes like Vesuvius or Fuji. Presumably this is partly because they are relatively unspectacular as landscape features and partly for the reason that their eruptive style seldom brings about major loss of life –

one of the prime requisites for a natural phenomenon to make press headlines! However, the volume of volcanic materials involved in the construction of large shield volcanoes necessitates them being ranked supreme.

Small shields, typified by Icelandic volcanoes such as Skjaldbreid and Kollota Dyngja, have grown simply as a result of overflow of fluid lava streams from a summit crater. Most such shields are less than 1 km in height, many being less than 100 m; their basal diameter, however, may be up to 20 times this. Icelandic shields, like those characteristic of regions such as the Snake River Plain, typically are situated within regions of crustal extension, either where lithospheric plates are separating or have separated and may be associated with flood lavas whose outpouring accompanied plate separation. Hot spot tracks leading away from the currently active centre under Iceland, lead back through northern Britain to East Greenland where there are extensive flood basalts erupted during the opening of the Atlantic about 60 million years ago.

The large shield volcanoes of Hawaii and the Galápagos Islands also are associated with mantle hot spots but are substantially larger. Those of Hawaii attain immense proportions; thus Mauna Loa, typical of the Hawaiian type, rises approximately 10 km from the Pacific floor and has a basal diameter of about 400 km. However, despite its volumetric immensity (425 000 km^3), the flank slopes are less than 6°, and in consequence it is not a striking landform. For a variety of reasons, not the least of which is the longer time over which they grow, the detailed geomorphology of large shields is generally more complex than small ones. Thus the former have plateau-like tops which often are surmounted by cinder cones, small subsidiary shields and complex collapse depressions (Fig. 9.1). A similar complexity characterizes the large volcanic shields of the Galápagos Islands which tend to have somewhat steeper flank slopes that lie within the range 10-34°. Their steeper profiles are in part the result of somewhat lower eruption rates than those typical of most Hawaiian volcanoes; however, their geological setting also is different, since the volcanic islands of the Galápagos lie closely south of an active spreading axis (the Galápagos Rift) that separates the Cocos and Nazca plates, whereas the active portion of the Hawaiian chain is firmly intra-plate in its structural setting.

In general, shield volcanoes are built from basalts or olivine basalts, but there are a few andesitic (Hayli Gub), alkaline (East African Rift) and ignimbritic (Vulsini) shields (Wood, 1977). Hawaiian eruptions are typically basaltic and while they may emanate from the summit region, much activity originates from fissures and rift zones, which in some cases are radial and in others approximately concentric with respect to the summit. Vigorous fire-fountaining usually accompanies the onset of eruptive activity, lava fountains often reaching heights of 500 m and temperatures of 1165°C (Swanson, 1973). Most lavas emerge as thin flows, averaging about 4 m thick, a very high proportion (80%) being emplaced via tubes and channels (Greeley *et al.*, 1976). The distribution pattern of flows associated with central vents active over lengthy periods tends to be radial, while fissure-related activity produces a more polarized distribution (see Fig. 4.3). The range in average eruption rate for Hawaiian shields is between 4 and 300 m^3 s^{-1} (average 60 m^3 s^{-1}) but during the initial stages of an eruption lava may emerge from vents at much higher rates; for instance 560 m^3 s^{-1} was recorded at the vent during the 1984 Mauna Loa eruption (Moore, 1987), while an eruption rate of over 1000 m^3 s^{-1} is implied by the 150-km-long flow associated with Trölladyngja in Iceland.

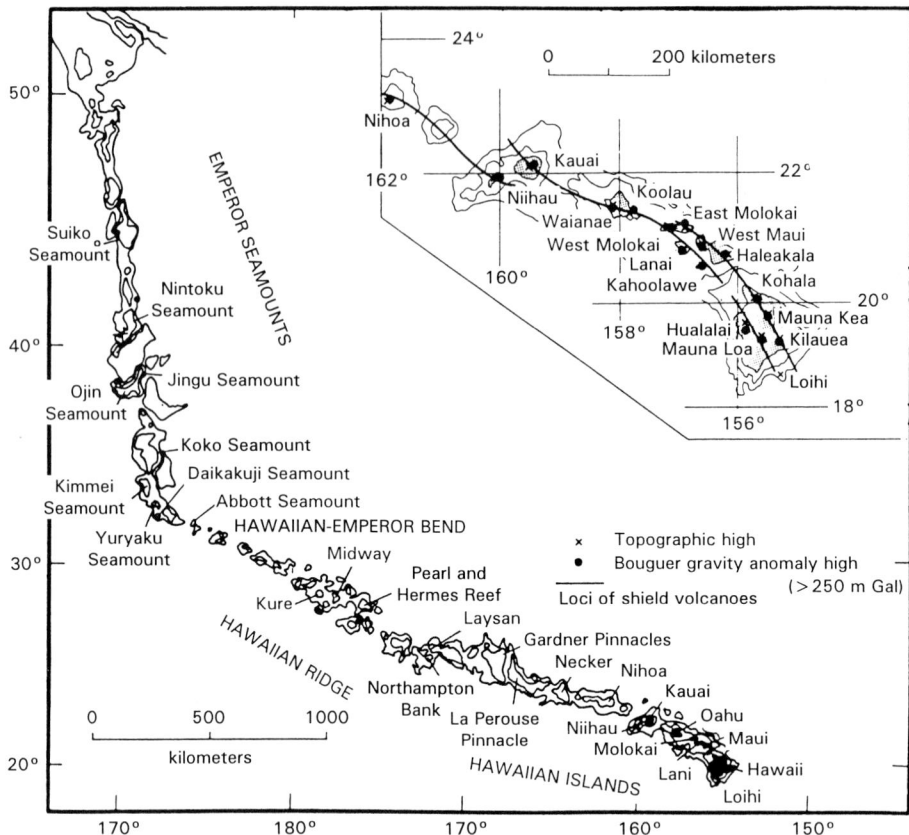

Fig. 9.2: Location of shield volcanoes in the Hawaiian Emperor chain. (After Clague and Dalyrimple, 1987.) The inset shows details of the relationships between topographic highs, Bouguer gravity anomaly highs and loci for principal Hawaiian islands. (After Jackson *et al.,* 1972.)

9.3   THE HAWAIIAN SHIELDS

The Hawaiian-Emperor volcanic chain, which stretches for 6000 km across the North Pacific, consists of at least 107 individual centres that have a total volume of about 1×10^6 km^3 (Fig. 9.2). The largely basaltic shields have grown on oceanic crust of Cretaceous age and their observable summits sit atop a broad ridge (the Hawaiian Ridge) in the ocean floor. The marked deflection observed in the trend of the Hawaiian and Emperor chains near 32° N reflects a major change in plate motion that occurred 43.1×10^6 years ago. Individual volcanic centres appear to be sited on short, sigmoidal and overlapping loci arranged *en echelon* in a clockwise sense in the Hawaiian chain (Fig. 9.2, inset). One

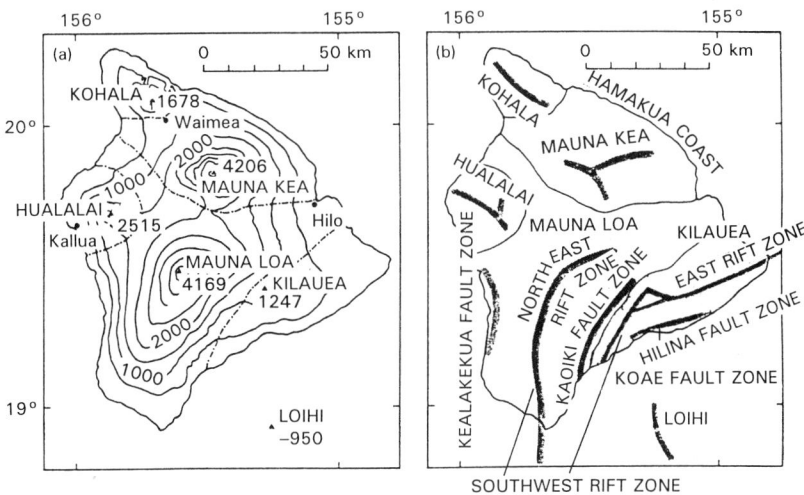

Fig. 9.3:(a) Island of Hawaii, showing major geographic features and Loihi Seamount.
(b) Major rift and fault zones of Hawaii. (From Peterson and Moore, 1987.)

possible explanation for this pattern is that it may have been the result of extensional strain resulting from tension within the Pacific plate.

The active shield volcanoes of the island of Hawaii, which lies at the southeastern end of the chain, are amongst the most closely studied volcanic structures on the Earth and a huge library of information relating to their eruptive history now exists. The island consists of five volcanoes with which are associated several prominent rift zones (Fig. 9.3). The volcano Kohala, which last erupted 60 000 years ago, is the most dissected, while Mauna Kea, last active 3600 years ago, is larger and has numerous cinder cones on its flanks. The giant shield Mauna Loa last erupted in 1984 and occupies more than half the area of the island; it has a summit caldera complex and several prominent rift zones associated with it. To its west lies the smaller shield of Hualalai, relatively steep-sided, undissected and last active in 1801, while adjacent to its southeast flank is Kilauea, the most thoroughly studied eruptive centre of the group. Kilauea has been active frequently in recent years, and has a summit caldera with several pit craters together with a series of recently active rift zones that intersect the island's coast (Plate 6). Finally, offshore 30 km south of Hawaii is the seamount, Loihi, an active and growing shield volcano whose summit currently is 950 m below the Pacific Ocean. The beautifully produced computer-drawn perspective diagram of the part of the Hawaiian chain between Oahu and Hawaii published by Fornari and Campbell (1987) nicely illustrates the overall physiography of this group of shields, bringing out the contrast between the older, steeper, submarine parts of the structures and the less steep, subaerial profiles of the visible coalescing shields (Fig. 9.4).

Due to the enormous load placed on the ocean floor by the volcanic constructs and the repeated withdrawal of magma from reservoirs beneath them, the Pacific lithosphere has

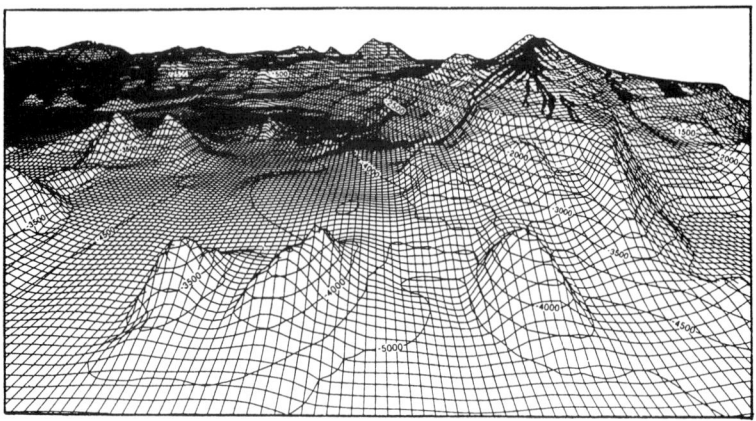

Fig. 9.4: Computer-drawn perspective diagram of the sea floor on the western flanks of the Hawaiian Ridge, looking north between Oahu (left-hand side) and Hawaii (right-hand side). Contour interval is 500 m and vertical exaggeration × 5. Produced by Dynamic Graphics Inc., Berkeley, California, and reproduced from Fornari and Campbell, 1987.

sagged and the Hawaiian Ridge subsided. Most of the shields have subsided between 2 and 4 km since reaching the ocean surface, and their bases have subsided rather more – between 5 and 8 km (Moore, 1987). The bulk of such subsidence typically seems to be accomplished about 1×10^6 years after the initiation of ocean floor activity at any particular locus. While an unknown amount of construction is achieved by dyke and other intrusions, calculations suggest that approximately one-half of all constructional activity is cancelled out by this subsidence.

Geophysical data suggest that the oceanic crust immediately below the Hawaiian chain is between 15 and 20 km thick – over twice the thickness typical of oceanic crust in adjacent regions. The greater thickness is believed to be a result of isostatic subsidence related to the transfer of crustal and mantle materials during volcano development. Because the lithosphere cools, thickens and subsides as it moves away from spreading axes and their attendant heat sources (Parsons and Sclater, 1977), it might be anticipated that this, together with volcanic loading, could account for the observed degree of subsidence. However, Detrick and Crough (1978) have shown that the amount of subsidence experienced by many islands and seamounts along the chain far exceeds that which could be accounted for by these processes alone, and to accommodate the surplus they invoke local thermal resetting of the lithosphere which, as it passes over a hot spot, renews the ageing process. The excess sagging is seen as one outcome of this renewed ageing of the lithosphere.

Radiometric ages for Hawaiian volcanoes obtained by McDougall (1964) have confirmed the early contention of Stearns (1946) that a volcano extinction sequence could be observed in moving westward along the Hawaiian part of the chain, and vindicated

Table 9.1: K-Ar ages for seven Hawaiian shield volcanoes (10^6 years). (After McDougall, 1964.)

Kauai	5.8 – 3.9
Waianae	3.5 – 2.8
Koolau	2.6 – 2.3
West Molokai	1.8
East Molokai	1.5 – 1.3
West Maui	1.3 – 1.15
East Maui	0.8
Hawaii (all centres)	‹1

Stearn's hypothesis that the main shield stage of each volcano was complete before the next one rose above sea level (Table 9.1).

All existing data suggest that Hawaiian shields are built up from the ocean floor in between 0.5 and 1.5×10^6 years (Jackson *et al.*, 1972). Calculations based on all available evidence indicate that volcanic propagation rates are 9.2 ± 0.3 cm y^{-1} along theHawaiian part of the chain and 7.2 ± 1.1 cm y^{-1} for the Emperor section. A best fit through all the age data gives 8.6 ± 0.2 cm y^{-1} for the chain as a whole. It also appears that there has been a slight increase in the rate of volcanic propagation during the last few million years.

It is generally accepted that volcanic activity in the Hawaiian Chain is related to a very active hot spot which has remained more-or-less fixed during the last 40×10^6 years but which, prior to this, between 65 and 40×10^6 years ago, lay in a more northerly latitude (Wilson, 1963; Morgan, 1972). Because the volcanoes are passively travelling away from this heat source on lithosphere which is slowly cooling, thickening and subsiding at the rate of about 0.02 mm y^{-1} (Clague and Dalyrimple, 1987), progressively older, extinct volcanoes are encountered as one moves westwards along the chain while, in contrast, the shields in the extreme east have not yet completed their growth cycle. The seismically active seamount, Loihi, situated southeast of the island of Hawaii, appears to be the latest product of activity over this long-lived hot spot.

9.3.1   Growth cycle of a typical shield

Hawaiian volcanoes erupt lavas of quite distinctive chemical composition during different stages in their life. Early studies by Stearns (1940), Macdonald and Katsura (1964) and Macdonald and Abbott (1970) established that several developmental stages characterized shield growth and these workers were able to document all but the very earliest (Fig. 9.5). More recently, geophysical studies of Loihi Seamount (Moore *et al.*, 1982) have allowed even the earliest stages of shield construction to be probed.

Most earlier schemes recognized a late shield-building stage during which caldera formation occurred and this development was often used as a marker in a volcano's evolutionary history; however, as long ago as 1884, Dutton suggested that caldera development was a recurring process during a shield's growth. In support of this view, Powers (1948) and Holcomb (1987) both have provided evidence for the existence of an

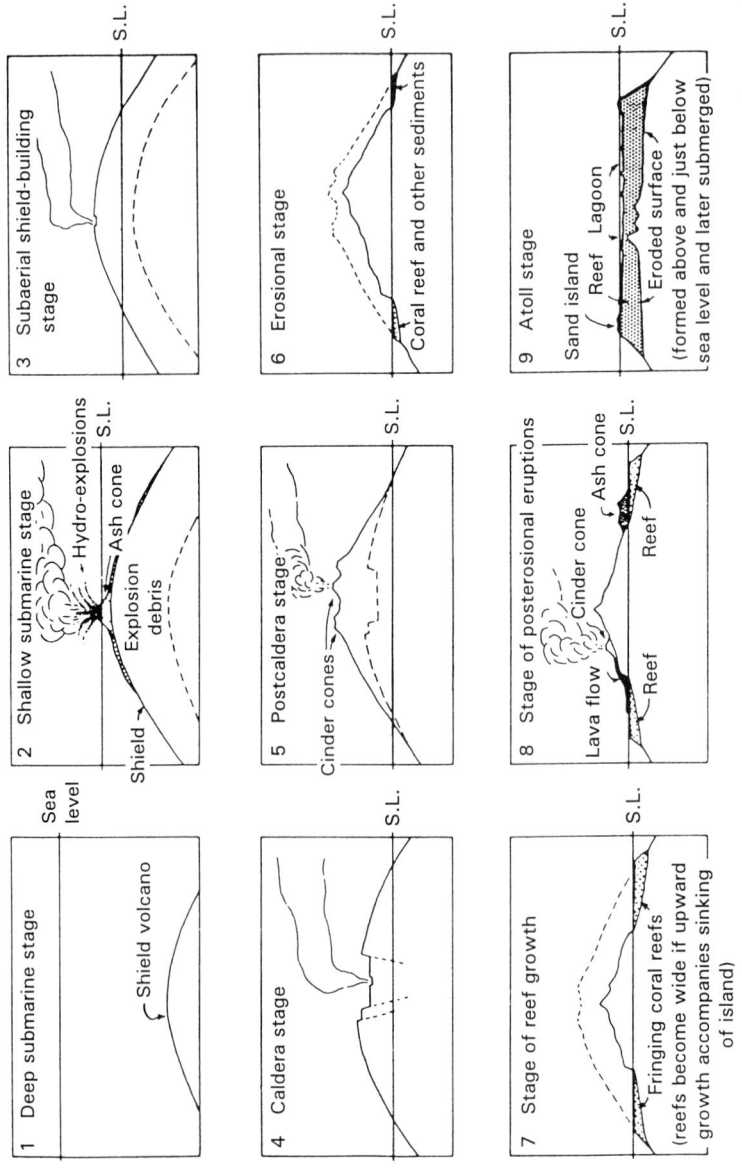

Fig. 9.5: Life cycle of a Hawaiian shield as depicted by Macdonald and Abbott (1970).

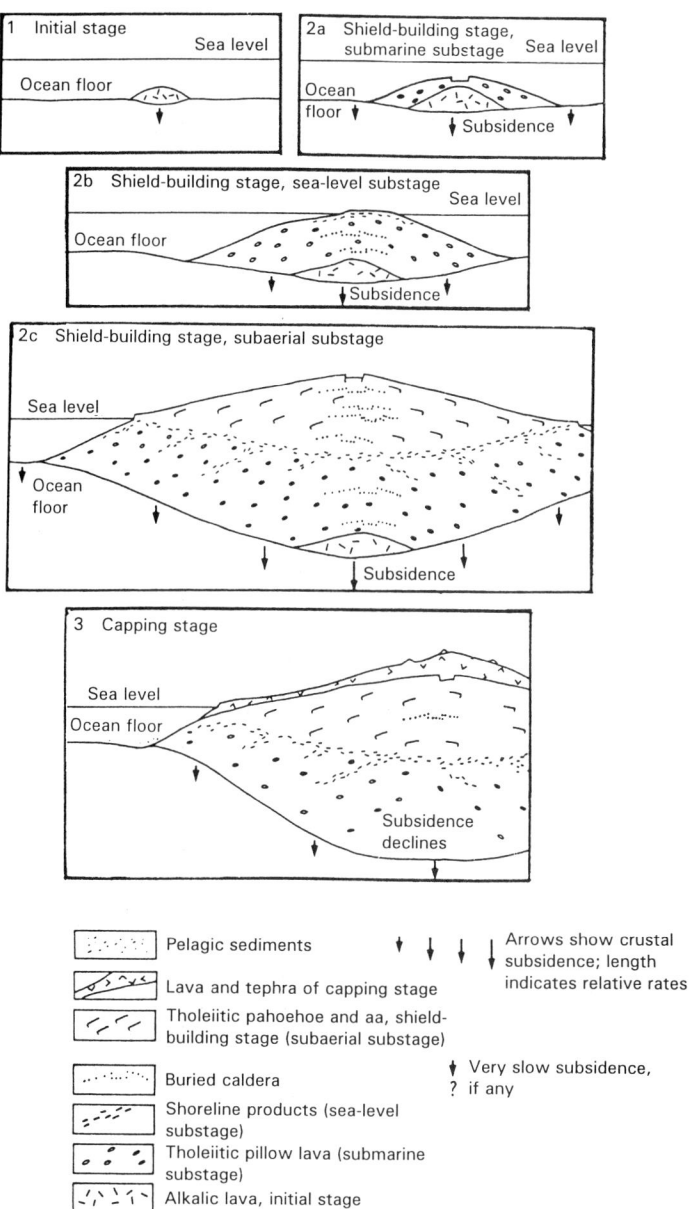

Fig. 9.6: Development stages for a typical Hawaiian shield. (After Peterson and Moore (1987).)

infilled and buried caldera beneath the present caldera on Kilauea. They believe a similar situation may pertain elsewhere on Hawaii. Furthermore, seismic studies of the Loihi Seamount have revealed that, although still a youthful submarine feature, it already has a 70-m-deep summit depression measuring approximately 2.8×3.7 km within which are two smaller pits (Malahoff, 1987). The flavour of this combined evidence is that not only do calderas form during the very early stages of shield construction, but also their development is not necessarily confined to a specific stage in the growth cycle.

A recent detailed study by Clague and Dalyrimple (1987) recognizes four main growth stages for Hawaiian volcanoes: (i) a submarine *pre-shield stage* dominated by low eruption rates and alkali basaltic lavas; (ii) a main *shield-building stage* during which between 95 and 98% of a volcano is constructed, usually in less than 1×10^6 years, and which is dominated by tholeiitic lavas erupted at high rates and often accompanied by caldera collapse; (iii) a *post-shield stage* during which mainly alkalic lavas are erupted, these often filling or partially filling the caldera, if present, or forming a cap to the volcano; such lavas account for about 1% of the total volume and are extruded at low eruption rates; (iv) an *alkalic rejuvenated stage*, whereby after perhaps as much as 1×10^6 years of quiescence and erosion, very small volumes of differentiated silica-poor lava may be erupted at very low rates from isolated vents. Subsequent stages involve further erosion and eventual atoll formation but since these latter stages have little relevance to shield growth on either Mars, Io or Venus, they are not discussed further here.

A more detailed scheme has emerged from the work of Peterson and Moore (1987), who have expanded on the above and part of whose proposed developmental history is shown in Fig. 9.6. Their growth stage 1 commences with the eruption of lavas from sea-floor fissures and vents, repeated activity gradually constructing a submarine lava pile built mainly from pillow lavas of alkaline composition. The resultant edifice has quite steep flanks, exemplified by the bathymetry of Loihi which shows slopes as steep as 45°. Linear rift zones may eventually develop during this stage, their trends being determined by local stress fields.

The early part of the main growth phase (stage 2) sees the rapid and repeated outpouring of predominantly tholeiitic basalt and picrite pillow lavas from both rifts and a summit vent or vents, these eventually building up a submarine shield whose flank slopes are within the range 10-20°. Because of loading by the growing volcanic pile, the subjacent lithosphere downsags, lengthening the time during which the shield summit emerges above sea level (now thought to be at least 1 million years). As it approaches sea level, the hydrostatic head declines sufficiently for boiling of the ocean water to occur as it interacts with vesiculating lava. Steam produced in this way disrupts the lava, often explosively, ejecting tephra and hyaloclastites which, together with coherent lava flows, contribute to the broad pile of mixed lava and pyroclastics that slowly emerges from the ocean. If upward growth exceeds the wave erosion rate, an increasingly large subaerial edifice evolves. (This stage has not been observed in Hawaii but has, of course, been seen at Surtsey in Iceland.)

Now that the island has been established, further repeated and rapid eruptions of tholeiitic basalts occur from both the summit and from major rift zones (Plate 7). Slopes on the subaerial part of the shield are generally considerably lower (3-10°) than in the submarine regions. Because growth is rapid, there is continued sagging of the lithosphere and the

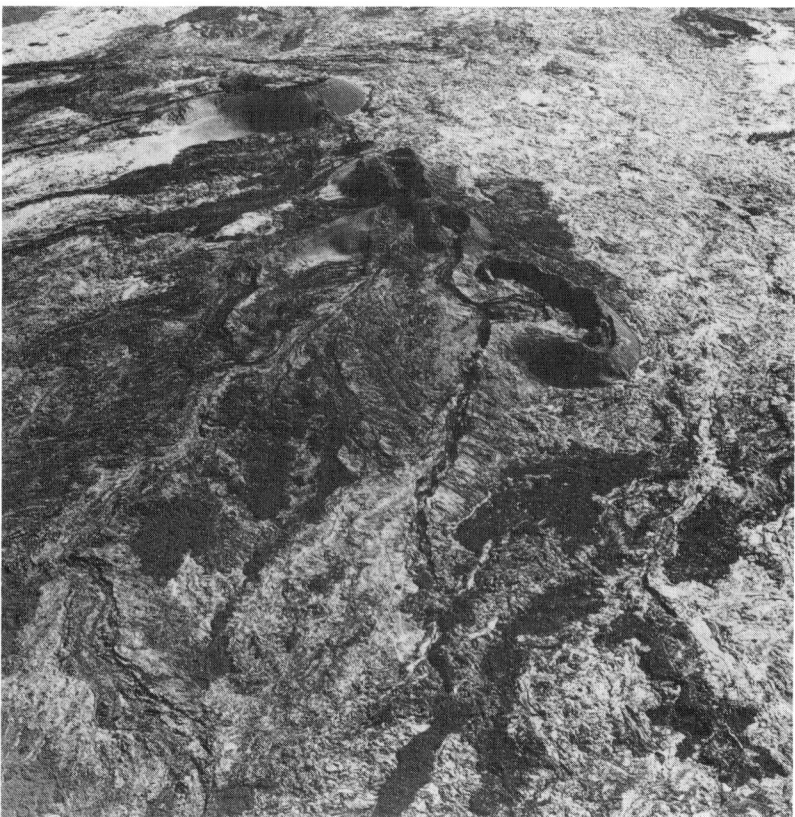

Fig. 9.7: Oblique view of cinder and spatter cones developed on the Southwest Rift Zone of Mauna Loa. Most of the cones are seen to be breached and probably grew, for the most part, while the lava channels in the foreground were active. Photograph by Ron Greeley, 1970.

earlier mixed lava/pyroclastic horizons become broadly synclinal. Where calderas develop, these may be repeatedly filled or emptied as active lakes occupy their interiors. Kilauea and Mauna Loa exemplify shields at this stage in the growth cycle.

Eventually a time is reached when magmatic differentiation yields more alkaline rocks, i.e. alkali basalts and transitional basalts; these may infill a caldera, if present, or form a lava capping to the shield (stage 3). In Hawaii, caldera formation has not been observed during this stage and it is likely that this is a function of declining magma volumes. Certainly the volume of lava erupted does decline at this point and further differentiation, producing rocks such as hawaiites, mugearites and trachytes, means that explosive activity increases. In response to this change in eruptive style a steeper-sided cap may grow, with slopes of up to 20°. As time elapses, eruptions become more widely separated by periods of inactivity, until eventually eruptions cease. Hualalai and Mauna Kea volcanoes are

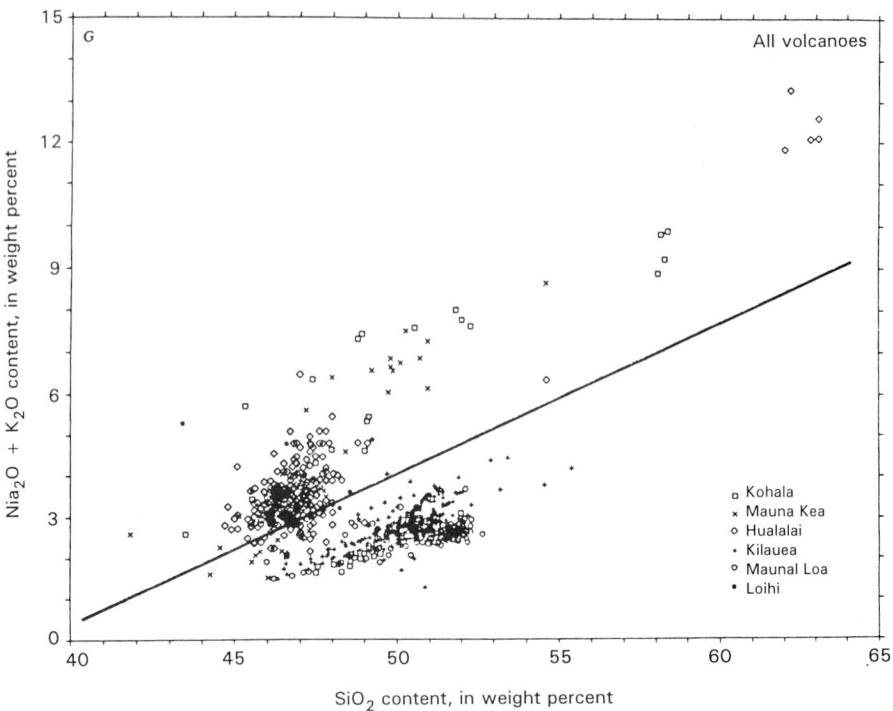

Fig. 9.8: Silica/alkalies plot for all Hawaiian shields with respect to the tholeiite/alkali basalt fields as defined by Macdonald and Katsura (1964). Modified after Peterson and Moore (1987).

both at this stage, although the latter has progressed much further along its evolutionary path than its neighbour. Subsequent developments follow the course identified by Clague and Dalyrimple and described above.

9.3.2   Petrological history of Hawaiian volcanoes

Each of the major Hawaiian shields is broadly similar to the others, although individual centres have reached different stages in the growth cycle. Most of the analyses of Hawaiian rocks are hypersthene-normative, so in terms of their overall chemistry, they can be said to be tholeiitic; more alkaline rocks remain subordinate. (The compositional spectrum can best be studied by referring to Appendix A-6 in BVSP, where a wide range of analyses is presented.) A plot of $Na_2O + K_2O$ versus SiO_2 for all the volcanoes on Hawaii is given here, and shows that the tholeiites and alkali basalts form a continuum (Fig. 9.8). Despite evidence that significant volumes of tholeiitic magma are stored beneath rift zones for long periods, seldom have fractionation processes produced rocks more silicic than basalt ($SiO_2 = <52\%$).

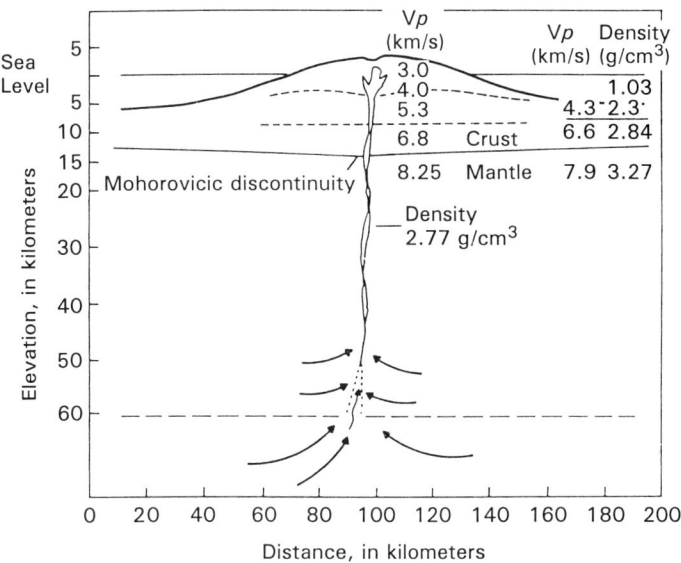

Fig. 9.9: A schematic cross-section through an idealized Hawaiian shield. V_p is the velocity of seismic P-waves. After Eaton and Murata (1960).

The majority of tholeiitic lavas are just quartz-normative, while many are also olivine-phyric. Rocks which are olivine-normative tend also to contain modal olivine, which occurs as a cumulus phase, and it has been shown that olivine fractionation, presumably under low to moderate pressures, has been the process controlling the chemistry of successive batches of tholeiitic melt extruded on Hawaii (Wright, 1971). Compared with MORBs, Hawaiian tholeiites are enriched in TiO_2, total iron (as FeO) and K_2O, but depleted in Al_2O_3 – a characteristic they share with all intra-plate basalts of this type, apart from those of the Galápagos Islands. Since such differences are not related to low-pressure fractionation, they are presumed related to source region differences.

The Hawaiian alkalic series includes alkali olivine basalts, hawaiites and mugearites (the dominant types), together with very subordinate trachytes, rhyolites and ankaramites. The most mafic members of this series (the ankaramites) are, like the tholeiites, enriched in TiO_2, FeO and K_2O but depleted in Al_2O_3, while the most evolved are enriched in alkalis and probably represent the liquid line of descent pursued by fractionating alkali basalt parent melt.

The relationship between the tholeiitic and alkali basalt series is still a contentious issue amongst petrologists. In terms of petrogenesis, low-pressure olivine fractionation evidently has played a major role in generating the compositions seen and much of the fractionation probably has taken place within shallow magma chambers; however, gravity segregation in relatively static lava lakes and flow differentiation must also have played a part. There are additional significant differences between tholeiites, particular on Kilauea, which appear to have occurred in response to high-pressure controls and/or other complex

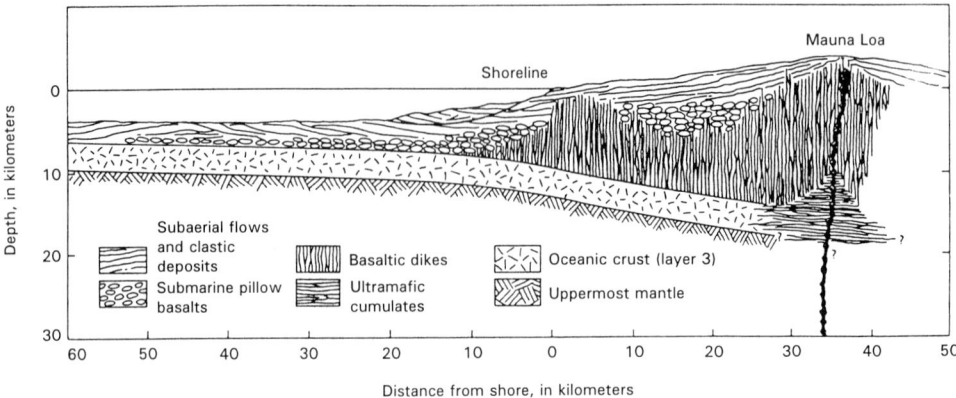

Fig. 9.10: Section to show the inferred geological structure of Mauna Loa shield, based on *P*-wave velocity and density data. Note the plethora of vertical basalt dykes and shallow magma chamber beneath the summit. (After Hill and Zucca, 1987.)

processes, as yet little understood. High-pressure crystallization studies in both natural and synthetic basalt systems (Kushiro, 1972; Presnall *et al.*, 1978) show that mantle peridotite can generate silica-undersaturated (alkaline) melts at a total pressure of greater than 12 kbar, and silica-oversaturated liquids (tholeiitic) at less than 12 kbar. This implies that the more alkaline melts are derived from a deeper (lherzolite) source than tholeiitic ones, the latter having been generated at depths of between 60 and 100 km (Jackson and Wright, 1970).

9.3.3   Genesis of Hawaiian-type shields

The generally accepted model for genesis of Hawaiian shield volcanoes remains that published by Eaton and Murata in 1960 (Fig. 9.9). Mantle rocks are partially melted at depths of between 60 and 170 km, in a region known as the Hawaiian Hot Spot. Ascent through the lithosphere takes place via discontinuous conduits and is aided by density contrasts between the melt and adjacent country rocks. Eventually the various magma bodies are supplied to a shallow reservoir or reservoirs situated between 3 and 7 km below the summits. When such reservoirs are filled, and magma pressure equals lithostatic pressure, there is upward and/or lateral emplacement of dykes to the summit or rift zones respectively; those which reach the surface form eruptions and add to the growing lava pile while the volcano's substructure is continually reinforced by a plethora of steeply inclined dykes (Fig. 9.10).

   Because high-volume, rapidly extruded eruptions deplete the magma reservoir quickly, the pressure within the reservoir quickly declines below the lithostatic pressure; such eruptions are thus of short duration. In contrast, relatively low-volume, more slowly erupted extrusions are sustained for much longer periods due to the resupply of more magma from depth.

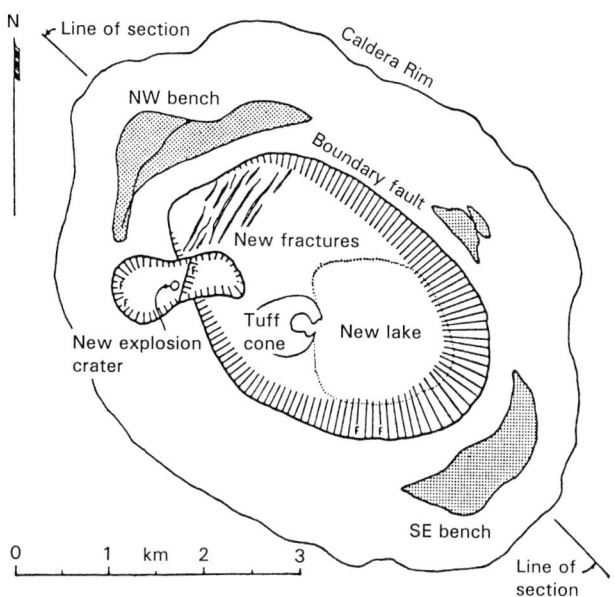

Fig. 9.11: Sketch map of the caldera of Fernandina, Galápagos Islands (above) and cross-section across the summit region (below). (After Simkin and Howard, 1970.)

9.4   CALDERA FORMATION AND OTHER SUMMIT ACTIVITY

On planets such as Mars, where there has been little erosion of volcanic structures, usually only the later stages of shield growth are displayed in their entirety and earlier phases can only be inferred from limited photogeological information used in connection with our experience in places like Hawaii and the Galápagos. It is relevant, therefore, to describe in a little more detail the activity which characterizes the summit regions of large terrestrial shields, since this usually is representative of their most recent activity.

As we have seen, caldera formation used to be considered symptomatic of a particular stage in a Hawaiian shield's evolution; however, recent evidence has suggested otherwise and it is now considered more likely that calderas form over most of a volcano's active life, at least prior to the alkali capping stage. Unfortunately few historic caldera collapse events been been documented; however, there is sufficient evidence to gain a feel for

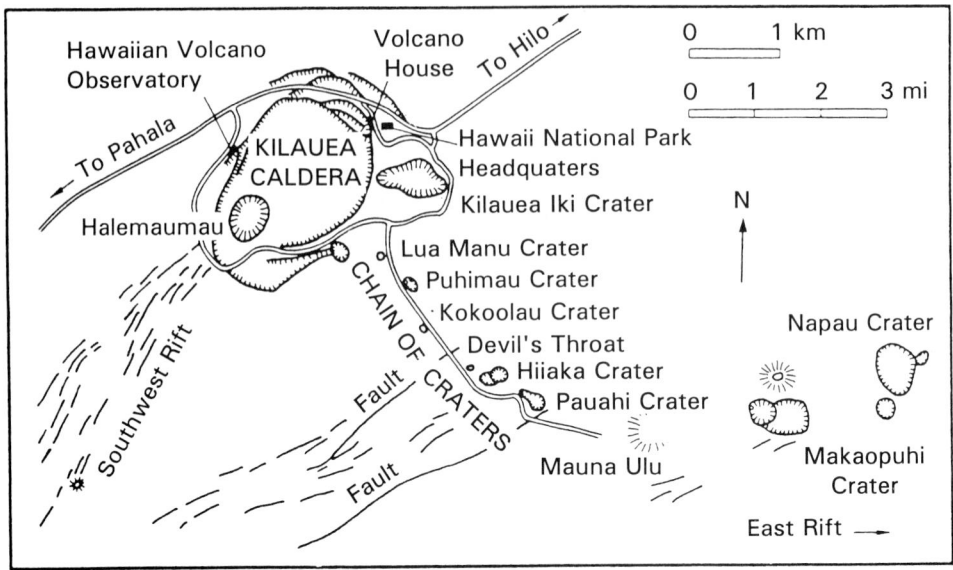

Fig. 9.12: Location map for features at and near the summit region of Kilauea volcano. (After Stearns and MacDonald, 1946.)

what is involved. Descriptive accounts of a number of infamous historic eruptions come to our aid here. For instance, it is believed that the largely submarine collapse of the Aegean volcano Santorini generated massive tsunamis that were responsible for devastating the Minoan civilization in 1470 B.C. Then, more recently, the mighty explosion that accompanied the 1883 caldera collapse of Krakatoa, west of Java, killed 36 000 people and expelled a huge cloud of ash that remained in suspension in the Earth's upper atmosphere for several years. Then again, another major collapse event – the 1912 Katmai eruption – is also documented and was a catastrophic event. None of these, however, involved summit collapse of a Hawaiian-style basaltic shield, each was a strato-volcano renowned for a more explosive style of activity. It is the more quietly-effusive Hawaiian shields that most closely resemble the large volcanoes of Mars and Venus.

Although little direct evidence of the events associated with caldera formation on basaltic shield volcanoes has been collected, there is at least some evidence from both Hawaii and the Galápagos Islands. The most direct evidence comes from one of the Galápagos shields, situated on Isla Fernandina, immediately west of the larger island depicted in Fig. 6.7. Here, on June 11th, 1968, a small seismic disturbance and large cloud of vapour was followed 4 hours later by generation of a much larger ash cloud which, in turn, was followed by a major explosion recorded at infrasonic stations throughout the hemisphere. Further seismic events were recorded for the next 10 days or so (Simkin and Howard, 1970). The violence of the explosion was such that ash expelled from the vent area fell at locations 350 km distant.

Prior to this major 1968 eruption the floor of the 4×6.5 km caldera lay 800 m below the rim, but subsequently it subsided in a series of short drops focussed along steeply inclined elliptical boundary faults (Fig. 9.11). Vertical displacements increased from zero in the northwest to 300 m in the southeast, yet despite such major movements, relatively little break-up of the caldera floor occurred and no fresh lava was erupted on to it.

The principal question posed by the Fernandina events is that of the whereabouts of the displaced magma. The only recorded contemporaneous effusion was a flow emplaced on the eastern flank during May 21, but the volume of this, and indeed of the ash erupted later, fall far short of the displaced volume; furthermore, no related submarine activity was recorded. It must be assumed, therefore, that the magma either must have been intruded as dykes into the volcano's substructure or withdrawn at greater depth by tectonic movements, perhaps along a suspected oceanic fracture zone.

Hawaiian calderas are usually elliptical and may contain subsidiary pits which, during the volcano's history, may be partially filled, completely filled or devoid of ponded lavas. The present caldera of Kilauea measures 4×3 km and ranges in depth from 120 m in the northwest to just a few metres at the southern rim. In the southwestern part of the caldera is an 800 m diameter pit, Halemaumau, which resides at the crest of a low shield built up on the caldera floor (Fig. 9.12).

A major explosive eruption is known to have occurred at Kilauea in 1790 but it is still not entirely clear whether the modern caldera subsided prior to this event or was partly associated with it (Holcomb, 1987). What is known, however, is that the summit of the volcano has experienced alternating phases of inflation and collapse over long periods and that the numerous pit craters have associated coarse tephra deposits that suggest a history of phreatic eruptions. In 1894 there was subsidence in Halemaumau and magma was withdrawn from that area. After a long period of inactivity the pit was filled to the brim with lava in both 1919 and 1921, and several flows spread out onto the main caldera floor. The pattern of almost continuous summit activity which was characteristic of the volcano during the nineteenth century and the early part of the twentieth, eventually ceased in 1924, when there were several violent phreatomagmatic eruptions and an ensuing collapse of the walls of Halemaumau, leaving a pit some 400 m deep. This change in eruptive behaviour appears to have followed a major submarine eruption from the East Rift Zone that coincided with summit collapse. It seems clear, therefore, that withdrawal of magma from a shallow reservoir beneath the summit of Kilauea was responsible for summit subsidence and that the magma withdrawn was both intruded into the volcano's substructure as dykes along the East Rift Zone, and extruded as basalt flows on this flank. Since 1952, episodic activity has continued at the summit.

Activity along rift zones is another characteristic of shields. Thus, along the upper parts of Kilauea's East Rift Zone there are numerous pit craters which appear to owe their

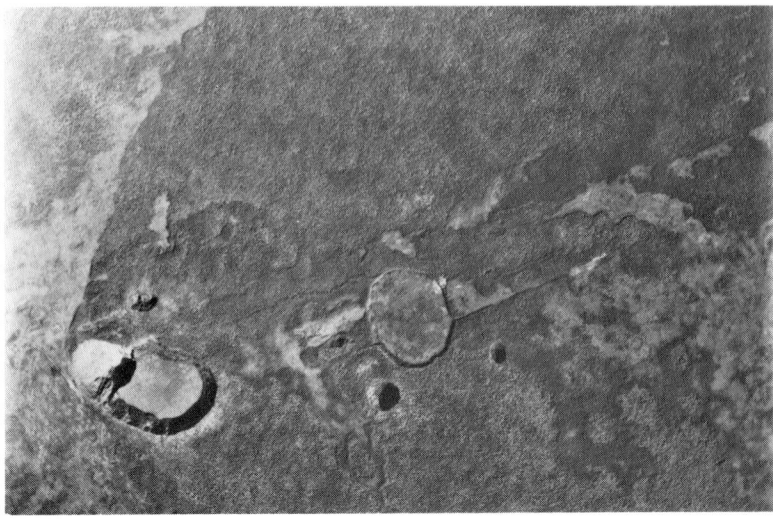

Fig. 9.13: Pit craters along Kilauea's East Rift Zone, as photographed in 1954. On the left is the 1 × 1.6 km pit, Makaopuhi. The deeper west pit inside it was subsequently flooded by lava flows from Mauna Ulu between 1969 and 1971 and then, in 1973, was completely filled with lava. Three kilometres east of Makaopuhi, on the same rift line, is the 800 m wide, 60-m-deep pit of Napau. It is transected by numerous faults and fissures aligned along the East Rift Zone. (U.S.G.S. Photograph HAI 9A.)

origin partly to magma stoping and partly to its withdrawal (Fig. 9.13). The latter is a response to eruption lower down the rift, causing the evacuation of a magma chamber beneath the surface and collapse of the crust immediately above it (Carr and Greeley, 1980). It is also possible in Hawaii to see sections eroded through rift belts associated with older volcanoes, such as Oahu and Lanai, which reveal dense swarms of near-vertical dykes. Orientation of active rift zones is evidently largely a function of local gravitational stresses since they form in those relatively unsupported sectors of shields that are not butted against adjacent volcanoes.

Summit eruptions usually are accompanied by shallow earthquakes that accompany the rise of magma towards the surface. Rift eruptions, in contrast, are characterized by less regular seismicity. During periods of prolonged rift activity the summit region deflates, presumably because magma is withdrawn sideways into reservoirs on the volcano's flanks. The forceful injection of magma into such zones may produce cracks and fissures at the surface and certainly results in intrusion of swarms of sub-parallel dykes. Such intrusions contribute significantly to volcanic construction and it is clear that while a shield may grow substantially by summit extrusions and dyke emplacement, flank growth also occurs and is responsible for the elongated morphology and low profiles of most Hawaiian shields.

10

Martian central volcanism

Martian shield volcanoes share many similarities with Hawaiian shields, being roughly circular or elliptical in plan, having relatively low profiles and summit caldera complexes, and being constructed, in the main, from low viscosity lavas. There are striking similarities, too, in the style of eruptive features, with flow scarps and frontal lobes, lava tubes and channels being widely recognizable, as well as small lava domes, spatter ridges and collapse depressions. They differ, however, in three important ways: firstly, they are very much larger; secondly, individual volcanoes were active over extremely long periods; and, thirdly, their global distribution is very different. The first and second of these imply a structural and thermal stability in the Martian lithosphere that was and is not matched by the Earth's, while the third evidently is a reflection of the absence of plate tectonics from Mars. Low profile patera volcanoes also exist on then planet; several of these appear related to phreatomagmatic activity.

10.1 DISTRIBUTION OF CENTRAL VOLCANOES

The global distribution of Martian central volcanoes is very uneven and is shown in Fig. 10.1. The quite different distribution pattern for Martian volcanoes compared with that for those of Earth is undoubtedly a reflection of the absence of plate tectonics on Mars; the large Martian shields are presumed to be related to a number of long-lived hot spots. The greatest number of volcanoes and also the youngest individual structure are located in the Tharsis province, the region of a major tumescence in the planet's crust (the *Tharsis Bulge*) whose crest is centred at 105°W 0° latitude. This is at a mean elevation of between 10 and 11 km above datum and is of continental proportions – it is roughly the size of Africa south of the River Congo – extending 4000 km from north to south, and 3000 km from west to east. Although the term "bulge" has been coined, in reality its slopes are low and it is really a broad, gentle rise. On the north flank, slopes range between 0.2° and 0.4°, but on the south side they are only about half this. This asymmetry affects also its extent, for not only is it steeper but it is also more extensive to the north of its crest, almost certainly an effect due to its straddling the line of dichotomy between the northern lowlands and the southern cratered highlands. Strangely, most of the large shield volcanoes are sited either near the crest of the bulge or on its northwestern flank; there are none on the southern flank.

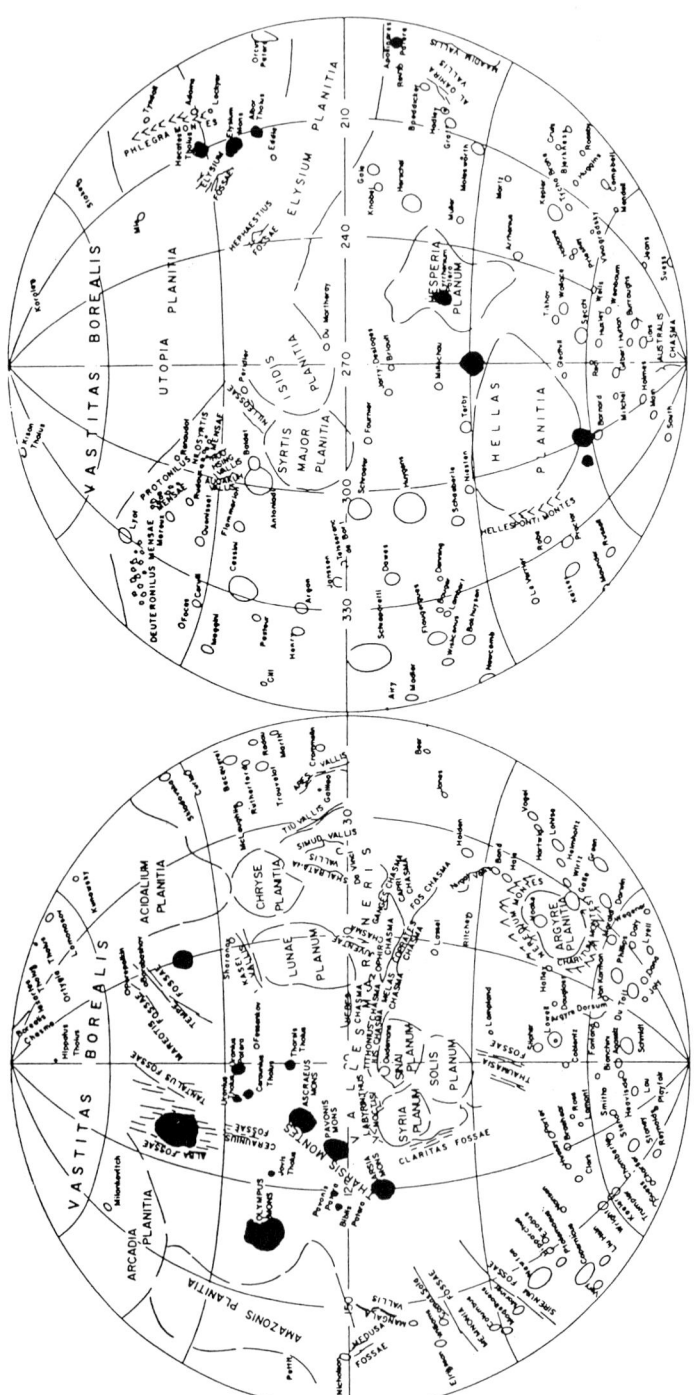

Fig. 10.1: Distribution of the larger central volcanoes on Mars. Individual centres are shown in solid black.

The most prominent volcano group is the Tharsis Montes, consisting of Arsia, Pavonis and Ascraeus Montes, spaced approximately 700 km apart and aligned in a southwest-northeast direction along the crest of the rise. Since major fractures pursue the same trend northeastwards and southwestwards, even beyond the Tharsis volcanic province, it can be assumed that they have developed along a major fracture zone, now buried by the products of Tharsis volcanism. Several smaller shields and steeper-sided tholi lie close to the continuation of this line to the northeast of Ascraeus Mons, while others lie to the east and west (Fig. 10.1). Twelve hundred kilometres northwest of Tharsis Montes lies Olympus Mons, the most spectacular of the Martian shields and also the youngest, while at a slightly greater distance north of Ascraeus Mons, on the extreme edge of Tharsis, lies the vast Alba Patera, a low-profile volcano with a striking family of circumferential fractures.

The shields of the Elysium region – Elysium Mons, Hecates and Albor Tholi – also sit atop a broad rise, but this is considerably smaller than that of Tharsis, with a diameter of approximately 2000 km and a mean height of 5 km. The nature of the Elysium volcanoes is somewhat different from those of Tharsis and there is evidence that their development may have involved not only lavas but also pyroclastic rocks (Malin, 1977).

Several major volcanic centres are located around the Hellas impact basin. Of these, Hadriaca and Tyrrhena Paterae are almost certainly mixed lava and ash volcanoes, while Amphitrites Patera has little vertical relief and a somewhat equivocal origin. The only other major volcano is Apollinaris Patera, which is located southeast of Elysium, at $10°S$ $185°W$, on the lowland hemisphere, just north of the line of dichotomy. Its morphology suggests an origin in both effusive and explosive activity.

10.2 CLASSIFICATION OF VOLCANO TYPES

The large volcanoes have been classified into three main types: (1) *shields*, which are built from thousands of individual flows, have summit calderas and overall low profiles characterized by steeper upper regions and gentler lower flanks; (2) *tholi* or dome volcanoes, which are similar to the former but have somewhat steeper slopes that may be a function of more viscous lava, lower eruption rates or pyroclast content; and (3) *paterae* which are of two kinds: (i) *lowland paterae* (e.g. Alba Patera, Uranius Patera), which are northern hemisphere lava shields characterized by extremely low profiles and complex summit calderas, and (ii) *highland paterae* (e.g. Tyrrhena Patera), which are located mainly in the southern hemisphere, have very low profiles and summit caldera complexes and also may be incised by channels (Plescia and Saunders, 1979). Most of subgroup (ii) are believed to be mixed lava and pyroclast edifices. A fourth group – volcano-tectonic depressions – may also be recognized, including certain suspected volcanic structures in Syrtis Major, in the region south of Hellas and amongst the volcanic plains around Tharsis (Plescia and Saunders, 1979), but little is known of these enigmatic features.

Justification for such a classification has been supported by the statistical data of Pike (1978) yet it has to be said that the groupings may be somewhat artificial; thus, many tholi may just be the steeper, upper parts of buried shield volcanoes, while some paterae may simply represent very low shields. Furthermore, some steeper tholi were

Table 10.1: Number of craters greater than or equal to 1 km in diameter per 10^6 km^2 for individual Martian volcanoes, compared with absolute ages derived from the chronologies of Neukum and Wise (1976) and Soderblom (1977). (After Plescia and Saunders, 1979.)

Volcanic centre	Number of craters $\geqslant 1$ km/10^6 km^2	Implied absolute age in 10^9 y (Soderblom, 1977)
Olympus Mons	27	0.03
Ascraeus Mons	110	0.1
Pavonis Mons	350	0.3
Arsia Mons	780	0.7
Apollinaris Patera	990	0.9
Biblis Patera1	400	1.3
Tharsis Tholus	1480	1.38
Albor Tholus	1500	1.4
Hecates Tholus	1800	1.7
Alba Patera	1850	1.7
Jovis Tholus	2100	1.95
Hadriaca Patera	2100	1.95
Elysium Mons	2350	2.2
Tyrrhena Patera	2400	2.25
Uranius Patera	2480	2.3
Uranius Tholus	2480	2.3
Ceraunius Tholus	2600	2.4
Ulysses Patera	3200	3.0
Tempe Patera	4300	3.4
Lunae Planum	2500	2.3

mis-classified before improved altimetric data became available. Notwithstanding this, the morphological distinctions described above have some value as they provide a framework within which discussion can proceed.

10.3   AGES OF CENTRAL VOLCANOES ON MARS

Absolute ages are unavailable for Martian rocks, but crater-counting methods have been used to establish a relative time-scale and also to arrive at possible absolute ages. The most complete chronology has been provided by Plescia and Saunders (1979), who have provided crater counts for each of the major constructs, divided into their types, with Lunae Planum as a datum (Table 10.1). This permits the sequence of shield activity at each of the centres to be established. They also derived absolute ages by inputting their data into the chronologies of Neukum and Wise (1976) and Soderblom (1977). It should be noted, however, that individual volcanoes did not form at an instant in time, but evolved often over very lengthy periods, episodes of activity being interspersed with periods of inactivity measurable in tens or hundreds of millions of years.

Because of the nature of the Neukum and Wise chronology, which extrapolates from large crater diameters down to 1 km size, rather than counting them, the ages derived by fitting crater data to their time-scale compresses most of the ages into a very early stage in Martian history. On this basis all shield activity, up to the stage when Arsia Mons erupted, was complete by 3.4×10^9 years ago. This is not consistent with thermal models. Therefore the natural predilection of Plescia and Saunders for the Soderblom time-scale is shared here, and, accepting this as at least a close approximation to reality, it can be observed that the construction of highland paterae spanned a period of about 1×10^9 years, beginning 3 $\times 10^9$ years ago. Subsequent centralized activity, with the growth of tholi and shield volcanoes, spanned the period 2.2×10^9 years (Elysium Mons) to 2.5×10^7 years (Olympus Mons). Such a time-scale means that the earlier phase of patera volcanism was predominantly Hesperian, with the major shield-building stage, as typified by the Tharsis region, spanning most of Amazonian time. A later detailed synthesis of Martian cratering by Hartmann *et al.* (1981) extends the cratering data to a wider range of Martian volcanic units, but inclusion of this data set would serve only to confuse an already complex issue. In general terms, it reaches similar conclusions to the work of Plescia and Saunders.

10.4   THARSIS AND ELYSIUM – GRAVITY AND TECTONICS

Gravity data derived from both the Mariner 9 and Viking Orbiters have allowed reconstruction of the Martian gravity field (see Carr, 1981). Low-resolution data from both sets indicate a major positive free-air gravity anomaly of about 500 mgal associated with Tharsis, complemented by 200-mgal gravity lows over Chryse and Amazonis (Sjogren *et al.*, 1975). High-resolution data shows large positive anomalies over Tharsis, Elysium and Isidis, and positive anomalies are also indicated for specific volcanic structures; thus Olympus Mons has the largest (344 mgal), but there is a 70-mgal positive anomaly associated with Alba Patera and one approaching 80 mgal over the Elysium shields (Sjogren, 1979). There is also a distinct correlation between gravity and topography which implies that topography can only be partially compensated, at least to shallow depths (Phillips and Saunders, 1975).

The Tharsis Bulge not only hosts Mars' major volcanic province but is also the focus for a widespread system of radiating fractures, nearly all of which are graben (Fig. 10.2). In northern Tharsis there are several intersecting sets, of differing ages, which indicate that there were episodic changes in the stress orientation; elsewhere, for instance in Memnonia, faults maintain a parallelism over considerable distance, implying uniform stresses affecting large regions. The sequence and timing of tectonic events has been studied by various workers (Wise *et al.*, 1979; Frey, 1979; Plescia and Saunders, 1982), with somewhat conflicting results. What is clear, however, is that fracturing reached a climax shortly following the decline in the impact flux, but that it continued with considerably less intensity until a much later date. Study of lava flows associated with Tharsis shields quite clearly indicates that while many faults affect them and thus are younger, others are definitely older; therefore it is clear that volcanism and tectonism continued at much the same time during the active life of the shield volcanoes.

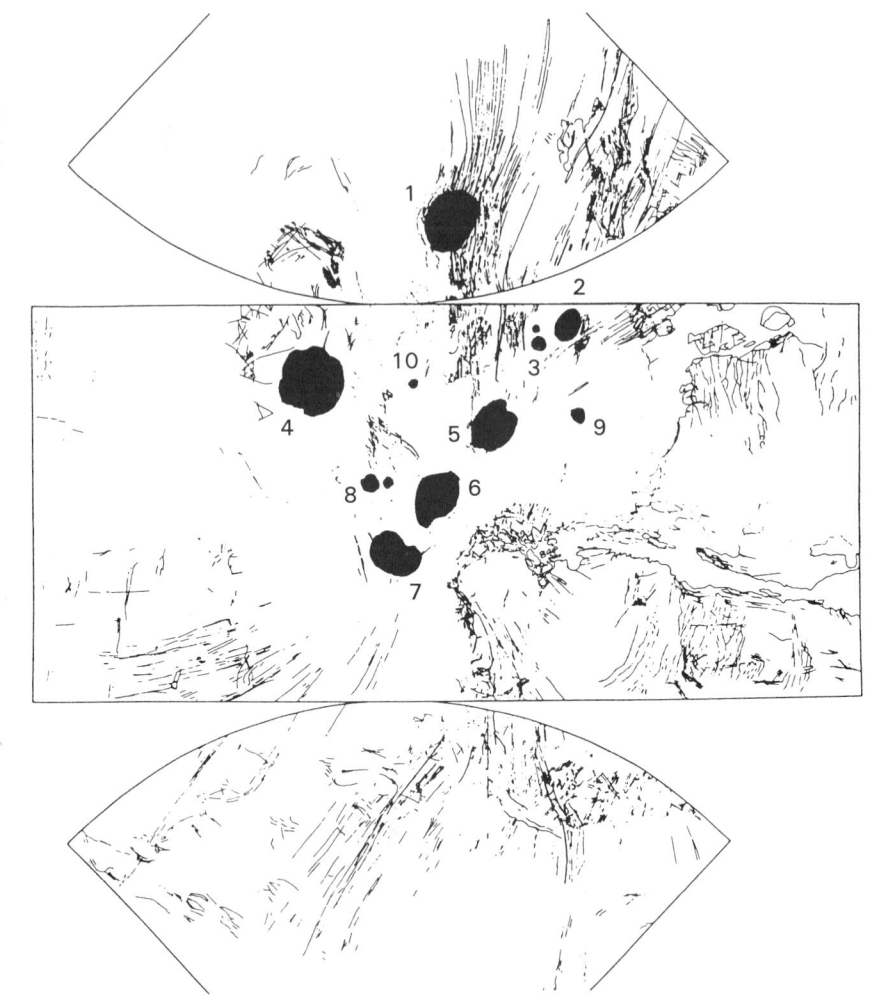

Fig. 10.2: Map of the western hemisphere of Mars between 60°N and 60°S showing the position of the main Tharsis volcanoes (black areas) and principal fractures and ridges. Volcanoes: 1. Alba Patera, 2. Uranius Patera, 3. Ceraunius Tholus, 4. Olympus Mons, 5. Ascraeus Mons, 6. Pavonis Mons, 7. Arsia Mons, 8. Biblis Patera, 9. Tharsis Tholus, 10. Jovis Tholus. (Modified from J.B. Plescia.)

Analysis of the pattern of fracturing by Plescia and Saunders (1980) suggests that the crest of the Tharsis bulge originally was located at 8°S 100°W, much closer to the western end of Noctis Labyrinthus than it now is, but that it shifted later to reside near the present shield of Pavonis Mons. The present distribution of fractures (and ridges) surrounding Tharsis has convincingly been shown by Phillips and Ivins (1979) to be due to the present topographic and gravity highs and not to be a function of crustal arching.

Fig. 10.3: Mosaic showing the summit region of Arsia Mons. The 120-km-diameter caldera has associated with it annular graben and from its backwalls hundreds of lava flows fan out down the flanks. Note that these form broad terraces close to the summit. Mosaic length 650 km. V.O. frames 052A02-08.

10.5   THE THARSIS VOLCANIC RISE

The Tharsis region of Mars shows a development of major volcanic structures and associated flowfields that is unique in the Solar System (Plate 1). It plays host to the tallest shield yet known, Olympus Mons, which rises over 23 km above the surrounding plains, and 27 km above mean datum. Also within its region is the huge patera volcano, Alba Patera, one of the most areally extensive central volcanic structures yet discovered. Tharsis is also a region of extensional faulting and major gravity anomalies.

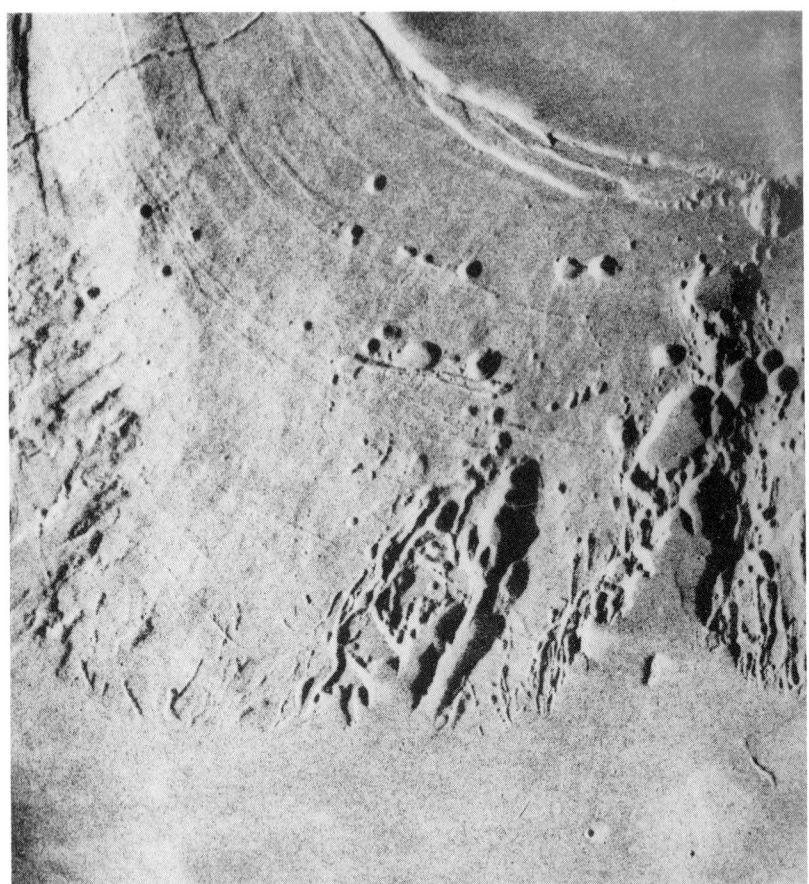

Fig. 10.4: Enlarged view of southwest flank of Arsia Mons, showing prominent embayment composed of coalescent pits and circumferential grabens. Frame width 115 km. V.O. frame 204A08.

10.5.1 The Tharsis Montes

The Tharsis Montes epitomize the large shield volcanoes of Mars. In overall morphology, albeit on a grander scale, they bear a striking resemblance to Hawaiian shields while in detail they have similarities with both Galápagos and Hawaiian volcanoes. By direct analogy, this imples that they grew as a result of the long-term eruption of large volumes of fluid, basalt-like lavas, considered by some workers to be of iron-rich composition (Carr, 1981). Of the three, the most northerly shield, Ascraeus Mons, rises to the greatest height (26 km) and has the largest relative height difference with respect to the surrounding plains (17 km); both Pavonis and Arsia Montes reach about 20 km above Mars datum. The main part of each construct is between 350 and 400 km in diameter and has a prominent summit caldera significantly larger than any known terrestrial shield caldera. However, flank slopes are relatively low, averaging less than 5°, and there is a tendency for the summit region and lower flanks to be less steep than the intermediate zone.

The prominent radial texture typical of each volcano's flanks is due to hundreds of narrow (generally less than 3 km wide) lava flows, many of which have apical channels; a large proportion of these can be traced upslope to the rims of the summit calderae, while a further significant proportion appear to have emanated from prominent embayments made up of numerous coalescing pits, and situated adjacent to the calderae in the southwest and northeast sectors of each shield (Fig. 10.3). Those near the summit of Arsia Mons are very large and striking, the embayment on the southwest, in particular, having spawned voluminous lavas onto that flank, producing a vast shoulder of fan-like flows described by Crumpler and Aubele (1978) as a parasitic shield (Fig. 10.4). This feature is prominent on radar profiles of the region (Roth et al., 1980). In general terms, successively older flows are exposed as the distance from the main shields increases (Scott and Tanaka, 1981). The disposition of flows of various ages is shown in the geological map (Fig. 10.5).

The summit caldera of Arsia Mons has a mean diameter of 120 km, is simple in structure and bounded by arcuate faults. Crater counts indicate the caldera floor to be much younger than the volcano's flanks. Suitably enhanced images of Arsia's caldera reveal that a line of low domes connects the embayments in each wall, suggesting eruptive activity associated with the major southwest-northeast fracture line continued after the latest caldera subsidence had taken place. However, as is the case with all three Tharsis Montes, little or no evidence of intra-caldera constructional volcanism is forthcoming. On the caldera rim are numerous graben whose spacing varies from between 1 and 12 km; these extend outwards from the rim for about 60 km (Fig. 10.4). The numerous narrow (0.5-1.3 km) lava flows which traverse the northwest slopes are older than these faults, transection relationships showing they were emplaced prior to fracturing of the caldera rim. The lavas themselves can be traced to a source in a major graben depression situated on the caldera rim, approximately 12.5 km from the edge of the current floor (Mouginis-Mark, 1981).

While large in terms of area, Arsia's caldera is relatively shallow compared with that of Ascraeus Mons, where the floor of the deepest pit lies 3.15 km below the rim (Mouginis-Mark, 1981). It is but one of eight major depressions that form the nested caldera complex of this volcano (Fig. 10.6). Study of the summit region by Peter Mouginis-Mark (1981) shows that some of the collapse events were preceded by major slumping of the caldera backwalls, these being approximately contemporaneous with the formation of circumferential graben. Originally the summit must have boasted several smaller pits, but the latest collapse saw the production of the large 40-km-diameter depression that now unifies the caldera complex.

Along the entire caldera rim it is possible to discern a plethora of narrow lava flows and, indeed, some sinuous channels, which radiate down the volcano's upper flanks. The flows are generally less than 1 km wide and between 10 and 20 km in length, while channels are between 100 and 200 m wide and up to 18 km long. The flows are estimated to be no more than 10 m thick (Schaber et al., 1978). Significantly, the crisp outlines of these flows and channels can be traced right up to the line of collapse marked by the caldera backwall, making it abundantly clear that effusive activity continued from the

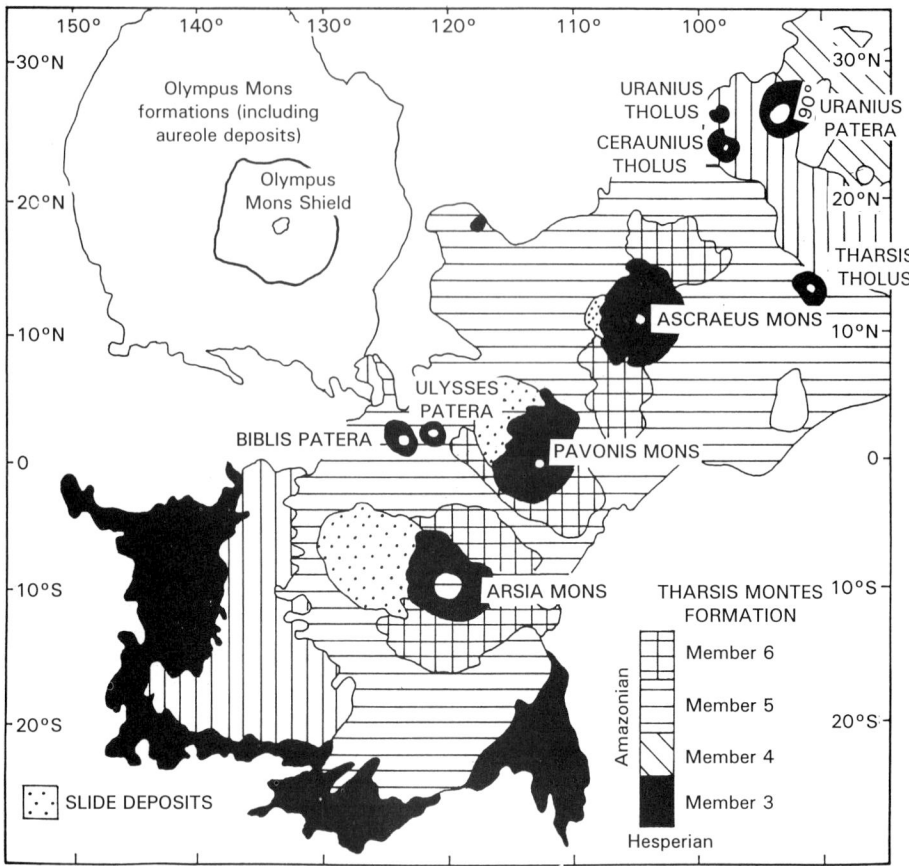

Fig. 10.5: Map showing main Tharsis volcanoes and associated volcanic units.

summit right up until the final collapse occurred and proving that late-stage explosive activity was not important in the shield-building process (Fig. 10.7). Source vents for flows are, however, not visible, even on high-resolution Viking images; neither can traces of the flows be discerned on slumped terraces within the nested caldera complex. This implies, firstly, that each collapse depression experienced resurfacing after its formation, and, secondly, that the source vents for the lavas originally were located further up the shield than they can now be traced.

Crater counts on the floor of Ascraeus Mons suggest it has a similar age to that of Pavonis Mons (Crumpler and Aubele, 1978), although little high-resolution imagery is available for the latter, making strict comparisons difficult. The latter has a single, 45-km-diameter caldera, about 4.5 km deep, surrounded by arcuate fault terraces that extend outwards and define a shallow summit depression approximately 100 km in diameter. Annular grabens also occur lower down the structure on the northeast, east and southeast slopes, commencing about 120 km from the summit (Fig. 10.8). These are similar to fractures developed around both Ascraeus and Arsia Mons, although their incidence is substantially lower on the former,

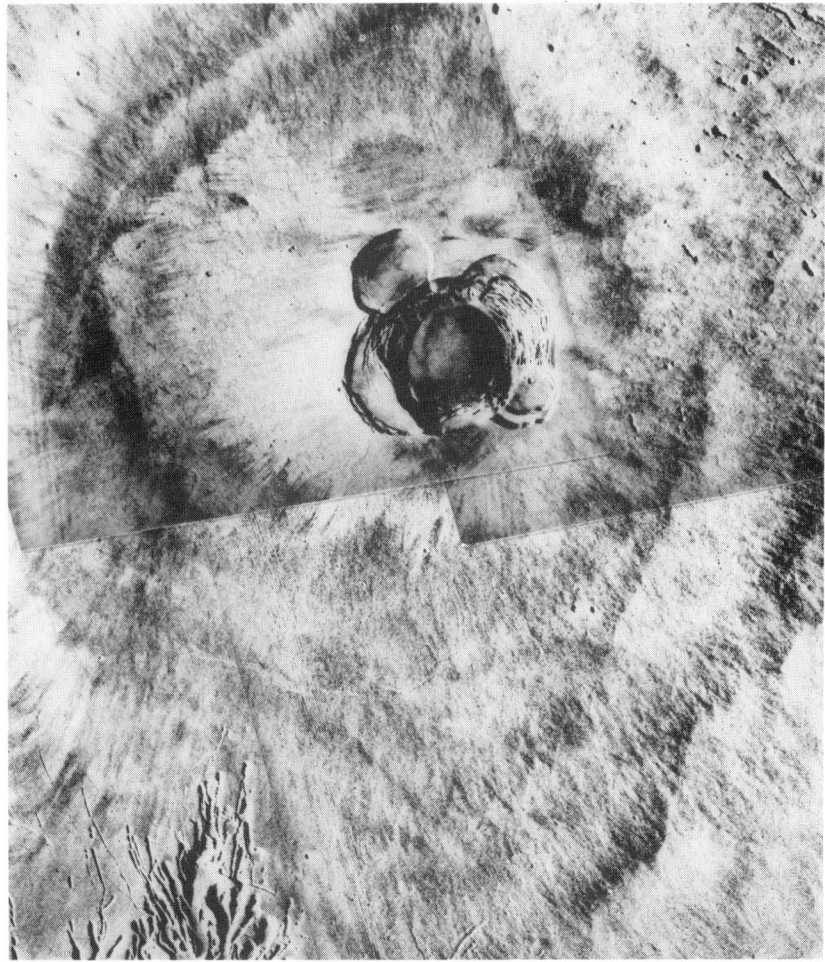

Fig. 10.6: Summit region of Ascraeus Mons, showing the nested summit caldera complex, radial flows and flank embayments. Frame width 210 km. V.O. frames 224A88-91.

which, instead, has a greater abundance of rimless concentrically arranged pits. Such pits do occur on the other two shields, but in lower abundance and often sited along the lines of prominent arcuate graben. Several sinuous rilles can be seen to cross these faults; these are considered to be major lava channels incised by turbulent flow and some of these apparently emerged from circumferential fractures, implying that the latter were the sources for the flows that subsequently were channelled down the volcano's flanks.

Such rheological studies as have been made of Tharsis Montes flows – flows associated with the main shield of Ascraeus Mons – indicate that they were of low yield strength and low viscosity, consistent in the main with basalt-like composition (Moore *et al.*, 1978; Zimbelman, 1985). However, ridged flows high on the shield of Ascraeus Mons appear to

Fig. 10.7: High-resolution mosaic of the south rim of Ascraeus Mons' summit caldera. The narrow summit lava flows are truncated by the caldera backwall, a characteristic they share with terrestrial lava shields such as Mauna Loa. Several flows are seen to be leveed, while others show apical channels. North is to the left. Frame width 25 km. V.O. frames 401B16/18.

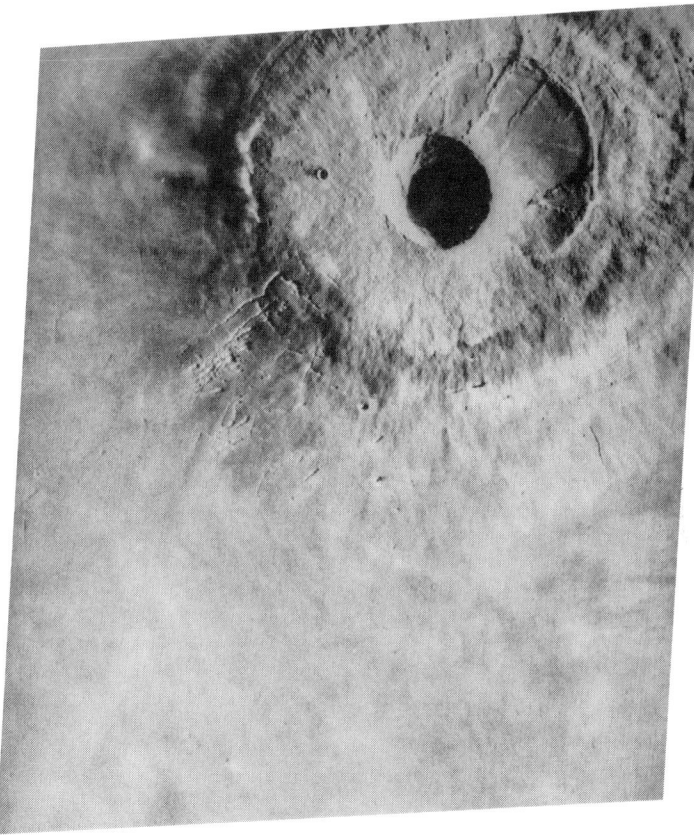

Fig. 10.8: The summit of Pavonis Mons, showing the 45-km-diameter caldera within its larger, shallow depression. Arcuate graben surround this. Note also the prominent wrinkle ridges that traverse the otherwise smooth, dark floor of this depression amd the embayments on the southwest flank. Frame width 300 km. V.O. frame 643A27.

have a rheology more closely akin to the basaltic-andesites of Arenal volcano, and may, therefore, represent more evolved magmas. The average effusion rates calculated for these flows range between 18 and 60 m^3 s^{-1} – towards the lower end of Hawaiian and Icelandic rates.

While a basalt-like composition is implied by the morphology of the Tharsis Montes flows, this is not to say that they are all exactly the same. On the upper parts of shield surfaces, out to radial distances of about 400 km, flows tend to be relatively narrow (less than 3 km), less than 150 km long (many being no more than 15 km in length) and often have a central channel, sometimes bounded by levees. Beyond this point they tend to widen (4-7 km wide) and become longer, some exceeding 400 km; central channels are common in this zone. At distances greater than 800 km, where regional gradients become lower, the flows broaden substantially and may be up to 50 km across at their terminations. These flows tend to lack channels and some are at least 650 km long.

Fig. 10.9: Mosaic showing landslide to the northwest of Arsia Mons. This huge lobe has a striated texture towards its margin, but is blocky inside this zone. Note that the striations cross both large sheet flows and small impact craters. Frame width 510 km. V.O. frames 042B9-13, 33-36.

When the length of flows on Arsia Mons is considered in terms of the height at which they emerged from the volcano, it is found that the greater the source height, the shorter the flows. This apparently is the case for all three shields, and indeed has been noted elsewhere in Tharsis (Cattermole, 1987, 1989), implying that it cannot be mere coincidence. Applying the basic laws of magma rise, it most likely is a direct reflection of the greatest altitude to which lava can be raised by the density contrast existing between the magma and the rocks in the source region. Confirmation for this comes from the concentration of shorter flows near the summits, where eruption rates must have been less than on the lower flanks – a response to the much greater vertical distance through which they needed to be lifted before eruption.

The stability of these massive constructs evidently wavered at some point in their growth, for massive landslides have modified their outlines – a characteristic they share with Olympus Mons. The most dramatic evidence for this comes from the west-northwest flank of Arsia Mons, where a lobate feature measuring 400 × 350 km extends outward from the base of the main shield. The dissected nature of the shield flanks above its upper termination suggests that there was a zone of detachment along which this side of the shield partly collapsed, slipping laterally as a gravity-assisted slide or series of slides that spread out over the lower plains.

The terrain within the slide unit is composed of hundreds of small hills, while at its outer edge are closely spaced ridges which produce a strongly striated effect (Fig. 10.9). A puzzling feature of this outermost, striated zone is that many striations crosscut impact craters and other small landscape features without in any way modifying their outline, as though having been superimposed upon them. The suggestion has been made that the slide formed while the landscape was mantled by ice (Williams, 1978). Not only would this have facilitated the flow of the debris, but its subsequent melting and removal would have superimposed the flow pattern onto the underlying topography. While purely speculative, this hypothesis is a plausible one and has interesting connotations for landform development in the Tharsis region at the time of volcano growth.

Taking all of the evidence together, it is clear that the three shields of Tharsis Montes followed a similar pattern of development. An initial stage of shield-building was achieved by the gradual accumulation of fluid lavas from both the summit area and from peripheral vents. After each shield had attained its maximum height, eruptive activity became concentrated along a major rift zone aligned in a southwest-northeast direction. Considerable lateral transport and supply of magma to vents and fissures along this zone over long periods of time resulted in repeated collapse of the shield summit region – as happens in Hawaii – forming major embayments and small satellite calderae in which some lavas ponded. Eruptions from these rift-aligned sources built out substantial shoulders on both the southwest and northeast flanks, from which emanated large numbers of major eruptions that disgorged high-volume flows for great distances over the surrounding plains. The great lateral extent of these flows implies that rates of extrusion were also high. Crater counts suggest that the construction of the main Arsia Mons shield terminated earlier than its two neighbours, but eruption both from southwest-northeast embayments and from within the calderae continued well into the more recent past (Crumpler and Aubele, 1978).

10.5.2   Olympus Mons

Of all the Martian volcanoes, Olympus Mons is unquestionably the most spectacular. Rising to a height of 27 km above Mars datum, and at least 23 km above the surrounding plains, it has a diameter of at about 600 km, giving it a volume over fifty times that of any terrestrial shield. If that was not enough, it is surrounded by a huge scarp that in places is 6 km high, while extending for between 300 and 700 km beyond the scarp base is a region of peculiar blocky terrain known as the *aureole* which stands our prominently in synoptic Viking images (Fig. 10.10).

In overall form the shield shares many similarities with the Tharsis Montes; it has a nested caldera complex, measuring 80 km across, which has suffered multiple collapse, the latest of these events having affected the southwest and northeast sectors. The deepest pit is 3 km deep. Despite the availability of very-high-resolution images of the summit region, evidence of intra-caldera activity is not forthcoming; the only features discernible on the floor are families of wrinkle ridges and arcuate graben, both landforms due primarily to tectonic forces (Fig. 10.11). In contrast to Ascraeus Mons, where the largest pit was formed last, here the reverse is true. Mouginis-Mark (1992) analysed high-resolution imagery of the summit caldera region, photoclinometrically, and showed that as much as 2.5 km of collapse took place within the 80×65 km caldera area, and also that the rim

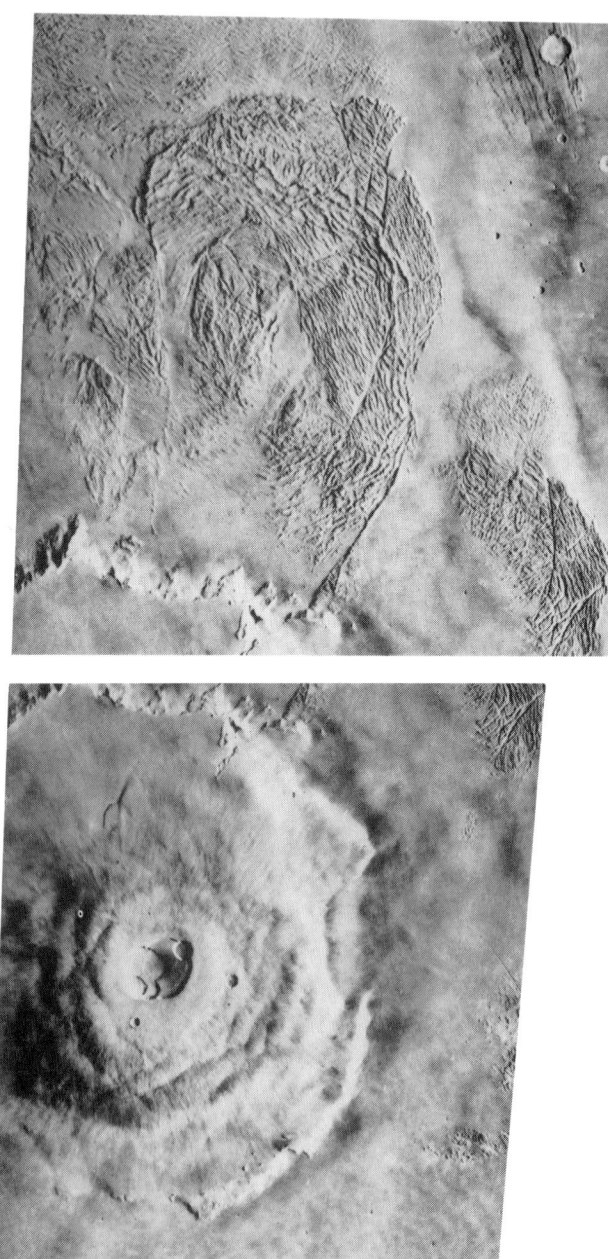

Fig. 10.10: Synoptic view of Olympus Mons, showing the summit caldera, terraced flanks, basal scarp and aureole. Length of photo pair 1200 km. V.O. frames 741A05 (above), 741A07 (below).

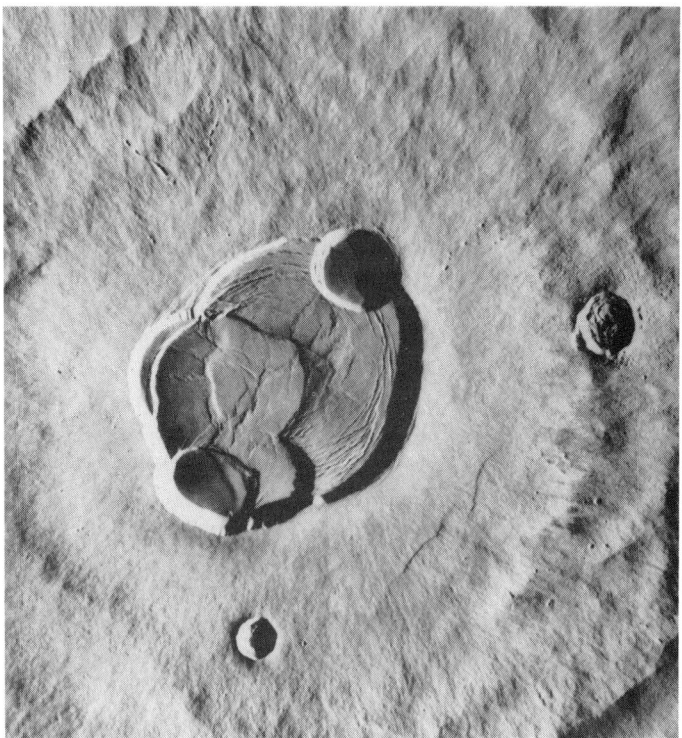

Fig. 10.11: The summit region of Olympus Mons, showing the nested caldera pits, radiating flow texture and broad flow terraces. Annular grabens occur on the floor of the oldest caldera component, while wrinkle ridges characterize the smaller, younger pits. Frame width 175 km. V.O. frame 890A68.

elevation varies by at least 2 km along its length. An 8-stage evolutionary sequence is recognized, commencing with catastrophic collapse of the largest depression and ending with the a 350-m displacement event to form the smaller, most southerly caldera. This evidently was a long-term process and was related, at least in part, to migration of the magma source within the volcano's substructure. Successive magmatic and collapse events led, in general terms, to development of extensional features towards the peripheries of the depressions and compressional ones close to the central regions. Interestingly, a rather abrupt transition occurs between the two regimes, at a distance of around 17 km from the caldera backwall. On the basis that ridges and graben within the oldest caldera were generated as a direct result of subsidence of the floor, and that this in turn was a manifestation of deflation of the underlying magma chamber, Zuber and Mouginis-Mark (1992) have been able to model the gross characteristics of the chamber. Thus, since the width of the chamber cannot have been significantly smaller than that of the caldera, and taking into account the distribution of tectonic features within the depression, they were able to infer that the depth of the magma chamber must have been at relatively shallow depth, i.e. $\leqslant 16$ km at the time of collapse.

The flank slopes are generally rather low (mean slope is 4°), but the shield has a somewhat sinusoidal profile since the middle part of the structure is steeper than either the summit region or the distal flanks. In addition, the surface of the volcano is terraced, there being a series of 15 to 50-km-wide flow terraces separated by distinct breaks in slope, and crossed by innumerable thin flows which have a roughly radial disposition. The flows themselves are difficult to see close to the caldera region but further away long, narrow flows (1-3 km wide), often with central channels and levee banks, are widely distributed, and except in the western sector, may be seen draping the face of the scarp before extending over the lower, flatter ground. The fact that these radial flows extend outwards for distances of about 350 km from the summit may mean that it is more realistic to consider 700 km as the real diameter of the volcano.

The scarp face itself transects a broad pedestal of pre-shield material about which little is known. This is transected by several faults, seen in section in the scarp backwall, which may be related to the generation of the scarp itself. Along the basal scarp there is evidence for several major landslides, particularly where the scarp has not been inundated by lava flows. These have a hummocky texture away from their margins with a ridge-and-trough fabric in peripheral locations. In some respects this texture is similar to that of the remarkable aureole, which extends in places for at least 1000 km from the shield summit and whose formation was either contemporaneous with or just subsequent to scarp development. This comprises a series of terrain blocks each of which is made up from distinctively textured, closely spaced ridges (Fig. 10.12). In broad terms, the inner edge of each block is embayed by younger flows or aeolian deposits, while the outer boundary is more scarplike. The overall impression gained in studying the whole aureole, is that the blocks form a complex of inwardly tilted prisms of intensely fractured terrain. Significantly, the number of superposed impact craters is very small, and, although smaller ones (less than 1 km) might be removed by surface creep or landslipping, the relative paucity of 5 km craters – which are less likely to be removed in this way – implies the feature must be young (Schaber et al., 1978). Another interesting point is that over the northwestern region is a positive free-air gravity anomaly of several tens of milligals (Sjogren, 1979).

Explanations for this unique but enigmatic landform are numerous and diverse. Carr (1973) suggested that the aureole is the remnant of an older shield volcano, while Blasius (1976) suggested it was an unroofed pluton. Neither of these early suggestions appears valid, since both would necessitate much higher rates of erosion than are observed on Mars. A rather different explanation stems from the work of King and Riehle (1974), who proposed Olympus Mons to have been a composite volcano, with a major input from ash generated during extensive explosive activity. The aureole blocks are considered by them to represent the eroded remnants of lava and tuff sheets, the latter being deposits of nuées ardentes. Morris (1979) also suggested construction from tuff sheets, but in his hypothesis these were erupted from several vents distributed around the main shield. Neither of these ideas seems entirely appropriate, however, since there is a total absence of evidence for any pyroclastic deposits anywhere on the Olympus Mons shield or, indeed, on the Tharsis Montes.

Hodges and Moore (1979) ingeniously proposed that Olympus Mons bore similarities

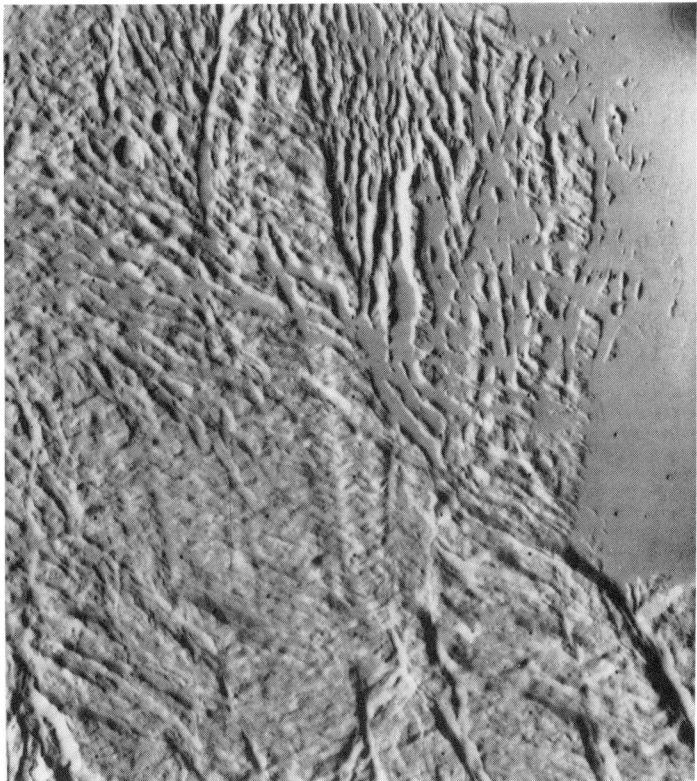

Fig. 10.12: Olympus Mons aureole texture. This consists of a series of blocks each composed of ridges interspersed with smoother plains units. Frame width 93 km. V.O. frame 043B20.

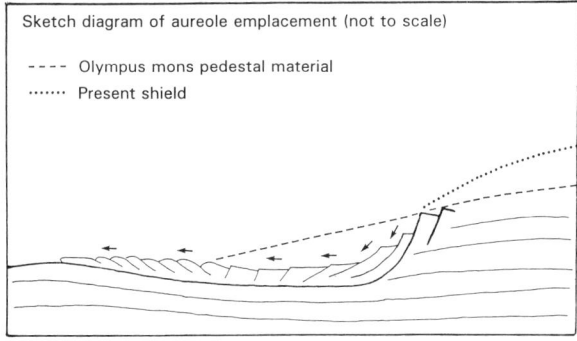

Fig. 10.13: Gravity-sliding mechanism for the generation of the Olympus aureole terrain. (After Lopes *et al.*, 1982.)

to certain Icelandic volcanoes which erupted under a cover of ice. According to this theory, the height of the basal scarp represents the thickness of an ice sheet which once covered the region, and beneath which eruptions took place. However, no evidence for subaerial activity within the confines of the aureole exists and, furthermore, neither is there any evidence for the ice sheet.

Various workers have considered the possibility of major thrusting or sliding. For instance, Harris (1977) suggested that the aureole lobes represent vast, gravity-assisted thrust sheets, an argument pursued also by Morris (1981). Carr *et al.* (1977) and Lopes *et al.* (1980) both have concluded the aureole to be the product of mass movement, the latter group invoking gravity-assisted rockslides on an immense scale (Fig. 10.13). Such an interpretation is supported by a comparison between the volume of material available before scarp formation and the present volume of the aureole materials. Furthermore, the greater extent of the aureole on the northwest side, where the shield slopes gently down to the surrounding plains (compared with the southeast, where it rises towards Tharsis Montes), clearly implies gravity control. Mega-sliding of this order could have been aided had there been a permafrost reservoir in the sub-shield pedestal. For this to have occurred the pedestal would, of necessity, have to have included a significant proportion of either brecciated or relatively porous material. The obvious candidate for this would be volcanic ash or tuff, for which, unfortunately, there is no evidence. Nevertheless, it is not beyond the realms of possibility that explosive activity may have preceded the building of the main Olympus Mons edifice; after all, there has been the suggestion that ash flows contributed to the evolution of Alba Patera, also on the northern side of Tharsis (Mouginis-Mark et al., 1988). Therefore, if the increased heat flow associated with the rise of magma to form the main shield had begun to melt sub-shield permafrost, this could provide a catalyst for the sliding out of the aureole blocks. While the author favours some kind of large-scale sliding mechanism, neither this nor any of the other suggestions explains the gravity data. At the present time, therefore, it has to be admitted that no single hypothesis is completely satisfactory and a solution to this enigma remains to be found.

10.5.3   Older Tharsis volcanoes

A number of smaller central volcanoes are situated within the Tharsis province. All are older than the Tharsis Montes and some may be as ancient as paterae in the heavily cratered highland hemisphere (Plescia and Saunders, 1979). At least two of these are somewhat steeper than the other shields, while one has a low profile and is very like the huge patera volcano Alba Patera; the remainder have the general characteristics of the younger Tharsis lava shields.

Northeast of Ascraeus Mons is a group of three volcanoes: Uranius and Ceraunius Tholi and Uranius Patera (Fig. 10.14). The two tholi have steeper slopes (5° and 7° respectively) than Uranius Patera (0.5°) and warrant separate description. Uranius Tholus has a diameter of 83 km and rises 3500 km above the surrounding lava plains. The flat-topped summit has a 14-km-diameter pit set towards the eastern edge of a larger but very shallow 32-km caldera, largely infilled with ponded lavas. Because several impact craters larger than 10 km are superimposed on its flanks this is believed to be a relatively old structure and certainly it, and the other two volcanoes, are significantly older than the

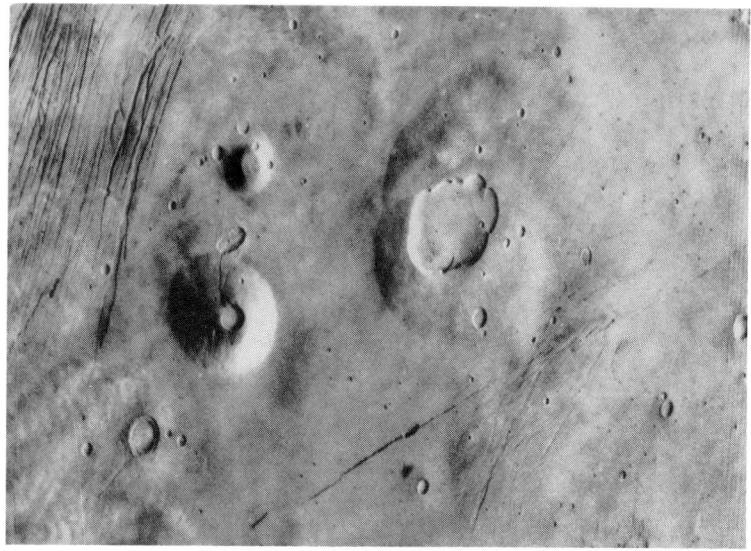

Fig. 10.14: Group of three volcanic shields situated east of Ceraunius Fossae (left). Uranius Tholus, the smallest, has a flattened summit region, while to its south is the larger, steep-sided Ceraunius Tholus. East of both is the volcano Uranius Patera, with a large summit caldera and much shallower flank slopes. Frame width 750 km. V.O. frame 759A73.

Hesperian-age flow plains which encroach upon them. Ceraunius Tholus is somewhat steeper and larger (130 km × 92 km) and has a 23-km summit caldera with vestiges of an older, shallower pit on its north side.

The surfaces of both volcanoes have a somewhat muted appearance and are finely striated, the most obvious striations being narrow channels with a radial trend. Generally these do not reside at the crests of ridged-profile lava flows but are incised into the volcano's flanks, although there are exceptions. Similar channels characterize the flanks of Uranius Tholus and appear to set these two rather steeper-sided structures apart from the rest in Tharsis. Indeed, their general appearance is more reminiscent of the Elysian volcano Hecates Tholus, discussed in a later section. Since several workers have invoked explosive activity in the latter's development, it may be that the rather different characteristics of these two Tharsis structures are a manifestation of at least some Plinian or similar eruptivity.

One particularly prominent 2-km-wide channel emerges from near the summit of Ceraunius Tholus and runs down the north flank into an impact crater at the base of the shield (Fig. 10.15). Carr (1974) suggested that this larger channel and several others like it may have been due to lava erosion, but Riemers and Komar (1979) argued that the general appearance of both coarse and fine channelling on both tholi is reminiscent of the terrestrial strato-volcano Barceno, in Mexico, which, during 1952-1953, experienced large-scale explosive activity that generated fast-moving density currents – base surges and nuées ardentes – which eroded arrays of channels in the volcano's flanks. The general

Fig. 10.15: JPL enhanced image of Ceraunius Tholus, showing the finely striated shield flanks and the prominent channels on the north side. The largest of these debouches into an elongate impact crater on whose floor it has deposited an apron of debris. Frame width 100 km. V.O. frame 516A24.

similarity in form and distribution between the Mexican and Martian channel arrays prompted their argument and, furthermore, they note that larger channels – up to 8 km long and between 1 and 2 km wide – are known to have formed by the same method on Asami volcano, Japan (Aramaki, 1956). Although magmas considerably more silicic than basalt were involved in the terrestrial channel-cutting episodes, this does not necessarily argue for silicic volcanism on Mars; both Francis and Wood (1982) and Mouginis-Mark *et al.* (1988) have shown that ash flow involving mafic magma is a perfectly reasonable Martian phenomenon.

Fig. 10.15 also shows a depositional apron at the foot of the shield where the wider channel debouches into the impact crater. From what this is made is arguable; it could be a stubby and rather viscous lava flow or an ash lobe whose perimeter has been oversteepened by subsequent erosion; it is unlikely to be a fluvially deposited sedimentary

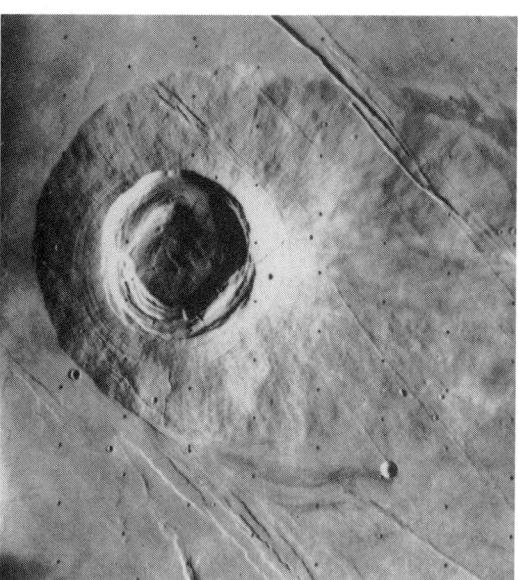

Fig. 10.16: Biblis Patera, a smaller shield west of Pavonis Mons. The 55-km-diameter caldera has associated circumferential graben, while the volcano's flanks are transected by regional Tharsis fractures. Frame width 195 km. V.O. frame 044B50.

deposit since no suitable catchment area would have been available at the summit of the tholus. It is not possible to discriminate between the two feasible alternatives. What is clear, however, is that the lobe, and therefore the cutting of the wide channel, post-dates the impact crater inside which the lobe resides, while this, in turn, is younger than the Upper Hesperian-age Tharsis Montes plains that encroach upon the shield. This is not the case with the fine-scale channels, which are embayed by the adjacent lava plains and clearly are an integral part of the main shield-building history. The same image also shows several partially coalescent pits near the summit caldera on the southwest flank of Ceraunius Tholus. These appear to be located on a slight tumescence very reminiscent of that seen on the Tharsis Montes, particularly Arsia Mons. This may indicate that the regional southwest-northeast fracture line was also operative when this tholus was being built.

Biblis Patera and Ulysses Tholus, situated west of Pavonis Mons, both have simple calderas slightly over 50 km in diameter with which are associated circumferential graben. The former has an elongate shape measuring 175 km × 105 km and rises 4 km above the surrounding plains (Fig. 10.16); the latter rises 3 km above the Amazonian-age lava plains and has a diameter of 91 km. Both have flank slopes approaching 4°. Their general appearance is very reminiscent of the summit region of Arsia Mons. The smaller, Jovis Tholus, is a 55-km shield with a 27-km-diameter summit caldera and is set somewhat apart from the rest, midway between Olympus Mons and Ascraeus Mons. Tharsis Tholus, east of Ascraeus Mons, is a trifle steeper (5°), rises 6 km above the plains and measures

Fig. 10.17: The 110 × 83 km nested caldera complex at the summit of Uranius Patera. Tilting of the northwest part of the floor appears to have preceded flooding of the central regions by ponded lavas. Frame width 270 km. V.O. frame 857A46.

155 by 120 km; it also has a caldera complex, 62 × 46 km across. As a group, these volcanoes bear close resemblance to the upper parts of the main Tharsis shields and, in consequence, are presumed to be partially buried, older structures that grew prior to the Upper Hesperian-Lower Amazonian activity which saw the gradual building of major centres along the crest of the Tharsis Bulge. Whether they ever grew to the vast dimensions of their successors is not known.

The third of the group of three Tharsis volcanoes whose description began this section is rather different from all of the others. The exposed part of Uranius Patera measures 202 × 184 km and has flank slopes of a mere 0.5°, which are traversed by well-defined, narrow and often leveed flows. These may be traced right up to the backwall of the nested (110 × 83 km) caldera (Fig. 10.17). The floor of this is crossed by several prominent wrinkle ridges while the northwest part of the floor surface is inclined towards the centre of the depression, suggesting it subsided prior to the emplacement of the more-or-less horizontal ponded flows which occupy the centre. The very large dimensions of the caldera when compared to the size of the volcano suggest that a large part of this has been buried by younger deposits and the close resemblance it bears to the summit region of Alba Patera, discussed in the next section, may mean that it was among the larger of the older structures.

10.5.4   Alba Patera

The Hesperian-Amazonian-age volcano Alba Patera has associated with it some of the most extensive and remarkable volcanic flow-fields found anywhere in the solar system. Greeley and Spudis (1981) set it aside from the others, describing it as ".... an apparently unique volcanic landform." Located on the northern extremity of Tharsis, its summit is located at 110°W, 40°N and lies at the centre of an oval ring-fracture zone measuring 550×400 km in diameter (see Fig. 6.9, p.109). Crater ages suggest that its oldest flows pre-date those of Tharsis Montes, that its peak of activity may have occurred around $1725 \pm 123 \times 10^6$ y, and that its most recent products post-date some of the lavas associated with Arsia Mons (Cattermole, 1989). On this basis its activity extended over a period of at least 1.5×10^9 y and may have spanned as much as 2.8×10^9 y.

The summit of Alba lies 7 km above Mars datum, and comprises a broad southwest-northeast-trending ridge with the summit calderas situated towards the northeast end. The mean slope value for the patera is only about 0.5°, yet the volcanic flows extend at least 1350 km from the summit, giving it a diameter of 2700 km, an area approaching 2×10^6 km^2 and a volume of at least 1.4×10^9 km^3. The extreme length of these flows, despite their having traversed gradients as low as 0.5° or less, implies both low viscosity and high volume for the individual flows. The flows themselves are of various kinds (Fig. 10.31) but can broadly be classified into three types: Group 1 – relatively narrow and often leveed flows radial with respect to the summit; Group 2 – longer, broader and often lobate sheet flows which extend much lower onto the shield flanks; Group 3 – massive tube- and channel-fed flows which are absent from near-summit locations but characteristic of the lower flanks (Cattermole, 1989). Quantitative analysis of these flows indicates that the large sheet and tube-fed lavas had low yield strength and viscosity and were erupted at very high rates, the latter, in particular, at rates sometimes in excess of 1×10^6 m^{-3} s^{-1} (Cattermole, 1987).

The summit region is the site of a double caldera complex which became the focus of extended effusive activity after an early period of flood lava activity in the general vicinity of the proto-shield (Fig. 10.19). Flows associated with the younger caldera partially bury the older, incomplete caldera depression. The former has at least five components, which together cover an area of 2065 km^2; transection relations indicate the size of the individual components to have decreased with time. The caldera is, however, very shallow, being at most 150 m deep, which contrasts markedly with the deeper pits on Ascraeus and Olympus Montes and indicates a significant difference in the dimensions of their subjacent magma reservoirs. As is the case with the Tharsis Montes, narrow and often leveed flows can be traced right up to the backwall of the caldera, indicating that Plinian-style activity was not important during the later stages of patera growth. Raitala (1989 and Raitala and Kauhanen

Fig. 10.18: Flow types characteristic of the southeast flank of Alba Patera. Prominent tube-fed flows are marked T–T´. The more northerly one is 120 km. long. Such flows retain much the same width along their course and have a low triangular cross-section. A 20-km-wide sheet flow with lobate margins and a flat upper surface is marked S–S´. This forms part of a massive flow-field complex that emerges from amongst the Alba ring fracture zone. Such flows often broaden substantially towards their distal ends. Frame width 85 km. V.O. frame 254S22.

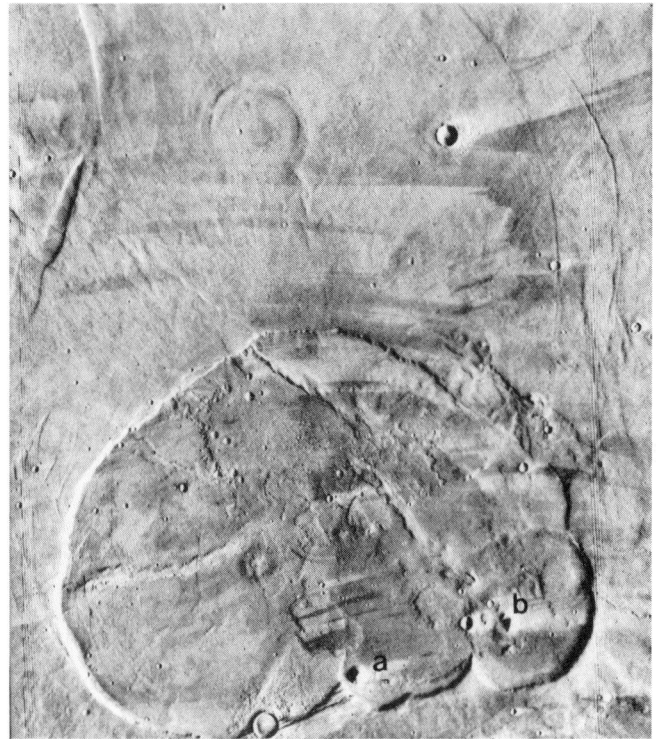

Fig. 10.19: The younger summit caldera of Alba Patera. This nested complex has five main components, the younger of which are located towards the bottom right of the image. The floor is at various levels and is traversed by prominent ridges composed of *en echelon* segments (r). Two prominent volcanic domes outcrop on the floors of the younger caldera pits (a and b). Note the plethora of narrow radiating flows along the northern rim and the narrow arcuate graben towards the top right. Frame width 73 km. V.O. frame 253S14.

(1989, in undertaking a tectonic analysis of the calderae and surrounding grabens, suggest that the central caldera zone is related to collapse into a relatively shallow reservoir, between 10 and 15 km deep (based on a caldera diameter/roof depth ration of 5.0). For the larger of the main calderae this depth might be expected to be closer to 25 km.

The general distribution of lava flows on the main part of the patera is radial about the summit but flows are scarce over the uplifted block of Ceraunius Fossae and in the northwest sector they are either absent or mantled by younger deposits (Fig. 10.20).

On the lower ground to the northwest, large fields of tube- and channel-fed lavas have a generally northnorthwest-southsoutheast strike and appear not to have been related to summit activity. Detailed geological mapping by the author (Cattermole, 1987, 1988, 1989; 1990) established there to have been three principal episodes of patera growth, (confirming

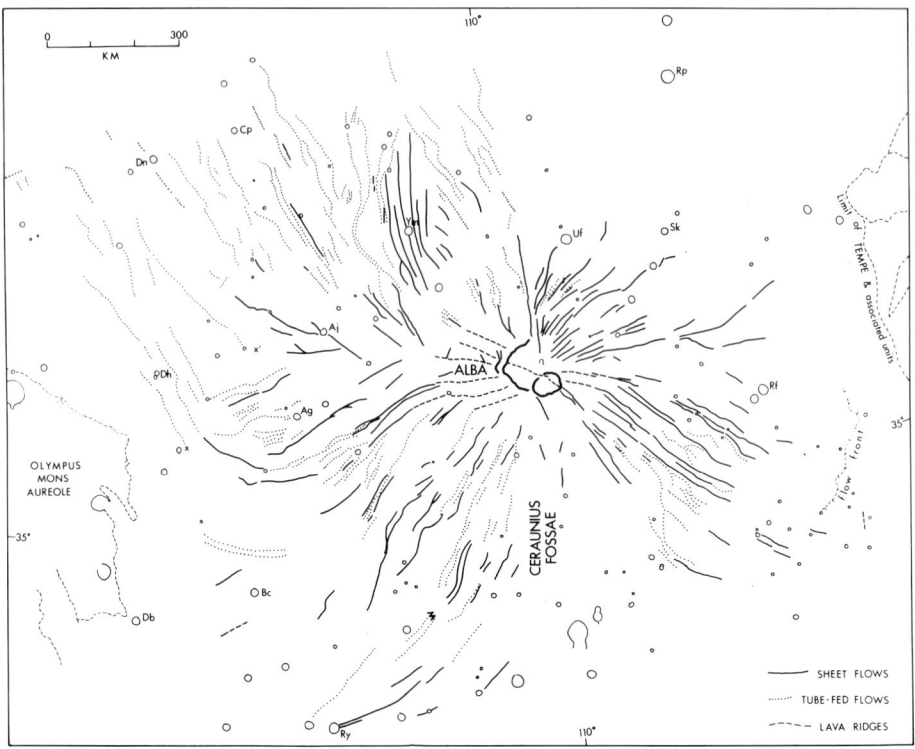

Fig. 10.20: Distribution of different flow types at Alba Patera volcano. While there is a generally radial pattern, massive tube- and channel-fed lavas to the west and northwest of the structure do not appear to have a summit-related origin.

earlier mapping by Scott and Tanaka (1980, 1986), together with generation of several large flow-fields of more restricted outcrop whose development may have been related to major fractures on the volcano's flanks (Fig. 10.21). The earliest phase in the development of Alba involved the widespread emplacement of fissure-fed flood lavas of Lower Hesperian age (Scott and Tanaka, 1986) which now occupy distal locations. Subsequently volcanism became more centralized, sheet flows and tube-fed lavas of large volumes being extruded, mainly from linear vents situated at or near the present summit, or from lower down the volcano's flanks. These flows were responsible for the majority of patera construction. Quantitative measurements of these flows indicate volumes at least an order of magnitude greater than most Hawaiian flows, sheet flows having volumes in the range 1-110 km^3 and tube-fed lavas achieving volumes as great as 3500 km^3 (Baloga and Pieri, 1985; Cattermole, 1987).

A further characteristic of Alba Patera is the existence of anastomosing channel networks, mainly on the northern flank and cut into relatively smooth-surfaced units from which obvious lava flow lobes are absent and which separate the two main stages of lava shield growth (Fig. 10.35). A recent analysis of these by Mouginis-Mark *et al.* (1988),

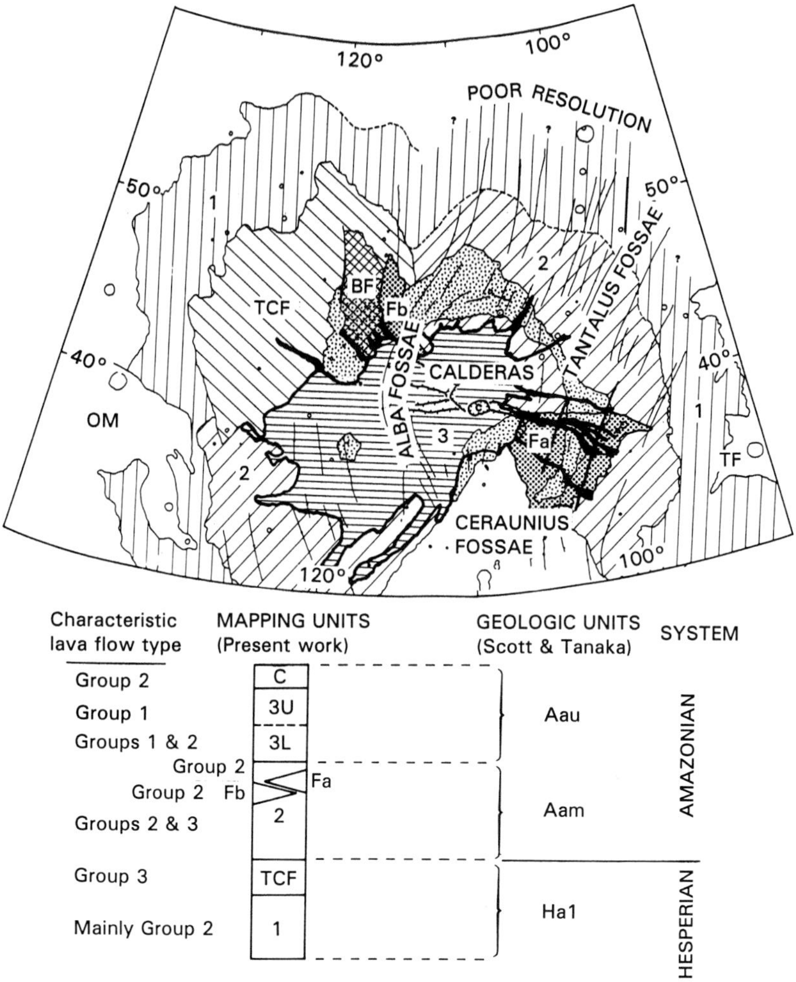

Fig. 10.21: General geological map of Alba Patera volcano (Cattermole, 1989)

combining photogeology with interpretation of thermal inertia data, led them to conclude, firstly that the channels were of fluvial origin and, secondly, that they had been incised by a process of sapping induced by the release of non-juvenile water within relatively unconsolidated deposits on the volcano's flanks. Lastly, after synthesizing their data, they also proposed that the fine-grained deposits themselves were of pyroclastic origin.

Pyroclastic materials may be dispersed either by eruption clouds emanating from a central vent, in which case the extent of the air-fall deposits will be a function of cloud height (Wilson and Walker, 1987), or laid down by ground-hugging pyroclastic flows whose generation has been discussed by Sheriden, (1979) and Malin and Sheriden (1982). Mouginis-Mark *et al.* (1988) discounted an air-fall origin on the basis that the channelled

Fig. 10.22: Anastomosing channel networks on the northern flank of Alba Patera. Note the well-developed dendritic drainage pattern.

deposits extended too far (500-600 km) from the potential caldera source region and could only have been dispersed that far by extremely high eruption clouds unlikely to form on Mars. In consequence, they appealed to long run-out pyroclastic flows as a means of dispersal that falls within the constraints imposed by Martian conditions. On the basis of the volume of material involved, they calculated that the total volume of volatiles that would need to be erupted to produce the observed flows would amount to around 5% of the mass of the current Martian atmosphere (Wilson and Mouginis-Mark, 1987).

If this conclusion is accepted – and there is no *a priori* reason why this should not be so – then eruption of the volatile-rich ash-flow materials can be shown to have post-dated the emplacement of high-volume sheet and tube-fed lavas on the lower flanks but to have preceded the final effusive phase of the volcano's evolution. As Mouginis-Mark and his co-workers note, the existence of pyroclastics within the lava pile certainly could have played a part in explaining its very low relief and also the form and distribution of the circumferential fractures. The development of the volcano might well, therefore, have followed an evolutionary path similar to that depicted in Fig. 10.23.

At a later stage, the extrusion of shorter, narrow and often leveed flows (volumes of 0.6-8.0 km^3) occurred from the summit itself, which experienced repeated episodes of

VOLCANIC EVOLUTION OF ALBA PATERA

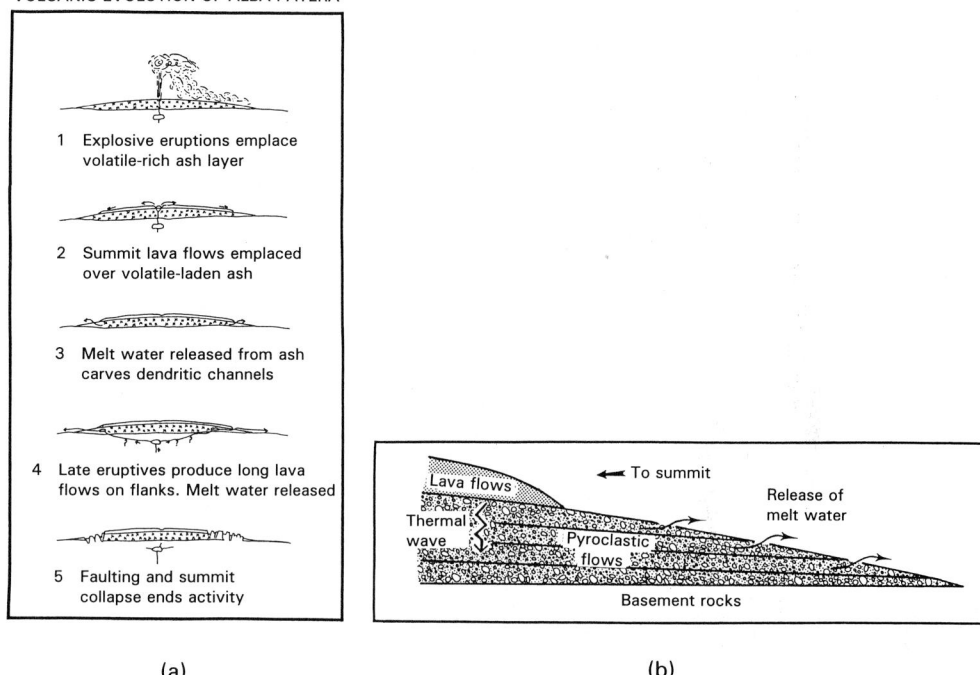

1   Explosive eruptions emplace
    volatile-rich ash layer

2   Summit lava flows emplaced
    over volatile-laden ash

3   Melt water released from ash
    carves dendritic channels

4   Late eruptives produce long lava
    flows on flanks. Melt water released

5   Faulting and summit
    collapse ends activity

(a)   (b)

Fig. 10.23: (a) Five-stage model for the evolution of Alba Patera. (b) Possible method by which channel networks were produced on smooth-textured flank deposits of Alba Patera by later-stage summit flows. It is assumed that the non-welded ash flows would have become charged with volatiles, following which summit activity generated lava flows that partially buried the earlier deposits. A thermal wave passing from the lava pile into the underlying volatile-enriched material would then generate meltwater and/or steam that eventually emerged at the surface along bedding planes between successive ash flow layers. (After Mouginis-Mark *et al.*, 1988.)

tumescence and collapse associated with the long-lived high-volume eruptive episodes. It is possible that the increased heat flow associated with this sub-terminal effusive activity provided the thermal energy necessary to melt permafrost locked within the ash deposits on the volcano's flanks, generating sufficient free-flowing water to either incise the channel networks or induce them by ground sapping (Fig. 10.24).

Finally, in the waning stage of activity, more viscous materials were erupted from a plethora of northnortheast-southsouthwest fissures, giving rise to *en echelon* spatter ridges

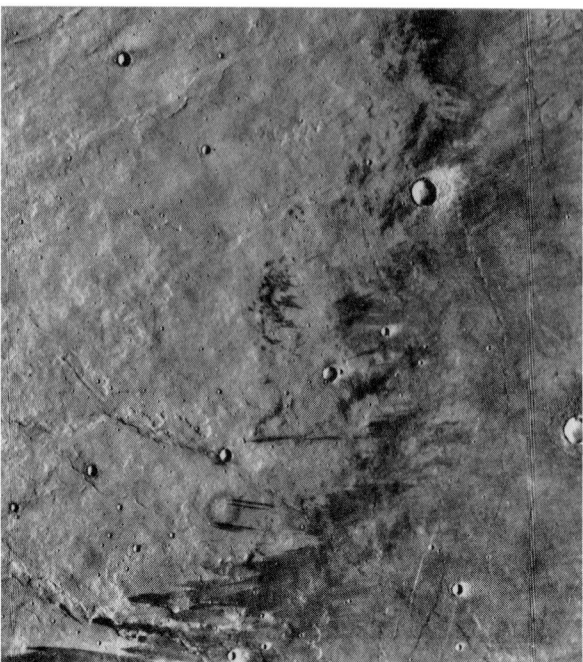

Fig. 10.24: Volcanic spatter ridge on the eastern flank of Alba Patera. Such ridges consist of linear ridge elements between 0.3 and 10 km long and between 100 and 500 m wide which may continue along the same trend for upwards of 150 km. Generally they are about 15 m high. V.O. image 253S44. Frame width 100 km.

up to 150 km long and between 100 and 500 m wide (Fig. 10.37). Similar structures are well known both from Hawaii and from the Snake River Plains (Carr and Greeley, 1980). Compared with their terrestrial analogues, the Martian features have lower aspect ratios (ratio of thickness to width), believed to be a reflection of the lower yield strength of the lavas, and are also about twice as wide. The latter is due to the greater width of fissures along which encrustation of spatter occurred and confirms the prediction of Wilson and Head (1983) that Martian fissure widths should be roughly twice those of Earth, and is entirely consistent with the greater volume of the flows observed on Alba volcano.

Because the Alba flows are so well-preserved and the summit calderas so well-imaged, they are readily mapped in detail, and in consequence, it is possible to consider certain volcanological implications. Firstly, it is clear from the flow morphology that Hawaiian-style effusive activity characterized the growth of the main shield of Alba. It is also apparent that the volumes of individual flow units gradually decreased with time. Now since caldera collapse can only take place if a sufficiently large void is created beneath the summit, and because subvolcanic reservoirs must be full prior to the onset of eruption (Blake, 1981), only during the extrusion of lava (or the injection of dykes) can such a space be created. As it is generally accepted that caldera dimensions are approximately proportional to the

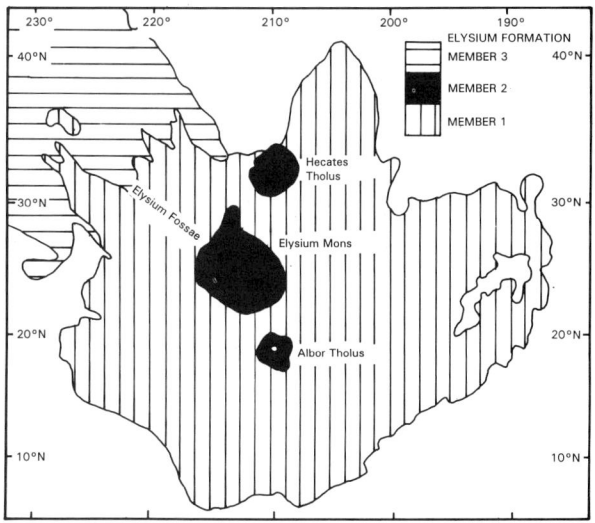

Fig. 10.25 Generalized geological map of the Elysium region of Mars.

size of the subjacent magma chamber (Koyanagi *et al.*, 1972; Wood, 1984), it is important to enquire whether or not the chamber volumes corresponding to the observed caldera pits can be accounted for by the volume of the circum-summit lavas. Since the volume of pre-erupted magma prior to collapse of the two more recent of Alba's pits must have been around 12 km³, calculations suggest that a chamber of radius 10.5 km and volume of 5000 km³ must have existed when the smaller caldera pits were generated; thus, assuming evacuation was achieved solely by effusive activity, less than 20 of the lower-volume summit flows would account for this. When the older, larger pits subsided, fewer than 50 sheet flows could have evacuated the 200 km³ volume implied by their greater dimensions. There is little doubt, therefore, that summit subsidence was achieved predominantly by effusive activity from or near the summit. Such is in accord with similar conclusions reached by Mouginis-Mark (1981) for Ascraeus Mons. Following on along the same line of argument, the greater volumes of the earlier sheet and tube-fed flows imply that progressively lesser volumes of magma were required to trigger eruption as time proceeded. The implication here is that the magma flux rate beneath the volcano must have been greater during the earlier stages of its construction, that is, during the early Amazonian epoch, when the main part of the structure was emplaced, than during the later stages.

The very large dimensions of Alba flows have no real parallel on the Earth and it is interesting, therefore, to consider the long-term magma supply rate to the volcano through geological time. If all of the present topography of Alba is due to constructional volcanism, the lava pile has a volume of 4.16×10^9 km³. If volcanism is considered to have occupied a period of at least 2.3×10^9 y, this gives an overall magma production rate of 2×10^{-3} km³ y^{-1}. Such a figure is significantly larger than present day rates of magma supply to the volcanic pile of Mauna Loa and Kilauea, given by Swanson (1973). Even assuming half of the patera was due to structural doming – which seems unlikely – lengthy periods of repose must have punctuated periods of major extrusive activity, since fluvial activity, structural readjustment and aeolian weathering are all known to have occurred. Calculations also show that several of the larger sheet- and tube-fed lavas on the lower flanks of Alba Patera must have released two orders of magnitude more thermal energy than the Earth's annual release due to volcanism, and one order of magnitude more thermal energy than the annual conducted heat flow of the Earth (Cattermole, 1989). This surprising conclusion leads inevitably to the implication that eruption of these very large volumes of magma must have constituted an extremely, if not the most, important, source of energy loss from the Martian interior over geologically short periods during the Hesperian and early Amazonian epochs.

10.6   THE ELYSIUM VOLCANIC RISE

There are three shield-like volcanoes in this region and they show significant differences from those of Tharsis. The lower Amazonian lavas associated with them cover an area of approximately 3×10^6 km^2 (Fig. 10.25). A 400-km-diameter fracture belt almost completely encircles the summit of Elysium Mons and outside this ring are numerous west northwest-east northeast-trending troughs, most with flat floors and the general appearance of graben. For reasons not entirely understood, these pass northwestwards into a series of branching channels which extend for several hundreds of kilometres further.

10.6.1   The Elysium shields
The largest of the Elysian group is Elysium Mons, which has a diameter of 170 km and a single caldera 12 km in diameter. Its summit lies about 9 km above the surrounding plains and it has a mean slope of 3.5° – comparable with slopes on large Tharsis volcanoes. The overall geometry is, however, asymmetric: there appears to be a main shield, approximately circular in plan, with the caldera sitting on the crest of a broad ridge that trends northwest-southeast across its crest. Viking imagery reveals a plethora of narrow volcanic flows on this main edifice, large numbers of hummocks up to 5 km across and linear channels (Fig. 10.26). Malin (1976) recognizes a second component in the volcano's structure – a broad northward-trending ridge about 200 km wide and 2 km high upon which there are numerous narrow flows and sinuous ridges, much like those visible on the upper flanks of Alba Patera. An earlier suggestion by Malin (1976) that Elysium Mons was in many ways similar to the terrestrial composite volcano Emi Koussi, in the Tibesti massif, can now be discounted. More recent altimetric data than was then available, reveals the cone to be far less steep than was originally believed, while his suggestion that silicic magmas may have been involved in its construction are not in accord with Viking Lander chemistry. Its overall morphology suggests it to be not unlike many Tharsis lava shields.

The other large Elysian structure, Hecates Tholus is more interesting and has been studied closely by Mouginis-Mark *et al.* (1982). Situated north of Elysium Mons, it rises about 6 km above the adjacent plains and is a low shield measuring 160×175 km. At the summit is a nested caldera complex measuring 11.3×9.1 km, with an estimated volume of 28 km^3 for the largest component. Unlike typical Tharsis shields, Hecates Tholus exposes

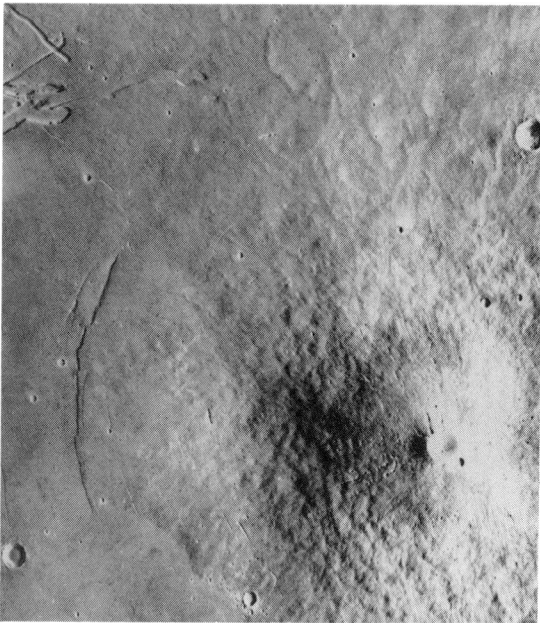

Fig. 10.26: (above) General view of Elysium Mons, showing the 14-km-diameter summit caldera, main shield and peripheral radial fractures. Frame width 265 km. V.O. frame 844A17. (below) Higher-resolution mosaic showing the hummocky texture typical of the shield flanks, together with several prominent channels north of the caldera and a long scarp on the eastern flank. Mosaic width 100 km. V.O. frames 541A44, 46.

Fig. 10.27: Summit region of Hecates Tholus showing the nested caldera complex, aligned pit rows, channels and generally rather smooth appearance of the circum-summit area. Frame width 60 km. Mariner 9 frame DAS13496298.

no lava flows, rather a complex of radial channels not unlike those seen on Ceraunius Tholus (Fig. 10.27). It was the existence of these channels which led Reimers and Komar (1979) to propose an origin in volcanic density currents. However, after studying the channel networks in detail, Mouginis-Mark and his co-workers showed that the anastomosing courses and dendritic tributary networks characteristic of the channels were unlike channels associated with volcanic density currents. Furthermore, the absence of channels from the summit region makes it difficult to believe their formation was associated with explosive activity focussed there. More plausibly, they suggest that the channels are fluvial in origin, having been incised into materials less coherent than the lava shield that these undoubtedly mantle.

The same group confirmed the earlier observations of Plescia and Saunders (1979) that variations in numbers of superposed impact craters implied several resurfacing episodes had modified the volcano's flanks; they also noted the relative paucity of small impact craters on the west part of the shield, over an area measuring 50×75 km. Indeed, the counts suggest the surface exposed there is younger than the youngest region of the Olympus Mons caldera (perhaps as young as 3×10^8 y). Very reasonably, they propose that this is a relatively young air-fall deposit which they estimate to be about 100 m thick and produced by an eruption cloud with a height approaching 70 km. Because a stable eruption column apparently was able to be sustained, the magma volatile content must have been about 1 wt % if the volatile were H_2O and more than 2 wt % if CO_2. Such a volatile component requires that the source magma must have originated at depths of greater than 50-100 km if the volatile was CO_2 and between 4 and 150 km if it was H_2O. While imposing a mantle source for CO_2-rich magmas, these estimates apply no such restrictions if water is the volatile, and allow for the possibility of absorption of permafrost or trapped groundwater as the rising aqueous magma neared the Martian surface. In order to reach such an altitude, model calculations suggest a volatile-charged Martian magma would need a mass eruption rate of around 10^7 kg^{-1} s^{-1}. This volcano provides a further example, therefore, of one in which phreatomagmatic volcanism has played an important role, apparently extending the period of such activity from the Hesperian into the Amazonian epoch.

Unfortunately, the resolution of available imagery allows little to be learned of the smallest of these volcanoes – the 30-km-diameter Albor Tholus – which is not discussed further here.

10.6.2   Apollinaris Patera

Apollinaris Patera has a rather isolated location southeast of the Elysium group, and is centred at 9°S, 186°W. Crater counts indicate an age roughly half that of Alba Patera (Plescia and Saunders, 1979). Its general appearance is that of a broad shield about 400 km in diameter with a 70-km summit caldera which has two different floor levels (Fig. 10.28). The volcano summit stands 5 – 6 km above the adjacent plains. The shield flank is strongly striated and is circumscribed on the west, north and east by a prominent scarp that transects the striated shield surface, presumed to be built from narrow lava flows. To the south, however, a large fan with its apex at the caldera rim extends for about 350 km and buries the scarp in that sector. On the surface of the fan is an array of what appear to be broad channels, some of which incise the caldera backwall. A recent survey of the volcano, based on analysis of digital elevation models and all available imagery (Thornhill et al., 1993 suggests that the edge of an old caldera is marked by the truncation of channels along a prominent morphological break 1 – 2 km below the present rim on the north and northwestern side of the structure. Subsequent caldera activity resulted in growth of a dome, followed by collapse to produce an inner caldera depression some 20 km smaller in diameter. The northwestern rim of the later caldera was coincident with the earlier one but the southeastern wall was displaced northwestwards by around 20 km, leaving a remnant of the dome within the southeastern part of the original depression. Subsequently the inner caldera has experienced a complex history of lava lake filling and evacuation.

Unfortunately the resolution of available Viking imagery does not allow for more detailed analysis and it is not possible to decide whether the fan-like deposit is built from lavas or from pyroclastic deposits cut by channels. However, Thornhil et al. prefer to intepret the fan as a flow field which developed from a point source breach in the caldera rim. The area of chaotic terrain to the south and west of the edifice may represent collapsed pyroclastic deposits, the collapse being triggered by removal of ground ice.

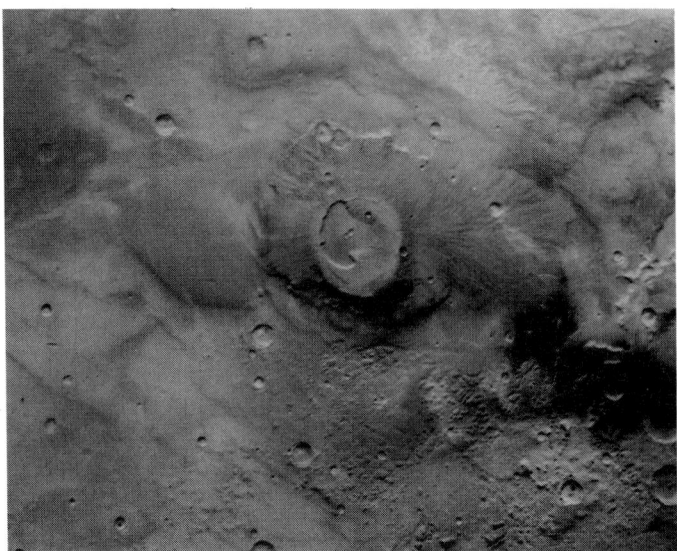

Fig. 10.28: View of the isolated central volcano Apollinaris Patera, with its 100-km-diameter caldera. The shield surface is truncated by a prominent scarp that is buried on the south side by a massive fan-shaped deposit that spills onto the adjacent lava plains. This could be a lava fan but might also represent channelled pyroclastics. Frame width 370 km. V.O. image 639A92.

10.7   HIGHLAND PATERAE OF THE HELLAS REGION

Several of the most ancient volcanic structures on Mars are located near to the borders of the Hellas Basin. Plescia and Saunders (1979) termed these highland paterae and, on the basis of crater counts, all appear to have formed during the same general phase in Martian history, 3.7-3.1 × 10^9 y ago. Potter (1976), Peterson (1977) and King (1978) made the first studies of these during geological mapping. Viking imagery suggests that some of these present evidence for the earliest explosive volcanism on the planet and Pike (1978) was constrained to suggest that Martian highland paterae were similar to terrestrial ash shields. Subsequently several investigators have applied models of pyroclastic flows under Martian conditions to deposits found in eastern Hellas, and it has been established that long runout pyroclastic flows can better explain the observed morphologies than ash fall deposits, and are quite capable of producing the very extensive channelled deposits seen. The inferred transition from explosive to effusive activity witnessed at both Tyrrhena and Hadriaca Paterae has important implications for Martian volcanic and climatic history.

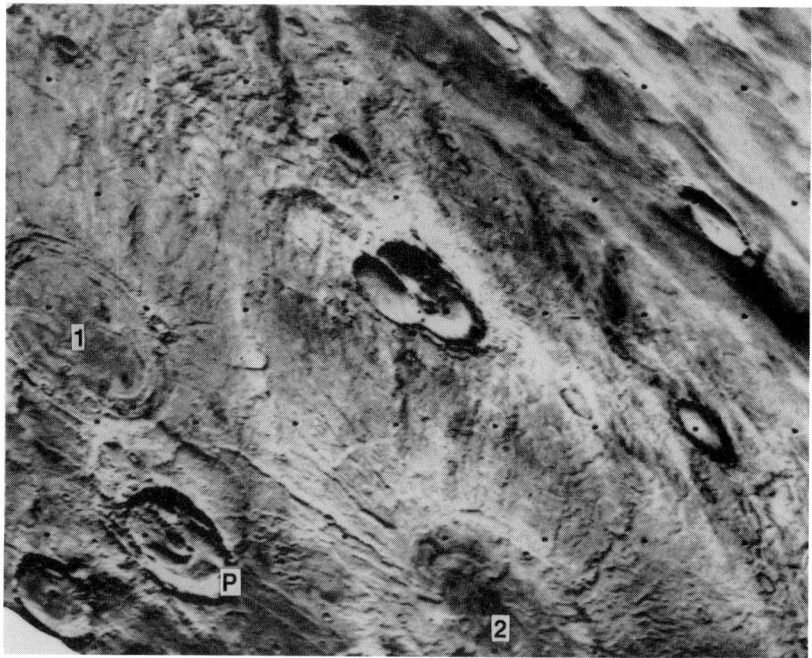

Fig. 10.29: Ancient volcanic ring structures close to the south rim of Hellas (north is to the left). Amphitrites Patera, (2) has a central depression with prominent radiating ridges and valleys, while to its south is another large ring feature, Peneus Patera (1), with concentric graben. Frame width approximately 800 km. V.O. frame 056B35.

10.7.1 Amphitrites Patera

The most equivocal of these ancient structures is Amphitrites Patera, situated close to the south rim of Hellas and comprising several 100-km-diameter ring structures with associated radiating ridges, some of which extend into the basin itself (Fig. 10.29).

The rings appear to have little, if any, vertical relief but have a broad structure that suggests some kind of volcanic origin.

10.7.2 Hadriaca Patera

Hadriaca Patera, located on the northeast rim of Hellas, is better defined and has at its summit an obvious flat-floored caldera 77 km in diameter (Fig. 10.30). The northern and eastern parts of the caldera backwall are rather indistinct and may show the effects of mantling by pyroclastic materials or of overflow by volcanic flows. On the eastern floor are several small hummocky features (<2 km across) which may represent late-stage volcanic constructs (e.g. domes or cinder cones) or be remnants of a flow or pyroclastic layer. Hadriaca appears to have little vertical relief – though more than Amphitrites Patera – but there is possibly as much as a 2 km height difference between the caldera and the foot of the channellized shield which extends 300 km from the caldera rim before merging

Fig. 10.30: Hadriaca Patera, an ancient volcanic structure on the northeast rim of the Hellas basin. The caldera is 77 km across and is the focus of radiating ridges and valleys rather reminiscent of the more mantled parts of Alba Patera. Several major channels are incised into the peripheral regions, two being visible at the bottom of the mosaic, and a third at the lefthand edge. V.O. frames 625A16. Frame width 200 km.

with the surrounding plains. To the southeast it is incised by the channels of Dao Vallis and is embayed by the low-lying plains that were laid down during the Hesperian and Amazonian periods. Close to the southwest backwall, a major scarp may be the eroded edge of a series of ponded flows associated with lava lake activity.

A recent study of this region by Crown and Greeley (1993) shows that the flank slopes range from 0.10° in the northeast to 0.47° in the south and west. The smooth, ridged and channelled morphology of the flanks – which appear to be devoid of lava flow lobes and scarps – is rather reminiscent of the mantled parts of Tyrrhena Patera, located 870 km to the northeast, and, to some extent, of Alba Patera. Such morphology indicates friable

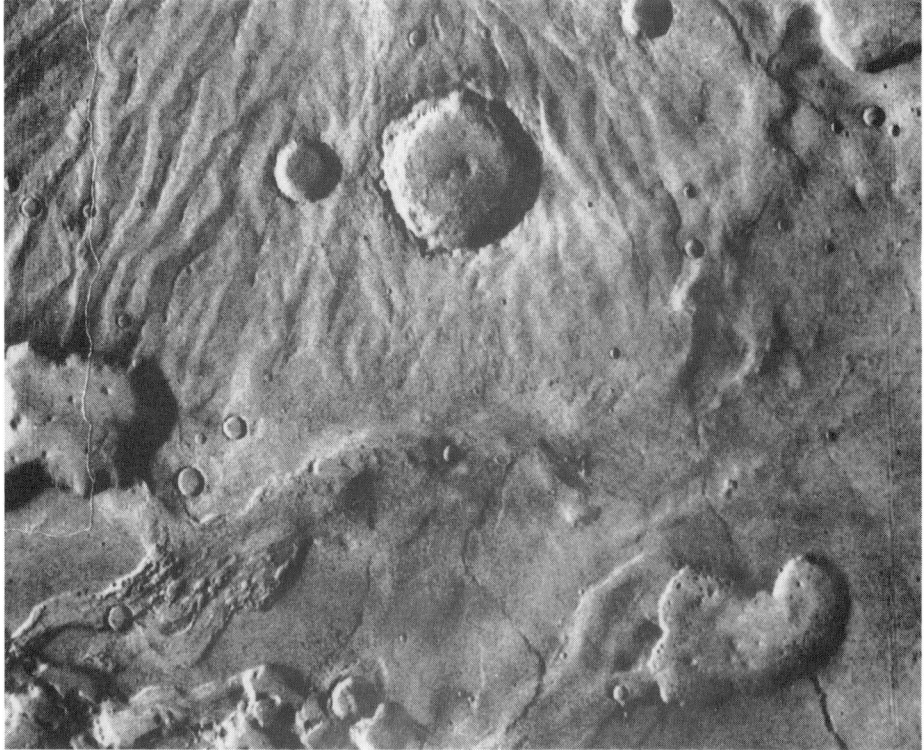

Fig. 10.31: Channels on the flanks of Hadriaca Patera. These have a radial pattern about the summit region. Stratification is exposed in some channel walls, while runoff channels may be seen within certain of the troughs. V.O. image 625A18. Frame width 160 km.

materials. Interestingly, a large channel, which begins its southwestward course in a large depression on the lower southeast flank, eventually merges with the floor of Hellas, some 800 km distant. The stratified and dissected deposits have been incised by outflow channels downslope. Layering is revealed on trough walls and occasionally v-shaped runoff channels occupy trough floors (Fig. 10.31).

The channels typically are 3 – 4 km in width and over 100 km in length. Zisk *et al.* (1992) indicate their depth to be between 200 and 300 metres. Typically they have flat bottoms and in this are dissimilar to lunar sinuous rilles which are U-shaped and are known to have been formed by evacuation of volcanic flows tubes. The channels on the flanks of Alba Patera generally occur on ridge crests, not on trough floors. Thus, channel morphology on Hadriaca Patera is more closely similar to that of channels on Tyrrhena Patera than to either of lunar rilles or the channels on Alba Patera. Further discussion of their origin will be left until Tyrrhena Patera has been described.

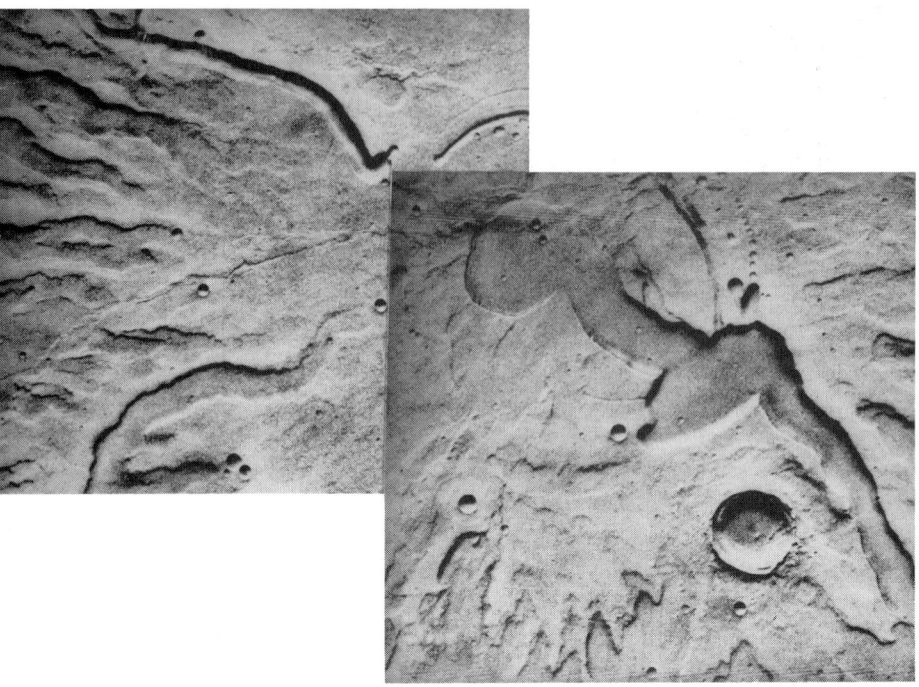

Fig. 10.32: The summit region of the eroded volcanic structure, Tyrrhena Patera (north is to the left). The central ring-faulted depression has been filled by later deposits and subsequently incised by volcano-tectonic depressions/channels. Mosaic width 140 km. V.O. frames 445A53, 54.

10.7.3   Tyrrhena Patera

The third of the major highland structures, Tyrrhena Patera, has received close study too, largely because it has benefitted from better image coverage. Located northeast of Hellas, it presents a markedly eroded appearance (Fig. 10.32). At the summit are two sets of ring fractures, the innermost defining a region 50 km in diameter inside which is an off-centre caldera depression. Leading off southwestwards from the caldera is a prominent broad channel which appears to be volcano-tectonic in origin; two others commence lower down the flanks. Greeley and Crown (1989), after a recent study of the region, recognized the presence of five geological units, of which the two older are a *basal shield unit* that extends 340 km to the south and over 300 km to the north and west of the summit caldera, and a slightly less old *summit shield unit*, which, on average, has a diameter of 200 km. The margins of both are embayed and dissected by younger units. It was these dissected, eroded deposits of higher albedo than their surroundings which were attributed by Pike (1978) and Greeley and Spudis (1981) to volcanic ash. Extensive *smooth plains* which differ morphologically from the adjacent ridged plains also are believed to be composed of volcanic ash. On the southwest side of Tyrrhena Patera is a younger, fan-shaped *southwest*

Fig. 10.33: High-resolution image of a part of the summit shield margin of Tyrrhena Patera, northwest of the volcano summit. Note the erosional scarps at the margin of the prominent channel towards the upper margin of the image and the faint lineations on the channel floor. Frame width 11.5 km. V.O. frame 794A01.

flank flow unit whose apex focusses on the caldera and which is composed of a plethora of narrow lava flows, often with levees and flow channels. At the summit there is *a caldera-filling unit*, believed to represent late-stage ponded lavas.

The radial texture and etched appearance of much of the basal and summit shield materials (Fig. 10.33) is interpreted by Greeley and Crown to be the result of erosion by water, wind and mass wasting of early ash deposits generated more or less contemporaneously by phreatomagmatic eruptions here and at a number of other centres surrounding Hellas. Such activity is seen as the result of eruptions through water-charged regolith that is believed to have been widespread in the southern hemisphere and whose former presence is indicated by the extensive fluvial channels incised into the Noachian-age plateau plains (Greeley and Guest, 1987). Subsequent to ash deposition, the degree of internal heating appears to have increased, whereupon flood lava eruptions emplaced the surrounding plains, which partially buried the flanks of the volcano. The sinuous channels that originate close to the summit may have continued to supply lavas to the lower flanks, after infilling most of the summit caldera.

10.7.4   Implications for early patera volcanism

Gulick and Baker (1990) have assessed the relative importance of contributions from lava, volcanic density flows and fluvial erosion in the genesis and modification of channels on Martian volcanoes. In most cases a combination of these phenomena appears to have been involved, with erosion by surface runoff and groundwater sapping dominating the latter stages of channel evolution. In the cases of both Tyrrhena and Hadriaca Paterae the channels clearly are erosional and have been cut into the rocks making up the volcanoes' flanks. These rocks are unlikely to be lava flows (their morphology and texture, and their susceptibilty to erosion are inappropriate) and more likely are either volcanic ash or ash flow deposits.

Applying the same criteria to Tyrrhena Patera that Mouginis-Mark *et al.* (1988) used on Alba Patera, Greeley and Crown conclude that the ash deposits cannot be air-fall in origin due to their very widespread dispersal (300-600 km from the caldera region). Ash-flow eruptions could, however, account for the observed distribution. Theoretical analysis of terrestrial ash-flow eruptions has indicated that large flows may have been emplaced at velocities of up to 300 m s^{-1} (Sparks *et al.*, 1978), while other theoretical considerations indicate initial velocities as high as 400-600 m s^{-1} may be possible (Sparks and Wilson, 1976). Their calculations suggest that initial flow velocities of between 325 and 450 m s^{-1} would be required to emplace the basal shield unit and an initial velocity of around 250 m s^{-1} could have emplaced the summit shield unit. If the smooth plains units are also accepted as being ash-flow deposits, initial velocities of around 650 m s^{-1} would be necessary. An ash-flow origin therefore seems highly plausible.

On the basis that eruption volumes were comparable with large terrestrial eruptions (10^2-10^3 km^3) – probably an underestimate from what we know about Mars – Greeley and Crown estimate that between 110 and 1100 basaltic eruptions could account for the entire edifice of Tyrrhena Patera. On the additional basis of calculations relating the rate at which subsurface water could accumulate to magma volume per eruption, they conclude that the volume of water required to drive the proposed explosive volcanicity could, in fact, accumulate rapidly (in tens of years) and that phreatomagmatic activity is quite consistent with the climatic changes suggested to have affected Mars during this early epoch (Clifford *et al.*, 1988). The fact that climatic conditions changed as time progressed may explain why ash eruptions do not appear to have characterized the younger shields of Tharsis.

Crown and Greeley (1993) apply gravity-driven flow models to Hadriaca Patera also and find that the distribution of the flank materials also can be most convincingly attributed to the emplacement of pyroclastic flows. That such materials were preferred to simple, ballistically-deposited, ash deposits is a reflection of the inability of non-welded ashes to support the steep-sided scarps typically found on Hadriaca. Both magmatic and hydromagmatic (involvement of volcanic activity with ice) eruption models were tested and found applicable. Assuming the former, the requisite eruption rates ($10^7 - 10^8$ kg s^{-1}), ejection velocities (400 m s^{-1}) and volatile contents (1.5 – 3.0% H_2O) are consistent with parameters derived for terrestrial Plinian activity; for the latter case, the requisite energy conversion efficiencies are comparable with experimental results, and the inferred permeability of the Martian regolith allows for the storage of large volumes of water and for its rapid transportation (flow rates $10^3 – 10^4$ m^3 s^{-1}). Thus either style of activity could transport the observed flank deposits to the requisite distance from their source vent or vents. The change from explosive to effusive style (flank deposits to caldera-filling flows) could be a manifestation of depletion of volatiles in the proximity of the centres with time, rather than to Mars-wide changes.

The observed stratification within the flank deposits of both Hadriaca and (to a lesser extent) Tyrrhena Paterae, is taken to indicate the possibility that welding has occurred, at least in the central regions of the flows. King and Riehle (1974) showed that welding and compaction of terrestrial silicic flows occur for emplacement temperatures of >600°C, while Riehle (1973) showed that minimum temperatures for welding of terrestrial rhyolitic flows are 625°C for 10-m-thick flows and 575°C for 40-m-thick flows. The distance for which welding could continue would be dependent upon the loss of heat in the eruption column (if the flow was generated by column collapse), and the loss of heat during flow emplacement. Modelling the flows of Hadriaca on the basis of the above work, Crown and Greeley (1993) showed that Martian pyroclastic flows would retain their heat for large distances (say, up to 275 km from the vent for flows of rhyolitic composition). For flows with eruption temperatures similar to terrestrial basalts, Martian pyroclastic flows would also retain their heat for large distances, certainly within the ranges required on both Hadriaca and Tyrrhena Paterae.

10.8   TEMPORAL SEQUENCE OF MARTIAN CENTRAL VOLCANISM

The earliest volcanic episode preserved in the Martian rock record is believed to have emplaced the Upper Noachian plateau plains and was succeeded by the extensive ridged plains of Lower Hesperian age, both believed to have been emplaced by flood lava eruption (see Chapter 8, also Tanaka, 1986). Centralized activity appears to have begun with the formation of the enigmatic structure Amphitrites Patera during the Lower Hesperian epoch, which was followed in Upper Hesperian times by generation of a small number of large, low-profile ash or mixed lava and ash volcanoes in the southern hemisphere near the borders of the Hellas impact basin, at Hecates Tholus in the region of Elysium, and in Syrtis Major Planum. The characteristic phreatomagmatic activity of this period was a manifestation of the rather different climatic conditions proposed to have been a feature of Mars at that time, as a result of which significant volumes of volatiles became entrained in magmas rising through the Martian subcrust.

Towards the close of Hesperian times, a further episode of flood volcanism began to emplace the volcanic plains of the lowland hemisphere and this was accompanied by the continuation of volcanic centralization in the region of Elysium at Albor Tholus, and in northern Tharsis, where activity became focussed on Alba Patera. While some explosive activity continued, there was a gradual change towards effusive volcanism, with the eruption of very large volumes of low-viscosity mafic magmas. The later stages in Mars' volcanic history saw the growth of vast shield volcanoes along the crest and margins of the Tharsis Bulge and the construction of Elysium Mons. The culmination of this uniquely Martian shield-building activity was the growth of Olympus Mons, some distance removed from the crest of Tharsis.

10.9   CONTROLS ON MARTIAN VOLCANISM

While the earliest central volcanism may have been controlled by deep-seated fracturing produced during the excavation of Hellas, this is less likely to have been the case for the activity in Tharsis and Elysium which must either been related to the growth of the major crustal upwarps which exist there or to processes which generated these. Considerable research has centred around trying to explain the Tharsis bulge. For instance, in an attempt to explain the gravity data for the Tharsis region, Sleep and Phillips (1979) proposed a model which assumed that the Martian crust to be thinner beneath Tharsis than elsewhere and the mantle less dense. (A lesser density might be expected in a region of active volcanism.) On the basis of such a model, they were able to achieve compensation under Tharsis at depths of only 300 km.

To constrain the thickness of the lithosphere beneath Tharsis, Comer *et al.* (1985) used elastic shell theory and analysed the effects of volcanic loading. They estimated the elastic lithosphere to be between 20 and 50 km thick beneath the regions surrounding Tharsis Montes, Alba Patera and Elysium Mons, while beneath Olympus Mons a minimum value of 150 km was derived. Their results confirmed, therefore, the apparent thinning of the Martian crust beneath the crest of both the Tharsis and Elysium bulges.

More recent finite element modelling was undertaken by McGovern and Solomon (1993) who note that the stresses imparted to the lithosphere by volcanic loads not only involve flexuring and faulting of that layer but also can influence magma transport in the substructure of an edifice. They note that, as in response to flexure in the lithosphere there are three regions where stresses become sufficient to cause faulting; these are (i) at the surface of the lithospheric slab supporting the volcano; (ii) near the bottom of the elastic lithosphere beneath the focus of the volcano; and (iii) on the upper flanks of the volcano. Modelling predicts that normal faulting should be dominant in (i); this is entirely consistent with the circumferential graben observed around each of the Tharsis Montes shields and of Alba Patera. It also applies to the encircling scarp at the base of Olympus Mons which can be viewed as a large-offset, listric normal fault. Should the volcanic edifice be modelled as detached from the underlying slab, the long wavelength northwesterly slope of Tharsis (on which the three shields are located) is predicted to add a NW/SE extensional stress. If this is superposed on the axisymmetric stress field it would engender radial extensional faults in the NE and SW quadrants and circumferential graben faults on the NW and SE flanks. What is observed at each of the three Tharsis Montes is that annular graben are observed on the northern and western flanks of the shields but not in the southeast quadrant. McGovern and Solomon suggest that the NW flanks of the volcanoes (Arsia Mons in particular) were able to slide down the northwest slope of the Tharsis Rise, but were buttressed on the opposite flank, preventing similar fault development.

The model also predicts strike-slip faulting should characterize the surface of the slab immediately beyond the volcanic load. The evidence for such fracturing around the Tharsis shields is, however, lacking. McGovern and Solomon surmise that because the earliest failure generated graben and normal faults, this relieved extensional stresses in their vicinity and subsequently acted as planes of stress relief for strains that accumulated later on. Because of this, they suggest strike-slip faults were prevented from forming. Furthermore, any potential early-formed strike-slip faults would occupy the lowest-lying peripheral ground, and would be most subject to burial by later volcanic deposits. Such deposits, added to which we might mention mass-wasted material, slumped debris and aeolian materials, would also bury circumferential faults in this region of the volcano.

The terraces which are so prominent a feature of the higher flanks of Olympus Mons have been interpreted as thrust faults (Thomas et al. 1990; the same interpretation is consistent with the above model, assuming the volcanic edifice is welded to the underlying lithospheric slab. The presence of circumferential graben high on the flanks of large shields is not consistent with this modelling; likely such faults are related to caldera-forming events.

McGovern and Solomon also note that the stress fields within large edifices have important implications for magma transport and eruption. As is well known, magma preferentially propagates through fractures that form normal to the direction of least compressive stress. Once flexure has occurred in a volcano that is growing (and still welded to the lithospheric slab beneath it) the principal stresses induced in the load rotate so that maximum compressive stress is, not as it was formerly, i.e. vertical, but horizontal. On this basis, horizontal propagation of magma would take place preferentially from the summit towards the periphery. This is entirely consistent with the observed evolution of the three main Tharsis shields. Compressive stresses will, of course, inhibit effusion of lavas high on a volcano's flanks; however, continued growth of an edifice by continual outpourings of magma from lower on the flanks imparts a growing load to the lithosphere and, in consequence, more flexure. Such flexure will have the effect of reducing the compressive stresses whereupon summit eruptions can more readily be accommodated.

The presence of linear rifts that bisect the Tharsis shields and the attendant eruptions concentrated along them, may well have been a response to the concentration of later-stage, radially-oriented normal faulting along a NE/SW strike by the overall northwestward slope of the Tharsis Rise on which the Tharsis Montes grew. Modelling predicts, however, that this could only be so for an edifice still welded to the lithospheric slab beneath, but for very low values of elastic plate thickness. More likely is the alternative that the rifts developed in response to radial rifting across which strain became concentrated due to the superimposed regional (slope-induced) stress field and movement along a plane of detachment along the downslope sector of each shield. This alternative has the additional advantage that rift formation would occur early on in a volcano's development which is more consistent with the chronology suggested by Crumpler and Aubele (1978) and, interestingly, gives us a scenario very similar to that known for Hawaii.

Various theories have been developed to account for the Tharsis Bulge (similar arguments would also apply to that in Elysium). Global shrinking, tectonic effects antipodal to the Hellas basin, and inhomogeneities in the Martian mantle are among those hypotheses to have been put forward (see Phillips, 1978). However, one of the more plausible hypotheses is that the Tharsis bulge is largely due to some form of mantle convection that, in actively uplifting the lithosphere, also produced widespread fracturing. Such convection could quite reasonably have been the result of radioactive heating, or of the existence of lower-density mantle material beneath Tharsis produced because of inhomogeneities inherent in the accretional process. However, such an explanation is not in accord with the findings of Phillips and Ivins (1979) that the present distribution of ridges and fractures around Tharsis are best explained as a result of stresses generated in the lithosphere by the presence of the bulge and not as a response to its uprise.

A very plausible alternative has been proposed by Solomon and Head (1982). These authors see Tharsis as a massive volcanic pile and not a tectonically produced dome. They suggest that because the Martian lithosphere was laterally inhomogeneous early in the planet's history, global and local stresses were caused preferentially where there was thin lithosphere, i.e. under Tharsis and probably Elysium. Once fractures had been propagated there, they would have provided channels of easy access for magma to reach the surface and there to be erupted. Once an elevated flow of thermal energy had been established in this way, the high heat flow would have maintained relatively thin lithosphere beneath the sites of active volcanism and would have had the effect of concentrating fracturing there too. This model scores over the above, particularly since it requires no abnormal chemical or dynamical properties to be sustained for lengthy periods in the Martian mantle. The origin of Tharsis does, however, remain a very contentious issue.

On the Earth, the most significant proportion of thermal escape since the formation of the crust is and has been associated with lithospheric plate boundaries. This has not been so on Mars, where the absence of plate tectonics has necessitated other means of heat escape. Its different lithospheric structure instead has led to the continued growth of central volcanoes above mantle hot spots which remained active through very long periods of time. As a result, Mars' early flood volcanism and the ensuing large-scale central volcanism provided possibly the most important means of planetary heat loss. The very large volumes of mafic magma extruded from the long-lived shields must be a direct manifestation of virtually continuous high heat-flow from a very active mantle beneath the two major volcanic provinces over a period of at least 2×10^9 years.

11

Central volcanism on Venus

We have seen in the previous chapter that both effusive and explosive activity have occurred on Mars. Surface conditions on Venus are, however, significantly different; thus, while theoretical considerations of vesiculation within magmas under Venusian conditions indicate that bubbles will form, especially if CO_2 is the primary magmatic volatile, the high surface atmospheric pressure (around 90 bars) and density would retard the coalescence of bubbles so as to make explosive volcanism very unlikely (Garvin *et al.* 1982). On Venus, therefore, effusive volcanism should dominate. Such volcanism is manifested in the extensive volcanic plains described in Chapter 6, but is also expressed in the development of central vent structures, including shields, domes and coronae, as well as larger volcanic rises.

The majority of steep-sided *domes* on the plains units are between 15 and 20 km in diameter and generally have a radar-dark summit region, often crowned by a summit crater. It might be suspected that these might be the source of the extensive plains units on which they occur; however, dome-like landforms have a very irregular distribution, in places clustering together in large groups, while elsewhere being more or less absent. Such a distribution suggests that their presence is not a prerequisite for the extensive volcanic plains surrounding them and that their own products may be more localized and perhaps more viscous than plains-forming magmas. The same may be true of the peculiar *coronae* whose distribution also is non-random and whose associated volcanic flows contribute only locally to plains development. While the extensive low viscosity flows associated with larger volcanic shields certainly add to the volume of basalt-like flows that are presumed to build the areally-extensive plains, they do so only locally and are entities in their own right and are often associated with rift zones and may be located on major volcanic rises. They thus are similar in some respects to the large shield volcanoes of Mars. All of these structures are an expression of centralized volcanic activity on Venus. This is now described.

11.1   DISTRIBUTION OF LARGE VOLCANIC STRUCTURES ON VENUS

There are on Venus numerous large volcanic constructs, generally less than 1 km high, whose diameters fall within the range 50 – 300 km. Most have summit calderae and several display a plethora of radar-bright radiating volcanic flows. These are the Venusian equivalent of the huge Martian shields already described. Many of the most strongly focussed centres of activity have constructed major *volcanic rises*, exemplified by Atla and Beta Regiones. Several of these are located at the intersection of major rift zones and, as such, share some of the characteristics of regions such as the East African Rift.

Shield structures range in size from a few kilometres across to several hundred. When the size-frequency distributions of intermediate and large volcanoes are plotted, it can be seen that there is a strong peak in the 20 – 40 km diameter range for intermediate structures, while the histogram for large edifices shows a broad flat top and a relatively small number of volcanoes with diameters exceeding 700 km (Fig. 11.1).

As a result of the Venera 15/16 missions, associated mapping revealed about 800 centralized volcanic structures between 20 – 100 km across and about 50 with diameters of 100 – 350 km. Head *et al.* (1992) termed these *intermediate* and *large* volcanoes respectively, a terminology that does appear to have some significance in terms of formational processes. A substantially larger tally of such structures exists now that Magellan has achieved global coverage. So far, 274 intermediate and 156 large volcanoes have been recognized; in addition, 86 calderae, 176 coronae, 259 arachnoids and 50 novae are known to exist. Their distribution is illustrated in Fig. 11.2 (a) – (f). The deficiency of major volcanic features from several lowland regions is likely due to an altitude-dependent influence of atmospheric pressure which dictates whether or not volatile exsolution will or will not take place. Also altitude will have an effect on the production of neutral buoyancy zones sufficient to form stalling zones occupied by magma reservoirs; the latter would favour outpouring of flood lavas at low altitudes and positive constructs at greater heights. It should be noted that there is a strong concentration of volcanic structures in the zone of Beta – Atla – Themis Regiones, an area covering approximately 20% of the surface along the equator, and in Alpha Regio, also in the equatorial regions.

11.2   MORPHOLOGY OF VENUSIAN SHIELD VOLCANOES

The initial distinction between Venusian shields is made on the criterion of size.

11.2.1   Intermediate-sized volcanoes

Some of the intermediate-sized structures have associated with them a sequence of radiating lava flows, usually with lobate margins that stand out from the surrounding dark plains by virtue of their radar-bright signature. These have been christened volcanoes of the *anemone type*, due to the petal-like flow terminations which are prominent in radar images. Many have summit depressions. The anemone-type centres are particularly distinctive, some having a more-or-less equidimensional summit caldera, others showing elongate or even-fissure-like depressions which clearly are related to regional fault patterns.

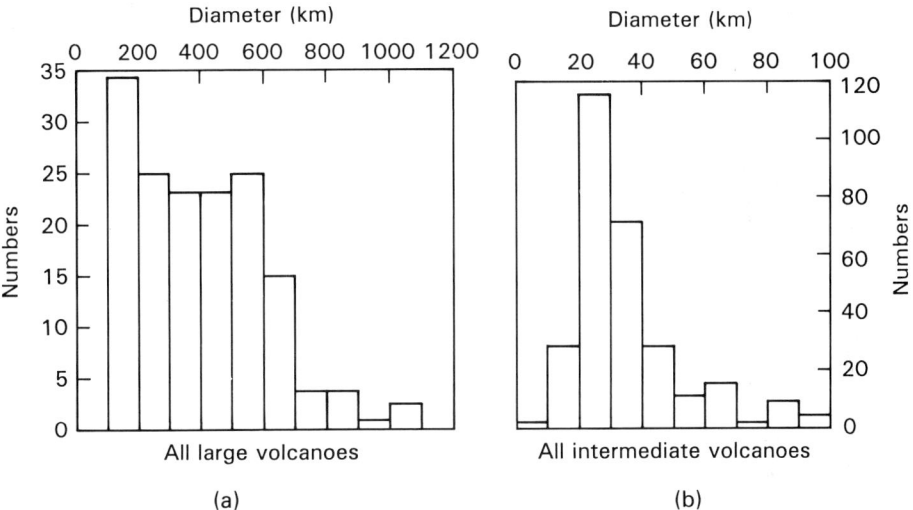

Fig. 11.1: Size distribution of intermediate and large volcanoes: (a) intermediate volcanoes; (b) large volcanoes. (After Head *et al.* (1992).)

Their distribution is definitely patchy, there being concentrations in the Beta-Atla-Themis-Imdr region and on the flanks of the equatorial highlands (Fig. 11.3).

Few outcrop in the lowlands, a pattern which may be due to burial in low-lying regions or a response to the paucity or even absence of neutral buoyancy zones near or below MPR.

11.2.2   Large volcanic shields

Large volcanoes are generally characterized by considerable positive relief and by extensive volcanic flow-fields and often have a summit depression of considerable size (Plate 8). Sometimes, however, the summit feature is not so much a depression as a corona-like structure which may account for one half of the radius. The flows associated with these larger structures typically have distinctive radar-bright lobate terminations and shield margins which contrast sharply with the surrounding fractured plains. Such structures often occur in groups which may be aligned along major fracture zones (graben). Such is the case with Ushas, Hathor and Innini Montes, located on a volcanic rise transected by north-south faulting, and situated between Lavinia and Navka Planitiae (Plate 12). Ushas Mons rises nearly 2 km from the surrounding plains and is the focus of a plethora of lobate volcanic flows which occupy a zone some 600 km in width. The more distal flows are cut by fractures but the summit flow-fields are not, suggesting that the edifice was formed over a considerable period. The summit itself appears to have no well-formed caldera, being radar-dark and with a number of short lava channels present.

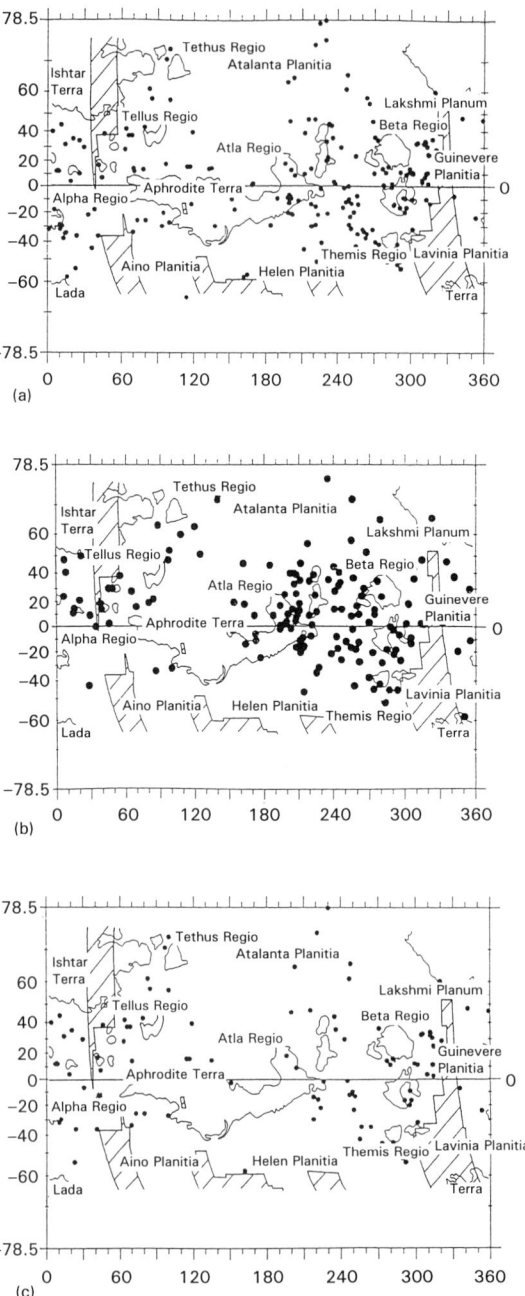

Fig 11.2: Global distribution of volcanoes on Venus. (a) Intermediate volcanoes; (b) large volcanoes; (c) steep-sided domes; (d) calderas; (e) coronae; and (f) arachnoids. (After Head *et al.* (1992).)

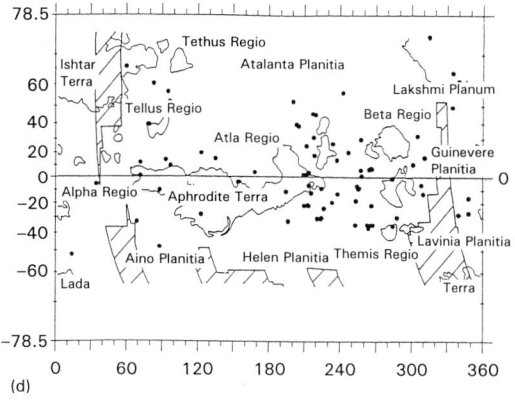

(d)

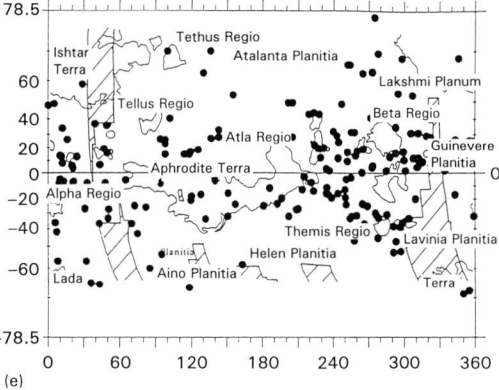

(e)

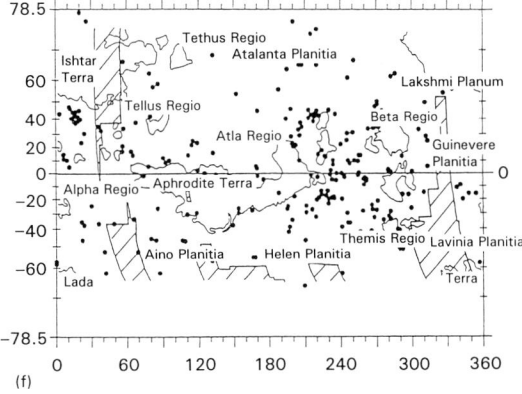

(f)

Fig. 11.3: Intermediate-sized volcanic structures in Atla Regio. Lava flows can be seen emanating from circular pits or linear fissures; many form petalate patterns. These have been called volcanoes of the "anemone" type. A volcanic collapse depression 20 × 10 km across outcrops near the centre of the image; this is drained by a lava channel approximately 40 km in length. Prominent north-south graben postdate the volcanicity. Magellan image P-38281.

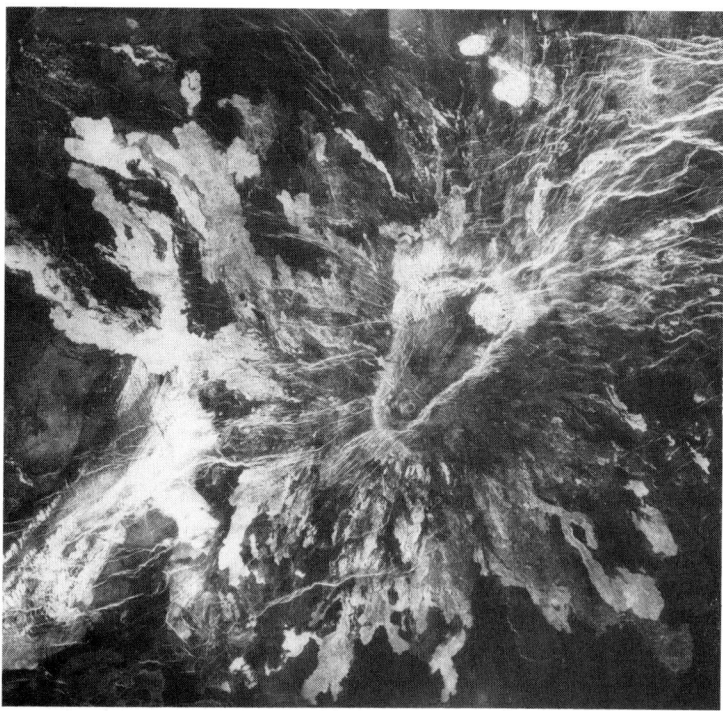

Fig. 11.4: Radar-bright lobate volcanic flows originate in a volcanic depression developed along a NE-SW fracture line in Phoebe Regio. Frame width 587 km. Magellan image P-40844.

Other major shield structures are Sif and Gula Montes, located in Western Eistla Regio, Sapas, Maat and Ozza Montes, located in Atla Regio, Rhea and Theia Montes in Beta Regio and Tepev Mons, in southern Bell Regio. Each has attendant volcanic flows associated with them which can be studied in detail on high resolution Magellan images (Fig. 11.4).

Very large topographic highs, which are measurable in thousands of kilometres but which have attendant volcanic structures upon them (such as the above) are called *volcanic rises*. The western part of Eistla Regio crowned by the high volcanic summits of Sif and Gula Montes – is typical of these. Atla and Beta Regiones are similar. They are not classed simply as volcanoes because processes other than volcanism may have contributed to their growth. These and their attendant shield edifices are described in section 11.3.

11.2.3   Coronae and related structures

Coronae were first identified on Venera 15/16 images and these, together with related arachnoids and novae (Janes *et al.*, 1992, range in diameter from 60 to over 2000 km (Stofan *et al.*, 1992). They show a variety of morphology but most coronae have an associated annulus of fractures between 10 and 150 km across. Novae tend to lack annular faults but to be dominated by a central dome with radial fracturing, while arachnoids are characterized

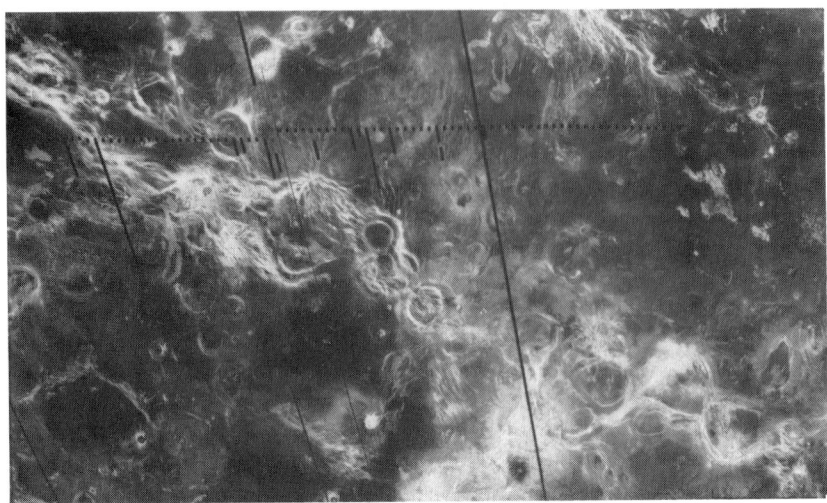

Fig. 11.5: The prominent chain of coronae running along Parga Chasma and Themis Regio. Frame width 4275 km. Magellan image C2-MIDRP 30S284;1.

by concentric fracturing beyond which strong radial faulting may extend for several radii. Head *et al.* (1992) identified 259 arachnoids and 50 novae. It is likely that the three types represent an evolutionary sequence generated by mantle upwelling. Many corona-like structures tend to occur in prominent chains which are related to major rifting. Such occurrences include the equatorial chain between Parga and Hecate Chasma and that which runs along the eastern flank of Lavinia Planitia. This indicates some interrelationship between corona formation and extensional stresses within the Venusian lithosphere.

Since coronae are among the most widespread of Venusian landforms they have been studied by several different groups, not least because an appreciation of their origins is fundamental to an understanding of the internal workings of Venus (Barsukov *et al.*, 1986, Pronin and Stofan, 1990 Stofan and Head, 1990, Squyres *et al.*, 1992, Stofan *et al.*, 1992). Their distribution is not entirely random (see Fig. 11.2), there being a clustering of such features around longitude 250°E, many being located between the longitudes of Beta, Phoebe and Themis Regiones, and near Atla Regio, while a distinct chain of coronae concentrates along Parga and Hecate Chasmata which extend eastward from Atla Regio (Fig. 11.5). Five main types can be recognized: (i) Concentric, (ii) Concentric-Double ring, (iii) Asymmetric, (iv) Radial/Concentric, and (v) Multiple. Of these, the first is the most common, about 50% of the total being of this kind. Concentric coronae also have the greatest size range (60 – 1060 km), while the unique feature, Artemis Chasma, measures 2600 km across. Differing degrees of volcanism can be identified with different types.

Tamfana, situated on the plains south of Alpha Regio, is a concentric corona with a raised rim, the opposite crests of which are 400 km apart. The rim rises around 700 m above the

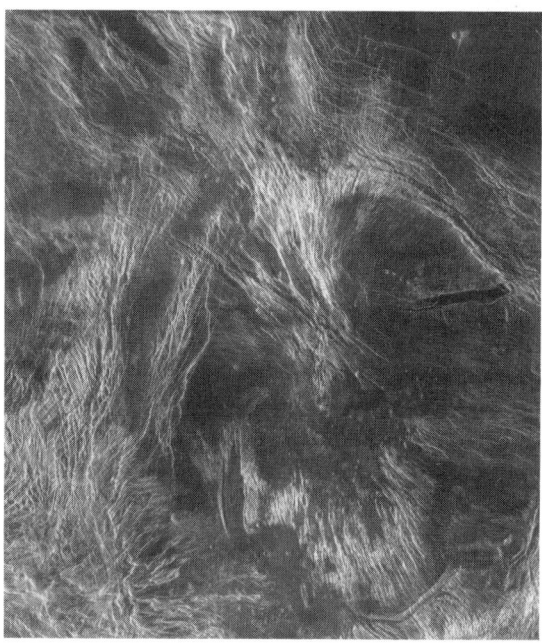

Fig. 11.6: The 400-km-diameter concentric corona, Tamfana. Note the radial and concentric fracturing and the well-developed moat which almost completely surrounds it. Magellan image F-MIDRP 35S003.1.

level of the surrounding plains from which it is separated by a continuous 500-m-deep moat, seen as the darker, smoother region beyond the radar-bright rim in Fig. 11.6. The rim has a steep and somewhat hummocky outer slope but its inner face is gentler. Most faults are paired to form long graben.

The corona interior has an area of 450 km^2 which in places lies 500 m below the plains and on which outcrop ponded lavas. The central part of the structure shows well-developed radial fracturing, appropriate to the initial updoming of the plains above a rising mantle plume. Lavas also have largely flooded the moat. Although some evidence may have been buried by later events, it appears that activity at Tamfana, like other coronae, commenced with uplift (and perhaps volcanic construction) of the rim and downwarping to form the encircling moat. This, in turn, generated concentric flexuring which was followed by the extrusion of highly fluid lavas, both on the interior and within the moat. Radial faulting apparently pre-dated volcanism within the moat region but in places also continued later.

Concentric/Double Ring coronae share many of the characteristics of the first group but are encircled by two distinct annular ridges and/or moats. Typically, individual annuli are separated by smooth plains, such structures typically being 60-70 km across. Thus far, 38

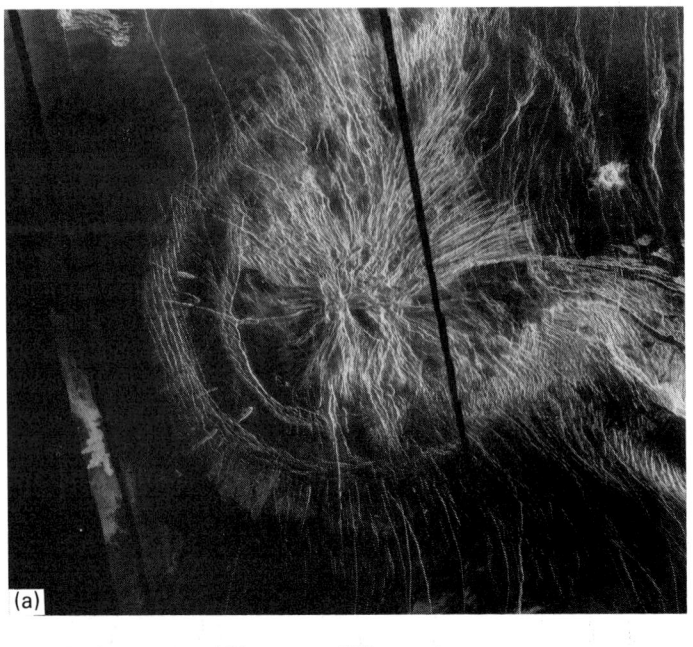

(a)

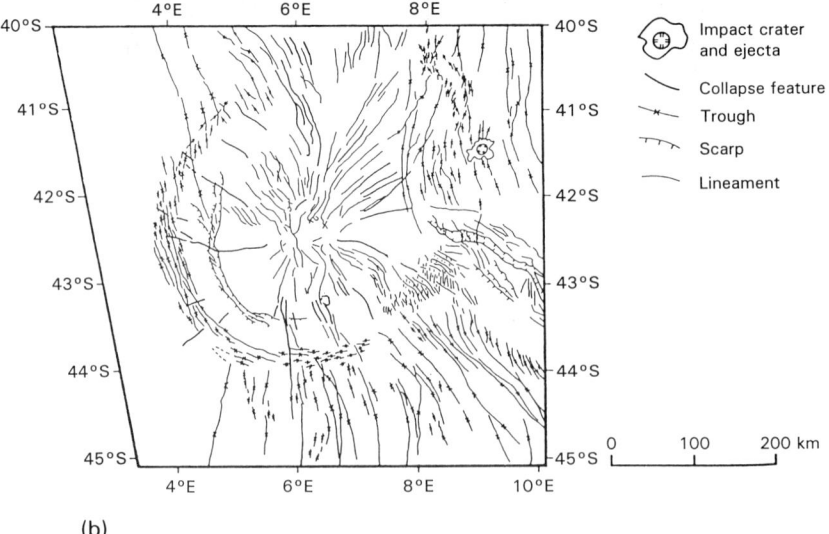

(b)

Fig. 11.7: (a) The corona structure, Selu, situated in Alpha Regio. This 350-km-diameter landform has a double annulus of fractures, open to the east. Note also the radial faults and the lobate radar-bright flows associated with the outer ring. Several lava channels can be seen, on the western flank. Fractures radial to the corona within the annular faults swing in strike to be parallel to the rift lying to the east and north; these may represent igneous dykes. Magellan image P-38156.(b) Sketch map of structural features associated with Selu. (After Squyres *et al.*, 1992.)

such structures have been identified, the majority of which occur in chains, especially in the region of Parga Chasma – Themis Regio. Radial/concentric coronae, on the other hand (17 mapped), exhibit an intense radial fracture pattern focussed at the corona centre, which is surrounded by concentric troughs and ridges. Fig. 11.7 shows a typical example, Selu, a broad plateau-like uplift whose summit lies 645 m above the surroundings; it has a diameter of 350 km and is located south of Alpha Regio, on the plains of eastern Lavinia Planitia.

It has a double annulus, the outer one spanning about 180° of arc, open towards the east, and characterized by troughs and grabens. The inner annulus is even less complete (90° of arc) and is characterized by outward-facing fault scarps. Individual graben are seldom longer than 100 km and typically are 4-6 km wide. In the east and north the rim is traversed by two intersecting sets of fractures, one set striking N-S and parallel to plains-crossing regional fractures.

The corona interior is extremely complex. The central region lies 500 m below the highest point (situated 50 km to the south of the focus) and is three-quarters encircled by higher topography. It is traversed by fractures, pits and pit chains, and partly covered by a small region of radar-dark plains. While fractures are radial within the annulus, beyond they turn into the pattern of rift faults that runs along the margin of Lada Terra. McGee and Head (1995) suggest that these may represent igneous dykes. The radial lineaments comprise small ridges, troughs and scarps spaced about 1-2 km apart, while larger features include steep-sided troughs which appear to be volcanic collapse depressions, a supposition supported by their association with co-linear chains of small pits with smooth floors.

Radar-dark plains outcrop at several locations within the interior, especially between the two annuli, and between the inner ring and the centre. Such deposits generally embay radially-fractured terrain but themselves are embayed by radially-trending collapse depressions and by the concentric faults associated with the two rings. In contrast, well away from the central regions, on the south and west flanks, radar-bright lava flows extend onto the adjacent plains, many inundating concentric faults.

The evolution of this structure appears to be as follows: early west-east extension, which generated N-S faulting, appears to have prevailed over much of the lifetime of Selu. Plains deposits which embay these early faults are themselves cut by the radial fractures that characterize the central regions; these are presumed to have formed in response to the uprise of the central plateau. Any radiating troughs in this region probably represent volcanic collapse events mimicking pre-existing lines of weakness. The concentric faulting was largely later and dominated by uplift, however, some subsidence does appear to have affected the innermost regions. Volcanism occurred throughout the development of Selu, although its importance may have diminished with time.

Asymmetric coronae, of which sixty have been identified, are marked by an asymmetry of form which produces annuli that range from sinuous to angular. Multiple coronae (35 mapped) consist of two or three linked structures which have a single encircling peripheral annulus; when these occur with other types they make for spectacular landscapes unlike anything known from the Earth. Several are located along ridge-and-trough belts to the east of Atalanta Planitia.

Synthesis of all the geological relationships observed indicates that radial fracturing consistently was the first stage in corona development. Such a phenomenon would accompany crustal uplift above a rising mantle diapir (Fig. 11.8(a)). Of course, not all

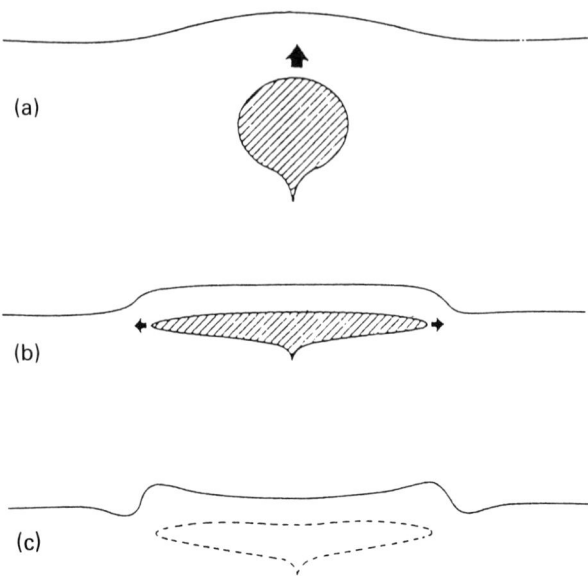

Fig. 11.8: Conceptual model for the growth of coronae. (a) Rising mantle diapir causes crustal flexing at surface. (b) Diapir impinges on underside of lithosphere, spreads laterally and flattens. (c) Diapir cools, allowing gravitational relaxation. (After Squyres *et al.* (1992).)

coronae show such fracturing at their cores; many instead have volcanic deposits in these regions, a characteristic which may be anticipated, since melts would have risen into the crust above the rising diapir as it hit the base of the Venusian lithosphere (Squyres *et al.*, 1992a; Janes *et al.*, 1992; Stofan *et al.*, 1992). Thus, at several coronae volcanism clearly accompanied radial faulting, since radial fractures are both buried by flows and also transect them, while some flows evidently emanated from radially-directed fissures. A small number of radially-fractured domes (e.g. Carpo and Mokos) further supports this hypothesis; these landforms probably represent the initial stages of corona development.

As the geodynamical modelling of Janes *et al.* (1992) shows, the evolutionary stage ensuing upon uplift may comprise viscous relaxation to a more plateau-like form. Such a plateau will be isostatically uncompensated. Plateau formation can be accompanied by a central subsidence, a topographic rim, an annular moat and concentric faulting. Volcanism may accompany all of this activity.

Modelling shows that a plateau-like landform will develop as the rising diapir impinges against the lithosphere, spreading radially and flattening out (Fig. 11.8(b)). Concentric fractures are to be expected on the inward-facing slopes of a moat, and along the crest and

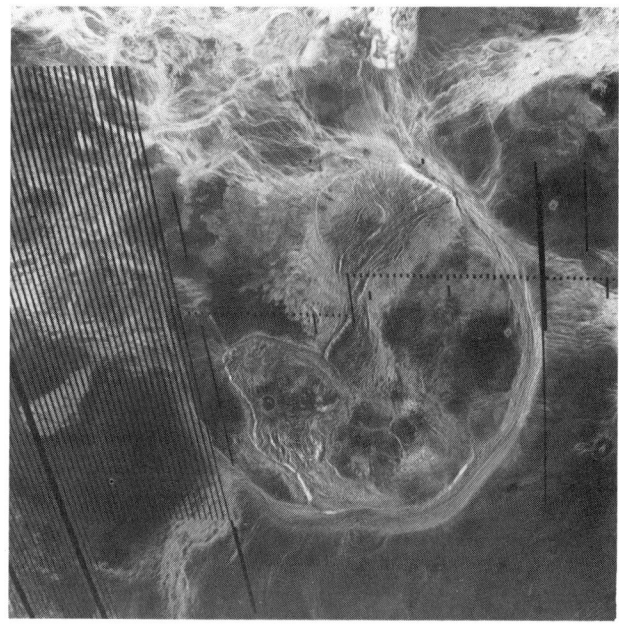

Fig. 11.9: Artemis Chasma. This 2600-km-diameter structure appears to be intermediate between a huge corona and a volcanic rise. Magellan image P-38857.

outward-facing slope of a rim, as thermoelastic cooling and gravitational relaxation of the thermally-supported plateau occurs. This sequence of events illustrates why dome and plateau formation precedes fracturing.

The later evolutionary stages of corona development involves continued volcanism, this often being so intense that it obliterates much or almost all the evidence of early radial fracturing. The sunken interior regions of coronae and the floors of moats may also be flooded by lavas, while the former may see the growth of shields or shield fields, steep-sided domes and collapse pits. Because the later stages of diapir activity involve the cooling and thinning of the diapir, and relaxation of the topography, volcanism will decline and the youngest volcanic flows tend to pond in topographic depressions. Such a pattern is dissimilar to that of volcanic shields, where late-stage effusive activity is characteristic. Exactly how much topography will remain after this late stage will depend largely upon the strength of the local lithosphere and any remanent thermal activity in the underlying.

A small number of coronae approach the size of volcanic rises; however, most lack the cross-cutting rifts common to most such rises and have less pronounced topography. Artemis Chasma is something of an exception, and, in measuring 2600 km across, is intermediate between the two. This structure, located south of Thetis Regio, has a structurally complex interior surrounded by a 150-km-wide circular annulus that shows strong deformation (Fig. 11.9). This annulus is characterized by graben, compressional ridges and other lineations whose origin is unclear. In places its floor lies 4 km deeper than the adjacent plains, while it is bordered by 4-km-high ridges which gives it a maximum

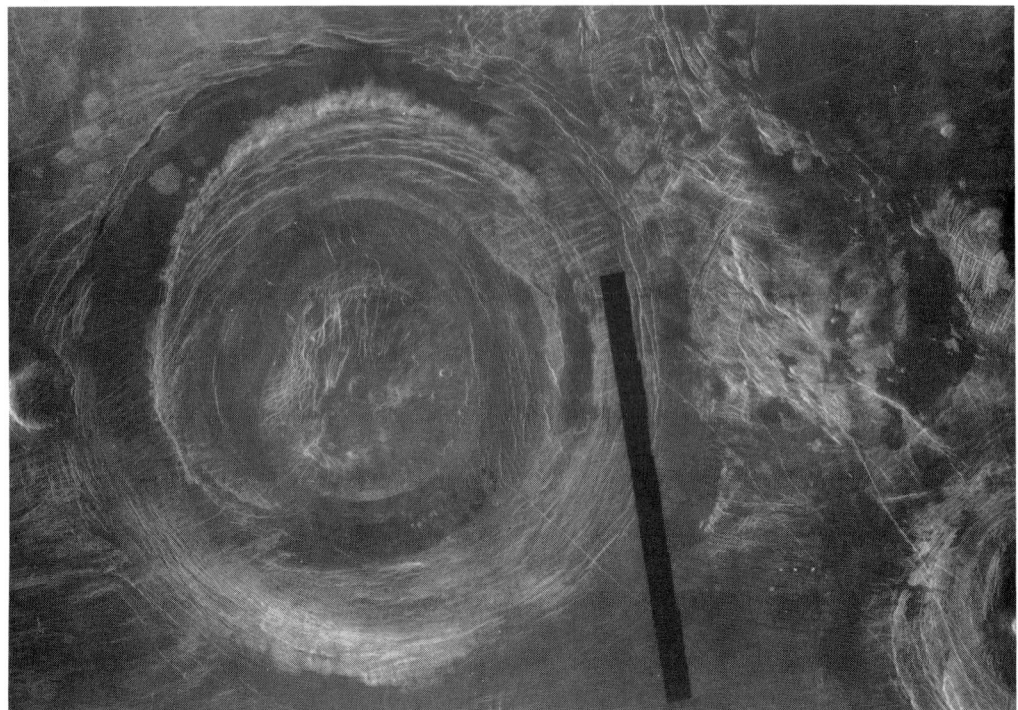

Fig. 11.10: The circular depression, Aramaiti, located south of Aphrodite Terra in Aino Planitia.. The structure is about 350 km across, having a depressed floors on which is located a 100-km-wide dome. An angular fragment of lighter plains which was not inundated by the younger radar-dark lavas, can be seen in the northwest sector. Concentric fracturing is strongly developed, while west-east fractures traverse the structure. Also visible within the interior are several small shields. Magellan image P-37946.

relative height difference of at least 7.5 km. Complex troughs cross the interior, some of which are indundated by radar-dark volcanic flows. However the interior does not lie at a higher level than the exterior plains – a major diference between it and a typical corona.

The origin of this feature remains an enigma. Evidently little vertical uplift occurred during its formation but equally there was considerable horizontal movement, both extensional and compressional. It has been suggested that some form of retrograde subduction might have played a part in its formation as well as that of other, smaller, structures in this region, i.e. Latona and Eigin (Sandwell and Schubert, 1992; Stofan *et al.* 1992.

Other corona-like landforms may have a different origin, although their slightly different morphology one suspects may be simply a function of their stage of modification by crustal relaxation. Some workers, e.g. Squyres *et al.* (1992) refer to these as *circular depressions* and see fit to assign them a separate identity; others (Head *et al.* 1992) classify them as coronae. They exhibit a concentric fault pattern, but this encircles a simple depression whose floor may lie 500-600 m below the surrounding plains and some regions of the floors may have been covered by lavas. Occasionally a very low rim surrounds this. On the floor of Aramaiti, a typical member of this group, is a distinctive 100-km-diameter dome which rises to the level of the exterior plain (Fig. 11.10). Squyres and his colleagues note that this structure essentially is the inverse of a typical corona, the raised rim taking the place of the depressed moat and the central dome taking the place of the central depression. While the downsinking of the depression generated concentric faulting, subsequent uplift of the dome was accompanied by radial fracturing on the floor of the structure; again, the reverse sequence to that typical of coronae. Such structures are more akin to calderas than coronae.

Arachnoids also are characterized by concentric and radial faulting but in their case the fracturing may extend outwards for several radii. Where several such structures are located in the same general area, the foci and the intersecting fracture sets gives the impression of spiders along cobwebs (hence the name). Their origin is undoubtedly similar to that of coronae, and Head *et al.* (1992) surmise that intrusion rather than extrusion may be dominant in the volcanism characteristic of these structures, accounting for the relative paucity of extrusive landforms in their vicinity (see Fig. 6.16).

11.2.4 Pancake domes and related features

Steep-sided volcanic domes are widespread on Venus. These often are flat-topped, with a circular summit depression and sometimes well-developed radial and concentric fracturing. Landforms with these characteristics have been christened aptly "pancake domes" (Fig. 11.11(a)). The 152 domes so far identified range in diameter from 10 to 70 km (most between 20 and 30 km), with a mean height of 700 m (maximum 1700 m). This means that the larger Venusian domes have volumes in excess of 100 km^3, implying magma volumes an order of magnitude greater than, say, the Laki fissures eruptions of Iceland, and more in keeping with terrestrial ash flow extrusions. Other such structures may exhibit a summit collapse crater and steep sides with scalloped or runnelled flanks (Fig. 11.11(b)). Elsewhere domes have been modified by tectonism, impact cratering, slumping and mass wasting.

Pavri *et al.* (1992) note that such domes are widely distributed but preferentially located near coronae and in plains regions adjacent to tessera massifs, and at altitudes near to or just below MPR. The latter may be a manifestation of their dependence upon the levels at which neutral buoyancy zones occur. The association with coronae suggests also their origins, like those of coronae, are connected with uprise of magma bodies of large volume above rising mantle plumes. This would favour the differentiation of more evolved (viscous) magma fractions which might be expected to generate such steep-sided edifices. Thus they could be envisaged as the Venusian equivalents of terrestrial rhyolite/dacite domes, however, while most terrestrial domes are emplaced episodically, their Venusian analogues appear to be single event products. It is possible also that they might be the (undisrupted) equivalents of terrestrial rhyolite ash-flow eruptions, the high atmospheric pressure experienced by a rising magma body on Venus inhibiting Plinian-style disruption

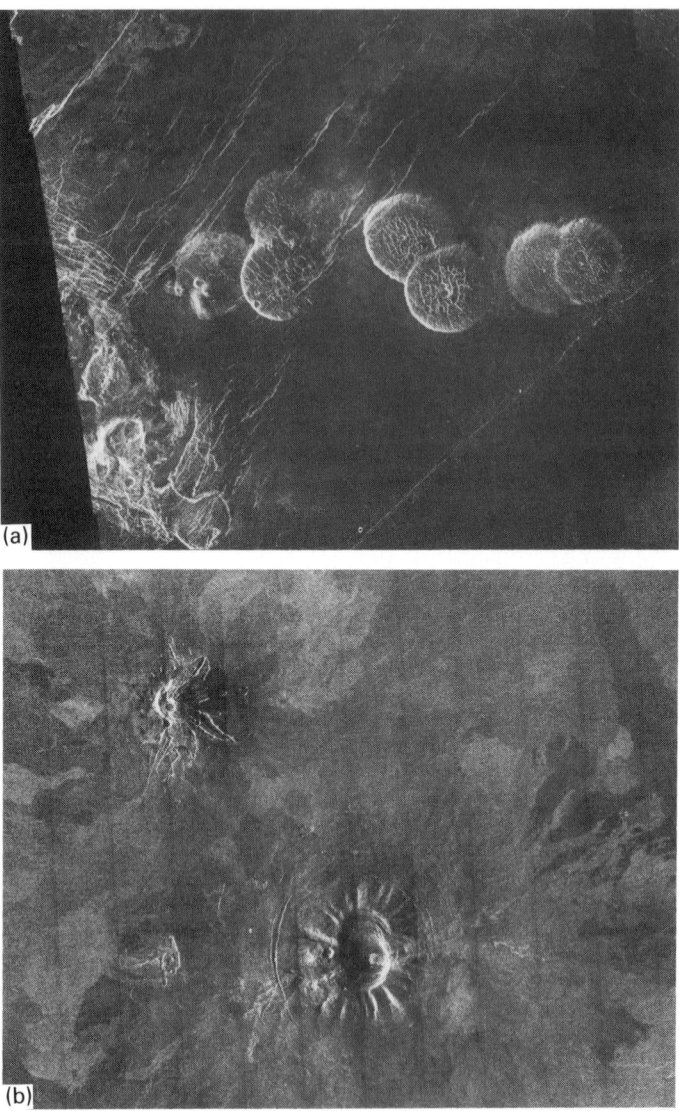

Fig. 11.11: (a) Steep-sided pancake domes in Tinatin Planitia. The largest dome measures 25 km across and rises about 1 km above the adjacent plains. All three domes have summit pits. An extrusion appears to have escaped from the central dome and flowed towards its southwesterly neighbour. Note the fractured carapaces and the collapse depressions on the central edifice. Frame width 250 km. Magellan image P-37125. (b) Unusual steep-sided dome located east of Beta Regio. The larger structure is 45 km across with a 20-km-diameter summit depression. The flanks of both domes are extensively gullied, presumably produced by either slumping or thermal erosion by lavas. The variable brightness of the flows on the adjacent plains indicates the considerable extent of the flows associated with dome-building activity. Frame width 240 km. Magellan image P-38811.

characteristic of terrestrial ignimbrite-generating events.

That magma upwelling is involved in their formation appears to be beyond dispute; furthermore, the probability that magma reservoirs on Venus will reach large volumes, favours large-scale differentiation within such reservoirs (Head and Wilson, 1992. Petrogenetic models which might explain the observed dome morphology were tested by Pavri *et al.* (1992) who noted that simple Newtonian models predicted the domes may have viscosities of $10^9 - 10^{12}$ Pa S (extremely high for silicates) and effusion rates more closely similar to terrestrial andesitic than basaltic eruptions. They were unable to achieve a precise match in aspect ratio for the Venus domes and therefore suggested such a model might not be appropriate. A Bingham-plastic model suggested yield strengths of between 10^4 and 10^5 Pa S for most domes and rheology calculations implied basaltic or more silicic magma was involved; however, the domes failed the simple height/radius test indicative of such a model, again suggesting the magma was not erupted as a simple Bingham plastic. A Bingham-plastic model with a temperature dependent yield strength produced a better match but even that was not entirely satisfactory. The uprise of such postulated viscous magma bodies could be either by chemical fractionation within a shallow reservoir alone, or by volatile enrichment within such a reservoir which would result in enhanced buoyancy and emplacement of highly vesicular lava at the surface. Whichever is the case, these features provide the strongest possible evidence that there has been magma fractionation on Venus, generating silicic magmas comparable to terrestrial andesites or dacite/rhyolites.

11.3 VOLCANIC RISES

The highlands of Western/Central Eistla, Beta, Atla, Tellus and Bell Regiones comprise a spectrum of large volcanic rises, the most complex of which is probably Atla Regio, a volcanic locus astride five converging rifts. Detailed examination of these regions shows that they differ in detail, particularly with respect to the extent and distribution of associated CRT (tesserae) and coronae. At Western Eistla and Atla, CRT is of minimal importance, however, around Beta tessera massifs are a primary terrain unit and arrayed along the periphery of the topographic rise. These units are consistently the oldest in each region. Coronae also vary in their distribution patterns. At Beta Regio they are arrayed mainly along the northern periphery and connected by lineaments to form chains. In Western Eistla they lie along a more-or-less linear belt crossing the highland massif. Coronae are not found on the higher parts of Atla Regio, being confined to the eastern end where they are located in rift zones. Each contains major shield volcanoes associated with extensive fluid, high volume, lava outpourings, as well as smaller volcanic structures such as shield fields and steep-sided domes. The large apparent depths of compensation, coupled with the above characteristics, strongly suggest that each is associated with a region of major mantle upwelling.

11.3.1 The volcanic rise of Beta Regio
Beta Regio is a broad volcanic rise located 6500 km west of the western end of Aphrodite Terra. It is a region of major extensive volcanism and rifting located at the intersection of

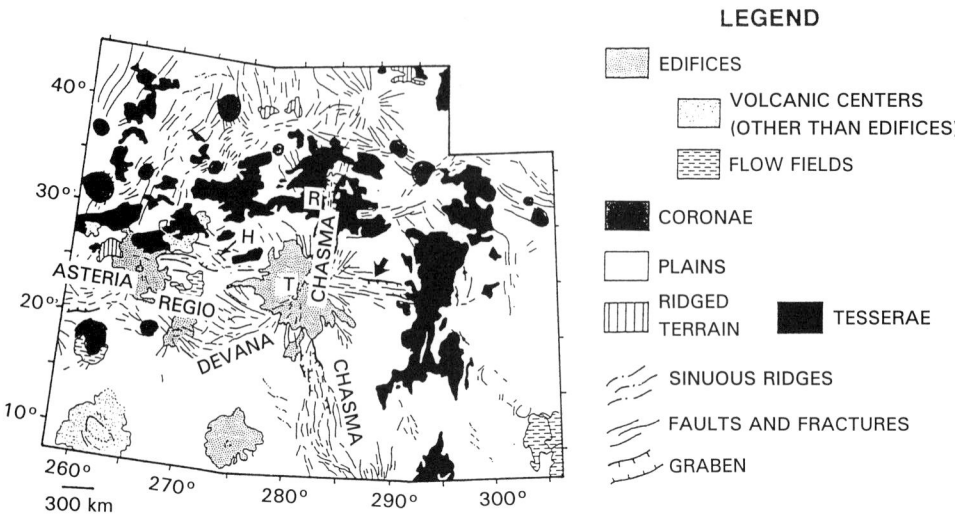

Fig. 11.12: General geological map of Beta Regio, showing volcanic edifices and structural elements. (After Senske *et al.,* 1992.)

several major tectonic lines. In this respect, Beta bears similarities to terrestrial rift volcanism that is plume-related (Dewey and Burke, 1974; not surprisingly, it had been compared with the East African Rift Valley (McGill *et al.*, 1981; Schaber, 1982; Campbell *et al.*, 1984; Stofan *et al.*, 1989). While the region is predominantly a rifted volcanic rise, extensive units of CRT outcrop in the neighbourhood of Rhea Mons and also on the eastern flank of the rise. This prompted Senske *et al.* (1991) to suggest that Beta Regio, rather than being a simple volcanic rise, actually is a zone of mantle upwelling that has severely disrupted an older tessera block.

One of the most prominent landscape features of Beta Regio is Devana Chasma, a rift valley which strikes N/S (Fig. 11.12). This cuts across an older system of W/E scarps related to an earlier extensional event. In the central region of the rift valley, faulting occurs over a broad region; towards the northern end, however, the surrounding region is composed of CRT, while in the central regions and to the south, the rift is surrounded by fractured plains with volcanic flows. The topography of the rift also shows a great deal of variation along its length: thus, while being a narrow 80-km-wide, 2-km-deep, trough near Rhea Mons, it becomes a broader 130-km-wide trough which has upstanding blocks of older crust in its centre. Then again, south of Theia Mons, in the belt between Beta and Phoebe Regiones, Devana Chasma is characterized by considerably greater relief (as much as 6 km) and more distinct flanking highs than are seen within Beta itself.

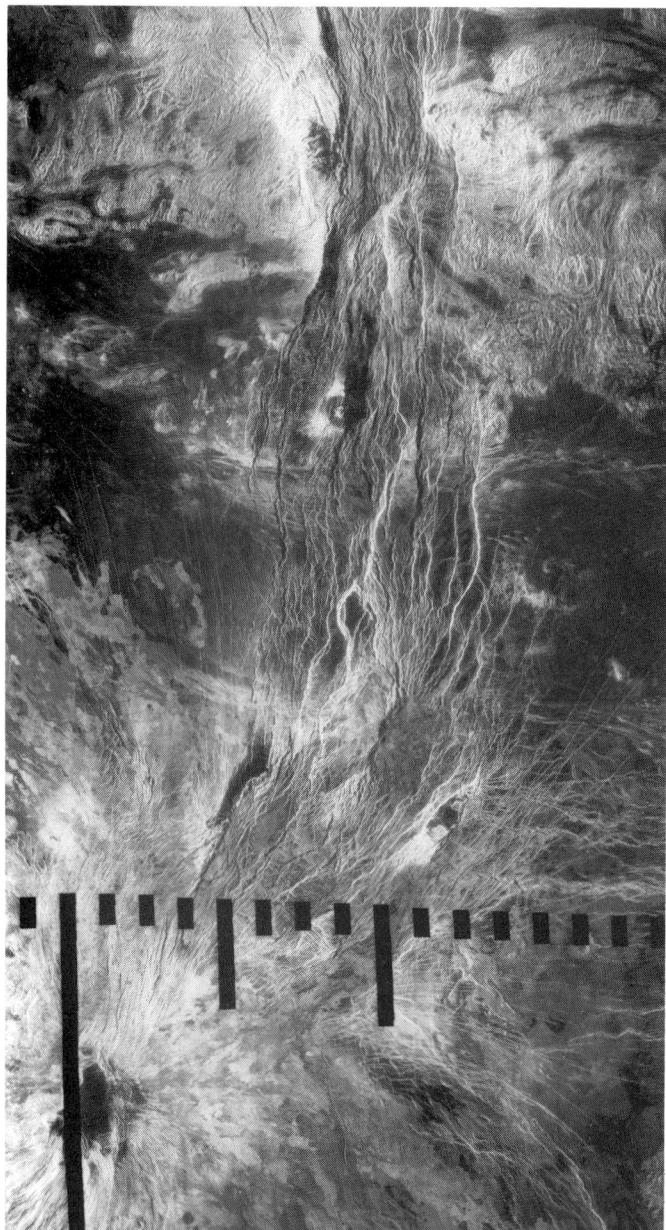

Fig. 11.13: The central part of Beta Regio. The 1050-km-long rift valley of Devana Chasma runs N/S. Theia Mons volcano is situated towards the bottom of the image, the rifted-apart tessera massif of Rhea Mons lies towards the top. Flows from Theia partially infill the rift floor. Magellan image P-41294.

The large volcanic shield of Theia Mons is superimposed on the rift, and itself is transected by younger faulting. Lying at the intersection of converging rifts, it is surrounded on all sides by volcanic flows and plains, the associated flows extending 1000 km from the summit (Fig. 11.13). Beyond and mainly to the east, west and north, isolated blocks of CRT (tessera) emerge above the plains surface. The western tessera is embayed by plains with shield fields and is transected by fractures related to adjacent coronae. The transection/ superposition relationships in the region clearly show that the tesserae blocks are the oldest crustal units exposed. Coronae are concentrated along the northern edge of the rise, being joined by lineaments that form chains.

On the northern flank of Theia, where Devana Chasma intersects the edifice, the rift narrows from 200 km to a width of 50 km. To the east, radar-bright and mottled lava flows partially infill the rift floor and evidently are younger than the faulting. At the summit of the volcano is a high volcanic plateau to the south of which is a radar-dark 75 km × 50 km caldera whose floor lies below the base of the mountain. As is the case with large Martian shields and paterae, the upper flanks of Theia are covered in relatively short, lobate, radar-bright flows that overlie longer flows that extend towards more distal regions.

Rhea Mons, which lies west of Devana Chasma, formerly was believed to be another shield, however recent images indicate that the topographic high, rather than being a caldera-topped volcanic construct with radiating lobate flows, is actually a region of disrupted tesserae. Although there are outcrops of smoother units around the summit region, which may be of volcanic origin (Fig. 11.13), those lobate radar-bright flank features previously interpreted as flows (Stofan *et al.*, 1989) are found to to be CRT. Consequently this mountain massif cannot now be considered primarily as a volcanic structure, rather an old deformed massif partially blanketed by volcanic products.

It appears that rifting in Beta Regio occurred at the same time as the main phase of volcanism centred upon Theia Mons; Beta Regio may represent a rifted tessera block, disrupted by a mantle plume (Senske *et al.* 1992). Identification of CRT fragments west of Beta and the detailed characteristics of the tessera terrain in northern and eastern Beta Regio appear to support this hypothesis. Any CRT formerly present in southern Beta Regio may have been covered by younger volcanic deposits. Theia Mons sits at the junction of several tectonic trends, and in this respect is similar to major volcanic centres at Eistla and Atla Regiones. This geometry is also similar to terrestrial rift-related volcanism, where faults tend to propagate and link up between regions of hot-spot volcanism. There is a large positive free-air gravity anomaly over Beta, with an apparent depth of compensation of at least 300 km. This is entirely consistent with the interpretation that this highland rise is located above a region of major mantle upwelling.

11.3.2 The volcanic rise of Atla Regio
Western Aphrodite comprises the plateau-like highland massifs of Ovda and Thetis Regiones which rise between 3 – 4 km above MPR; eastwards of latitude 140°E is a much lower zone of troughs and elongated ridges, including Diana and Dali Chasmata, which extends for 5000 km. Eastern Aphrodite is built largely from the massif of Atla Regio whose highest points, the summits of the volcanoes Ozza and Maat Montes, attains elevations of 9 km above MPR (Fig. 11.14).

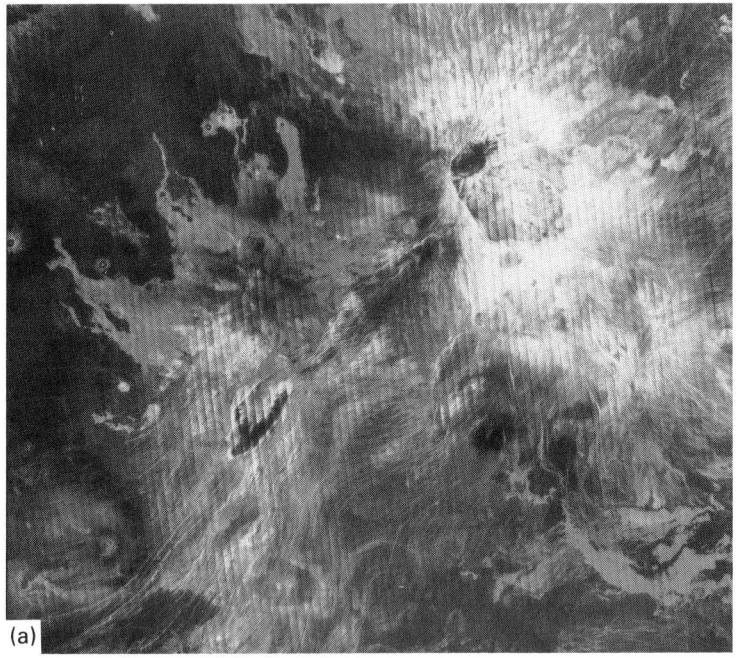

(a)

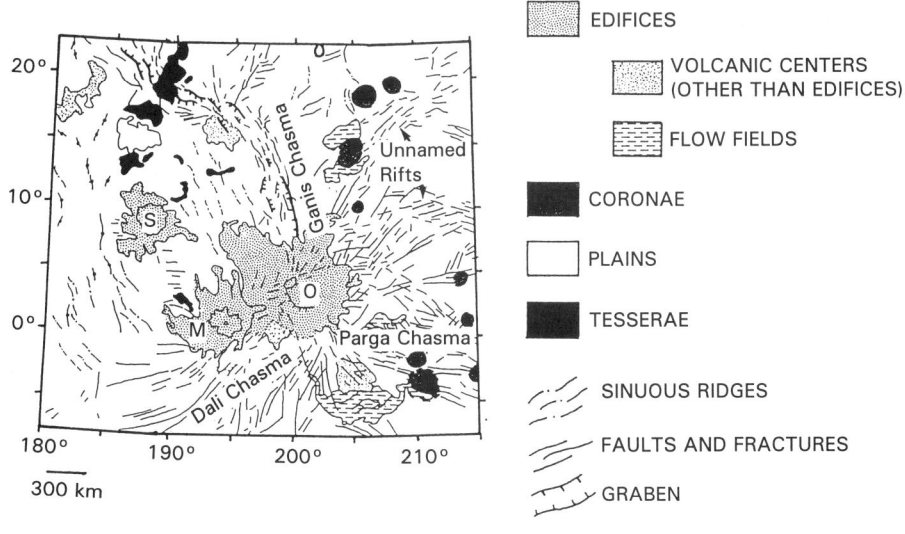

(b)

Fig. 11.14: (a) The volcanic rise of Atla Regio. Ozza Mons is situated within the radar-bright area close to the centre, while Maat Mons lies to the southwest. Both have a radar-dark summit. Magellan image C1-MIDRP 00N197;1. (b) Geological sketch map of the same region. (After Senske *et al.* (1992).)

Atla Regio lies at the eastern end of Aphrodite Terra, being a broad highland dome, with several major volcanic centres and lying at the intersection of major rifts (Fig.11.7 Measuring 1500 × 2500 km, most of the rise lies between 2 and 4 km above MPR but reaches over 7 km in places. Three main rift zones – Ganis, Parga and Dali Chasmata – radiating out from the centre, the location of Ozza Mons; one extends northeastwards towards Beta Regio, another tectonic junction. Geoid and gravity anomalies for Regio both are high (120 m and 130 mGal respectively) and are centred over Ozza Mons which, like Maat Mons, stands out clearly on Plate 8. Estimates of the depth of compensation for this area range from 200 km to 250 km (Smrekar and Phillips, 1991; Phillips *et al.*, 1991).

Maat Mons is the highest of Atla's shield volcanoes, rising 9 km above MPR, and being 300 km in diameter. Lava flows and radial graben extend for hundreds of kilometres from the summit. It is not located at the end of a major rift, rather it is situated on the northern flank of Dali Chasma. The lavas on its southeast flank appear to infill the rift floor, while those the north and northwest apparently were extruded down the raised rim of the fault and ponded in the plains below. The summit is marked by a region of radar-dark material 100 km in diameter, within which can be seen pits, flows and small domes; there is also a striking field of small domes located just north of the summit. It is bounded to the south and west by concentric, 5-10 km-wide graben.

Ozza Mons is located NE of Maat and rises to 7.5 km. While it shares many of Maat's characteristics, fracturing is by no means the dominant characteristic of this volcano. There is no summit caldera, rather a radar-dark plateau feature measuring 50 km × 100 km, and rising 1.5 km above its surroundings, seen just below the centre of the image. A cluster of small volcanoes, domes and associated lava flows which lie north of the summit are responsible for partially burying Ganis Chasma, giving a clear indication of the age relationships in this area. To the south lies a subsidiary volcano which also is elongated along a NE-striking line. Although many fractures are radial, many also tend to be aligned NE and are continuous with the faulting that connects Atla with the region of chasmata to the southwest. There is very obviously a close connection between the siting of volcanic centres and the regional extensional faulting.

Sapas Mons, another large shield volcano, is situated close to the western margin of Atla Regio. This is the foreground feature in Plate 8. The radar-dark plains surrounding Atla contain several highly-embayed tessera blocks. This is true also of the region surrounding the northern termination of Ganis Chasma. These areas of CRT appear to be the local basement, a relationship they hold also in the vicinity of both Ishtar and Beta Regiones. However, older CRT units are much more widespread within Beta Regio than at Atla.

Despite the abundance of volcanism, fracture systems are the most prominent features of Atla and extend outwards over the surrounding area. Ganis Chasma connects Maat Mons with Nokomis Montes to the northeast. This 1.5-km-deep feature is approximately 1000 long and as wide as 300 km in places and evidently is a major rift zone consisting of sets of overlapping graben. Volcanic flows both crosscut and are transected by the fault scarps. The broad similarity between Atla and Beta is obvious, a feature which has led several groups to consider Atla a region of mantle upwelling (Senske *et al.*, 1992; Solomon *et al.*, 1992; Bindschadler *et al.* 1992). Interestingly, coronae are absent from the highest part of Atla, being preferentially concentrated in the eastern part, where they are found along rift lines, assumed to be foci of mantle upwelling along lines of extension.

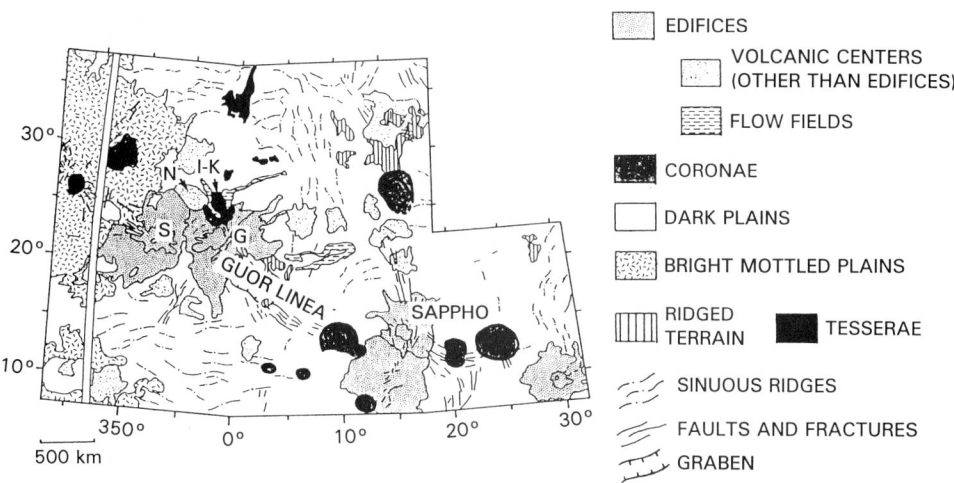

EDIFICES

VOLCANIC CENTERS
(OTHER THAN EDIFICES)

FLOW FIELDS

CORONAE

DARK PLAINS

BRIGHT MOTTLED PLAINS

RIDGED
TERRAIN   TESSERAE

SINUOUS RIDGES

FAULTS AND FRACTURES

GRABEN

Fig. 11.15: Geological sketch map of Eistla Regio. The highland massifs are joined by fracture belts. The volcanic shields of Sif, Gula and Nissaba Montes are labelled S, G and N respectively. The corona Idem-Kuva is labelled I-K. (After Senske *et al.* (1992).)

11.3.3   The volcanic rises of Eistla Regio

Eistla Regio lies astride the WNW extension of a line continuing the trend of western Aphrodite. It is a series of broad crustal rises, each several thousand kilometres across, which trend WNW/ESE. Western and Central Eistla are broad volcanic rises traversed by rift zones, with large volcanic structures occupying the higher ground; these tend to have high radar backscatter (Senske, 1990; Senske *et al.*, 1992. The region is marked by strong positive gravity anomalies, and estimated compensation depths of between 100 and 200 km have been reported (Smrekar and Phillips, 1991; Grimm and Phillips, 1992). Eastern Eistla, which was not imaged during the first Magellan mapping cycle, is dominated by the 525 km × 370 km corona structure, Pavlova.

The volcanic rise of *Western Eistla* measures 3200 km × 2000 km and is dominated by the large volcanoes of Sif and Gula Montes (Fig. 11.15). Gula Mons is the larger of the two, and is approximately 400 km × 250 km, broadly elliptical in planform and rises 4.6 km above MPR and 3.2 km above its surroundings. The flank slopes are at angles of between 0.25° and 1.4°; thus it has the profile of a large shield volcano. Volcanic flows radiate down its flanks and extend outwards for up to 300 km (Plate 10). Their radar signature ranges from radar-dark to radar-bright, the latter contrasting strongly with the surrounding plains shown on the above image. Sif Mons is smaller, and is an isolated mountain that rises 3.4 km above MPR with a local relief of around 2 km and a diameter

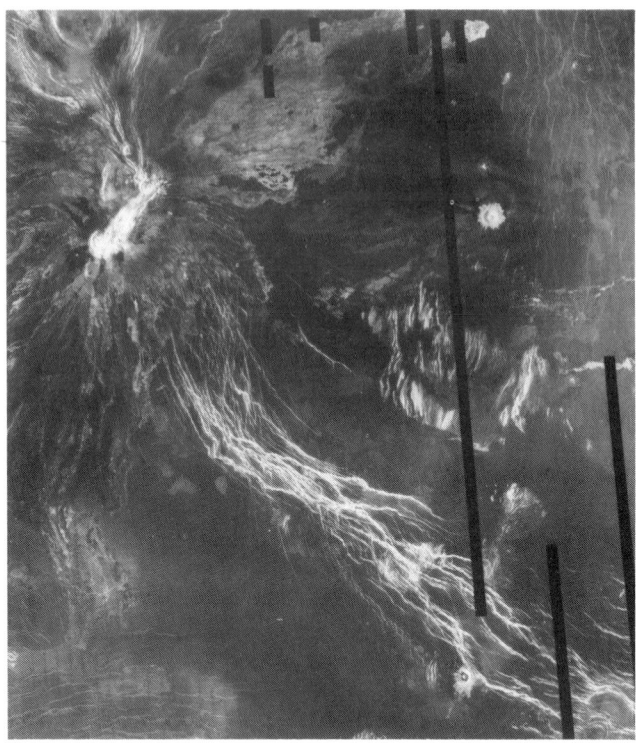

Fig. 11.16: Guor Linea Rift and Gula Mons. The rift extends southeastwards from Gula and is in a region of thinned crust. Frame width 700 km. Magellan image C1-MIDRP 15N009;1.

of 200 km. It has a prominent summit caldera southeastwards from which run a line of 3-10-km-diameter collapse pits. North of both Sif and Gula Montes and roughly midway between them is a further volcanic edifice, Nissaba – the roughly circular feature with intermediate radar signature due west of the most northerly of the Gula flows. This is a broad volcanic dome and transection relationships suggest it to be older than the corona, Idem-Kuva, located to its east and between the two bright flows north of Gula Mons.

Mapping of the associated volcanic rocks by Senske and his colleagues (1992) shows that both Sif and Gula expose three assemblages of lava flows: to the north of Sif, two phases of eruption are separated by an episode of extensional fracturing. The earlier phase is dominated by flood eruptions that generated radar-dark flows whose source cannot be discerned; these are overlain by mottled flows which may embay earlier volcanic landforms. The post-fracturing volcanism saw the effusion of mottled and radar-bright flows that form well-defined flow-fields whose sources can often be tracked down. Thus, the westernmost flow-field appears to originate in a large dome. Other late-stage flows, particularly those to the east of Sif, exhibit flow channels and may inundate graben. The longest of such flows is 110 km in length and less than 1 km wide. An even younger sequence of smooth mottled volcanic units embays the previous unit and is interpreted by Senske *et al.* (1992) to represent late-stage flank eruptives associated with shield summit construction. This type of sequence is very reminiscent of several Martian shields and paterae.

Both shields, but Gula Mons in particular, are also the focus of an array of extensional fractures; the latter crossing the volcano's flow apron and onto the adjacent plains. Fracturing is also widespread to the northwest of Gula Mons, where two prominent coronae, each 250 km in diameter, have associated families of both concentric and radial fractures. This is taken as an indication of lithospheric extension to the northwest of the Gula structure. The most prominent fracture set in this region is that of Guor Linea which extends southeast of Gula Mons for at least 1000 km (Fig. 11.16). This feature lies astride a topographic ridge that joins Western and Central Eistla. Many of the Guor Linea faults are located along the walls and floor of what essentially is a linear 50-75-km-wide rift running along the axis of a broad linear rise. Such an association clearly shows that Guor Linea is a rift system which has developed in response to local extension and thinning of the lithosphere (Solomon *et al.*, 1991; Grimm and Phillips, 1992).

The main line of the Guor Linea rift is interrupted by the massive edifice of Gula Mons. This implies that, in general terms, volcanic construction postdates the main rifting; however, some fractures transect Gula flows, indicating that extensional deformation also occurred after the main volcanic episode. The main fractures associated with Gula are radial and most are found to the southwest, northwest and east of the volcano. Sif Montes-related radial fractures are found to the north, northwest and west of Sif, but are less prominent than those of Gula Mons.

Extending around the rise are extensive regions of *domed plains*, several areas being characterized by large numbers of small volcanic domes. Grimm and Phillips (1992) quote frequencies of 400 per million square kilometres. The emplacement of such plains appears largely to have predated volcanism associated with Sif Mons, but some clusters of domes are superimposed on later fracturing. The formation of such units implies a strong local concentration of volcanic activity but with only limited surface access of partial melts to the Venusian surface.

Central Eistla Regio was originally called Sappho; Magellan imagery reveals that this region is actually built from two volcanic structures that sit atop broad volcanic rise. Sappho Patera is a 300-km-diameter volcanic edifice marked by concentric fractures and a 200-km-diameter summit caldera. This makes it one of the largest calderas on the planet. To its south is another volcanic structure, Anala, which is less well-defined but rises 1 km above MPR (Fig. 11.17). Flow channels have been identified on the high flanks of the latter. Major flow fields extend outwards for 900 km from these two centres, that of Anala overlapping parts of Sappho's, giving a clue to the broad age relationship. There are also several partially buried coronae flanking the main rise. The latter are older than the shields (McGill, 1994) and were formed very soon after the emplacement of the regional plains.

Fractures on this rise strike mainly N/S, whereas those to the east and west strike perpendicular to these. NNW-trending fractures are cut by the flow apron, but faults striking just east of north extend all the way to Sappho. At Sappho itself, a patera structure, the

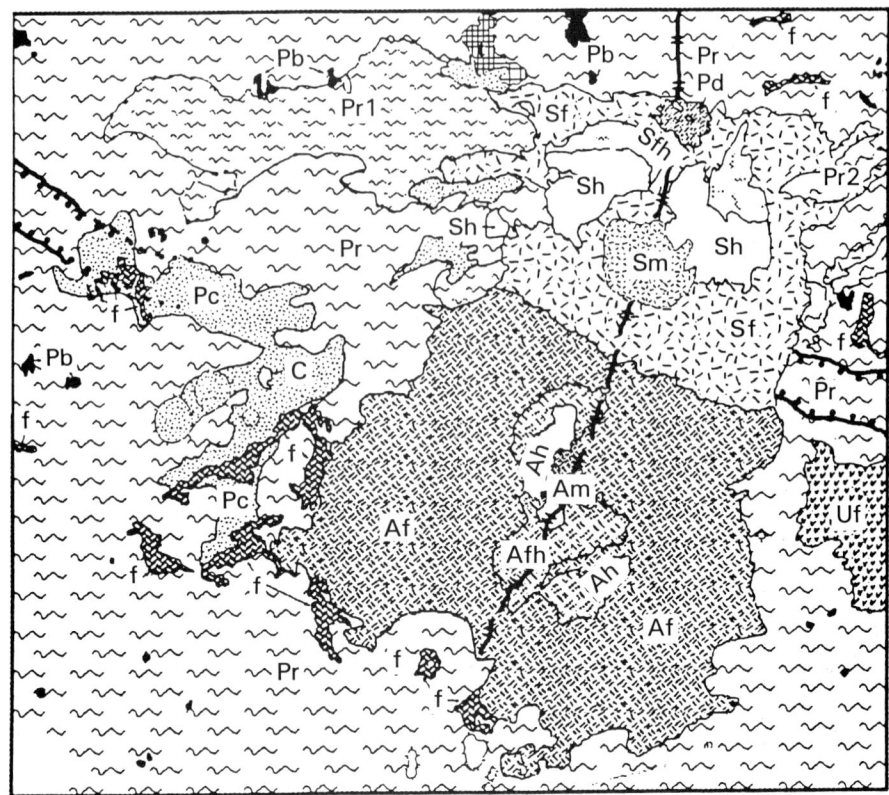

Fig. 11.17: Geological sketch map of Central Eistla Regio. (After McGill, 1994.)

fractures are deflected around the periphery of the caldera, while NNE to NE-trending fractures join the two volcanic edifices. Another family of fractures trends approximately W/E and is buried by Sappho (and may also be offset beneath it). The Guor Linea rift stops some 900 km west of Sappho and is truncated by a corona on the western flank of the Central Eistla rise.

In terms of the age relationships of the four major volcanic centres, it appears that Sif is younger than Gula, while Anala is younger than Sappho. Although no direct evidence can effect correlations between West and Central Eistla, Grimm and Phillips (1992) note that volcanic styles in Central Eistla appear to be more mature than those in the West, suggestive of greater age there. While Senske *et al.* (1992) feel that volcanic construction largely postdated uplift, it does seem that the fracture systems which typify the area both crosscut and are inundated by volcanic deposits on both rises, indicating that volcanism and tectonism were broadly coeval. McGill (1994), after detailed mapping of the region, concludes that the sequence of formation of the major volcanic landforms was : (i) regional plains, (ii) coronae, (iii) shield volcanoes.

Pioneer-Venus gravity data has been utilized by Grimm and Phillips (1992) to determine the 2-dimensional gravity anomaly for this region. The vertical gravity data is shown in Plate 14. From the gravity and topography they derived mass anomalies on two internal horizons: the first lies 20 km down and represents the density contrast experienced at the crust-mantle boundary; a deeper boundary was found at 200 km, and is presumed to lie where lateral density variations owing to mantle circulation are greatest, i.e. beneath the thermal boundary layer at the base of the lithosphere. They interpreted the deeper set of anomalies, and then solved for the mantle flow that would be driven by such anomalies. On the basis of their chosen flow model, both western and central Eistla are seen as sites of strong mantle upwelling, this being stronger beneath Gula Mons than Sif. The zone of upwelling also extends to the southeast beneath Guor Linea as a less intense "saddle" with about one-half to one-third the strength of the main plumes. This implies that, rather than indicate simple passive extension beneath Guor, the stresses are being applied by the actively-convecting mantle. The solution for the shallow set of anomalies suggests that the crust beneath western Eistla, particularly Guor Linea, is thinner than average for the region; however crustal thinning is not indicated for Sappho Patera, indeed, some solutions suggest it may have been somewhat thickened.

Grimm and Phillips postulate that Western Eistla, Guor Linea and the plains to the north are related to crustal thinning (or shallow cooling), while Central Eistla and the southern plains are regions of crustal thickening (or shallow heating). Interestingly, the same workers find that maximum crustal thickening/shallow heating occurs beneath Heng-O corona, situated to the south of Western Eistla; however, various trial solutions for the gravity data usually produced no anomaly for the lower layer. This prompts the observation that, if coronae are sites of mantle upwelling, as many suspect, then Heng-O must no longer be active.

The large gravity anomaly of Western Eistla indicates that the rise overlies a site of strong mantle upwelling and crustal uplift; this is where thinning has occurred. In contrast, the larger component of crustal compensation and somewhat weaker flow-related density anomaly at Central Eistla may imply that the crust has been thickened magmatically, in which case mantle upwelling may currently be waning. Perhaps, therefore, Central Eistla sits above a more mature plume than that underlying the western rise.

11.3.4   The volcanic rises of Bell and Phoebe Regiones

Bell Regio is a broad crustal dome 1500 km across situated to the north of Eastern Eistla Regio and, in many respects, is somewhat similar to it. There are two rises joined by a family of N/S-trending rifts (Jannle *et al.*, 1987). The northern one hosts the corona Nefertiti and numerous volcanic flows, the southern contains the volcano, Tepev Mons, and several smaller volcanic centres. Northern Bell Regio is lower than its more southerly counterpart, and is the site of less intensive volcanism. The elongated 500×225 km corona, Nefertiti, has numerous radiating flows as well as being defined by concentric faults (Jannle *et al.*, 1988). Tepev Mons rises 5 km above its surroundings and consists of two foci elongated in a W/E direction. The eastern component is a volcanic peak with a radar-dark signature; western Tepev has a shallow 40-km-diameter summit caldera with a radar-dark floor. Numerous volcanic flows radiate out from both. Another smaller depression with a pancake

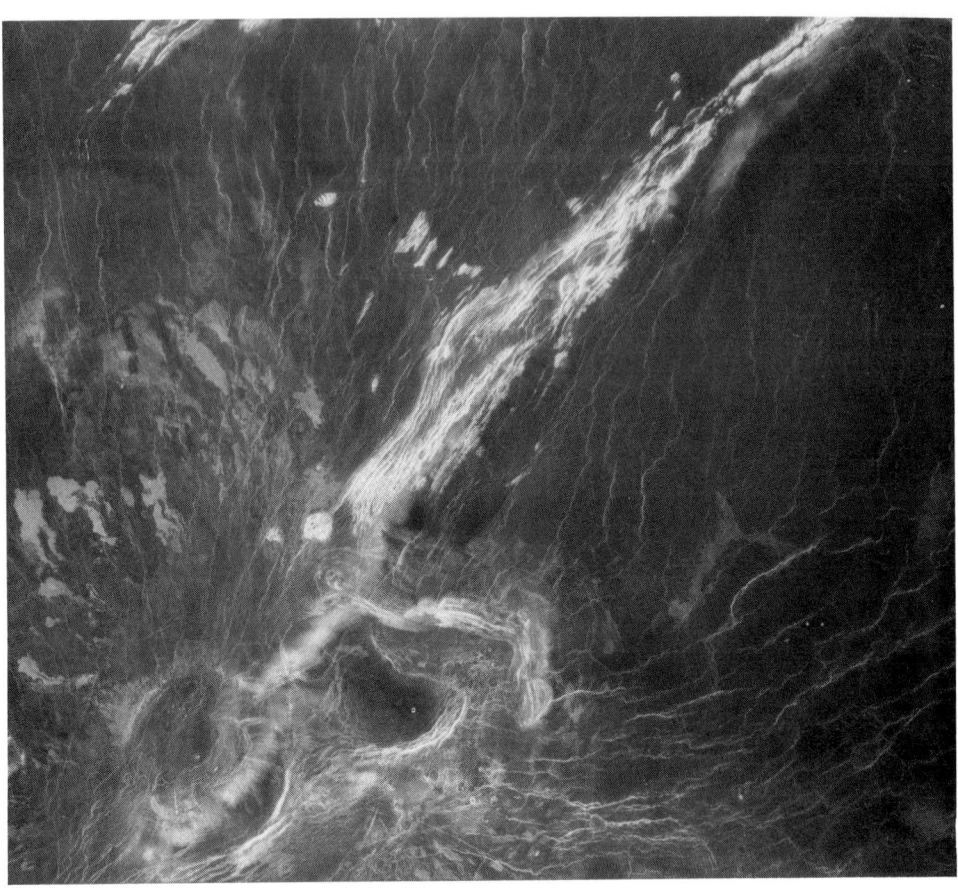

Fig. 11.18: Shield volcano straddling the rift system extending southeastward from Bell Regio. The summit of the volcano lies 2 km above the surroundings and comprises an intensively fractured and mildly convex plateau with a depression and plethora of radar-bright, lobate, volcanic flows, domes and small pits. Frame width 580 km. Magellan image P-40844.

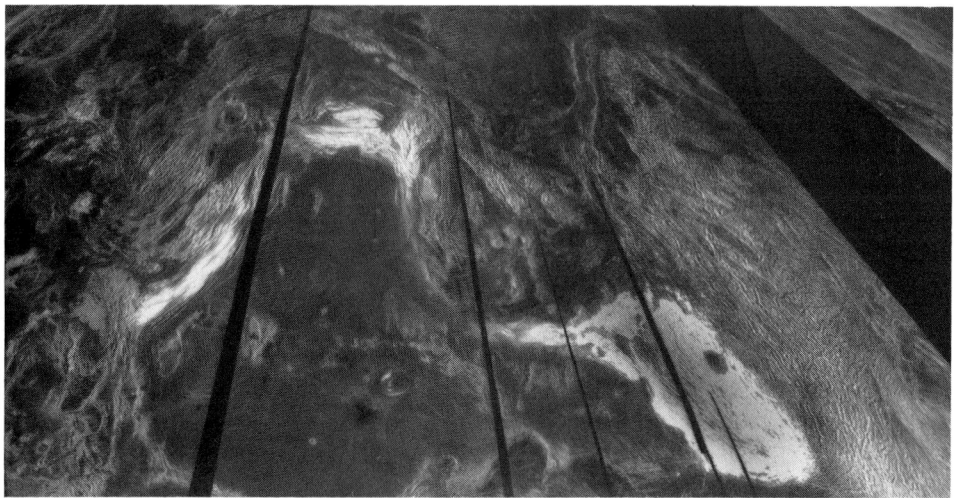

Fig. 11.19: Magellan image map of the volcanic rise of Ishtar Terra. Magellan image C2-MIDRP. 60N332;2 (part of).

dome lies to the southeast of the summit, together with prominent NW/SE fractures. Roughly due east of Tepev Mons is another volcanic focus with which are associated numerous radar-dark flows, pit chains and other collapse features. A small family of concentric graben is located further south; this seems to be associated with the same eruptive focus.

To the southwest of Southern Bell Regio is a series of tessera blocks; further small blocks outcrop to the south. Within them are two main sets of lineations: one NW/SE, the other NE/SW; sometimes there is a third, W/E set. In all cases the NE/SW set is the youngest and, in places, the areas between the faults are inundated by volcanic flows associated with the volcanic rise. As elsewhere, this implies that CRT is the oldest terrain in the region, predating volcanism, and possibly being the initial stage of deformation associated with an evolving mantle plume. Smrekar and Phillips (1991), from Pioneer gravity data, derive an apparent depth of compensation beneath Bell Regio of between 100 and 200 km, that is, very similar to that beneath Western Eistla Regio.

Phoebe Regio takes the form of an elongate plateau lying at the south-southwestern end of the Devana Chasma rift system. In some respects it is like the northern part of Beta Regio, apparently comprising a tessera massif upon which is superimposed a major volcanic edifice. The volcanic structure has a modest gravity high associated with it, and straddles a narrow rift fault (Fig. 11.18).

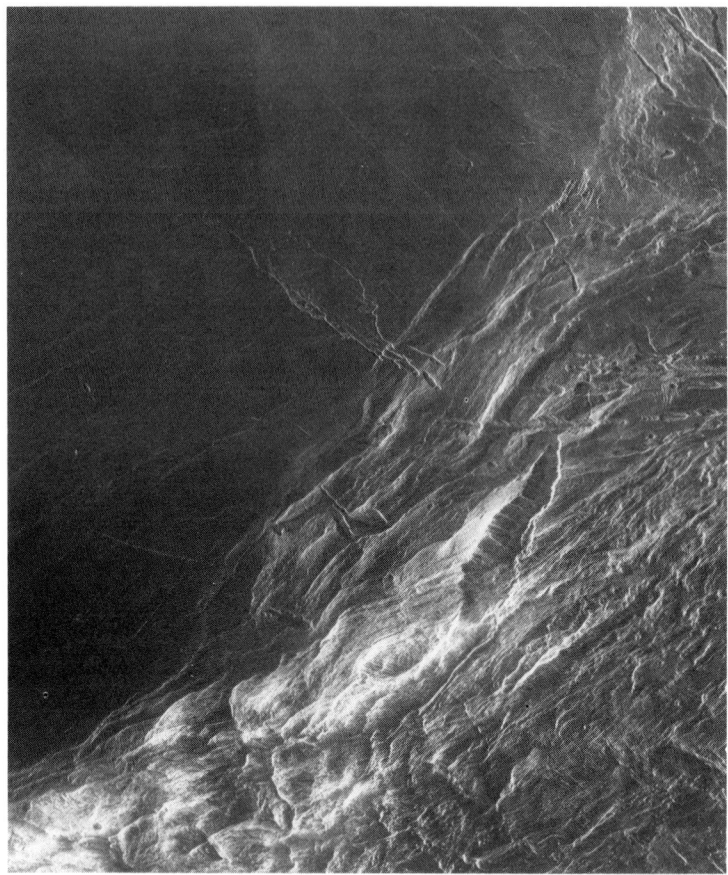

Fig. 11.20: The boundary between Danu Montes and the radar-dark plateau of Lakshmi Planum. Lava channels and collapse depressions are incised into the inward-facing slopes of the encircling mountains. The longest such channel extends 75 km onto the plateau surface. Magellan image P-38335.

11.3.5   Western Ishtar Terra

Ishtar Terra is unique amongst the elevated regions of Venus, for although it has some of the characteristics of other plateau-shaped highlands, its perimeter stands several kilometres above the interior; this does not happen elsewhere. Western Ishtar comprises the huge plateau of Lakshmi Planum, tilted slightly towards the south, which is wrapped around by mountain belts and bounded by steep scarps; east of it rises the mountain massif of Maxwell Montes (Fig. 11.19). Plateau-like regions of tessera (CRT) lie exterior to the bounding mountain belts of Lakshmi Planum, and lie at a lower elevation. Large positive anomalies in gravity (55 mGal) and geoid (60 m) are located to the south of Maxwell Montes, well away from the centre of Lakshmi Planum. It is difficult to assess how significant this offset may be.

Lakshmi Planum; has been studied in considerable detail (Kaula *et al.*1992. Two major volcanic structures, Sacajawea and Colette with their associated annular fractures dominate the central regions. East of Sacajawea Patera lies CRT which is heavily embayed by volcanic flows; superposition/transection relationships indicate that the latter consistently are younger than the former. Smooth plains cover large areas, particularly in the lower, southern, sector of Lakshmi, and in places are incised by curvilinear troughs, many of which trend roughly parallel to either the plateau margins or neighbouring mountain belts. The evidence for volcanism within the mountain belts of Ishtar is variable. Lava channels can be seen to have flowed down onto the interior plateau surface from Danu Montes, but not elsewhere (Fig. 11.20) Volcanic landforms found near the northeast arm of Danu Montes include partly-coalescent collapse pits and graben which trend either NW-SE or E-W, and transect those thrust faults which are present. This implies that volcanism postdated much of the orogenesis. Volcanism is, however, much more widespread on the exterior slopes, where all of the mountain belts display a plethora of lava flows, sometimes with collapse depressions, spreading down their flanks.

In the cases of Danu and Freyja Montes, and Vesta Rupes, the presence of aligned collapse depressions and pits indicates that the rise of magma was facilitated by crustal extension. Indeed, extensional stresses clearly affected all of the belts, but the highest incidence of extensional features is to be found along Danu Montes and Vesta Rupes, while it is also evident that they affected the plateau margins. Smrekar and Solomon (1992 interpret the sets of narrow closely-spaced features which characterize these regions of high gradients, as graben and normal faults. Since many of the fault sets strike perpendicular to the downslope direction, the most likely explanation for their existence is gravitational spreading. The fact that, in several of the belts, extensional structures lie parallel to what appears to be the direction of shortening, implies also that this process was ongoing while mountain building was in progress. This kind of relationship can be observed in the Earth's Himalayan chain, and is a natural consequence of the tendency for thickened, elevated crust to spread under the ubiquitous influence of gravity.

11.4   CONTROLS ON CENTRALIZED VOLCANISM

It should by now be clear that one characteristic Venus shares with the other terrestrial planets is the development of basaltic volcanism. Extensive volcanic flooding is witnessed in the lowland areas, while volcanic shields on a variety of size scales, domes and lava channels are distributed widely. The development of features such as lava channels, collapse pits and flat-topped lobate flows suggests that plains-forming eruptions were both of high volume and low viscosity. The steeper-sided edifices may be a manifestation of eruptions involving more evolved, viscous magmas. Many regions of radar-dark plains within the highlands also exhibit tell-tale features of volcanism, showing that here, too, volcanic activity has occurred. Frequently this can be connected with extensional deformation.

The distribution of Venusian volcanic features is not widely confined to linear zones, as on the Earth, although rift-related volcanism is apparent; rather there is a concentration along the equatorial latitudes, particularly in the region of Beta-Atla-Themis Regiones which covers 20 percent of the area of the planet, and to the south of Alpha Regio, along zones of extensional faulting. Furthermore, there is a deficiency of features such as coronae in the lowlands, which most likely is attributable to a combination of elevation-dependent eruption conditions and partial or complete burial by younger volcanic units. In the time period represented by the present Venus surface, volcanic activity has occurred, at one time or another, virtually everywhere, i.e. it is one of the dominant geological phenomena on Venus.

11.4.1   Structure of interior and lithosphere of Venus

The similarity in mass and density of Venus and the Earth has led to the assumption that the Venusian interior, in terms of both gross structure and density distribution, is similar to Earth's (Phillips and Malin, 1983). However, because of the very elevated surface temperatures, the lithosphere of Venus ought to be significantly more buoyant than Earth's. This is important, since it would militate against active subduction and terrestrial-style plate recycling. Consequently, in the absence of a method of removing thermal energy from the interior by plate tectonics, mantle plume and hotspot activity ought to be more vigorous on Venus than on our own planet.

Despite some similarities between the equatorial highlands of Venus and rifted sea-floor regions on Earth, explored by several workers as potential analogues (Brass and Harrison, 1982; Kozak and Schaber, 1989; Head and Crumpler, 1990), the altimetric data provided by Magellan have not provided confirmatory evidence for rapidly-spreading oceanic-type plate activity. In fact the recent realization that the largest contiguous highland region, Aphrodite Terra, is actually built from a series of roughly circular volcanic rises and equidimensional tessera massifs, laced with a preponderance of quasi-circular structural patterns and associated with major plains- and flood-style volcanism, has weakened such an idea. Furthermore, gravity data indicate that the highlands are supported by low-density roots that penetrate to depths which, in some instances, must lie well below the lithosphere and show that the topography is being dynamically supported. Such a scenario means that the topography is maintained by thermal buoyancy in the mantle, that is, by upwelling mantle plumes. Such "hotspots" would either supply direct support or do so indirectly by the heating and consequent expansion of the overlying lithosphere. Nevertheless, there is some evidence that differential movements have caused transform-style offsets along the highland belt, but these cannot be along the lines of those which characterize mid-oceanic rises.

The fact that both isolated volcano-tectonic features, for example coronae and arachnoids, and the large-scale volcanic rises are characteristic of Venus, makes it difficult to envisage any other process capable of generating the observed physiography than widespread plume activity. Such activity, well known from terrestrial experience, has the capacity to generate lithospheric swells, major rifting and associated volcanism. This is exactly what is seen on Venus. Theoretical considerations point to the fact that if coronae are produced above terrestrial-style hot-spots, picritic magmas could typify their cores, while tholeiitic magma would form at the cooler margins. Should plumes on Venus by 100-150°C hotter than their terrestrial counterparts, however, melts generated in the core regions of rising plumes could approach basaltic komatiite composition, with olivine-tholeiite towards the margins (Hess and Head, 1990).

Direct observational tests of mechanical and thermal models for the Venusian lithosphere are crucial if an understanding of the planet's dynamical history is to be reached. One way to estimate the thickness of the elastic lithosphere is from the flexural response to lithospheric loads (a technique which has been used with some success for the Tharsis Bulge on Mars). The pre-Magellan study of Head (1990) who investigated the relationships between the northern plains and Ishtar Terra, identified possible lithospheric flexure where the northern plains appear to underthrust the Ishtar massif along Uorsar Rupes. Using the Venera 15/16 topographic profiles, Head estimated an elastic lithosphere of thickness 11-18 km, which is consistent with a thermal gradient of 15-25 K km^{-1} and a heat flow of 50-70 mW km^{-1} (Solomon and Head, 1990). This result is comparable with expected values of global thermal gradient and heat flow (Phillips and Malin 1984). Since the northern plains stand close to the modal elevation of the planet, this is also consistent with the prediction that heat loss on Venus can be estimated by scaling from Earth and that most of the thermal transport through the outer layers of Venus is by conduction.

It is likely that Venus has a basalt-like crust overlying a peridotitic mantle. This being and it is generally agreed we are dealing with the following layered configuration: (i) strong upper crust, (ii) ductile lower crust, (iii) strong upper mantle, and (iv) ductile mantle at greater depth (Banerdt and Golombek, 1988). The high degree of correlation between long-wavelength topography and gravity already noted suggests that the planet lacks a lower mantle low-viscosity zone like the Earth's, in which case mantle convective motions should readily couple with the overlying lithosphere, producing recognizable tectonic and topographic patterns at the surface. While the crust itself may be quite strong, because the ambient temperature lies close to the point where crustal rocks would begin to lose their strength, the strong upper layer may be quite thin, say a few kilometres or less. Since the surface temperature on Venus falls with elevation at the rate of 8 K km^{-1} (Seiff *et al.*, 1980), the highest regions may be 80 K cooler than the MPR. Suppe and Connors (1992) have suggested that the relief of Venusian fold belts, which shows a remarkable linear dependence upon absolute elevation, may reflect an absolute elevation dependence of the depth of the brittle-plastic transition, which may be controlled by an isostatic coupling of elevation, lithosphere thickness, and geothermal gradient.

Using pre-Magellan data, Sandwell and Schubert (1992), arrived at an elastic lithosphere thickness of 10-25 km and a thermal gradient of 10-25 K km^{-1} for the same area of Uorsar Rupes and the circumferential fractures associated with the nearby corona, Eithinoha. For the very large Heng-O and Artemis coronae, and the smaller Latona structure, the same workers derived thicknesses of between 30 and 60 km, which give correspondingly lower thermal gradients of 3-9 K km^{-1}. This corresponds to values of conductive heat loss approximately one half that of the expected average planetary value. They invoke a combination of differential thermal subsidence and lithospheric subduction to explain the observed features in this region. Grimm and Phillips (1992), making the assumption that faulting in Western Eistla Regio was generated in response simply to flexural loading, arrived at an estimated value of 50 km for the elastic lithosphere, at the time of graben formation. This equates with a thermal gradient of 33 K km^{-1}.

The larger values for thickness pose something of a problem to current ideas about the thermal structure of the planet, since the derived thermal gradients are suprisingly low, particularly for the coronae. On current thinking, these ought to be sites of high heat-flow; however, as Solomon *et al.* (1992) point out, the lower lithosphere and asthenosphere

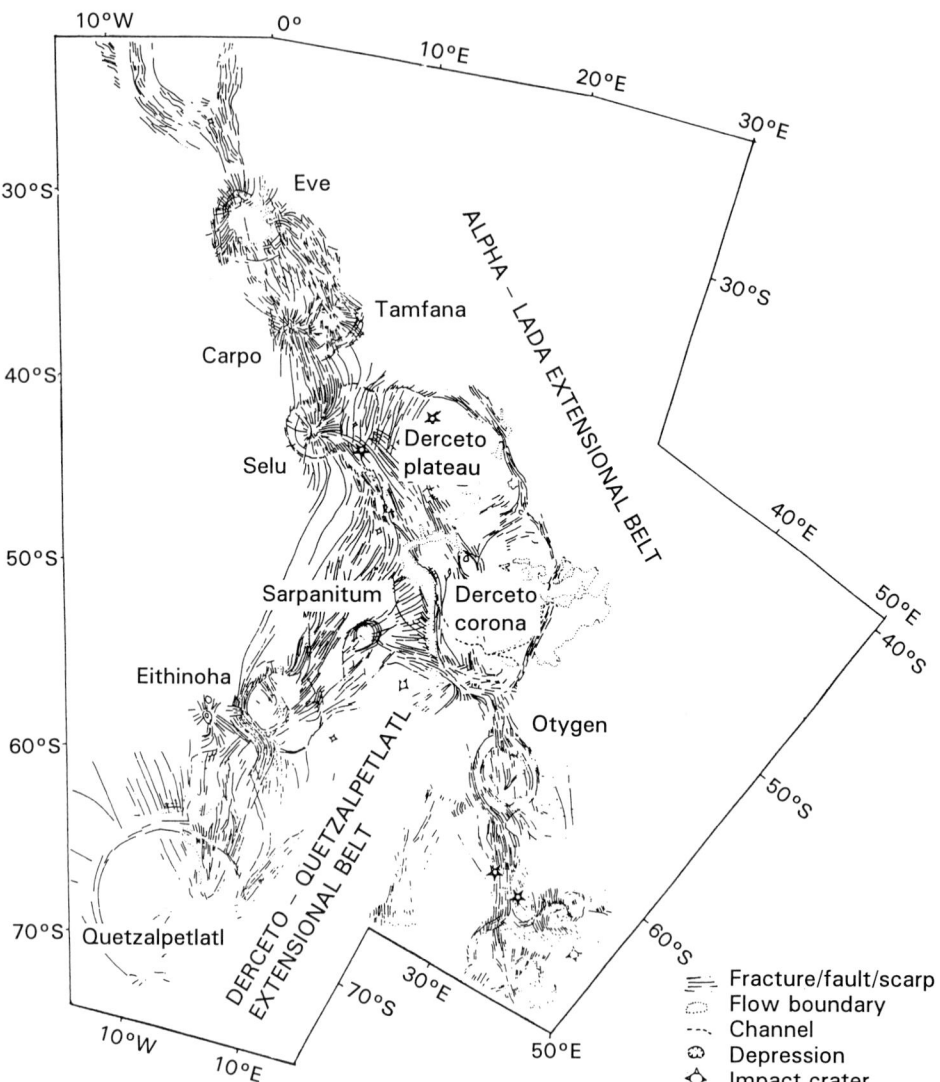

Fig. 11.21: The extensional belts of northern Lada Terra. The generalized structural map was prepared by Baer *et al.* (1994) from hand-mosaicked C1 MIDRs.

may be viscoelastic, while the Venusian crust and mantle are believed to be anhydrous. Both would have a fundamental effect upon the calculations, and as a result, values for thermal gradient could be much higher.

11.4.2   Magma uprise and plume activity on Venus

There is a general consensus that volcanic rises are sites of active mantle upwelling and that their associated volcanism occur in response to pressure-release partial melting in the underlying mantle. Such volcanic rises, although much larger than most coronae-type structures, do show a size overlap with them. Thus Artemis Chasma and Heng-O coronae, 2600 km and 1060 km across respectively, compare closely with the 1500-km-wide rise of Bell Regio; most coronae, however, seldom exceed 400 km diameter. It is also apparent that most volcanic rises show a greater diversity of volcanic features and greater amounts of uplift and extension (i.e. major rifting) than typical coronae.

It may also be observed that coronae and volcanic rises tend to cluster along major rift zones which, by implication, are lines of extensional tectonics. Senske and Head (1992) suggest that major rifting, for instance, that associated with the corona chains of Parga and Hecate Chasmata, where the altitude difference between plains and adjacent highland margins locally exceeds 6 km, may be the result of gravity sliding or slope failure. However, for rifts where there are not such large topographic differences, mantle downwelling may be a more likely instigator of extensional stresses. This may apply to the 6000-km-long extensional belt which runs from Alpha Regio to Lada Terra and includes the coronae Eve, Tamfana, Carpu, Selu, the Derceto volcanic plateau and Otygen; this strikes in a northwesterly direction and intersects obliquely a 1600-km-long belt that includes Derceto and Quetzalpetlatl (Fig. 11.21). Baer *et al*, (1994) note that the formation of the extensional belts overlapped in time but that extension has affected each of the coronae involved. In places volcanism originated within the zones of extension but elsewhere activity originated at corona sites and flowed down into the low parts of the faulted zones. In no single case was any of the coronae formed after cessation of the regional extensional faulting. Evidently crustal extension and corona formation must have been interrelated, in some cases coronae having determined the surface expression of regional faulting, in others coronae having been influenced by the concentration of regional stresses.

The above rift zones lie along the margin of the plain of Lavinia Planitia, which has been interpreted as a region of mantle downwelling (Bindschadler *et al.*, 1992). Because of the predicted absence of an asthenosphere on Venus there will be strong coupling of mantle flow to the lithosphere, thus, as McGee and Head (1995) note, during the early stages of its evolution a region of downwelling mantle will result in subsidence, compressional hoop strains and peripheral extensional radial strain (Fig. 11.22). It is these strains that likely are responsible for the rifting observed in Lada Terra/Alpha Regio. The reason that rifting became localized along only one boundary of Lavinia Planitia could be that in the early stages of downwelling strain would be expected to be relatively small. This being so, they would be largely influenced by regional stresses and inhomogeneities in lithospheric strength. Thus, the likelihood of Lada Terra being a region of relatively weak, thickened, crust, is strong, therefore stresses would tend to be concentrated along this zone. Alternatively this focussing might simply be related to a difference in inherent

(a)   PRIOR TO ONSET OF DOWNWELLING

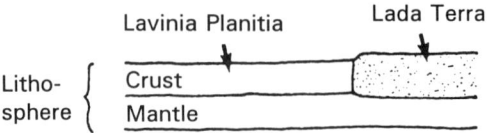

(b)   AFTER DOWNWELLING INITIATED

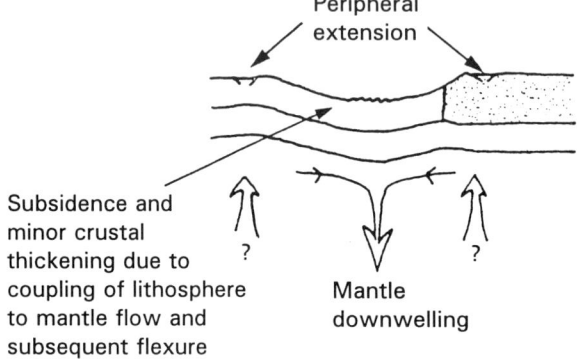

(c)

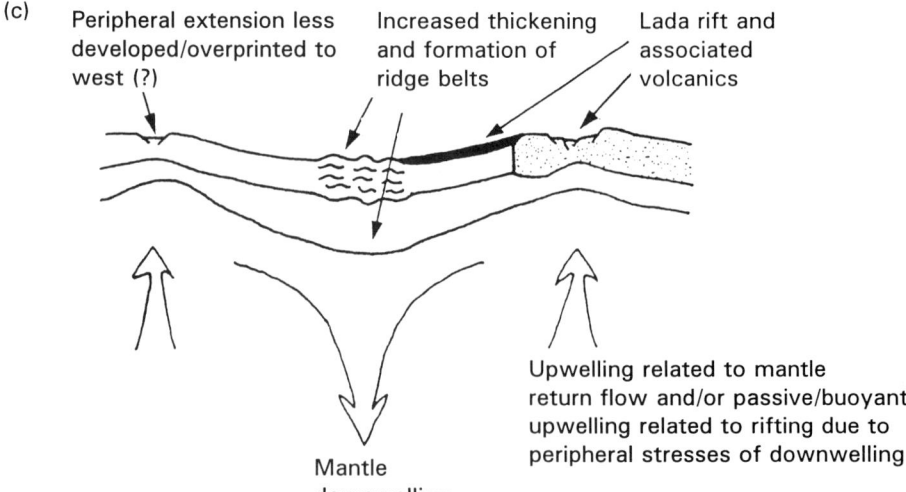

Fig. 11.22: Cartoon illustrating possible mantle downwelling beneath Lavinia Planitia and extensional origin for Lada rift and its related volcanism. (After Magee and Head, 1995.)

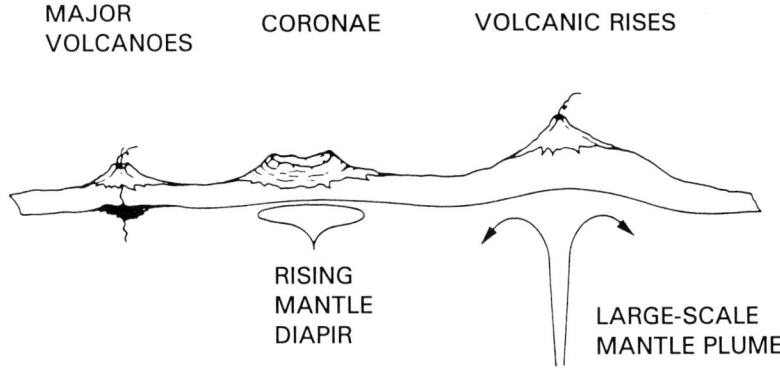

MAJOR
VOLCANOES   CORONAE   VOLCANIC RISES

RISING
MANTLE
DIAPIR   LARGE-SCALE
                                     MANTLE PLUME

Fig. 11.23: Cartoon illustrating different scales of mantle upwelling on Venus. Shield volcanoes develop above relatively large high-level magma reservoirs, while major volcanic rises form above major upwellings where there is long-lived high heat flow. Coronae sit somewhere between these two extremes. (After Stofan *et al.* 1992.)

strength of the lithospheric slabs beneath Lada and Lavinia respectively.

This poses the question of how magmatism is connected with the tectonics of this region. In the case of simple rifts, magmatism will be related to an uprising mantle plume; the Lada rift is somewhat different. Clearly volcanism is related to rifting, with major centres being roughly 630 km apart; this suggests that mantle material rose upwards in response to lithospheric extension. The production of magma by such a process is believed to be dependent upon the amount of extension in the lithosphere and the temperature of the rising mantle material (White and McKenzie, 1989) Terrestrial studies indicate that melting will commence when the ratio of final to original area (β) of the lithosphere reaches around $\beta=2.5$. For smaller degrees of extension, i.e. $\beta=1.2$, melting will occur only if the upwelling mantle is at least 150°C hotter than normal (Mohr, 1992). Is sufficient extension recognizable in the Lada rift region; for instance, could it generate the 2×10^4 km³ volume of lava extruded at Mylitta Fluctus? Estimates of the amount of extension suggest that, *under terrestrial conditions*, this is unlikely since a mantle temperature of 1480°C would be required, i.e. at least 150°C hotter than normal (McGee and Head, 1995). However, convection models for Venus studied by Phillips and Malin (1983) lead us to believe that Venusian mantle temperatures would be at least 200°C higher than on Earth It is thus conceivable that passive rifting above a rising plume of normal temperature could generate the amount of volcanism observed.

Another possibility is that volcanic activity in this region was related to actively rising "plumelets", each associated with a corona or major flow-field, but not specifically associated with the rifting. Magee and Head note, however, that in this context both coronae and volcanic centres associated with the rift zone are of much higher volume than those located away from it (i.e. mean flowfield within rift = 1.9×10^5 km², as opposed to 1.8×10^4 km² outside bounds of rift). Furthermore there is a strong concentration of coronae along the rift zone. It seems unlikely, therefore, that a large number of individual plumelets could have randomly risen along this line of extension.

The third possibility is that the volcanic activity observed along this rift is related to a reverse upflow set in motion by mantle downwelling beneath Lavinia Planitia. This would be a similar situation to the mantle counterflow recognized on Earth where delamination takes place. Similar arguments apply to this idea as have been discussed for passive upwelling, and the effectiveness of such a model would be strongly dependent upon mantle temperatures. Assuming that the diameter of each corona is roughly correlated with the scale of its associated mantle plume, Magee Roberts and Head (1993) estimate that upwelling in the Lada Rift region must have commenced at depths of at least 170 – 510 km.

If major rises and coronae originate in similar mantle processes, which is widely accepted, then the implication is that mantle upwelling activity within Venus occurs on different scales. The large apparent depths of compensation associated with volcanic rises and the very extensive volcanism imply that they are connected with much larger-scale mantle upwellings than are coronae. Since most coronae are smaller than rises, it may be reasonably assumed that coronae represent a smaller scale of upwelling, or weaker upflow of material, or flows which are shorter-lived along the lines depicted in Fig. 11.23. The fact that there is a major concentration of volcanic rises, large shield volcanoes and coronae in the equatorial belt between Atla and Beta Regiones, and along two principal tectonic trends, i.e. Parga and Hecate Chasmata, which are characterized by extension, volcanism and uplift, suggests that both large- and small-scale plume activity originates in the mantle along hot regions encircling the core-mantle boundary. Unfortunately, at present, there is no way of knowing how long-lived are plumes beneath major rises. The fact that along some volcanic zones, e.g. Parga – Hecate Chasmata, corona spacing is closer and the coronae more abundant than in the case of the Lada Rift, may be a reflection of differences in the depth of melting at the various locations, the depth being shallower in the former case.

Calculations by Stofan *et al.* (1992) suggest that most coronae can be generated by mantle diapirs 75-100 km across. They also can be produced with a temperature difference of around 300 K from the mantle temperature (Janes *et al.*, 1992). On this basis, and bearing in mind that there are approximately 300 coronae on Venus, they calculate that the approximate heat loss rate through coronae is 4×10^{-3} of the estimated mantle heat production through global volcanism for Venus (Solomon and Head, 1991). Thus mantle plumes related to corona production probably account for but a small proportion of the heat transfer to the surface of Venus, a condition which is comparable with the estimates for terrestrial plumes, which generally are believed to account for around 10 percent of the Earth's internal heat loss (Sleep, 1990). This is consistent with the belief that the Venusian mantle is heated mainly from within by radiogenic sources.

McGill (1994) after a detailed investigation of Eistla Regio, confirms that it was formed by one or more major mantle upwellings, comparable in size to terrestrial "superplumes". He suggests that coronae and volcanic shields were offsprings which may have been generated by convective instabilities within the superplumes, or by growth from multiple sites of enhanced partial melting in the Venusian mantle. He also surmises that the change from corona formation to the smaller shield volcanoes accompanied a gradual thickening of the lithosphere with time. This would appear to contradict the expected thinning of the lithosphere due to the long-term high heat flow, but can be explained away by assuming that it was a response to rapid cooling of the crust which ensued from the last burst of global cooling and global resurfacing believed, by some workers, to have affected the planet.

Koch (1994) uses a somewhat different approach, using a boundary integral method and solving for the motion of the plume head and for the topography, geoid and stress field at the surface. This involves taking into account the geoid to topography ratio (GTR) for different features. She observes that when a plume head approaches the Venusian surface, stresses in the overlying fluid cause it to thin below the surface; during spreading, the arched surface passes through various stages that correspond to observed Venusian landforms. The initial stage sees the rising plume head generating a broad topographic dome, with a large geoid and attendant radial extensional faults. Subsequently, the topography subsides, becoming more plateau-like and the dominant stress pattern becomes a zone of concentric extension encircled by a broad belt of concentric compression. Beta, Atla and Western Eistla Regiones all have large GTRs and may be the expression of an early stage in plume development. The flatter-topped structures, i.e. Ovda and Thetis Regiones – which have lower GTRs – may be later stage features. Such a model would envisage novae, arachnoids and coronae as representative of a temporal developmental series. Her modelling implies initial plume head diameters of about 1000 km, and predicts that highland rises evolve on a timescale of around 10 Ma, while smaller-scale uplifts evolve in a 100 Ma timescale.

11.4.3   Magma reservoirs and Neutral Buoyancy Zones

What is not known for Venus the relative contribution of intrusion and extrusion within the crust. On Earth, extrusive rocks account for between 10-20% of the total. Several factors argue that for Venus this proportion may be lower: firstly, high-density crust appears to be more extensive on Venus; secondly, the high surface pressure inhibits vesiculation within melts rising towards the surface. The latter would have the result of magma reaching neutral buoyancy zones at shallower levels than on Earth. This phenomenon may account for the apparent abundance of fractures and graben associated with volcanic units: such fractures may represent dykes.

The implications for the production of volcanic landforms on Venus of magma reservoirs and neutral buoyancy zones has been examined by Head and Wilson (1992). Because the high atmospheric pressure inhibits volatile exsolution on Venus, it also generally hinders the formation of neutral buoyancy zones and shallow magma reservoirs. Thus, for magma ascending and erupting near to or below MPR there should not be stalling and the melts would erupt at high volumes and effusion rates. Similar magmas rising 2 km above MPR would behave rather differently; roughly one half would ascend directly to the surface while the other half would stall in neutral buoyancy zones. The general tendency on Venus is for such stalling zones to form with increasing magma gas content, and to appear at shallower depths than on Earth. With increasing height above MPR, Venusian stalling zones become deeper.

There is a general tendency for most Venusian central volcanoes to reach fairly modest heights. This can be explained by the behaviour of neutral buoyancy zones as volcano height increases. A magma chamber centre will of necessity be at greater depth relative to the summit as the structure grows, due to the major change in atmospheric pressure with increasing altitude. As a stalling zones and magma reservoirs will stay in the substructure of a typical shield for longer periods than they would on Earth. The consequent lower rate of vertical migration implies that reservoirs would tend to stabilize and undergo lateral growth, becoming significantly larger than their terrestrial counterparts. On Venus, large stalled magma bodies would lead to multiple and more widely dispersed vents and large volume individual eruptions.

Another effect of large reservoir size is that in encourages formation of positively buoyant fractions of melt through processes such as differentiation and volatile exsolution. In shallow locations, nonbuoyant material undergoing volatile exsolution will require higher gas bubble content to generate eruptions than on the Earth. However, when this gas-enriched magma reaches the Venusian surface, it is more likely to retain its bubbles than to undergo explosive disruption. One of the important consequences of this phenomenon is the higher potential on Venus for effusion of lavas that have high gas bubble contents, giving rise to more viscous rheologies for relatively basic melts than would be encountered on the Earth. This leads to the prediction that dykes and shallow intrusions should be commonoplace and that pancake domes and stubby flows are, in effect, Venusian equivalents of terrestrial ignimbrites, i.e. very high gas bubble content, high-volume, high-discharge rate.

11.4.4   Age of volcanism and resurfacing on Venus

The complex geology and relative paucity of impact craters imposes severe limitations on stratigraphic studies. With regard to the resurfacing rates due to volcanic activity and tectonism, crater distribution studies suggest that it has been episodic and spatially inhomogeneous. Initial crater statistics (based on the tally of only 1333 craters) appear to indicate that the mean age of the surface of the planet is between a few hundred million to one billion years. The observed crater population leads to three possible end-member models for the resurfacing of the planet: (i) catastrophic resurfacing, (ii) "leaky planet" global resurfacing, and (iii) regional resurfacing. These have been discussed in detail by Phillips *et al.*, 1992).

The catastrophic resurfacing model considers the observed crater tally as a production population and invokes an episode of rapid resurfacing between $500 - 300 \times 10^6$ y ago, of sufficient depth to obliterate the pre-existing crater record, and provide a pristine surface on which the current production population accumulated (Schaber *et al.*, 1992). Such a model views subsequent volcanism as minimal in both volume and extent. In support of this idea, Schaber and his colleagues cite the relatively small number of impact craters highly modified by volcanic processes. However, what is not clear is both the volume and method of emplacement of many of the plains units found on Venus which cannot yet be directly attributed to volcanicity. Where they are not obviously related to specific volcanic foci thay may have formed in response to catastrophic events, or equally well may have been related to extrusion of variable volumes of lavas from different loci and at different times, i.e. they were formed sequentially, over the last few hundred million years.

The "leaky planet" scenario sees volcanism as having occurred relatively uniformly in time, and having taken place almost everywhere at the same time. By this method the Venusian crust would be thickening gradually with time, and impact craters constantly being either modified or obliterated. However, the lack of impact craters in intermediate to advanced stages of volcanic burial tends to militate against this theory.

The extent of volcanic deposits which are easily mappable over specific regions individually cover areas in excess of 125 000 km², furthermore emplacement of volcanic units took place serially, probably on both local and regional scales. Such a process means that although individual volcanic episodes occurred in sequence, over long periods of time these could have contributed to general resurfacing of the whole planet, or, at least, very large parts of it. This has sometimes been described as the "cookie-cutter" model, an analogy whose meaning will not be lost, even on non-cooks! It envisages resurfacing as having proceeded by the infilling of topographic lows at different times with the occasional obliteration of impact craters as an integral part of the same process. The preponderance of large numbers of different types of volcanic feature related to mantle instabilities on the scale of a few hundred kilometres (e.g. coronae, shield fields, large shields, arachnoids etc.), suggest that this is true, and that magmatism on Venus is linked with pressure-release partial melting associated with plumes and hotspots. The fact that the total volume of volcanic deposits predicted by the model is greater than that observed does not invalidate it; as has been observed above, large areas of plains units undoubtedly have been deposited by processes as yet not understood.

Recent Monte Carlo simulations of equilibrium resurfacing models have resulted in seventeen times more embayed impact craters than actually observed, or unobserved non-random crater distributions for resurfacing areas between 0.03% – 100% of the surface (Strom *et al.*, 1994). Such constraints imposed by the cratering record suggest that Venus may suffered a global resurfacing event approximately 300×10^6 y B.P., after which volcanic resurfacing diminished greatly, as did tectonism. The present crater population is envisaged as having accumulated since that time.

As we have observed, the entire Venusian surface has been strongly affected by by volcanism and tectonism; however, a mere 2.5% of impact craters are embayed by lava and only 12% affected by strong deformation. Such a situation occurs on only one other solar system body, Neptune's strange moon, Triton. Global resurfacing has been invoked to explain conditions there too. The observation strongly suggests that Venus was globally resurfaced in relatively recent times (approximately 300×10^6 y ago), and that the modifying event took a few tens of millions of years. Strom *et al.* (1994) suggest it probably ended rather abruptly, i.e. in <10 m.y.). The same Monte Carlo simulations imply that between 4 – 6% of the surface has been volcanically modified since that time, and that the lava production rate during this period was around 0.01 – 0.15 km³ y⁻¹. This is about 3 – 33 times less than Earth's current rate of interplate volcanism. The most recent activity on Venus apparently has been concentrated in the region of Beta-Atla-Themis Regiones, where the greatest number of modified craters is to be found. Furthermore, high values of emissivity have been recorded for volcanoes in that region, implying relatively recent activity.

One suggestion for the global resurfacing of Venus suggests that because of the high ambient temperatures at the surface, an oscillatory convective regime operates and did so throughout much or most of Venus's geological history (Arkhani-Hamed *et al.*, 1993).

This resulted in episodic global resurfacing, planetary cooling, and a change in the convective regime from oscillatory to quasi-steady state. Turcotte (1993), on the other hand, suggests that episodic recycling of the crust occurred. In his model, a stable lithosphere, gradually thickening with time and lacking plate tectonic activity, results in lower heat flux at the surface and a consequent mean increase in internal temperature. The latter could lead to enhanced mantle convection that triggered rapid plate tectonic activity and rapid resurfacing. A third proposal invokes phase transitions within the Venusian interior to cause catastrophic convective episodes (Steinbach and Yuen, 1992).

11.5 EPILOGUE

Episodic or regional resurfacing events appear to have occurred on all of the larger terrestrial planets. It has been proposed that "superplumes" have caused accelerated mantle convection and heat loss on the Earth (Garzanti, 1993) and that currently such a structure is rising below Western Europe (Hoernle *et al.*, 1995) . The most widely documented of these occurrences happened at the Cretaceous-Tertiary boundary and lasted about 40×10^6 y. There also is considerable evidence that volcanic activity peaked during the Hesperian and Early Amazonian periods. This is the same period during which enhanced volcanism has been invoked to account for the melting of subsurface ice and generated the outflow channels (Tanaka and Chapman, 1992). Herrick and Parmentier (1994) suggest that the planet may have experienced episodic large-scale overturn of its mantle with periods of the order of $10^7 - 10^9$ years It should be no surprise, therefore, that such a phenomenon should be discussed with respect to Venus.

12

Volcanism on Io

During early 1979, just before the Voyager 1 and 2 fly-bys of Jupiter, Stan Peale, Pat Cassen and Ray Reynolds published a paper in *Science* which predicted the occurrence of active volcanism on Jupiter's Galilean satellite, Io (Peale *et al.*, 1979). This surprising suggestion followed their calculations relating to the tidal stresses produced within the body of the satellite by being pushed and pulled between the gravitational fields of Jupiter and another Jovian moon, Europa. They estimated that about three orders of magnitude more heat could be generated by these tidal stresses than was released by long-lived radioactivity. In consequence, internal activity in the form of volcanism could very well result. Later in that year, after Voyager had passed by Io, inspection of certain images at the Jet Propulsion Laboratory (JPL) revealed an odd bright feature associated with the limb of Io (Morabito *et al.*, 1979). Subsequently this was confirmed to be a massive eruption plume which was generating material at a prodigious rate.

Since that time several more such plumes have been observed and we now know that volcanic material is generated on Io so quickly that it is capable of covering itself in a 100-m-thick layer of volcanic material every one million years! This rapid resurfacing process accounts for the unexpected absence of impact craters revealed on Voyager images. Johnson *et al.* (1979) have estimated that at least 10^{10} tons of volcanic material are produced every year and that Io is currently emitting between 1 and 1.5 W m^{-2} tidal energy as a result of volcanic activity. Because of the presence of sulphur and its compounds in the torus surrounding the moon, its presence among the volcanic products was strongly suspected.

12.1 SURFACE COMPOSITION OF IO

Io is roughly the same size and density as Earth's Moon, a fact which leads to the inevitable conclusion that Io too is composed predominantly of silicate material. However, the surface of the moon is more red than any other object in the solar system and analysis of its overall spectrum indicate that it matches most closely that of sulphur (Nelson *et al.*, 1982). The surface colours of Io as seen on most image-processed Voyager images include various shades of red, orange, yellow and brown, plus almost black and almost white areas (Plate

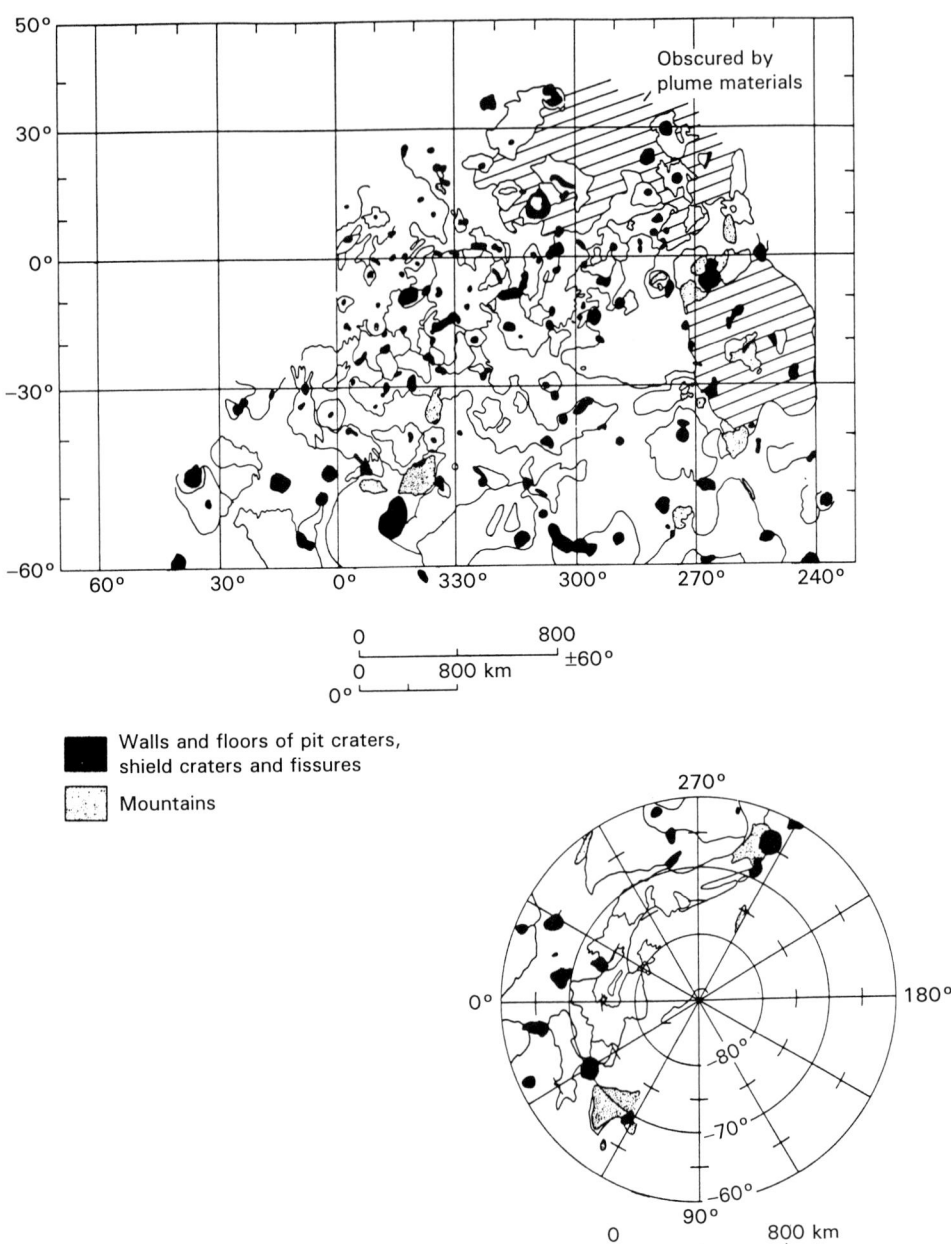

Fig. 12.1: Preliminary geological map of Io prepared from moderate resolution imagery from the Voyager 1 mission (March 1979). The upper map is in Mercator projection, the lower in polar stereographic. (After Schaber, G.G. (1980), figure 1a, with permission.)

12). All of these can be matched by those observed when molten sulphur is allowed to cool undisturbed from above its melting point in laboratory simulations. Sulphur at various temperatures or anhydrous mixtures of sulphur allotropes with SO_2 frost and sulphurous salts of sodium and potassium therefore are prime candidates for the surface composition (Fanale *et al.*, 1979).

While there has been wide support for the view that its surface is mantled in sulphur and/or its compounds and that a part of Io's volcanic activity involves the effusion of sulphur-rich flows (Smith *et al.*, 1979), there are several unresolved problems with this idea which lead to considerable uncertainty. One of these attaches to a characteristic of sulphur itself, whose natural colour versus temperature scheme is significantly modified when only small amounts of impurities (less than 0.2%) are incorporated (Sagan, 1979; Sill and Clark, 1982). Furthermore, Young (1984) has argued that the colouration seen on most of the Voyager imagery used in earlier modelling for sulphur and its allotropes is not truly representative of the true colours of Io, which is much more a yellowish-green than an orange-red, and he therefore asserts that earlier colour-match modelling may be invalid.

Carr and his colleagues (Carr *et al.*, 1979) have expressed the view that both silicate and sulphur volcanism is characteristic of Io and that the near-surface consists both of silicate and sulphur deposits. More recently Carr (1986) has reviewed the literature relating to this subject and assessed the relative roles of sulphur and its compounds and silicates in driving Ionian volcanism. This will be discussed further after the geomorphological characteristics of the satellite's surface have been described.

12.2   SURFACE FEATURES ON IO

During the Voyager 1 and 2 fly-bys about 35% of the Ionian surface was photographed at a resolution of about 5 km, the coverage mainly focussing on the equatorial and southern polar regions. The best resolution was achieved in an equatorial belt between 275° and 360°W, where features about 0.5 km across were resolved. Initial geological mapping of Io was completed by Masursky *et al.* (1979) and Schaber (1980, 1982), while a shaded relief and surface markings map was published by the U.S.G.S. in 1987 (Map I-1713). A detailed geologic map of the Ra Patera area of Io by Ron Greeley, Paul Spudis and John Guest became available as recently as 1988 (Map I-1949).

Three principal kinds of geological units can be recognized: (i) mountain material, (ii) plains materials and (iii) vent materials. Their distribution as mapped by Schaber (1980) from the Voyager 1 data is shown in Fig. 12.1. Of these three main types of material, the vent and plains materials are considered related to volcanic activity, but while most of the mountain materials may be volcanic, some could represent older crustal products either unrelated or not directly related to volcanism. Embayment relationships suggest that the mountain massifs are relatively old with respect to both the plains and vent materials; furthermore they, of all the landforms, show modification by both tectonic and erosional processes.

Fig. 12.2: The mountain massif of Boosaule Montes, showing the adjacent large vent structure and surrounding layered and intervent plains deposits with low shields, calderas and rifts. Frame width 800 km. Centred at 6.5°S., 252.0°W. (Voyager frame 125JI+000.)

12.2.1   Mountain massifs

The mountain material comprises less than 2% of the area mapped and has considerable vertical relief. Commonly it is mantled with high-albedo deposits, probably sublimates, represented by white, yellow and orange-brown hues on processed imagery. These frequently appear to have been generated from vent areas nearby. Individual mountain massifs may extend laterally for upwards of 100 km, while the highest massifs rise over 9 km above their surroundings. While the colour of the material suggests it is mantled by sulphurous deposits, the very significant relief that it supports militates against the massifs being wholly of sulphur or its compounds (Masursky *et al.*, 1979; Clow and Carr, 1980). One of the most prominent mountains comprises three massifs Boosaule Montes, which abut on a large unnamed vent structure centred at 6°S, 267°W (Fig. 12.2). A similar vent characterizes another massif, Euboia Montes. The common association of steep-sided pits with the massifs, suggests that at least some, if not all, have a volcanic origin. The

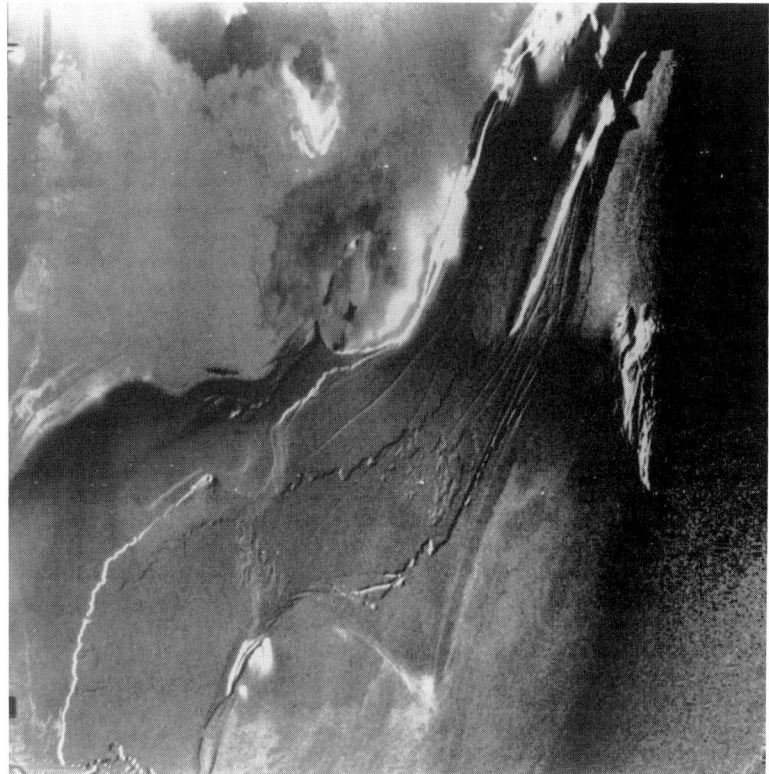

Fig. 12.3: The region of Nemea Planum, showing layered plains units with linear rifts and eroded scarps. Frame width 800 km. Centred at 71.4°S.,265.8°W. (Voyager frame 141JI+000.)

relationship may mean that the massifs are rim deposits of the large vents, but they also could be older volcanic massifs into which younger structures have been incised. What is clear is that their considerable relief, their ability to maintain steep slopes bordering the pits and their often fractured character implies that they are constructed of materials with a relatively high yield strength. This militates against them being constructed solely from sulphur or its compounds and seems to imply they are composed largely of silicate rocks (Clow and Carr, 1980). On this basis it appears most likely that they represent outcrops of relatively old volcanically generated silicate subcrust mantled in sulphurous sublimates.

12.2.2   Plains and intervent flows
Layered plains units are the predominant landforms on Io and are estimated to account for about 40% of the mapped area. On Voyager images they range in hue from black and white to various shades of red and yellow; polar plains tend to be dark brown to black

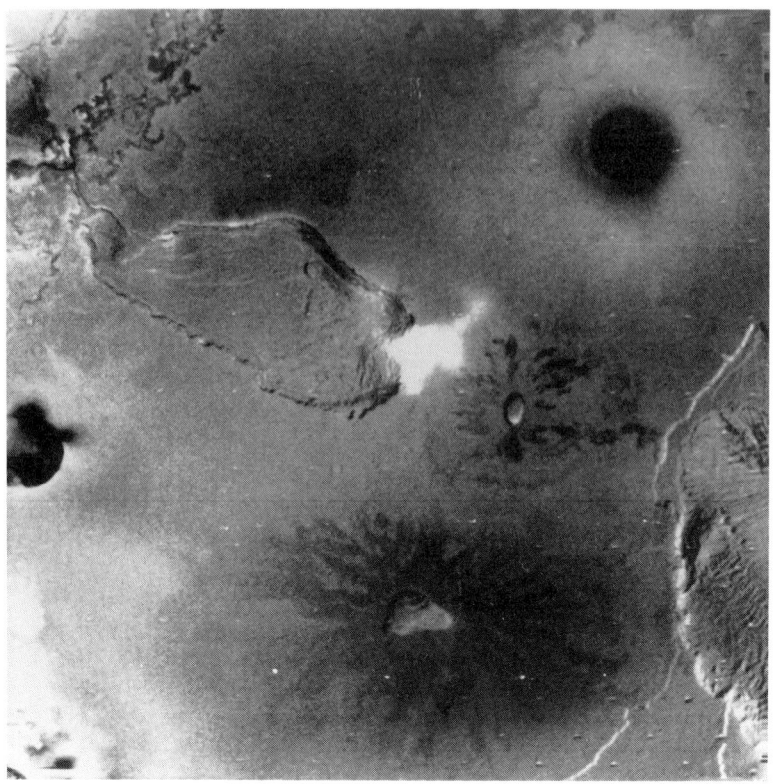

Fig. 12.4: Prominent white fumarolic deposit at base of Iopolis Planum layered plains unit.
To its west is Maasaw Patera with its prominent 2 km-deep caldera and dark, radiating flows.
The prominent massif at the southwest corner of the image is Euboea Montes. Frame width
505 km. Centred at 38.4°S., 329.5°W. (Voyager frame 139JI+000.)

(Masursky *et al.*, 1979). Schaber (1980) recognized three principal types of plain in the
region he mapped from the Voyager 1 mission. *Intervent plains*, which cover about 40%
of the mapped area, generally have smooth surfaces with regionally consistent, intermediate
albedo (Fig. 12.3). Such materials are believed to represent fallout from volcanic plumes
or lava flows interbedded with fumarolic deposits from pit craters, low shields (paterae)
or fissures. *Layered plains* cover about 9% of the mapped region and form extensive,
smooth, flat surfaces bounded by 150 to 1700 m high escarpments, often transected by
numerous graben faults (Fig. 12.3). Even more convincing evidence that erosional processes
may have operated is provided by the third recognized unit, the *eroded layered plains*,
which largely are restricted to the south polar region and account for less than 1% of the
area mapped.

The general similarity between the intervent and layered plains units suggests a similar origin, although the latter appear to show more positive evidence of headward erosion and wall slumping along the prominent fault-controlled scarps. The significant relief developed within the plains again reinforces the conclusion that silicate rocks compose the subcrust, but the variations in surface colour shown on Voyager images within the various plains units, suggest they may be veneered in sulphur or sulphur compounds.

McCauley *et al.* (1979) have described an erosional process in which they envisage SO_2-rich plains having slowly retreated due to sapping, rather like the sapping of CO_2 or water envisaged in the formation of Martian box canyons. Escarpments would thus represent lines of easiest release along which liquid SO_2 or elemental sulphur would escape under artesian pressure. Their calculations suggest that if crustal fracturing occurred SO_2 would be driven towards the surface and that, at the triple point for SO_2, part of the fluid would crystallize and the system would expand, forming SO_2 vapour. On reaching the surface the partly crystallized mixture would be released energetically with velocities approaching 350 m s^{-1}. Such release could take place from either a fissure or a vent and would allow material to be ejected as far as 70 km from the source area. It is possible that the widely distributed high albedo patches seen on Io's surface were formed by such a mechanism (Fig. 12.4).

12.2.3   Vent materials

Over 300 vent and vent-related structures have been identified on Io. The largest of these are most frequently located in equatorial latitudes, many attaining diameters in excess of 250 km; those in higher latitudes generally are smaller (less than 100 km). The randomness of their distribution sets them apart from terrestrial, Venusian and Martian volcanoes, apparently indicating that hot spots on Io are not controlled by well-established, patterned convection cells.

While vent structures account for only 5% of the total surface area of Io, associated flows cover wide areas, forming low-profile shields or *paterae*. Some flows are greater than 700 km in length, implying very fluid materials and high effusion rates. The central structures themselves range from huge low-relief shields to small pits just a few kilometres across. There are also some smaller discoid structures such as Apis Tholus, and more complex volcanic structures like Kibero Patera. As is the case with terrestrial and Martian volcanic structures of this type, calderas crown the summit regions of many Ionian shields and paterae, these sometimes being complex and bearing witness to prolonged phases of summit collapse.

Low shield volcanoes, more aptly termed paterae, are typified by Ra Patera, which has a central ovoid dark region measuring 25×40 km, presumed to be a vent, about which are distributed long, narrow, radial or subradial flows (Fig. 12.5). While altimetry is not available for Io, the trend of the flows is taken to indicate sufficient central height for radial flow to extend over a wide area; however, as has been shown on Mars, flank slopes need be only about 0.5° for this to occur in fluid lavas, implying the summit may rise only 1 km above the surrounding plain. Greeley *et al.* (1988) have mapped seven major flow complexes on the flanks which cover an area that measures 760×480 km. This includes relatively narrow flows, some of which appear to have been tube-fed, as well as broader sheet flows.

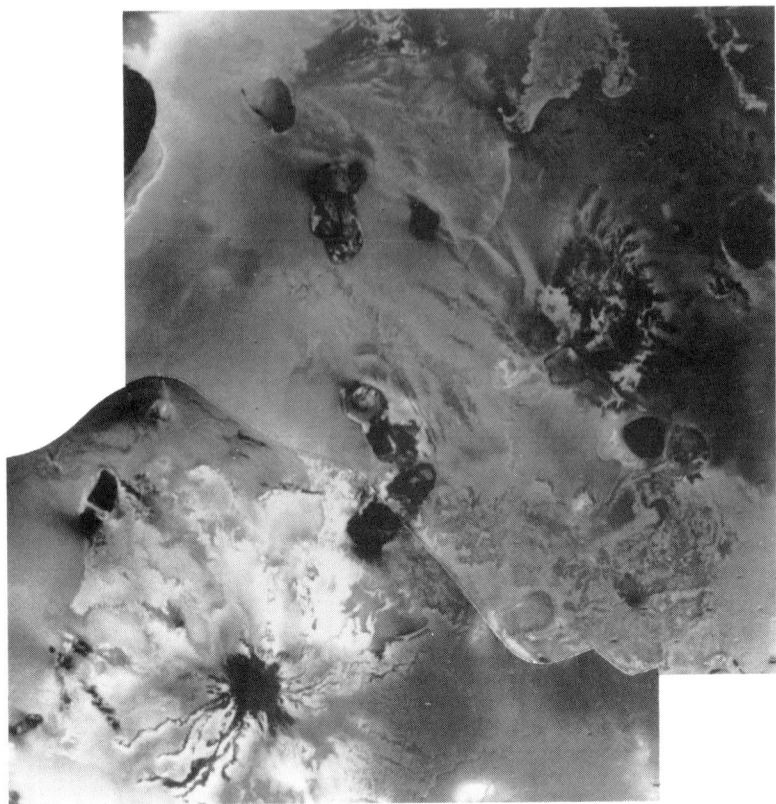

Fig. 12.5: The prominent Ionian low shield, Ra Patera is seen towards the lower margin of this mosaic. Note the radiating, dark, volcanic flows with lobate terminations and the dark floor of the summit caldera. To its east lies the breakout shield, Kibero Patera, where dark tongues are seen emerging from the surrounding complex flows. Frame width 950 km. Centred at 8.9°S., 303.1°W. (Voyager frames 047J1+100 and 049JI+000.)

Temperatures measured at hot spots like this range up to 600 K, which certainly is consistent with sulphur volcanism. Furthermore Pieri *et al.* (1984) have found colour changes along the lengths of flows associated with Ra Patera that are suggestive of transitions of sulphur through various allotropes as it cools. Sulphur, which melts at the low temperature of about 115°C, is very fluid, and even if present on the surface to the extent of only a few per cent, could be mobilized into local sulphur flows and temporary lava lakes. As is the case with most summit calderas, that of Ra Patera is floored with low albedo materials which may be molten sulphur.

Relative height measurements made at Maasaw Patera are significant here. This is a similar structure to Ra Patera, having a well-defined summit caldera which measures 25 × 40 km and several different floor levels, like terrestrial shields. Using a new photometric

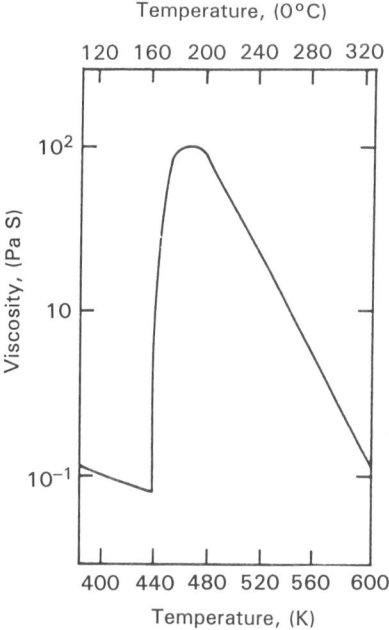

Temperature, (0°C)

120 160 200 240 280 320

Viscosity, (Pa S)

10^2

10

10^{-1}

400 440 480 520 560 600

Temperature, (K)

Fig.12.6: Plot of viscosity versus temperature for pure sulphur. The melting point is 119°C, but the viscosity minimum is at 160°C. The colours become progressively lighter with decreasing temperature, from black above 200°C to yellow below 160°C. (After Meyer, B. (1977). *Sulfur, Energy and Environment*. Elsevier, Amsterdam.)

technique, D. W. G. Arthur (1980) obtained a maximum depth of 2180 m for the summit pit. This order of wall height, if typical of most Ionian calderas, is unlikely to be sustained in sulphur or its compounds alone, and while many surface flows may be composed of sulphur, it has to be assumed that the main bulk of the paterae volcanoes is composed of silicate material, perhaps interbedded with sulphur-based flows and pyroclastic rocks.

The relatively low temperatures recorded at Ionian hot spots have been levelled as an argument against the notion of silicate magmatism on the surface of Io. However, recent broadband infrared studies on an active (basaltic) lava lake on Kilauea, Hawaii (Gradie *et al.*, 1988), have shown that effective radiating surface temperatures may be as low as 526-644 K during quiescent phases of the lake's activity, and that the higher-temperature component (i.e. the hotter lava beneath, exposed where there are breaks in the cooler crust) may account for as little as 0.05 to 0.1% of the surface area. Based on these observations, there seems little to prevent a conclusion that ponded silicate lavas could be important components of active Ionian calderas. This supports the calculations presented by Carr (1986).

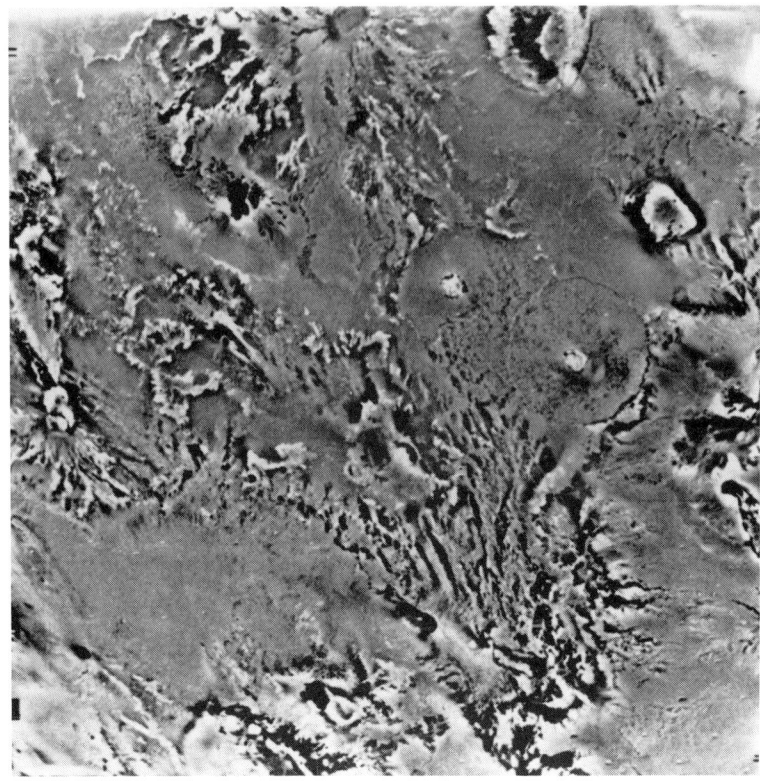

Fig. 12.7: The two disk-shaped shield volcanoes, Apis and Inachis Tholi, 100 and 140 km in diameter respectively. These may represent shields built from rapidly-erupted lavas which spread out evenly in all directions from the central vent, or may be composed of pyroclastic materials. Frame width 925 km. Centred at 12.4°S., 348.3°W. (Voyager frame 071JI+000.)

Other volcanic constructs are more complex. Kibero Patera is one such structure. This is characterized by a large central mass with rather rugged relief associated with which is a 40-km-diameter caldera. A complex of compound flows emanates from its eastern side and this has numerous peripheral lobate tongues (Fig. 12.5). While the tongues could be relatively viscous silicate flows, Greeley and his colleagues suggest they may be representative of a unique process related to sulphur volcanism (Greeley *et al.*, 1988; Fink *et al.*, 1983; Greeley and Fink, 1984). Fig. 12.6 shows that from its melting point at around 119°C, up to a temperature of about 160°C, liquid sulphur exhibits a steady decrease in viscosity. Above 160°C, however, it shows a very abrupt viscosity increase of near four orders of magnitude, due to polymerization effects. Above 190°C, its viscosity decreases again, as it depolymerizes. With this behaviour in mind, the central mass of Kibero Patera is believed to have formed when predominantly sulphur material was erupted in a high-

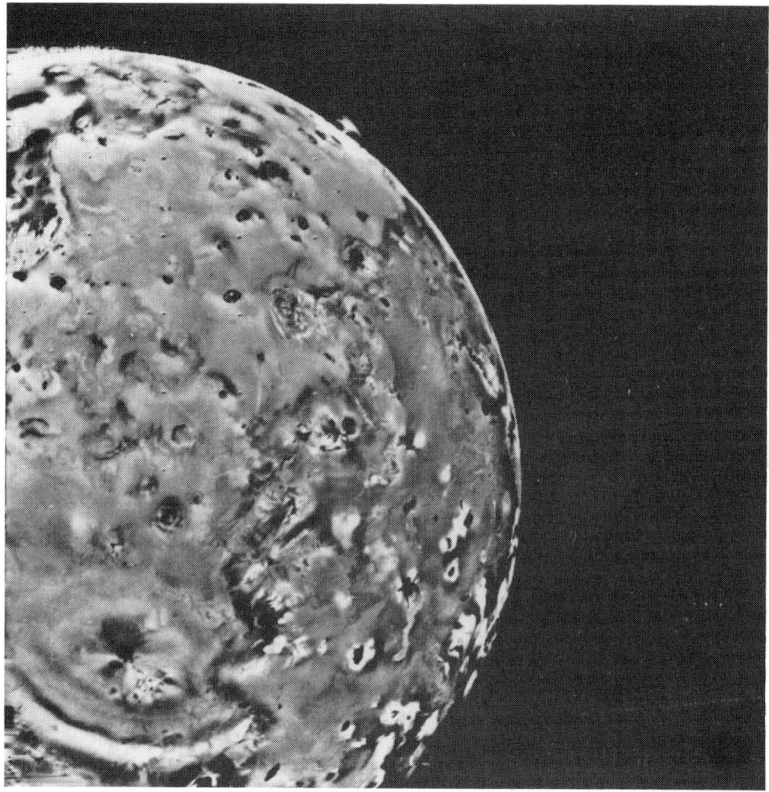

Fig. 12.8: A prominent umbrella-shaped plume may be seen rising from the limb of Io, silhouetted against the dark sky background. The plume source, Pele, is seen in plan towards the bottom of the picture. Note the dark central region and the surrounding heart-shaped "halo". Frame width 3200 km. Centred at 16.7°S., 193.2°W. (Voyager frame 1107JI+001.)

viscosity state (i.e. greater than 200°C). As it cooled below about 160°C the sulphur, because it naturally experiences an abrupt decrease in viscosity, effectively liquefied. Subsequent eruptive pressure then caused breakouts of the high-viscosity crust to occur, forming the observed tongue-like features. Such volcanic structures informally have been termed *breakout shields*.

In addition to paterae and breakout shields, there are also a number of more regular disk-shaped structures, such as Apis and Inachus Tholi (Fig. 12.7). Each is about 180 km across, has a prominent central pit and a well-defined periphery. Individual flows are not visible on Voyager images, even at high resolution, and therefore it has to be assumed that these are built either from very fluid lavas that spread out evenly on all sides of the vent or from symmetrically-distributed fall-out deposits from eruption clouds.

12.3   VOLCANIC PLUMES

During the Voyager 1 mission, nine active eruption plumes were observed rising above the surface of Io (Strom *et al.*, 1979). These reached heights of between 70 and 280 km, implying vent ejection velocities ranging from around 500 to 1000 m s^{-1} (Smith *et al.*, 1979; Strom *et al.*, 1979). Eight of these were active at the time of the Voyager 1 encounter and there is hard evidence that recent activity has occurred at these and other centres. In particular there were several areas of anomalously high surface temperatures detected, indicating rates of volcanic activity very much higher than those typical of the Earth Hanel *et al.*, 1979). All of the active plumes, bar one, were concentrated in the equatorial belt between 30°N and S. This may be taken to imply that the surfaces in this zone are younger than those nearer the poles. The largest plume measured 1000 km wide and 280 km high.

Most plumes comprise a highly symmetrical umbrella-shaped cloud with a central fountain, which, in the case of the largest plume, was about 35 km across at its base. When viewed in plan, plume sources are characterized by a dark central region consisting of radial to sub-radial jet-like streaks that may extend outwards for 150 km, a surrounding circular or irregular bright aureole and a very diffuse outer region, often heart-shaped, surrounding the bright aureole (Fig. 12.8). The two most widely documented plume sites are named Pele and Loki, both of which are associated with caldera depressions.

Some plumes were observed to remain approximately the same size over a number of days, while others showed changes over a span of 4 or 5 h (Strom *et al.*, 1979). Thus in the four months separating the two Voyager probes, the two calderas Aten and Surt had become very much darker than first imaged by Voyager 1, while major changes had also occurred at Loki (McEwan and Soderblom, 1983). The same workers also note that more than one type of plume occurs on Io. Thus just one plume was very diffuse and lacked the umbrella shape and columnar fountain of the others; it measured 100 km high and 210 km across in visible light but in UV light appeared twice that size. Close inspection revealed it to have two components: a central core seen in visible light and a fainter envelope only discerned in UV. A possible interpretation could be that the central region was composed of pyroclastic materials, while the "UV-scattering halo" was an envelope of extremely fine-grained dust, or even gas, which strongly reflected at the shorter wavelengths.

Strom and colleagues (1979) note that some plume sources are located on fracture lines. However the symmetry of the eruption clouds indicate that pipe-like vents or short fissures are the main source. Since the eruptions evidently are driven by internal heating, they presumably are generated by a mechanism akin to violent geyser activity, involving SO$_2$ Kieffer, 1982)

12.4   THE ROLE OF SILICATES AND SULPHUR ON IO

Pure sulphur has several allotropes with characteristic colours that transform from one to the other with changing temperature. Thus liquid sulphur at temperatures above 250°C is black but with falling temperature this changes through red to orange and yellow Meyer, 1965). Rapidly quenched sulphur apparently has a translucent whitish appearance

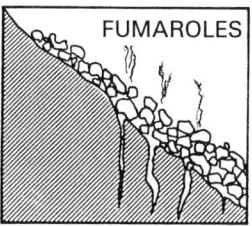
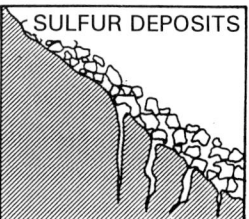
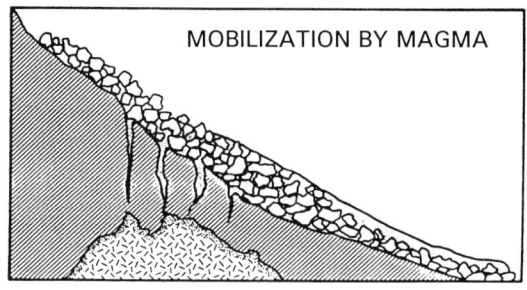

Fig. 12.9: Diagram to show the proposed origin of a secondary sulphur flow, based on studies of the 1950 Mauna loa sulphur flow. Fumarolic sulphur is deposited near to the surface and accumulates within talus. Subsequent magmatic heating melts the original sulphur, liquefying it and mobilizing into a flow. (After Greeley *et al.*, 1984.)

regardless of its original temperature. Some work has been undertaken in temperature mapping of Ionian volcanic flows on the basis of known colour transformations but, as has been discussed above, this has to be viewed with extreme caution since (i) processing of Voyager images does not necessarily faithfully reproduce true colours (Young, 1984), and (ii) even small amounts of impurities change the colour schemes significantly.

Modelling experiments by Fink *et al.* (1983) suggest that, once mobilized, sulphur flows could proceed for great distances. They also showed how crustal overturning, resurfacing and fracturing would be common on sulphur flows, which also would exhibit development of surface ridging on the centimetre scale. This may give to such flows a geomorphological signature that allows discrimination from silicate lavas. Currently, however, image resolution for Io is insufficient to discern such small-scale features and use of this potential diagnostic criterion must await future missions. Fracturing and break-up of the upper surfaces of sulphur flows will affect the colouration of such flows, particularly where molten sulphur from the flow interior rises up. Since colour has been widely used as a criterion for diagnosing sulphur at the surface of Io and its temperature, clearly it is an interesting field of study, which eventually may have applications in unravelling the volcanic history of the moon.

In an attempt to provide concrete data on actual sulphur flows, on Io, Greeley *et al.* (1984) visited Hawaii and studied the 1950 sulphur flow on Mauna Loa. This is located along the Southwest Rift Zone, which was active during 1950. It comprises a series of *en echelon* fractures that have been sites for basaltic eruptions. Secondary sulphur deposits, resulting from continuous fumarolic activity, are widespread along the fracture lines and are distinctive by virtue of their yellow to white colour. The 1950 sulphur flow is a fan-shaped body 27 m long and 14 m wide, with a thickness range from 0.1 to 0.45 m. Inspection by the field team showed there to be a 30-cm flow tube with a 5-cm thick roof. The whole was emplaced as a single cooling unit.

The nature of the field evidence on the 1950 flow indicates that the original sulphur was emplaced by fumarolic activity which accumulated between talus composed of basaltic rocks. Subsequently this was mobilized by heat generated during the 1950 basaltic eruptions (Fig. 12.9). Chemical analysis of the flow material indicated 99.9% pure sulphur with just a trace of calcium. The yellow hue implies an absence of S_3 or S_4 allotropes, since laboratory studies reveal that if the pure substance is heated above its melting point but below 420 K and allowed to cool without disturbance, it reverts to a lemon yellow colour on cooling. On the other hand, sulphur heated to above 450 K retains a brownish hue for some time afterwards, due to the continued presence of S_3 and S_4 allotropes (Gradie and Moses, 1983). The Mauna Loa flow therefore must have been produced above its melting point (389-398 K) but below 420 K.

Sulphur flows have also can be studied in island arc locations, particularly in Indonesia (Cattermole, 1990). At Kawah Mas crater, high on the slopes of Gunung Papandayan, West Java, fumarolic sulphur precipitates in a variety of colours, ranging from orange-yellow to yellowish-white. Within the fumarole field are a number of active vents in which molten dark brown sulphur forms small pools that spill over onto the crater floor to form modest sulphur flows (Plate 13). One flow active during 1990 reached a length of 4.5 metres, a width of 0.6 metres and flowed within well-defined levees. The vent orifice was lined with dark brown sulphur while, with increasing distance from the vent, the sulphur changed from orange-red to rich lemon yellow. This proved to be 99.7% pure sulphur.

Given the proven presence of sulphur compounds on Io, then secondary sulphur deposits are very likely. Indeed the whitish to yellow deposits seen frequently associated with some scarps and fissures support this contention. Primary sulphur deposits are, however, another matter, for several reasons. The assertion of Clow and Carr (1980) that sulphur alone cannot support the high caldera walls and scarps seen on Io, together with the shield-like nature of the volcanic constructs on the satellite – so typical of basaltic terrains on the Earth and Mars – provide strong circumstantial evidence that silicate volcanism must have occurred and been the major product of energy release. In support of this, the very high resurfacing rates required to explain the absence of impact craters cannot be accomplished by sulphur deposition, assuming that sulphur flows on Io are of comparable thickness to those of the Earth (10-45 cm).

Silicate magmatism would provide a mechanism for mobilizing fumarolic and plume-generated sulphur deposits which, because of the very low viscosity of molten sulphur, would spread widely but thinly over the surface. The large dimensions of caldera structures on Io imply a strong crust of at least 10 km thickness; indeed geophysical considerations indicate that it is probably more than 20 km thick (Carr, 1986). As a consequence, a significant amount of the tidal energy should be dissipated below the lithosphere and transported towards the surface as silicate melt. The level of the thermal anomalies known

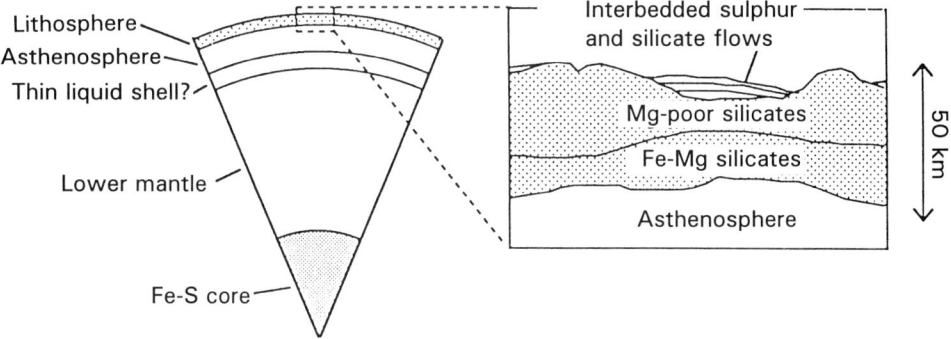

Fig. 12.10: Possible internal structure of Io. The detail of the lithosphere shown on the right-hand side supposes a simatic layer equivalent to the terrestrial asthenosphere, overlain by interleaved silicate and sulphur flows, forming the crust.

to occur and detected by the Voyager IR interferometer experiment support the contention that high rates of silicate volcanism must have occurred. Indeed the elevated emission rates observed at Ionian hot spots imply eruption rates much higher than are typical of terrestrial eruptions (4000 m^3 s^{-1} may be necessary to explain the Loki emission of 10^{13} W, (Cassen *et al.*, 1982)). It also seems eminently likely that Io is a strongly differentiated world with a basaltic lithosphere. Since magma reaches the surface of a planet largely as a result of differences in pressure exerted at the magma source by the rock column and the magma column (Eaton and Murata, 1960), it can be calculated, assuming an all-basalt lithosphere and reasonable quantities for surface density (3.0 g cm^{-3}) and coefficient of volume expansion (9.6 × 10^{-5}), that a volcano on Io cannot grow higher than 1.8 km if the lithosphere is 30 km thick, and no higher than 3.6 km if it is twice as thick (Carr, 1986). This is consistent with what is observed at the surface, the general paucity of high-relief shield volcanoes such as those of Mars, Venus and the Earth being most marked.

On the assumption that the major driving force behind volcanism is silicate magmatism, the known volatile materials such as sulphur and its compounds, as well as the sodium and potassium-rich materials believed to be present due to their detection in the Io torus, would be continually mobilized. The vaporization of these volatiles would provide the materials for the observed plumes, fumarolic deposits and flows, and for the Ionian torus. Davies and Wilson (1988) have shown that the thermal interactions between silicate sill-like intrusions and sulphur compounds mean that for a given depth of intrusion, there is a maximum amount of sulphur that can be mobilized by a sill, and also a minimum thickness of sill that will yield that much sulphur. They conclude that the processes involved in eruption of hot sulphur through a cool sulphur crust are complex. The style of the silicate volcanism on Io appears to have been somewhat reminiscent of that observed in the Snake River Plains, with extensive flat-lying plains units presumably generated by high-volume fluid basalt-like lavas, as well as low-profile shields and constructs associated with fissures. However, this has been tempered by highly explosive plume activity and the basalt (?) succession must be intercalated with substantial sulphur-rich pyroclastic and flow deposits.

12.5   INTERIOR STRUCTURE OF IO

It is clear that Io is not entirely molten inside, as once was predicted. More likely, it has differentiated to form an iron-sulphur core enclosed in a silicate mantle, a fact implicit in the rapid resurfacing rate (1 cm yr^{-1} if all the flows are silicate and 10 cm yr^{-1} if they are all sulphur – to which must be added 1 cm yr^{-1} from plume sources). While the mantle must be largely solid it seems inevitable, in the light of the style of activity seen, that the upper mantle (equivalent to Earth's asthenoshpere) must be convecting and that there may indeed be a liquid shell immediately beneath the solid lithosphere. The latter could be up to 50 km thick (Fig. 12.10).

13

Volcanism on icy satellites

Icy satellite studies relatively recently have been brought together in a series of excellent volumes published by the University of Arizona Press, Tucson. These are entitled: *Planetary Satellites* (1977) edited by Joseph Burns, *The Satellites of Jupiter* (1982), edited by David Morrison, and *Satellites* (1986), edited by Joseph Burns and Mildred Matthews. The interested reader cannot better be served in a quest for insight into these strange worlds than by perusing the pages of these three compendia. A simpler summary will be found in Ron Greeley's excellent title *Planetary Landscapes*, published by Allen and Unwin in 1985. The most recent and readily accessible book is David Rothery's excellent book *Satellites of the Outer Planets* (1992).

13.1 ICE AND ROCK LITHOSPHERES AND ASTHENOSPHERES

The outer planet moons generally have relatively low masses and therefore do not segregate nickel-iron cores; it is the silicate phases which tend to concentrate in the innermost regions. Having said that, the larger satellites may contain small amounts of Ni, Fe and S. The principal geological activity takes place in the outer layers – the lithosphere and asthenosphere – which may be made of pure water ice, water ice contaminated with compounds of N_2 (nitrogen), NH_3 (ammonia), CH_4 (methane), CO (carbon monoxide), or of water and rock mixtures. The distinction between the rigid lithosphere and ductile asthenosphere on such a world will be dictated by the mass of the moon, the make-up of the materials concerned and the thermal gradient. At some depth the ice will begin to deform and become capable of solid-state creep, at which point the transition from lithosphere to asthenosphere will have been reached. This depth will be the level at which the temperature has reached roughly 0.6 the melting temperature of the ice concerned. On Europa, for instance, this would imply a lithosphere about 30 km thick; on Callisto, however, it would be at least 500 km in thickness. Many of the smaller moons, particularly those of Saturn, could be lithospheric throughout their masses (Consolmagno and Lewis, 1978).

The effect of contaminating water ice with silicate rock fragments, is to increase the rigidity of the mixture at a given temperature. This dictates that a rock-ice lithosphere will tend to be thicker than a pure ice on, given the same conditions. The effect of adding, say, ammonia (NH_3) – which has the potential to account for up to 20% of outer planet moons – is also to make it more rigid at lithospheric depths. However, as the contaminated ice is subjected to greater temperature and pressure, it undergoes partial melting, generating a liquid of NH_3. $2H_2O$ composition when $T° = >176$ K. This leaves a residue of pure water-ice with an intergranular fluid of the ammonia hydrate melt. This would be equivalent to Earth's Low Velocity Layer.

13.2 CRYOVOLCANISM

The term volcanism implies the activity of molten materials (magmas); these eventually crystallize to form solid rocks or volcanic glasses. General usage of the term covers the activities of silicate magmas, however, with the advent of fly-by missions past the outer planets, the meaning of volcanism has become widened to include discussions of surface units deposited by sulphur (Io) and by ice and ammonia compounds (other moons of outer planets). To date, the study of *cryovolcanism* (ice-volcanism) is still in its infancy and only a very restricted literature exists. It may involve not only aqueous materials developed near the surface but also silicates from the deeper interiors of these ice and rock bodies. Interesting discussions may be found in Johnson and Nicol (1987), Lunine (1989) and Stevenson (1982) .

The physical and chemical make-up of the outer planet satellites has been reviewed by Lewis (1971). The same author has shown how ammonia hydrates are likely constituents of these icy satellites (Lewis, 1972), possibly comprising up to 20% of their total mass. The particular interest in this likelihood is that the presence of ammonia in water-ice lowers the melting point of a eutectic mixture (approx 35% by mass NH_3) to about 174 K. Such a melting temperature is low enough to be attained by radioactive heating alone inside icy satellites with radii in the range 500 to 1000 km. This provides a process whereby resurfacing could be achieved on the appropriate-sized moons of both Saturn and Uranus. In deriving equations of state for ammonia-water liquids, Croft *et al.* (1987) discuss how the density of ammonia-water eutectic liquid at its melting temperature (174.4 K) is about 0.941 g^{-1} cm^3. Since the solid density is about 0.937 g^{-1} cm^3 for a mixture of water-ice or monohydrate and about 0.963 g^{-1} cm^3 if it consists of monohydrate and dihydrate, the liquid is very nearly neutrally buoyant, and possibly negatively buoyant, with respect to the solid. Therefore ammonia-water volcanism can occur within sufficiently warm, undifferentiated satellites within the outer planet systems. They suggest that in the larger satellites, such as Titania, where temperatures would exceed the melting point within most of the interior, the surface layer of ammonia hydrates might become quite thick – perhaps tens of kilometres. Continued injection of such a thick crust by liquid of eutectic composition would become increasingly difficult due to lack of buoyancy, so much so that magmatic activity might have to become largely plutonic in the later stages of development. This would reduce the likelihood of surface flooding but would increase the likelihood of surface tectonics.

Ice in the asthenosphere of icy worlds is capable of flow at rates sufficient to maintain convection, but what about eruption of subsurface melts? The partial melting of water-ice contaminated with ammonia (NH_3. $2H_2O$) would be slightly less dense than the surrounding

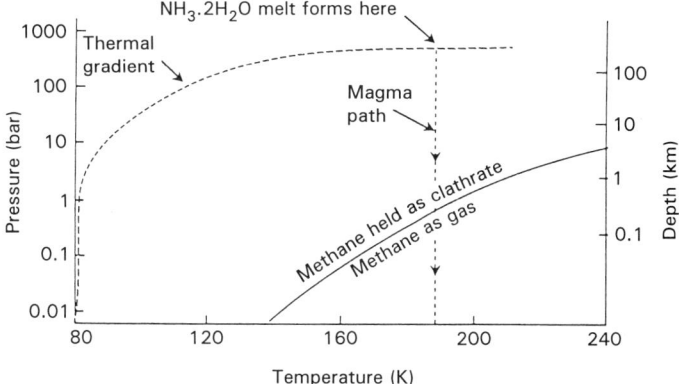

Fig. 13.1: A possible thermal gradient inside an icy satellite, showing the depth at which a $NH_3 \cdot 2H_2O$ melt could form, and that this might follow during rapid ascent without significant cooling. The solid line shows the conditions under which a methane clathrate decomposes to H_2O-ice and methane gas. (After Rothery, 1992.)

solid ice, and certainly less dense if the latter contained any rock clasts. It would therefore tend to rise towards the surface and, if a sufficiently large volume had formed, might be able to reach the surface and erupt before it froze (at 176 K). (Smaller volumes would freeze below the surface.) The viscosity and yield strength of such a melt, under the low gravity conditions found on icy moons, would allow the melt to behave rather like fluid basalt does on the Earth. Should some crystallization occur during uprise, however, then the yield strength and viscosity would become more like dacite or rhyolite. Thus there is scope for a variety of different cryovolcanic processes that potentially could produce different types of landform on ice-rock moons.

Melt might also be carried to the surface by gases mobilized within the interior of an icy moon. For instance, small amounts of ammonia in water could vaporize as the load pressure diminished towards the surface. Alternatively, the passage of an $NH_3 \cdot 2H_2O$ magma through ice containing, say, methane or some other volatile substance in clathrate form, could cause the clathrate forming the walls of the fissure to break down into H_2O-ice and gas (Fig. 13.1). Both of these processes has the potential to generate an explosive eruption akin to a terrestrial geyser.

13.3   THE JOVIAN SATELLITES

As has been shown in the previous chapter, the most volcanically active body in the entire solar system is Io, one of the inner moons of the giant planet, Jupiter. Its style of activity is related not to ice or other light elements and compounds, but to interactions between silica and sulphur stirred by huge tidal stresses. The same large tidal stresses experienced by the other Galilean moons of Jupiter ensure that they also have undergone fairly severe

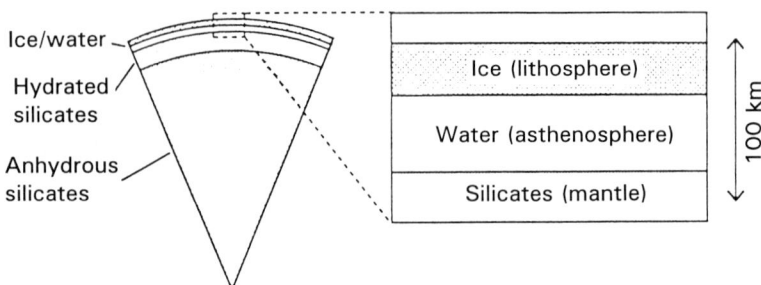

Fig. 13.2: Possible interior structure of Europa. The icy lithosphere is likely to be at least 30 km thick. (After Rothery, 1992.)

disruption, as a result of which materials from within have emerged at the surface, modifying their ancient cratered crusts. The intrusion of such icy materials comes under the general scope of volcanism.

13.3.1 Europa

Europa has a diameter of 3138 km and was the least well imaged of Jupiter's moons. While there is little to suggest that there is current geological activity in the form of volcanism, it must be noted that the surface is very flat and has a very low incidence of impact craters, far less than that of the Moon's maria, and therefore must have been continuously resurfaced. Viscous relaxation alone could not account for the paucity of craters and some form of endogenic process must be invoked. Europa's density is 3.57 g cm^{-3} – slightly less than Earth's Moon – and to account for this it is necessary to assume that 5% water is mixed in with the silicate component (Golombek and Banerdt, 1990). When tidal and radiogenic heat considerations are taken into account, it is likely that the lower part of the enveloping water layer is liquid, possibly forming a subsurface water "ocean". The water layer could be 100 km thick, underlain by a 1200-km-thick silicate mantle amd 850-km-thick sulphur-iron core (Greeley, 1987) (Fig. 13.2).

The icy surface is transected by dark curvilinear ridges which break it up into polygonal blocks. There are also regions of brownish and greyish mottled terrain with a hummocky texture and rough surface; these appear to be younger than the more reflective icy plains. Several workers have suggested that brown and grey plains units are the result of endogenic activity, the icy crust having been repeatedly fractured both by tectonic forces and by the forceful intrusion of slushy, water-rich materials from within (Lucchitta and Soderblom, 1982; Morrison, 1983). It is not inconceivable that these units represent regions underlain by warm ice that contains silicate dust, behaving like shallow intrusions, or that the material has actually been extruded onto the Europan surface. The prominent dark bands obvious in all synoptic images of the moon have been discussed by Finnerty *et al.* (1980), who proposed that they are the result of the infilling with aqueous fluids and silicates of fractures opened by global expansion (Plate 18). The intrusive bodies thus would most closely resemble volcanic dykes.

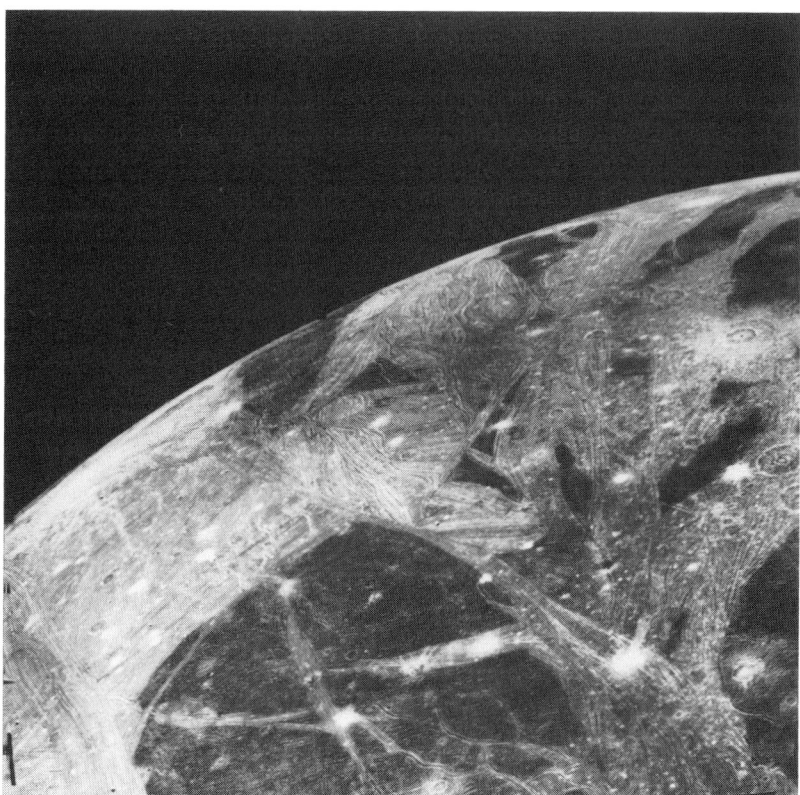

Fig. 13.3: Voyager image of Ganymede showing the two principal terrain types: (i) dark cratered terrain, and (ii) grooved, light terrain. VO1 image 358J2-1.

13.3.2   Ganymede

Ganymede is the largest satellite in the Solar System, having a diameter of 5262 km and a density of 1.93 g cm^{-3}; it must be composed of at least fifty percent water ice. According to Greeley (1987), it may have a 500-km-thick icy lithosphere above a 900-km-thick ice/silicate mantle, with a 1250-km-radius silicate core beneath. Ganymede has a crust which has been severely deformed and which some observers have seen as the closest analogue to Earth's plate structure. Mapping indicates Ganymede to consist of large patches of dark, cratered terrain, heavily furrowed and considered to be remnants of the ancient crust, and lighter, grooved terrain which accounts for about 60% of the imaged surface (Fig. 13.3). The latter generally appears to have formed at the expense of the former and its formation, in particular, is believed by many to be related to the widespread crustal extension and normal faulting which are likely to have released subsurface fluids.

The light terrain shows a wide range of impact crater densities and spectacular tectonic deformation. It also frequently shows embayment relationships which suggest it was emplaced as low viscosity volcanic flows, although restricted areas of this higher-albedo material appear to have been deposited subaerially, rather like pyroclastic deposits (Murchie and Head, 1988a). The dark materials also show embayment relationships with older materials and are observed to partially or completely bury older, furrowed materials (Allison and Clifford, 1987). They too show a wide range of crater densities, implying a range of ages for material with similar albedo. It appears, therefore, that the lower albedo units are also cryovolcanic. A recent synthesis by Murchie (1989) describes a complex history of tectonism and volcanism, suggesting that dark-mantle volcanism ended approximately 3.8×10^9 years ago and was followed by a different style of tectonism and light-mantle volcanic activity.

One explanation of the difference in albedo of the various materials is that the darker units are older, having been darkened by time-dependent processes (Shoemaker *et al.*, 1982); however, crater densities of different dark materials do not show a consistent relationship with albedo (Murchie and Head, 1988b). An alternative suggestion is that the albedo variations largely are controlled by substrate properties, the regolith being an intimate mixture of different materials upon which frost develops on a silicate-rich lag (Spencer, 1987). Various possibilities exist which could vary the albedo: thus it might be a function of differences in the proportions of silicates entrained in the ice lavas, of variations in the composition of the silicates thus entrained, or of differences in the salts or low-melting point volatiles in the lavas. After careful consideration of each of these possibilities, Murchie and Head (1988) concluded that old, low-albedo materials erupted as a relatively cool, silicate-bearing, near eutectic H_2O-NH_3 liquid and that the younger, higher-albedo materials erupted as higher-temperature, silicate-bearing, H_2O-rich liquid. The much higher ablation rate of low-pressure ammonia monohydrate compared with pure water-ice is invoked in the formation of darker, more silicate-rich lag on the older NH_3-enriched volcanic material.

13.3.3   Callisto

Callisto is the least dense of the Galilean moons and is more heavily cratered than Ganymede. It has a diameter of 4800 km and a density of 1.83 g cm^{-3}. It had been thought that Callisto's surface was a relict of the period of late heavy bombardment and that ice-volcanic deposits were absent. However, Stooke (1989) has suggested that certain hummocky materials of slightly higher albedo than their surroundings, particularly at the foot of the outward-facing scarps north of the large impact basin Valhalla, and less heavily cratered plains units in a few other locations, may be volcanic in origin. Higher-resolution imagery will be required before such claims can be substantiated, but evidently the possibility of ammonia-based volcanism cannot yet be totally ruled out on this world.

13.4   THE SATELLITES OF SATURN AND URANUS

At the present time analysis of potential volcanic complexes or plains units on these distant moons is in its infancy. Stooke (1989a) has reported a possible 30-km-diameter volcanic mound on Mimas and also suggests global resurfacing may have affected this

Fig. 13.4: Voyager-1 mosaic covering the Saturn-facing partion of the leading hemisphere of Dione. The most prominent graben trough is located at the upper left; this is called Latium Chasma. The largest impact crater is Aeneas, with a diameter of 200 km. Image P-38811.

moon. The same author (1989b) reports a possible volcanic complex on the trailing hemisphere of Mimas. Surprisingly, since it is located on a very tiny moon, a corona structure imaged on the Uranian satellite Miranda has been suggested as having a cryovolcanic origin (Croft, 1988). If such proves to be the case, then similar ripple-like structures imaged on Enceladus may have a similar origin. At this stage such interesting proposals must be considered highly speculative and future missions will certainly be required before much progress can be made in unravelling the possible volcanic history of these outer worlds. The reader who is interested in more deeply investigating the geology of these bodies is referred to the bibliography given in the introduction to this chapter.

13.4.1   Dione and Tethys
Dione has a diameter of 1118 km and a density of 1.44 g cm^{-3}. Of all of Saturn's moons it is the one which shows the greatest signs of geological activity. It has a highly reflective, cratered, icy surface crossed by streaks and troughs which appear to be graben (Fig. 13.4).

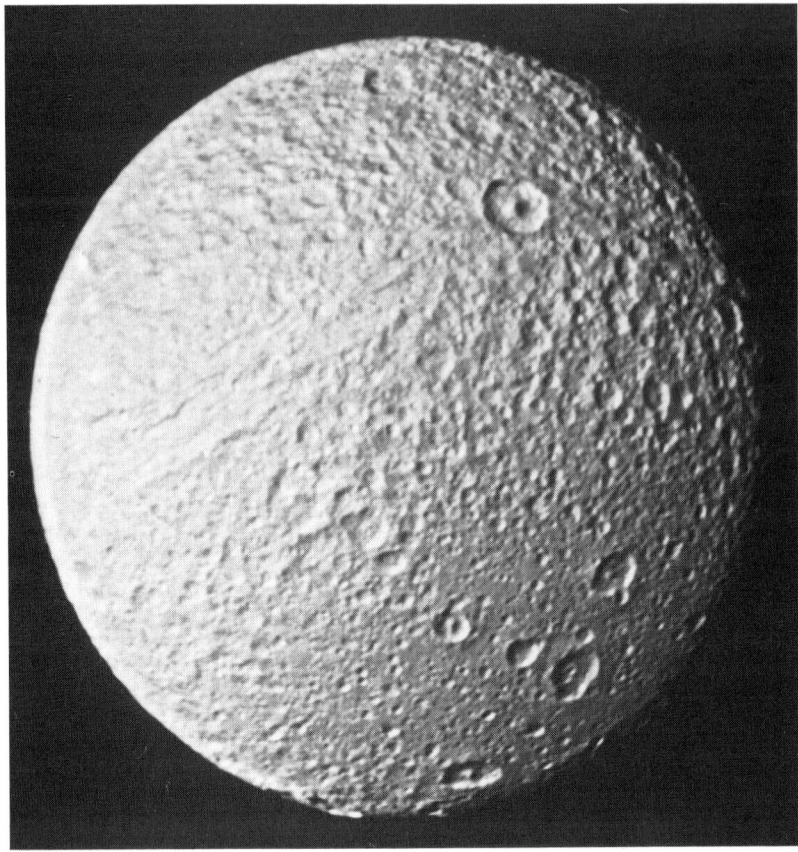

Fig. 13.5: Voyager 2 synoptic view of one hemisphere of Tethys. Note the lightly-cratered plains to the bottom right of the image. The feature, Ithaca Chasma, can be seen running diagonally across the image from the prominent impact crater Telemachus (top right).

1-km-high arcuate scarps are common and indicate that this moon has a complex tectonic history. As to volcanism, there is considerable dispute. Lightly-cratered plains regions on the moon evidently are relatively youthful and may represent areas of once heavily-cratered terrain which have been inundated by eruptions of $NH_3 . 2H_2O$ melts. The extensive nature of these plains indicates (should this explanation prove to be valid) that such melts were very fluid. An alternative explanation sees the plains deposits as having been generated by explosive eruptions of methane or ammonia ice. Both are highly speculative ideas and it will require higher resolution imagery to reach any firmer conclusions regarding this issue (Moore, 1984).

Tethys is virtually the same size as Dione, but is slightly less dense. The densely cratered

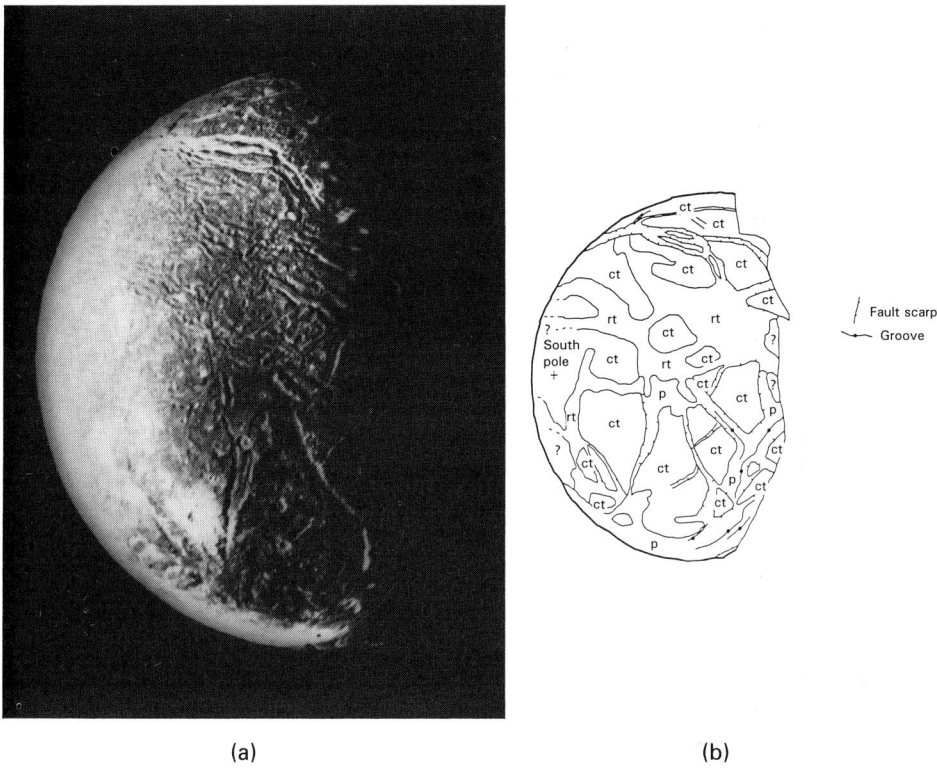

(a)   (b)

Fig. 13.6: (a) Voyager-2 mosaic of Ariel, showing prominent grooves floored by relatively smooth plains units. Image P-29520. (b) Sketch map of geological units (*ct*, cratered terrain; *rt*, ridged terrain; *p*, plains). (After Rothery, 1992.)

surface gives way to more lightly cratered plains on the trailing hemisphere (Fig. 13.5).

It is possible that these light plains units have been resurfaced by NH_3. $2H_2O$ eruptions, a suggestion based on the low rims of several impact craters located on the plains units.

13.4.2 Ariel and Titania

Saturn's 1160-km-diameter satellite, Ariel, was imaged at relatively high resolution by Voyager-2, although it imaged a mere 35% of the surface. Fig. 13.6 shows that the moon consists of three distinct geological units: cratered terrain, ridged terrain, and plains.

Prominent amongst the landforms are chasmata which appear to be graben faults, some, apparently, with transform movements. Many of these clearly have been flooded by some

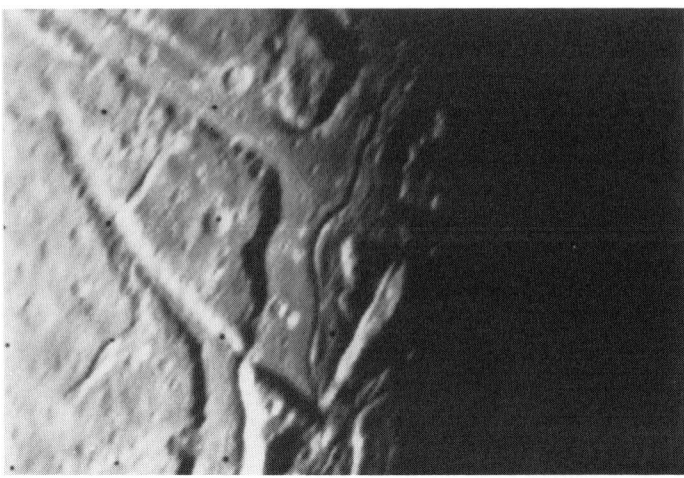

Fig. 13.7: A Voyager-2 image of a part of Brownie Chasma, showing the 3-km-wide "rille" of Sprite Vallis.

kind of cryovolcanic flow (Fig. 13.7) and some modelling work has been undertaken to establish what the nature of this material may be (Schenk, 1991).

Suffice it to say that this is somewhat inconclusive, bearing in mind the resolution of the imagery and the incomplete coverage of the Voyager flyby. However, candidates for such material include crystal-free melts of NH_3. $2H_2O$, and water-ice contaminated with ammonia, methane, carbon monoxide or nitrogen. Whatever the nature of such flows, photogeological evidence such as that presented in Fig. 13.9, strongly suggests that flow material may have moved downslope through tubes, some of which subsequently collapsed (to form medial valleys) while others remained as ridges while the surrounding flow surfaces subsided (to form medial ridges).

Titania is considerably larger than Ariel (1580 km diameter) but shares several of its terrain characteristics. Unfortunately the rather poor resolution of images obtained by Voyager precludes detailed mapping of its surface. Having said that, regions with substantially lower crater densities than the rest of the surface, may show evidence for cryovolcanic resurfacing.

13.4.3 Miranda

Voyager images of the 472-km-diameter moon, Miranda, are spectacular, showing huge arcuate scarps, swaths of strongly grooved terrain, and regions of cratered terrain. The complexity of the geology and great amplitude of fault scarps on this relatively small body, have led many workers to invoke break-up and re-accretion to explain the observed physiography. Approximately one half of the moon consists of the hummocky cratered unit, the crater rims being oddly subdued; this contrasts strongly with the other, which consists of grooved "coronae" on which crater rims are far sharper and whose bounding ridges may have altitudes exceeding 1 km (Fig. 13.8).

The coronae themselves may have begun to develop around major impacts that followed

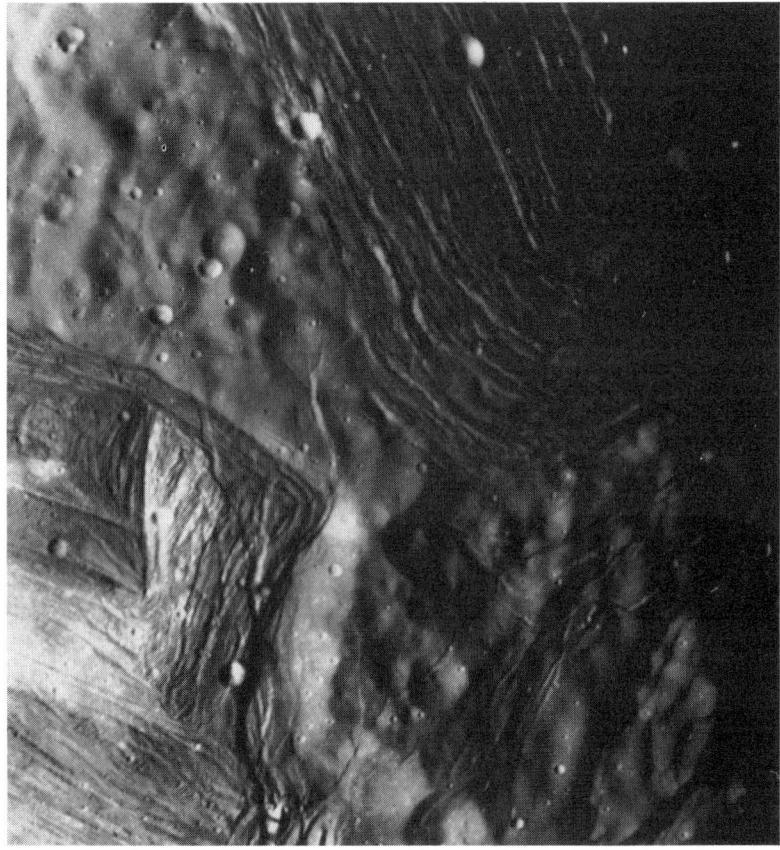

Fig. 13.8: Voyager-2 mosaic of Miranda, showing Arden Corona (left), Inverness Corona (lower centre) and Elsinore Corona (right limb).

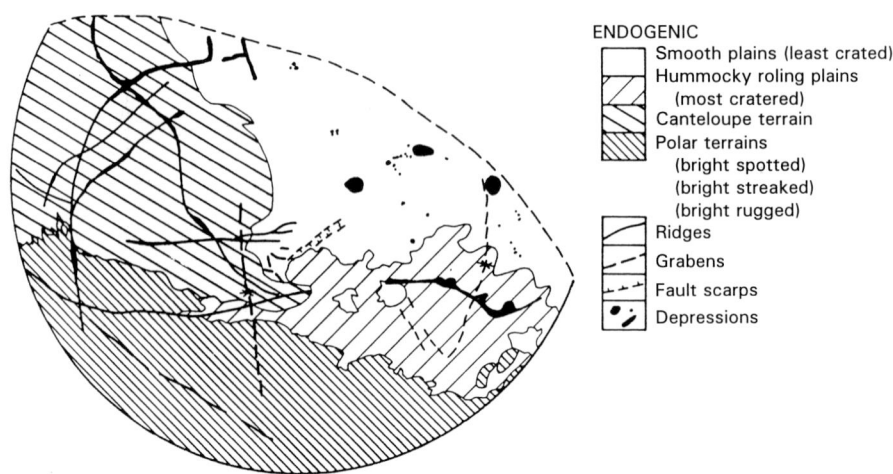

ENDOGENIC
Smooth plains (least crated)
Hummocky roling plains
    (most cratered)
Canteloupe terrain
Polar terrains
    (bright spotted)
    (bright streaked)
    (bright rugged)
Ridges
Grabens
Fault scarps
Depressions

Fig. 13.9: Physiographic sketch map of Triton. (After Smith *et al.*, 1989, fig. 31.)

re-accretion of the icy world, tectonic weaknesses controlling subsequent uprise and escape of cryovolcanic products (Janes and Melosh, 1988). The prominent bounding ridges are generally assumed to represent fracture-controlled extrusions of "warm" ice or nitrogen-water compounds (Schenk, 1991; Jankowski and Squyres, 1988).

13.5   SATELLITES OF NEPTUNE

Images transmitted by the Voyager 2 spacecraft reveal a plethora of features on the giant Neptunean moon Triton which indicate cryovolcanism has been (and possibly still is) an important geological process. One British newspaper reported Bruce Murray as saying: "My God, it looks like Hawaii!". The first images released showed Triton to be characterized by hues of pink and blue. The pinkish colours possibly come from radiation-damaged methane ice which has been spewed out by Triton's peculiar "volcanoes" (Plate 19). The blue colouration is almost certainly representative of surface ice crystals. The remaining 7 moons are small and little is known about them.

13.5.1   Triton
Triton has a diameter of 2705 km and a surface temperature of 38 ± 4 K. Its density is 2.08 g cm^{-3} which is very similar to Pluto. It must be assumed that Triton has a greater rock to ice ratio than either Ganymede or Callisto. Its size dictates that it was large enough to become differentiated into a silicate core, probably about 2000 km in diameter, and an icy mantle 350 km thick. According to Smith *et al.* (1989) water ice must be the primary

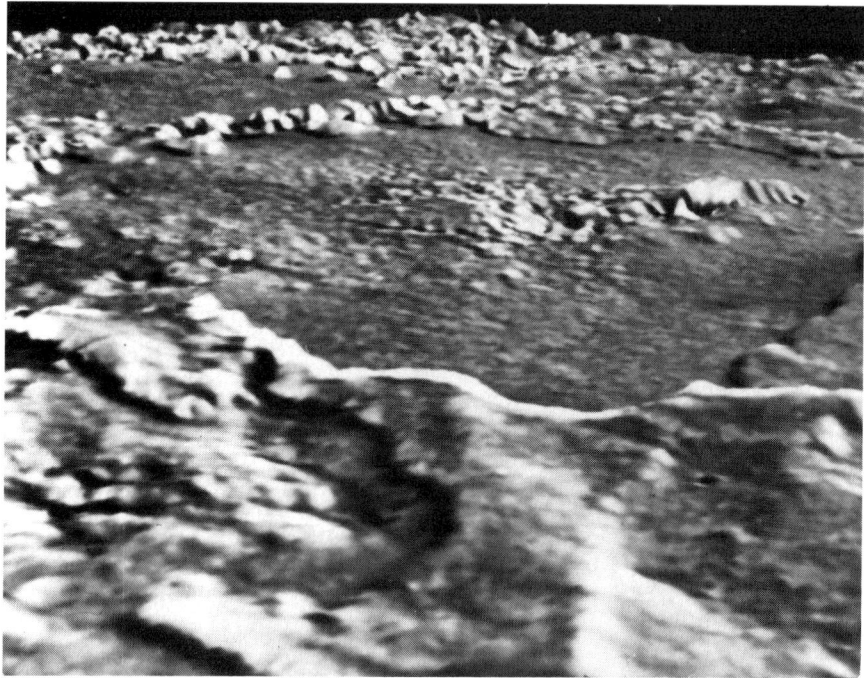

Fig. 13.10: Smooth plains, showing a major depression and overlapping layers towards the bottom of the image. The crater is 15 km in diameter. A possible volcanic channel lies east of this crater.

component of the near-surface layer, this being overlain by a veneer of nitrogen and methane ices and their derivative compounds. It is unique among the outer planet moons in having a retrograde motion, a characteristic which probably arose after impact with another body. Such an event would require that the tidal energy produced must be dissipated, so providing a major source of heat that is assumed to have led to global melting and differentiation.

Voyager-2 images revealed that albedo patterns on Triton comprise two units: (a) a brighter polar unit which blankets much of the southern hemisphere; and (b) a slightly less reflective and redder plains unit extending approximately northwards from the equatorial regions Plate 19). However, these albedo patterns do not correlate well with topography, and must represent a blanketing layer over the underlying physiographic units. The more reflective polar unit is believed due to N_2 and CH_4 frost; in the southern hemisphere concentrations of these elements are found only in tiny pockets too small to be resolved by the Voyager imaging system. Only 40% of the surface was imaged; all of this appeared to be young, heavily-cratered surfaces being absent. A simple physiographic map is shown in Fig. 13.9.

The three principal terrain types are (i) smooth plains (ii) hummocky plains, and (iii) canteloupe terrain. The smooth plains clearly overlie the other two units and on some images lobate scarps at its edges suggest emplacement as rather viscous flows, probably of crystallized ammonia-water mixtures (Fig. 13.10).

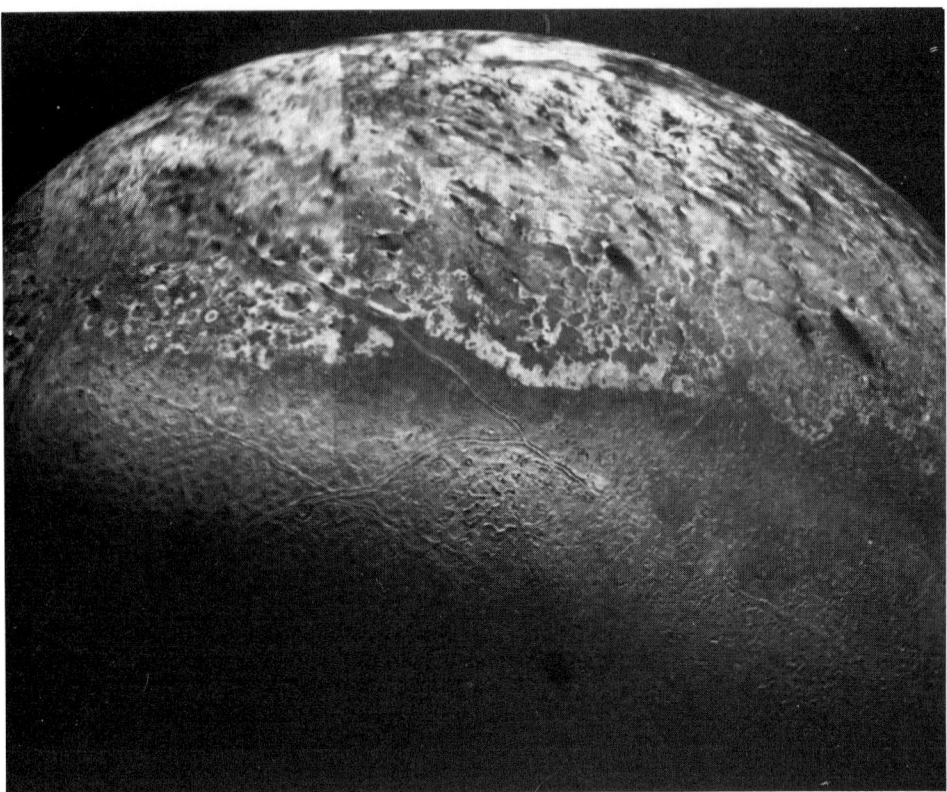

Fig. 13.11: Hummocky terrain, showing a smooth-floored depression with central rough unit and several narrow channels, probably evacuated volcanic tubes or fissures.

While most typical plains are at relatively high elevation, similar units are found at lower levels. These occupy four flat-floored depressions up to 200 km across (Fig. 13.12). The one shown above is typical, being circumscribed by steep scarps which in places are terraced. To what extent these are the result of volcanic construction or caldera collapse, or of erosion, is not clear. However, the presence of rather rugged surfaces, together with pits near the centre of the depression, suggests these to be the most recent volcanic deposits.

Hummocky terrain also appears to represent cryovolcanic material. It is within this unit that the majority of impact craters are located, their density being comparable with that of lunar maria plains. Within it are located dome-like landforms, and what appear either to be lava channels or tubes, or narrow graben. One prominent ridge-like feature has apical pits along it and may represent a flow crest with underlying tube (Fig. 13.11).

The oddest terrain type is the canteloupe unit. Consisting of a meshwork of ridges that separate near-circular depressions up to 25 km across (Fig. 13.12), it is unique in the solar system. It is not at all clear what is the nature of the depressions but what does seem

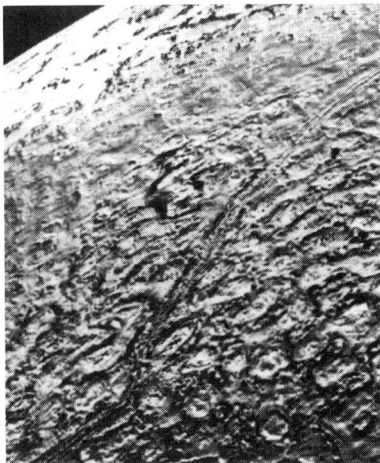

Fig. 13.12: Canteloupe terrain on Neptune's largest moon, Triton. The image is 450 km wide.

obvious is that the intervening zones have been the sites of viscous extrusions that gave rise to a net-like pattern.

While there is no positive evidence that volcanic processes currently are active in generating any of the above terrains, Voyager-2 did image certain features in the southern polar region which appear to be volcanic in nature. These were in the form of dark plumes which erupted more-or-less vertically to a height of 8 km and then were streaked out horizontally by the prevailing winds, streaking out for distances of up to 100 km. They may represent geyser-like eruptions of nitrogen compounds, set up either by solar heating, or by endogenic activity, e.g. intrusion of relatively warm icy lavas below the polar ice which vaporizes the nitrogen gas from the ice. Either way, these seem to represent the most positive evidence for current cryovolcanic activity on this fascinating moon.

Appendix 1

NOTES ON PLANETARY GEOCHEMISTRY

1. Lunar geochemistry

The chemistry and petrology o lunar rocks has been summarized in Chapter 8. It has been established that anorthositic rocks dominate the highland regions, while Fe- and Ti-bearing basalts dominate the maria. The lunar highland crust has been much brecciated by impacts but is the best studied example of a primary planetary crust in the solar system. Petrographical studies indicate it to be complex in detail, principally due to brecciation and shock, but it is geochemically rather simple.

There is a broad consensus that the highland crust was precipitated from an early magma ocean that formed during the accretion of our satellite. Three ingredients can be recognized: the first is a large volume of plagioclase feldspar cumulate rocks (ferroan anorthosite) which rose by flotation in the anhydrous primary magma ocean. A Sm-Nd closure age (based on sample 60025) of 4440 ± 20 m.y. provides a lower limit for plagioclase precipitation and crustal formation. The residual liquid (the second component) from this process (KREEP) crystallized about 4350 m.y. BP, indicating that about 90 m.y. was required for the whole to precipitate. A third component (the Mg-suite of rocks) appears to represent plutonic bodies intruded into the anorthositic crust soon after it crystallized. Since the Moon formed 4560 m.y. ago, the highland crust was in place very quickly, indicating that melting followed very closely upon the Moon's formation.

The crust, naturally, overlies the upper mantle from which the subsequent mare basalts were fractionated. The fact that in these mare rocks Eu is depleted relative to other REEs, has generally been taken to imply that Eu was depleted prior to crystallization and flotation of plagioclase feldspar. The geochemical characteristics of the Moon are shown in Table A.1.

The recent Clementine mission has allowed scientists to identify major compositional provinces on the Moon. As expected, these include the basaltic flows of the maria, young and fresh impact craters and the huge South Pole-Aitken Basin on the farside. At present this data, together with medium- and high-resolution imagery, is in the infancy of its analysis. In due course doubtless this will provide new insights into the evolution of the Moon and its geochemical characteristics.

Table A.1: The major element composition of the Moon

	Bulk Moon	Primitive Mantle	Highland Crust	Mantle following Crust extraction
SiO_2	43.4	44.4	45.0	44.3
TiO_2	0.3	0.31	0.56	0.28
Al_2O_3	6.0	6.14	24.6	4.1
FeO	10.7	10.9	6.6	11.4
MgO	32.0	32.7	6.8	35.6
CaO	4.5	4.6	15.8	3.36
Na_2O	0.09	0.09	0.45	0.05
K_2O	0.01	0.01	0.07	0.003
Cr_2O_3	0.60	0.61	0.06	0.67
MnO	0.15	0.15	-	-
(Fe,FeS)	2.3 (core)			

Data from Taylor, S. R. (1982).

2. Martian geochemistry

Analysis of fines by Viking Landers

The Viking 1 and 2 Lander probes both successfully obtained geochemical data. While it was disappointing that the gas-chromatography experiment showed a singular lack of organic molecules in the Martian regolith, the XRF hardware analysed scoop samples from both of the Lander sites. Twenty-two samples were retrieved, several being of sufficiently large volume to allow for XRF analysis. The geochemical character of these "fines" is shown in Table A.2. The samples from both locations contained abundant Si, together with significant amounts of Mg, Al, Ti, Ca and S. The relative proportions of these elements, however, were unlike any known terrestrial rock type and it is to be assumed that the samples represent admixtures of different silicate rock fractions. Since the gas chromatography experiment detected about 1 wt% H_2O, this amount of water is doubtless bound into the collected samples. These analyses are likely to represent a planetwide dust average.

Phobos results

The Russian Phobos mission obtained gamma-ray measurements of K, U and Th, plus major element data over a broad region encompassing Lunae Planum and Xanthe Terra. The data are consistent with basaltic rocks (Table A.3)

Table A.2: Chemical composition of Martian fines.

Major elements (wt%)	Viking 1	Viking 2
SiO_2	44.7	42.8
Al_2O_3	5.7	
Fe_2O_3	18.2	20.3
TiO_2	0.9	1.0
MgO	8.3	
CaO	5.6	5.0
K_2O	<0.3	<0.3
SO_3	7.7	6.5
Cl	0.7	0.6

Trace elements (p.p.m.)		
Rb	<30	<30
Sr	60±30	100±40
Y	70±30	50±30
Zr	<30	30±30

SNC meteorites: Martian crust and volatiles

Original attempts to estimate the Martian volatile inventory were based on comparisons between the Earth and Mars, mainly using ratios of noble gases; however, results of such analysis were always at odds with requirements (they were far too small). Modern evidence comes from what once was considered an unlikely source: meteorites. The groups of 8 stony meteorites known as the SNC group (Shergotty, Nahkla and Chassigny) share at least two unique characteristics: (i) they have a remarkably youthful crystallization age $(1.3 \times 10^9$ y) and (ii) they show an embedded gas content which belies their origins.

The youthful age indicates their origin in a body which remained active well beyond the time that the Moon and asteroids ceased to suffer major geological activity. Mars seems to be the only other body that fits these requirements. Confirmatory evidence that this is so is derived from the isotopic composition of the SNCs, which is remarkably close to that at the Viking Lander sites. This similarity extends to an observed enrichment in ^{15}N, a characteristic that is otherwise unique to Mars. When normalized to the abundances of such refractory elements as Si and La, SNCs are several times richer in moderately volatile elements such as Na, K and Br than those inferred for the Earth. This appears to indicate that Mars, as suspected, is a volatile-enriched planet.

Of the 10 separate lithologies found in the SNCs, 8 are cumulate rocks. The remaining two show signs of crystal fractionation. There is some indications also that the shergottites bear a resemblance to terrestrial komatiites; certainly they are enriched in pyroxenes. Since most SNCs are relatively finely crystallized and have a glassy mesostasis, evidently they solidified at fairly shallow depths. This could be taken to indicate that shallow differentiated igneous intrusions are common on Mars.

Table A.3: Phobos 2 analysis compared with Viking 1 data.

Element	Phobos 2	Viking 1
O	48 ± 5	50.1 ± 4.3
Mg	6 ± 3	5.0 ± 2.5
Al	5 ± 3	3.0 ± 0.9
Si	19 ± 4	20.9 ± 2.5
S		3.1 ± 0.5
Cl		0.7 ± 0.3
Ti	1 ± 0.5	0.51 ± 0.2
K	0.3 ± 0.1	0.25
Ca	6 ± 3	4.0 ± 0.8
Fe	9 ± 3	12.7 ± 2.0
U	$(0.5 \pm 0.1) \times 10^{-4}$	
Th	$(1.9 \pm 0.6) \times 10^{-4}$	

Igneous crustal materials

It has long been suspected that certain areas of low-albedo materials might be more closely representative of crustal rocks than the altered bright regolith and dust. Adams and McCord (1969) were the first to demonstrate that the $0.35 - 1.1$ µm spectral characteristics of dark regions observed telescopically were consistent with rocks akin to terrestrial olivine-basalt that had been oxidized in the laboratory. More recent observations and laboratory modelling have significantly improved our knowledge of these potential Martian crustal rocks.

Laboratory studies by Singer (1980 of naturally occurring oxidized or weathered basaltic rocks, as well as synthesized powders, has isolated one specific condition that can generate the particular spectral characteristics of these dark regions. This is a thin Fe^{3+} rich coating on a dark substrate. This is closely similar to what is obtained spectrally by analysing a Mauna Kea basalt coated with palagonite.

Evidence for a predominantly mafic crust on Mars comes from Earth-based spectral reflectance data, meteorite data, from the measured high density of the Martian mantle (McGetchin and Smyth, 1978), and from the Viking lander XRF experiments (Toulmin *et al.*, 1977). Extensive data regarding spectral characteristics in the region $0.35 - 1.1$ µm have been analysed by Huguenin *et al.* (1978) as representing a variety of basaltic to ultrabasic compositions, including abundant olivine in some locations, as well as pyroxenes. Subsequently, Huguenin (1987) suggested that hydrolysed olivine is present widely in the Martian dust. A reinterpretation by Singer (1980) confirmed the abundance of clinopyroxene but indicated that there was no conclusive evidence that olivine was a constituent in the low albedo materials. The dominant pyroxenes appear to have low to moderate Ca and with quite high Fe contents.

In the light of all information currently to hand, it would appear that the Martian crust is dominated by basalt-like basic rocks, many of which are relatively little altered; the relative importance of ultramafic material is still a matter of some debate. Recently it has been suggested that mafic volcanic ash may also be present (Geissler *et al.*, 1990).

Table A.4: Venera and Vega measurements of U, Th and K.

Probe	U(× 10⁻⁴ wt%)	Th(× 10⁻⁴ wt%)	K(wt%)	K/U(× 10⁴)
Venera 8	2.2 ± 0.7	6.5 ± 0.2	4.0 ± 1.2 1.82	+1.65 / -0/85
Venera 9	0.60 ± 0.16	3.65 ± 0.42	0.47 ± 0.08 0.78	+0/47 / -0.27
Venera 10	0.46 ± 0.26	0.70 ± 0.34	0.30 ± 0.16 0.65	+1.65 / -0.46
Vega 1	0.64	1.5	0.45	
Vega 2	0.68	2.0	0.40	

3. Venusian geochemistry

The major element chemistry of Venera and Vega samples has been discussed briefly in Chapter 7. The earlier Venera probes all landed in the equatorial plains and yielded results that indicated rocks of basaltic composition. Venera 13 and 14, however, sampled the higher-standing regions of Aphrodite Terra (considered by some as potential regions of lower-density "continental"-like crust); these too indicated rocks ranging in composition from ocean-floor basalt to alkali-basalt.

Abundances of U, Th and K measured at the Venera 8, 9 and 10 sites resemble terrestrial rocks rather than either lunar rocks or meteorites; this is true also of the Vega 1 and 2 analyses (Table A.4). The data for the Venera 8 site has been interpreted as more closely akin to terrestrial continental rocks; however, analytical accuracy is not terribly high, and this may be a spurious interpetation.

The data indicates basalt-like rocks and accords with the morphology of flows and landforms observed on Venus; these have terrestrial analogues in most instances. More evolved magmas or viscous basaltic magmas may also occur, giving rise to steep-sided domes and related landforms.

Appendix 2

SOURCES OF PLANETARY DATA

The main source of planetary data is NSSDC. The address is as follows:

National Space Science Data Center (NSSDC)
World Data Center A for Rockets and Satellites,
National Aeronautics and Space Administration,
Goddard Space Flight Center,
Greenbelt, Maryland 20771,
U.S.A.

There are also a number of regional imagery centres, both within and outside of the U.S.A. NSSDC will be pleased to provide details of these.

There presently are three ways to electronically access data from NSSDC. Limited data are available from selected menu options of the NSSDC Online Data and Information Service (NODIS). A larger data bank is permanently accessible via NSSDC's ANONYMOUS account for DECnet/COPY or Internet/FTP access. Both NODIS- and ANONYMOUS- accessible data are on magnetic disk. Thirdly, larger amounts of data are accessible from NSSDC Data Archive and Dissemination System (NDADS) consisting of optical disk (12" WORM) jukeboxes.

A brochure is available from NSSDC's Co-ordinated Request and User Support Office (CRUSO; 301-286-6695). This identifies and explains all NODIS options.

NSSDC also distributes much high interest data on CD-ROMs. The planetary community, more than any other, has committed more of its data to this medium than any other. With CD-ROMs, NSSDC distributes DOS-compatible and MAC-compatible software which typically retrieves and displays data. NSSDC periodically issues a paper report identifiying data currently available via CD-ROM. The most recent is *NASA Space and Earth Science Data on CD-ROM* (NSSDC/WDC-A-R&S 93-01). An online version of this is available through NODIS.

Hardcopy planetary data from various space missions is also available from NSSDC. Current catalogues are available on disk, fiche and film. Colour transparencies generally are dealt with by commercial organizations outside NASA. Again, NSSDC will provide details of these.

References

Note: In the following list of references the entry **BVSP** refers to *Basaltic Volcanism on the Terrestrial Planets*, published by the Lunar and Planetary Institute in 1981 under the abbreviated title: *Basaltic Volcanism Study Project 1976-1979,* and edited by T. R. McGetchin, R. O. Pepin and R. J. Phillips.

Adams, J. B. and McCord, T. B. (1969) Mars: interpretation of spectral reflectivity of light and dark regions. *J. Geophys. Res.,* **74**, 4851-4856.

Adams, J. B. and McCord, T. B. (1973) Vitrification darkening in the lunar highlands and identification of the Descartes material at the Apollo 16 site. *Proc. Lunar Sci. Conf. 4th*, 163-177.

Adams, J. B., Pieters, C. and McCord, T. B. (1974) Orange glass: Evidence for regional deposits of pyroclastic origin on the Moon. *Proc. Lunar Sci. Conf., 5th.*, 171- 186.

Alfvén, H. and Arrhenius, G. (1976) *Evolution of the Solar System.* NASA SP-345.

Allison, M. L. and Clifford, S. M. (1987) Ice-covered water volcanism on Ganymede. *J. Geophys. Res.,* **92**, 7865 - 76.

Anders, E. and Grevesse, N. (1989). *Geochim. Cosmochim. Acta,* **53**, 199, Table 2.

Anderson, I. (1984) The restless volcanoes of Venus. *New Scientist,* 23 February 1984, 22.

Aramaki, S. (1956) The 1783 activity of Asama Volcano, Parts I and II. *Japan J. Geol. Geogr.,* **27**, 191-229; **28**, 11-32.

Arkhani-Hamed, J. (1993), Schaber, G. G. and Strom, R. G. (1993) Constraints on the thermal evolution of Venus inferred from Magellan data. *J. Geophys. Res.,* **98**, 5309 - 5315.

Arthur, D. W. G. (1980) Precise Mars relative altitudes (abstract) In *Reports of Planetary Geology Program - 1978-1979.* NASA TM-80339.

Aubele, J. C. and Slyuta, E. N. (1990) Small domes on Venus: characteristics and origin. *Earth, Moon and Planets,* **50/51**, 493 - 532.

Aubele, J. C., Head, J. W., Slyuta, E. N. and Basilevsky, A. T. (1988) Characteristics of domes on Venus and a comparison with terrestrial cinder cones and oceanic volcanic edifices. *Lunar Planet. Sci. 19th,* 3-4.

Baer,G., Schubert, G., Bindschadler, D. L. and Stofan, E. R. (1994) Spatial and temporal relations between coronae and extensional belts, northern Lada Terra, Venus. *J. Geophys. Res.,* **99**, 8355-8370.

Baker, V. R., Komatsu, G., Parker., T. J., Gulick, J. S., Kargel, J. S., and Lewis, J. S. (1992) Channels and valleys on Venus: preliminary analysis of Magellan data. *J. Geophys. Res.*, **97**, 13421-43.

Baksi, A. K. and Watkins, N. D. (1973) Volcanic production rates: Comparison of oceanic ridges, islands and the Columbia River Plateau basalts. *Science* **180**, 493-496.

Baldwin, R. B. (1963) *The Measure of the Moon*. Chicago Univ. Press.

Baloga, S.M. (1987) Lava flows as kinematic waves. *J. Geophys. Res.*, **92**, 9271-9279.

Baloga, S. M. (1987) A review of quantitative models for lava flows on Mars. *MEVTV Workshop on Nature and Composition of Surface Units on Mars. LPI TM-88-05*, 17-19.

Baloga, S. M. and Pieri, D. C. (1985) Estimates of lava eruption rates at Alba Patera, Mars. *Rept. Planet. Geol. Geophys. Program-1984*, 245-247. NASA TM-87563.

Baloga, S. M. and Pieri, D. C. (1985) Estimates of lava eruption rates at Alba Patera, Mars. *Report Planet. Geol. Geophys. Program - 1984, NASA TM-87563*, 245-247.

Baloga, S.M. and Pieri, D. (1986) Time-dependent profiles of lava flows. *J. Geophys. Res.*, **91**, 9543-9552.

Barsukov, V. L., Basilevsky, A. T., Kuzmin, R. O., Pronin, A. A. and many others (1984) Geology of Venus as revealed by analysis of radar images obtained from Venera 15 and 16. *Geokhimiya*, **12**, 1811-1820 (in Russian)

Barsukov, V. L, Volkov, V.P. and Khodakovsky, I. L. (1982) The crust of Venus: theoretical models of chemical and mineralogical composition. *Proc. Lunar Planet. Sci. Conf. 13th., J. Geophys. Res.* **87**, A3-9.

Barsukov, V. L., Surkov, Yu. A., Dimitriev, L. and Khodakovsky, I. L. (1986) Geochemical study of Venus by landers of Vega-1 and Vega-2 probes. *Geokhimiya*, 275-289. *Lunar Planet. Sci. Conf. 13th*, Houston.

Basilevsky, A. T. and Head, J. W. (1988) The geology of Venus. *Ann. Rev. Earth Planet. Sci.*, **16**, 295-317.

Basilevsky, A. T., Pronin, A. A., Ronca, L. B. and Kryuchkov, V. P. (1986) Styles of tectonic deformation on Venus: Analysis of Venera 15 and 16 data. *Proc. Lunar Planet. Sci. Conf. 16th, J. Geophys. Res.* **91**, D399-411.

Bell, A. F. and Hawke, B. R. (1981) Lunar dark haloed impact craters: origin and implication for early mare volcanism. *J. Geophys. Res.*, **89**, 6899 - 6910.

Belton, M. J. S. et al. (1992) The Galileo solid-state imaging experiment. *Space Sci. Rev.*, **60**, 413 - 455.

Bence, A. E., Grove, T. L. and Papike, J. J. (1980) Basalts as probes of planetary interiors: constraints on the chemistry and mineralogy of their source regions. *Precambrian Res.,* **10**, 249-279.

Best, M. G. (1982) *Igneous and Metamorphic Petrology*, 44-59, Freeman, San Francisco.

Birch, F. (1965) Energetics of core formation. *J. Geophys. Res.*, **70**, 6217-21.

Birch, F. (1969) In: *The Earth's Crust and Upper Mantle*, 18ff. Monograph 13, Amer. Geophys. Union.

Blake, S., (1981) Volcanism and the dynamics of open magma chambers. *Nature,* **289**, 783-785.

Blake, S. (1984) Magma mixing and hybridisation processes at the alkalic, silicic, Torfajökull central volcano triggered by tholeiitic Veidivötn fissuring, south Iceland. *J. Volcanol. Geotherm. Res.,* **22**, 1-31.

Blasius, K. R., (1976) *Topical studies of the geology of the Tharsis region.* PhD dissertation, CALTECH.

Bowen, N. L. (1928) *The Evolution of the Igneous Rocks.* 333pp. Princeton, New Jersey. (Also Dover Publications)

Boyd, F. R. and Meyer, H. A. O. (eds.) (1979) Kimberlites, Diatremes and Diamonds: Their Geology, Petrology and Geochemistry. Amer.Geophys.Union.,*Proc. 2nd. Int. Kimberlite Conf.,1.*

Booth, B. and Self, S. (1973) Rheological features of the 1971 Mt.Etna lavas. *Phil.Trans.Roy.Soc.Lond.,Ser.A,* **274**, 99-106.

Brass, G. W. and Harrison, C. G. A. (1982) On the possibility of plate tectonics on Venus. *Icarus,* **49**, 86 - 96.

Bultitude, R. J. (1976) Eruptive history of Bagana volcano, Papua New Guinea, between 1882 and 1975. In: *Volcanism in Australasia,* R. W. Johnson (ed.). Elsevier.

Burns, J. A. (ed.) 1977 *Planetary satellites.* University of Arizona Press, Tucson.

Burns, J. A. and Matthews, M. S. (eds.) *Satellites.* University of Arizona Press, Tucson.

Cadogan, P. (1981) *The Moon - Our Sister Planet.* Cambridge, 1981.

Cameron, A. G. W. (1973) Accumulation processes in the primitive solar nebula. *Icarus,* **18**, 407-450.

Cameron, A.G.W. (1978) The primitive solar accretion and formation of the planets. In: Dermott S.F. (ed) *The Origin of the Solar System.* Wiley, New York, pp 49 - 74.

Campbell, B. A. and Campbell, D. B. (1992) Analysis of volcanic surface morphology on Venus from comparison of Arecibo, Magellan and terrstrial airborne radar imagery. *J. Geophys. Res.,* **85**, 8271 - 8281.

Campbell, D. B., Head, J. W., Harmon, J. K. and Hine, A. A. (1984) Venus: Volcanism and rift formation in Beta Regio. *Science,* **226**, 167-169.

Carlson, R. W. and Lugmair, G. W. (1979) Sm-Nd constraints on early lunar differentiation and the evolution of KREEP. *Earth Planet. Sci. Lett.,* **45**, 123-132.

Carmichael, I. S. E., Nicholls, J., Spera, F. J., Wood, B. J. and Nelson, S. A. 1977) High temperature properties of silicate liquids: applications to the equilibration and ascent of basic magmas. *Phil. Trans. Roy. Soc. Lond., Ser. A.,*

Carmichael, I. S. E., Turner,F. J. and Verhougen, J. (1974) *Igneous Petrology.* 739pp. McGraw-Hill, New York.

Carr, M.H. (1974b) The role of lava erosion in the formation of lunar rilles and Martian channels. *Icarus,* **22**, 1-23.

Carr, M. H. (1981) *The Surface of Mars.* Yale, 118-122.

Carr, M. H., (1984) In *Geology of the Terrestrial Planets.* (M. H. Carr, ed.), NASA SP-469.

Carr, M. H. (1986) Silicate volcanism on Io. *J.Geophys.Res.,* **91**, 3521-3532

Carr, M. H. and Greeley, R. (1980) *Volcanic features of Hawaii.* NASA SP-403.

Carr, M. H., Masursky, H., Strom, R. G. and Terrile, R. J. (1979) Volcanic features of Io. Nature, *280*, 729-733.

Cassen, P. M., Peale, S. J. and Reynolds, R. T. (1982) Structure and thermal evolution of the Galilean satellites. In: *The Satellites of Jupiter*, Ed D. Morrison, University of Arizona Press, Tucson.

Cattermole, P. J. (1976) The crystallisation and differentiation of a layered intrusion of hydrated alkali basalt parentage parentage at Rhiw, North Wales. *Geol. J.,* **11**, 45-70.

Cattermole, P. J. (1986) Linear volcanic features at Alba Patera, Mars: probable spatter ridges. *J.Geophys. Res.,* **91**, E159-165.

Cattermole, P.J. (1987) Sequence, rheological properties and effusion rates of volcanic flows at Alba Patera, Mars. *J. Geophys. Res.,* **92**, B553-560.

Cattermole, P. J. (1989) Volcanic flow development at Alba Patera, Mars. *Icarus.* **83**, 453 - 493.

Cattermole, P. J. (1990) Unlikely twins - Io and Java. *Astronomy Now*, **4**, No.12, 21 - 27.

Chapman, C. R. and Jones, K. L. (1977) Cratering and obliteration history of Mars. *Ann. Rev. Earth Planet. Sci.* **5**, 515-540.

Chester, D.K., Duncan, A.M., Guest, J.E. and Kilburn, C.R.J. (1985) *Mount Etna: The anatomy of a volcano.* Chapman and Hall, London.

Cigolini, C., Borgia, A. and Casertano, L. (1984) Intercrater activity, aa-block lava, viscosity and flow dynamics: Arenal volcano, Costa Rica. *J. Volcanol. Geotherm. Res.,* **29**, 155-176.

Clague, D. A. and Dalyrimple, G. B. (1987) *Volcanism in Hawaii*, USGS Prof. Paper 1350, 5 - 54

Clark, S. P. Turekian, K. and Grossman, L. (1972) Model for early history of the Earth. In Robertson, E. C. (ed.) *The nature of the solid Earth.* McGraw-Hill, New York, 3-18.

Clifford, S. M., Greeley, R. and Haberle, R. M. (1988) Evolution of Climate and Atmosphere. *EOS,* **69**.

Clow, G. D. and Carr, M. H. (1980) Stability of Sulfur slopes on Io. *Icarus,* **44**, 729-733.

Colton, W. E., Howard, K. A. and Moore, H. J. (1972) Mare ridges and arches in southern Oceanus Procellarum. *Apollo 16 Preliminary Science Report*, NASA SP-315, 29.90-29.93.

Colwell, R. N. (ed.) (1983) *Manual of remote sensing*, vols 1 and 2, 2nd ed. American Society of Photogrammetry, Falls Church.

Comer, R. P., Solomon, S. C. and Head, J. W. (1985) Mars: Thickness of the lithosphere from the tectonic response to volcanic loads. *Rev. Geophys. Space Phys.,* **23**, 61-92.

Consolmagno, G. J. and Lewis, J. S. (1978) The evolution of icy satellite interiors and surfaces. *Icarus*, **34**, 280 - 293.

Cooke, R.S.J., McKee, C.O., Dent, V.F. and Wallace, D.A. (1976) Striking sequence of volcanic eruptions in the Bismarck volcanic arc, Papua New Guinea in 1972-74. In: *Volcanism in Australasia.* R.W.Johnson (ed.) Elsevier.

Cotton, C. A. (1952) *Volcanoes as landscape forms.* Whitcomb and Tombs (reprinted 1969, New York: Hafner)

Cox, K. G., Bell, J. D. and Pankhurst, R. J. (1979) *The Interpretation of the Igneous Rocks*, 332-395. Allen and Unwin, London.

Crisp, J. and Baloga, S. M. (1989) A model for lava flows with two thermal components. Submitted to J. Geophys. Res.

Croft, S. K. (1988) Miranda's Inverness Corona interpreted as a pyroclastic complex. *Lunar Planet. Sci. 19th*, 225-226.

Crumpler, L. S. and Aubele, J. C. (1978) Structural evolution of Arsia Mons, Pavonis Mons and Ascraeus Mons, Tharsis region of Mars. *Icarus, **34**, 496-511.

Curtis, C. D. (1964) Applications of crystal field theory to the inclusion of trace transition elements during fractional crystallisation. *Geochim. Cosmochim.Acta,* **28**, 389-402.

Daly, R. A. (1933) *Igneous Rocks and the Depths of the Earth*. Chapters xi and xii. McGraw-Hill, N. York.

Dawson, J. B. (1971) Advances in kimberlite geology. *Earth Sci. Rev.,* **7**, 187-214.

Dawson, J. B. (1989) Personal communication.

van Diggelen, J. (1951) A photographic investigation of the slopes and heights of the ranges of hills in the maria of the Moon. *Bull. Astron. Inst. Netherlands,* **11**, 283-289.

DeHon, R. A. (1974) Thickness of mare material in the Tranquillitatis and Nectaris basins. *Proc. Lunar Sci. Conf. 5th*, 53-59.

DeHon, R. A. (1979) Thickness of the western mare basalts. *Proc. Lunar Planet. Sci. Conf. 10th*, 2935-2955.

DeHon, R.A. and Waskom, J.D. (1976) Geologic structure of the eastern mare basins. *Proc. Lunar Sci. Conf. 7th*, 2729-2746.

Delano, J. W. (1980) Apollo 15 red glass: chemistry and liquidus phase relations. *Lunar and Planetary Science XI*, 210-212.

Delano, J. W. and Taylor, S. R. (1980) Composition and structure of the deep lunar interior. *Lunar and Planetary Science, XI*, 225-227.

Detrick, R. S. and Crough, S. T. (1978) Island subsidence, hot spots, and lithospheric thinning. *J. Geophys. Res.,* **83**, 1236-1244.

Dewey, J. F. (1975) Finite plate implications; some implications for the evolution of rock masses at plate margins. *Amer. J. Sci.* **275A**, 260-284.

Dowty, E., Keil, K., Prinz, M. and Takahashi, H. (1976) Meteorite-free Apollo 15 crystalline KREEP. *Proc. Lunar Sci. Conf. 7th,* 1833-1844.

Dutton, C. E. (1884) Hawaiian volcanoes. *USGS Ann. Rept. 4th*, 75-219.

Eaton, J. P. and Murata, K. J. (1960) How volcanoes grow. *Science,* **132**, 925-938

Ehlers, E. G. and Blatt, H. (1980) *Petrology*. 732pp. Freeman, San Francisco.

Elder, M. (1976) *The bowels of the Earth*. London: Oxford.

Elsasser, W. M. (1963) Early history of the Earth. In: Geiss, J. and Goldberg, E. (eds.) *Earth Science and Meteoritics*. North Holland, Amsterdam, 1-30.

Elston, D. P. and Willingham, C. R. (1969) *Five-day mission plan to investigate the geology of the Marius Hills region of the Moon*. Astrogeology 14, USGS open-file report, 55p.

Fanale, F. P., Brown, R. H., Cruikshank, D. P. and Clark, R. N. (1979) Significance of absorption features in Io's IR reflectance spectrum. Nature, *280*, 761-763.

Faure, G. (1977) *Principles of Isotope Geology*. 464pp. Wiley, New York.

Faure, G. and Powell, J. L. (1972) *Strontium Isotope Geology*. 464pp. Springer-Verlag, Berlin.

Fedotov, S. A. (1975) Mechanism of magma ascent and deep feeding channels of island arc volcanoes. *Bull. Volc.,* **39**, 1-14.

Fedotov, S. A. (1977) Mechanism of deep-seated magmatic activity below island arc volcanoes and similar structures. *Int. Geol. Rev.,* **6**, 671-680.

Fedotov, S. A. (1978) Ascent of basic magma in the crust and the mechanism of basaltic fissure eruptions. *Int. Geol. Rev.,* **20**, 33-48.

Fielder, G. (1961) *The structure of the Moon's surface*. Pergamon, London.

Fielder, G. (1965) *Lunar Geology*. Lutterworth, London.

Fielder, G. and Wilson, L. (1975) *Volcanoes of the Earth, Moon and Mars*. Elek, London.

Fink, J.H. (1980) Surface folding and viscosity of rhyolite flows. *Geology,* **8**, 250-254.

Fink, J.H. and Zimbelman, J.R. (1985) Field measurements of 1983-4 lava flows at Kilauea and Mauna Loa volcanoes. *Lunar Planet. Sci. 16th.,* 238-239.

Fink, J. H. and Zimbelman, J. R. (1986) Rheology of the 1983 Royal Garden basalt flows, Kilauea volcano, Hawaii. *Bull. Volcanol,.* **48**, 87-96.

Fink, J. H. and Zimbelman, J. R. (1988) *Abstr. Lunar Planet. Sci. Conf.,* 327 - 328.

Fink, J. H., Malin, M. C., D'Alli, R. E. and Greeley, R. (1981) Rheological properties of mudflows associated with the spring 1980 eruptions of Mount St.Helens volcano, Washington. *Geophys. Res. Lett.* **8**, 43-46.

Fink, J. H., Park, S. O. and Greeley, R. (1983) Cooling and deformation of sulfur flows. *Icarus,* **56**, 38-50.

Finnerty, A. A., Ransford, G. A., Pieri, D. C. and Collerson, K. D. (1980) Is Europa surface cracking due to thermal evolution? *Nature,* **289**, 24-27.

Florensky, C. P. and many authors (1977) The surface of Venus as revealed by Soviet Venera 9 and 10. *Bull. Geol. Soc. Amer.* **88**, 1537-1545.

Florensky, C. P. and many authors (1983) Venera 13 and 14: sedimentary rocks on Venus? *Science* **221**, 57-59.

Fornari, D. J. and Campbell, J. F. (1987) Submarine topography around the Hawaiian Islands. In *Volcanism in Hawaii*, USGS Prof. Paper 1350, 109-124.

Foxworthy, B. L. and Hill, M. (1982) *Volcanic eruptions of 1980 at Mount St.Helens. The first 100 days*. Geol. Surv. Prof. Paper 1249.

Francis, P. and Wood, C. A. (1982) Absence of silicic volcanism on Mars: implications for crustal composition and volatile abundances. *J. Geophys., Res.,* **87**, B9881-9889.

Frey, H. (1979) Thaumasia: a fossilized early-forming Tharsis uplift. *J. Geophys. Res.,* **84**, 1009-1023.

Fyfe, W. S. (1973) The generation of batholiths. *Tectonophysics,* **17**, 273-283.

Gaddis, L. R. and Pieters, C. M. (1985) Remote sensing of lunar pyroclastic mantling deposits. *Icarus,* **61**, 461-489.

Gaddis, L. R., Adams, J. B., Hawke, B. R., Head, J. W., McCord, T. B., Pieters, C. M. and Zisk, S. H. (1981) Characterization and distribution of pyroclastic units in the Rima Bode region of the Moon. *Lunar Planet. Sci., XII*, 321-323.

Garvin, J. B., Head, J. W. and Wilson, L. (1982) Magma vesiculation and pyroclastic volcanism on Venus. *Icarus,* **52**, 365-372.

Garvin, J. B., Head, J. W., Zuber, M. T. and Helfenstein, P. (1984) Venus: the nature of the surface from Venera panoramas. *J. Geophys. Res.* **89**, 3381-3399.

Garzanti, E. (1993) Himalayan ironstones, "superplumes" and the breakup of Gondwana. *Geology,* **21**, 105 - 108.

Gault, D. E. and several authors (1968) *Surveyor 7 mission report,* chapter 9. JPL Techincal Report 32-1264.

Geikie, A. (1903) *Textbook of Geology.* MacMillan, London.

Ghose, M. C. (1976) Composition and origin of Deccan basalts. *Lithos* **9**, 65-73.

Gifford, A. W. and El-Baz, F. (1978)) Thickness of mare flow fronts. *Lunar Planet. Sci. IX*, 382-384.

Goettel, K. A. (1980) Density of the mantle of Mars. *Lunar Planet. Sci. 11th.*, 333-335.

Goins, N. R., Dainty, A. and Toksöz, M. N. (1977) The deep seismic structure of the moon. *Proc. Lunar Planet.Sci. Conf.,8th*, 471-486.

Golombek, M. P. and Banerdt, W. B. (1990) Constraints on the subsurface structure of Europa. *Icarus*, **83**, 441 - 452.

Golombek, M. P. and McGill, G. E. (1981) Rio Grande Rift: Active or passive? *Conference on Planetary Rifting,* 99-102, LPI Contrib. No. 457, Houston.

Gradie, J. and Moses, J. (1983) Spectral reflectance of unquenched sulfur. *Lunar Planet. Sci. Conf. 14th.,* 255-256.

Greeley, R. (1971) Lunar Hadley Rille: consideration of its origin. *Science,* **172**, 722-725.

Greeley, R. (1971) Observations of actively forming lava tubes and associated structures, Hawaii. *Mod. Geol.,* **2**, 207-233.

Greeley, R. (1976) Modes of emplacement of basalt terrains and an analysis of mare volcanism in the Orientale Basin. *Proc. Lunar Sci. Conf., 7th*, 2747 -2759.

Greeley, R. (1982) The Snake River Plain, Idaho: Representative of a new category of volcanism. *J. Geophys. Res.* **87**, 2705-2712.

Greeley, R. (1985, 1987) *Planetary Landscapes.* Allen and Unwin, London.

Greeley, R. (1976) Modes of emplacement of basalt terrains and an analysis of mare volcanism in the Orientale Basin. *Proc. Lunar Sci. Conf. 7th*, 2747-2759.

Greeley, R. (1994). *Planetary Landscapes.* 2nd edition. Chapman and Hall, London.

Greeley, R. and Crown, D. A. (1989) Volcanic geology of Tyrrhena Patera, Mars. Unpublished manuscript.

Greeley, R. and Guest, J.E. (1987) *Geologic Map of the eastern equatorial hemisphere of Mars.* USGS Map I-1802B.

Greeley, R. and Spudis, P. D. (1978) Mare volcanism in the Herigonius region of the Moon. *Proc. Lunar Planet. Sci. Conf. 9th*, 3333-3349.

Greeley, R. and Crown, D. A. (1989) Volcanic geology of Tyrrhena Patera, Mars. Volcanic geology of Tyrrhena Patera, Mars. *J. Geophys. Res.,* **95**, 7133 - 7149.

Greeley, R. and Fink, J. H. (1984) Sulphur volcanoes on Io? *Astronomy Express* **1**, 25-31.

Greeley, R. and King, J. S. (eds.) (1977) *Volcanism on the Eastern Snake River Plain, Idaho.* NASA CR-154621.

Greeley, R. and Schneid, B. D. (1991) Magma generation on Mars: Amounts, rates, and comparisons with Earth, Moon and Venus. *Science*, **254**, 996 - 998.

Greeley, R. and Spudis, P. D. (1981) Volcanism on Mars. *Rev. Geophys. Space Phys.*, **19**, 30-41.

Greeley, R., Kadel , S. D., Williams, D. A., Gaddis, L. R., Head, J. W., McEwen, A. S., Murchie, S. L., Nagel, E., Neukum, G., Pieters, C. M., Sunshine, J. M., Wagner, R. and Belton, M. J. S. (1993) Galileo imaging observations of lunar maria and related deposits. *J. Geophys. Res.*, **98**, 17183 - 17205.

Greeley, R., Spudis, P. D. and Guest, J. E. (1988) *Geologic Map of the Ra Patera area of Io*. USGS Map I-1949.

Greeley, R., Storm, D. and Wilbur, C. (1976) Frequency distribution of lava tubes and channels on Mauna Loa Volcano, Hawaii. *Geol. Soc. Amer. Bull. Abstr.*, **8**, 192.

Greeley, R., Theilig, E. and Christensen, P. (1984) The Mauna Loa sulfur flow as an analog to secondary sulfur flows(?) *Icarus,* **60**, 189-199.

Greeley, R., Wilbur, C. and Storm, D. (1976) Frequency distribution of lava tubes and channels on Mauna Loa volcano, Hawaii. *Geol. Soc. Amer. Abstr.*, **6**, 892.

Green, D. H. and Ringwood, A. E. (1970) Mineralogy of peridotitic compositions under upper Mantle conditions. *Phys. Earth Planet. Int.*, **8**, 359-371.

Green, J. and Short, N. M. (1971) *Volcanic landforms and surface features: a photographic atlas and glossary*. New York, Springer.

Gregg, T. K. P. and Greeley, R. (1993) Formation of Venusian Canali: Considerations of Lava Types and their Thermal Behaviours. *J. Geophys. Res.*, **98**, 10 873 - 10 882.

Grimm, R. E. and Phillips, R. J. (1994) Gravity anomalies, compensation mechanisms, and the dynamics of western Ishtar Terra, Venus. *J. Geophys. Res.*, **97**, 16035-16054.

Gulick, V. C. and Baker, V. R. (1990) Origin and evolution of valleys on Martian volcanoes. *J. Geophys. Res.*, **95**, 14,325 - 14,344.

Grossman, L. (1972) Condensation in the primitive solar nebula. *Geochim. Cosmochim. Acta*, **36**, 597-619.

Grossman, L. and Larimer, J. , (1974) Early chemical history of the solar system. *Rev. Geophys. Space Phys.*, **12**, 71-101.

Grout, F. F. (1945) Scale models of structures related to batholiths. *Amer. Journ. Sci.*,**243A**, 260-284.

Guest, J. E. and Greeley, R. (1977) *Geology on the Moon*. Wykeham, London and Basingstoke.

Guest, J. E. and Murray, J. B. (1976) Volcanic features of the nearside equatorial lunar maria. *J. Geol. Soc. Lond.*, **132**, 251-258.

Hanel, R. and several authors (1979) Infrared observations of the Jovian System from Voyager 1. *Science,* **204**, 972-976.

Hapke, B. , Danielson, G.E., Klassen, K. and Wilson, L. (1975) Photometric observations of Mercury from Mariner 10. *J. Geophys. Res.* **80**, 2431.

Harker, A. (1904) The Tertiary Igneous Rocks of Skye. *Mem. Geol. Surv. U. K.*, Chapters xi, xii and xiii.

Harris, P. G. (1974) Volcanic liquids, their origin and nature. *Science Progress,* **61**, 515-533.

Harris, P. H. (1957) Zone refining and the origin of potassic basalts. *Geochim. Cosmochim. Acta,* **12**, 195-208.

Harris, S. A. (1977) The aureole of Olympus Mons, Mars. *J. Geophys. Res., 83, 3099-3107.*

Harrison, C. G. A. and Rooth, C. (1976) The dynamics of flowing lavas. In *Volcanoes and Tectosphere.* Ed H. Aoki and S. Iizika, 103-113. Univ. Tokyo, Tokyo.

Hartmann, W. K. (1972) *Moons and Planets.* Wadsworth, Belmont, California.

Hartmann, W. K. (1973) Martian cratering.4. Mariner 9 initial analysis of cratering chronology. *J. Geophys. Res.* **78**, 4096-4116.

Hartmann, W. K. (1978) Planet formation:mechanism of early growth. *Icarus,* **33**, 50-61.

Hartmann, W.K. (1983) *Moons and Planets.* 2nd ed. Wadsworth, Belmont, 509 pp.

Hartmann, W. K. and many authors (1981) Chronology of planetary volcanism by comparative studies of planetary cratering. BVSP, chapter 8

Hartmann, W. K. and 12 authors, (1981) Chronology of planetary volcanism by comparative studies of planetary cratering. BVSP, 1050 ff. Pergamon, New York.

Hawke, B. R. and Bell, J. F. (1981) Remote sensing studies of lunar dark halo craters: Preliminary results and implications for early volcanism. *Proc. Lunar Planet. Sci. Conf. 12th,* 655 - 678.

Hawke, B. R. and Head, J. W. (1978) Lunar KREEP volcanism: Geologic evidence for history and mode of emplacement. *Proc. Lunar Planet. Sci. Conf. 9th,* 3285-3309.

Hawke, B. R., MacLaskey, R. D., McCord, T. B., Adams, J. B., Head, J. W., Pieters, C. M. and Zisk, S. H. (1979) Multispectral mapping of the Apollo 15-Apennine region: The identification and distribution of regional pyroclastics. *Proc. Lunar Planet. Sci. Conf. 10th,* 2919-2934.

Head, J. W. (1974) Lunar dark-mantle deposits: possible clues to early mare deposits. *Proc. Lunar Sci. Conf. 5th,* 207-222.

Head, J. W. (1976) Lunar volcanism in space and time. *Rev. Geophys. Space Phys.,* **14**, 265-300.

Head, J. W. (1981) Lava flooding of ancient planetary crusts: Geometry, thickness and volumes of flooded lunar impact basins. *The Moon and Planets,* **26**, 61-88.

Head, J. W. (1990) Processes of crustal formation and evolution on Venus: an analysis of topography, hypsometry and crustal thickness variations. *Earth, Moon and Planets,* **50/51**, 25-55.

Head, J. W. and Gifford, A. (1980) Lunar mare domes: Classification and modes of origin. *Moon and Planets,* **22**, 235-258.

Head, J. W. and McCord, T. B. (1978) Imbrian-age highland volcanism on the Moon: the Gruithuisen and Mairan domes. *Science,* **199**, 1433-1436.

Head, J. W. and Wilson, L. (1979) Alphonsus-style dark-halo craters: Morphology, morphometry and eruption conditions. *Proc. Lunar Planet. Sci. Conf. 10th,* 2861-2897.

Head, J.W. and Wilson, L. (1981) Lunar sinuous rille formation by thermal erosion: eruption conditions, rates and durations. *Lun. Planet. Sci. 12th,* 427-429.

Head, J. W. and Wilson, L. (1981) Ascent and eruption of basaltic magma on the Earth and Moon. *J. Geophys. Res.* **86**, 2971-3001.

Head, J. W. and Wilson, L. (1983) A comparison of volcanic eruption processes on Earth, Moon, Mars, Io and Venus. *Nature,* **302,** 663-669.

Head, J. W. and Wilson, L. (1986) Volcanic processes and landforms on Venus: theory, prediction and observation. *J. Geophys. Res.,* **91,** B9407-9446.

Head, J. W. and Wilson, L. (1992) Magma reservoirs and neutral buoyancy zones on Venus: implications for the formation and evolution of volcanic landforms. *J. Geophys. Res.,* **97,** 3877 - 3903.

Head, J. W., Bryan, W. B., Greeley, R., Guest, J. E., Schultz, P. H., Sparks, R. S. J., Walker, G. P. L., Whitford-Stark, J. L., Wood, C. A. and Carr, M. H. (1981) Distribution and morphology of basalt deposits on planets. In: *Basaltic Volcanism on the Terrestrial Planets.* Pergamon, N. York.

Head, J. W., Campbell, B. B., Elachi C., Guest, J. E., McKenzie, D., Saunder, R. S., Schaber, G. G., Schubert, G. (1991) Venus volcanism: initial analysis from Magellan data. *Science,* **252,** 276 - 288.

Head, J. W., Crumpler, L. S. and Aubele, J. C. (1992) Venus volcanism: classification of volcanic features and structures, associations, and global distribution from Magellan data. *J. Geophys. Res.,* **97,** 13153-13197.

Head, J. W., Murchie, S., Mustard, J. F., Pieters, C. M., Neukum, G., McEwen, E., Greeley, R., Nagel, E. and Belton, M. J. S. (1993) Lunar Impact Basins: New data for the western limb and far side from the first Galileo flyby. *J. Geophys. Res.,* **98,** 17149 - 17181.

Head, J. W., Peterfreund, A. R., Garvin, J. B. and Zisk, S. H. (1985) Surface characteristics of Venus derived from Pioneer-Venus altimetry, roughness and reflectivity measurements. J. *Geophys. Res.* **90,** 6873-6885.

Head, J. W. and several authors (1978) *Mare Crisium: The View from Luna 24.* Pergamon, New York, 43-74.

Heiken, G. H., McKay, D. S. and Brown, R. W. (1974) Lunar deposits of possible pyroclastic origin. *Geochim. Cosmochim. Acta,* **38,** 1703-1718.

Henderson, P. (1982) *Inorganic Geochemistry.* Pergamon, Oxford.

Herrick, D. L. and Parmentier, E. M. (1994) Episodic large-scale overturn of two-layer mantles in terrestrial planets. *J. Geophys. Res.,* **99,** 13,153 - 13,197.

Hess, P. C. and Head, J. W. (1990) Derivation of primary magmas and melting of crustal materials on Venus: some preliminary petrogenetic considerations. *J. Geophys. Res.,* **97,** 13153 - 97.

Higgins, G. H. and Kennedy, G. C. (1971) The adiabatic gradient and the melting point gradient in the core of the Earth. *J. Geophys. Res.,* **76,** 1870-1878.

Hildreth, W. (1979) The Bishop Tuff: evidence for the origin of compositional zoning in silicate magma chambers. *Geol. Soc. Am. Sp. Paper 180,* 43-75.

Hill, D. P. and Zucca, J. J. (1987) Geophysical constraints on the structure of Kilauea and Mauna Loa volcanoes and some implications for seismomagmatic processes. In: *Volcanism in Hawaii,* USGS Prof. Paper 1350, 903-917.

Hodges, C. A. (1973) Mare ridges and lava lakes. *Apollo 17 Preliminary science Report,* NASA SP-330, 31.12-31.21.

Hodges, C. A. and Moore, H. J. (1979) The subglacial birth of Olympus Mons and its aureole. J.Geophys. Res., *84,* 8061-8074.

Hoernle, K. et al. (1995) *Nature*, **374**, 34.

Hofmann, A. W. and Hart, S. R. (1975) An assessment of local and regional isotopic equilibrium in a partially molten mantle. *Yb. Carnegie Inst. Wash.*, **74**, 195-210.

Holcomb, R. T. (1987) Eruptive history and long-term behaviour of Kilauea Volcano. In: *Volcanism in Hawaii*, USGS Prof. Paper 1350, 261-350.

Holweger, H. (1988) in *The Impact of Very High S/N Spectroscopy on Stellar Physics* (ed G. Cayrel de Strobel and M. Spite), p.411, International Astronomical Union.

Housley, R. M. (1978) Modelling lunar eruptions. *Proc., Lunar Planet. Sci. Conf. 9th.*, 1473-1484. by centrifuged models. Academic Press, London and N. York.

Howard, K. A. and Muehlberger, W. R. (1973) Lunar thrust faults in the Taurus-Littrow region. *NASA SP-330*, 31-22 to 31-25.

Howard, K. A., Carr, M. H. and Muehlberger, W. R. (1973) Basalt stratigraphy of southern Mare Serenitatis. *Apollo 17 Prelim. Sci. Rept.*, NASA SP-330.

Hoyle, F. (1946) On the condensation of the planets. *Monthly Notices Roy. Astron. Soc.* ,**106**, 406-422.

Huang, S. S. (1973) Extrasolar planetary systems. *Icarus,* **18**, 339-376.

Hulme, G. (1973) Turbulent lava flow and the formation of lunar sinuous rilles. *Mod. Geol.,* **4**, 107-117.

Hulme, G. (1974) The interpretation of lava flow morphology. *Geophys. J. Roy. Astron. Soc.,* **39**, 361-383.

Hulme, G. (1976) The determination of the rheological properties and effusion rate of an Olympus Mons lava. *Icarus,* **27**, 207-213.

Hulme, G. and Fielder, G. (1977) Effusion rates and rheology of lunar lavas. *Phil. Trans. Roy. Soc. Lond., Ser A,* **285**, 227-234.

Ip, W. -H. (1974) Planetary accretion in jet streams. *Astrophys. Space Sci.,* **31**, 57-71.

Irvine, T. N. (1979) Rocks whose composition is determined by crystal accumulation and sorting. In H. S. Yoder Jnr, (ed): *The Evolution of the Igneous Rocks: 50th Anniversary Perspective* 244-306.

Irving, A. J. (1977) Chemical variation and fractionation of KREEP basalt magma. *Proc. Lunar Sci. Conf. 8th*, 2443-2448.

Ito, K. and Yamada, H. (1982) Stability relation of silicate spinels, ilmenites and perovskites. *Adv. Earth Planet. Sci.,* **12**, 405-419.

Jackson, E. D., Silver, E. A. and Dalyrimple, G. B. (1972) Hawaiian-Emperor Chain and its relation to Cenozoic circumPacific tectonics. *Bull. Geol. Soc. Amer.,* **83**, 601-618.

Jaeger, J. C. (1968) Cooling and solidification of igneous rocks. In: *Basalts.The Poldervaart treatise on rocks of basaltic composition, vol. 2*, 503-536 (ed. H. H. Hess and A. Poldervaart), Interscience, New York.

Jakobsson, S. P. (1972) Chemistry and distribution pattern of recent basaltic rocks in Iceland. *Lithos* **5**, 365-386.

Janes. D. M., Squyres, S. W., Bindschadler, D. L., Baer, G., Schubert, G., Sharpton, V. L. and Stofan, E. R. (1992) Geophysical models for the formation and evolution of coronae on Venus. *J. Geophys. Res.*, **97**, 16,035 - 16,067.

Janes, D. M. and Melosh, H. J. (1988) Sinker tectonics: an approach to the surface of Miranda. *J. Geophys. Res.*, **93**, 3127 - 3143.

Jankowski, D. G. and Squyres, S. W. (1988) Solid-state ice volcanism on the satellites of Uranus. *Science*, **241**, 1322 - 1325.Johnson, A. M. (1970) *Physical Processes in Geology*. Freeman, San Francisco.

Jankowski, D. G. and Squyres, S. W. (1988) Solid-state ice volcanism on the satellites of Uranus. *Science*, **241**, 1322 - 1325.

Jannle, P., Janssen, D. and Basilevsky, A. T. (1987) Morphologic and gravimetric investigation of Bell and Eistla Regiones on Venus. *Earth, Moon and Planets*, **39**, 251 - 273.

Jannle, P., Janssen, D. and Basilevsky, A. T. (1988) Tepev Mons on Venus: morphoplogy and elastic bending models. *Earth, Moon and Planets*, **41**, 8282 - 8294.

Johnson, A. M. (1979) Field methods for estimating rheological properties of debris flows. (Unpublished manuscript, University of Cincinatti).Johnson, T. V. Cook, A. F., Sagan, C. and Soderblom, L. A. (1980) Volcanic resurfacing rates and implications for volatiles on Io. *Nature*, **280**, 746-750.

Johnson, A. M. and Hampton, M. A. (1969) Subaerial and subaqueous flow of slurries. *Unpublished final report, U.S.G.S. Contract 14-08-0001-10884*. Branner Library, Stanford University, California.

Johnson, A. M. and Hampton, M. A. (1969) Subaerial and subaqueous flow of slurries. *Unpublished final report, U.S.G.S. Contract 14-08-0001-10884*. Branner Library, Stanford University, California.

Johnson, M. W. and Nicol, M. (1987) The ammonia-water phase diagram and its implications for icy satellites. *J. Geophys. Res.*, **92**, 6339 - 49.

Johnston, D. H. and Toksöz, M. N. (1977) Internal structure and properties of Mars. *Icarus*, **32**, 73-84.

Jones, K. L. (1974) Evidence for an episode of martian crater obliteration intermediate in martian history. *J. Geophys. Res.* **79**, 3917-3932.

Karlstrom, T. N. V., McCauley, J. F. and Swann, G. A. (1969) *Preliminary Lunar Exploration Plan of the Marius Hills Region of the Moon*. Astrogeology 5, USGS open-file report, 42p.

Kaula, W. M. (1975) The seven ages of a planet. *Icarus*, **25**, 1-15.

Kaula, W. M., Fanale, F. P. and Anderson, D. L. (1981) Implications of basaltic volcanism for the evolution of planetary bodies. In BVSP, 1238ff.

Kaula, W. M. and several others. (1974) Apollo laser altimetry and inferences as to lunar structure. *Proc. Lunar Sci. Conf. 5th*, 3049-3058.

Kaula, W. M., Bindschadler, D. L., Grimm, R. E., Hansen, V. L. Roberts, K. M. and Smrekar, S. E. (1992) Styles of deformation in Ishtar Terra and their implications. *J. Geophys. Res.*, **97**, 16,085 - 16,120.

Kawada, K. (1977) The system Mg_2SiO_4-Fe_2SiO_4 at high pressure and temperature and the Earth's interior. Ph.D. thesis. Univ. Tokyo.

Kennedy, G. C. and Higgins, G. H. (1973) Temperature gradients at the core-mantle interface. *The Moon*, **7**, 14-21.

Kiefer, W. S. and Murray, B. C. (1986) Mercury's smooth plains. *Abstr. Mercury Conference*, Tucson.

Kieffer, S. W. (1982) Ionic volcanism. In *Satellites of Jupiter* (Ed. D. Morrison) 647-723, University of Arizona Press, Tucson.

King, E. A. (1978) *Geologic map of the Tyrrhena Patera quadrangle of Mars*. USGS Map I-1073.

King, J. S. and Riehle, J. R. (1974) A proposed origin of the Olypmus Mons escarpment. *Icarus, 23*, 300-317.

Koch, D. M. (1994) A spreading drop model for plumes on Venus. *J. Geophys. Res., 99*, 2035 - 2052.

Krauskopf, K. B. (1948) Mechanism of eruption at Paricutin volcano, Mexico. *Geol. Soc. Amer. Bull., 59*, 711-732.

Kuiper, G. P. (ed.) (1952) *The Atmospheres of the Earth and Planets*. 2nd ed. Chicago University Press, 306-405.

Kuiper, G. P. (1959) The exploration of the Moon. In: *Vistas in Astronautics, 2*, 273-313. Pergamon, London.

Kushiro, I. (1972) Effect of water on the composition of magmas formed at high pressures. *J. Petrol., 13*, 311-342.

Lambert, I. B. and Wyllie, P. J. (1972) Melting of gabbro (quartz eclogite) with excess water to 35Kbar, with geological applications. *J. Geol., 80*, 693ff.

Langmuir, C. H., Vocke, R. D. Jnr., Hanson, G. N. and Hart, S. R. (1978) A general mixing equation with applications to Icelandic basalts. *Earth Planet. Sci. Lett. 37*, 380-392.

Leake, M. A., (1981) PhD. Dissertation, University of Arizona.

Lipman, P. W. (1980) Cenozoic volcanism in the western United States: implications for continental tectonics. In *Studies in Geophysics: Continental tectonics, 161-174*. Nat. Acad Sci. Washington.

Lipman, P. W. amd Banks, N. G. (1987) Aa flow dynamics, Mauna Loa 1984. In: *Volcanism in Hawaii*, USGS Prof. Paper 1350, chapter 57.

Lofgren, G. E., Donaldson, C. H. and Usselman, T. M. (1975) Geology, petrology and crystallization of Apollo 15 quartz-normative basalts. *Proc. Lunar Sci. Conf. 6th*, 79-99.

Lopes, R. and Guest, J. E. (1982) Lava flows on Etna, a morphometric study. In *The Comparative Study of the Planets* (eds. A. Coradini and M. Fulchignoni) Reidel, Dordrecht.

Lopes, R. M. C. and Kilburn, C. R. J. (1987) The planimetric development of aa flow-fields on Mt.Etna, Sicily. Hawaiian Symposium on How Volcanoes Work (paper)

Lowman, P. D. (1976) Crustal evolution in silicate planets:implications for the origin of the continents. *J. Geol., 84*, 1-26.

Lowman, P. D. (1978) Crustal evolution in the silicate planets. *Naturwissenschaften, 65*, 117-124.

Lucchitta, B. K. (1973) Photogeology of dark material at the \Taurus-Littrow region of the Moon. *Proc. Lunar Sci. Conf. 4th*, 149-162.

Lucchitta, B. K. (1976) Mare ridges and related highland scarps - result of vertical tectonism? *Proc. Lunar Sci. Conf. 7th*, 2761-2782.

Lucchitta, B. K. (1977) Topography, structure and mare ridges in southern Mare Imbrium and northern Oceanus Procellarum. *Proc. Lunar Sci. Conf. 8th*, 2691-2703.

Lucchitta, B. K. and Schmitt, H. H. (1974) Orange material in the Sulpicius Gallus formation of the southwestern edge of Mare Serenitatis. *Proc. Lunar Sci. Conf. 5th*, 223-234.

Lucchitta, B. K. and Soderblom, L. A., (1982) The geology of Europa. In: Morrison, D. (ed) *Satellites of Jupiter.* University of Arizona Press, Tucson, pp 521 - 555.

Lunine, J. I. (1989) The Urey Prize Lecture: Volatile processes in the outer solar system. *Icarus*, **81**, 1 - 13.

Maalöe, S. (1985) *Principles of Igneous Petrology.* 374pp. Springer-Verlag, Berlin.

Maalöe, S. and Aoki, K. (1977) The major element composition of the upper mantle estimated from the compositions of lherzolites. *Contrib. Mineral.Petrol.,* **63**, 161-173.

Malin, M. C. (1979) Mars: evidence of indurated deposits of fine materials. Abstract *NASA Conf. Publ.* **2072**, 54.

Malin, M. C. (1980) The lengths of Hawaiian lava flows. *Geology*, **8**, 306-308.

Marsh, B. D. (1978) On the cooling of ascending andesitic magma. *Phil,Trans. Roy. Soc. Lond.,Ser. A.,* **288**, 435-443

Marsh, B. D. and Carmichael, I. S. E., (1974) Benioff zone magmatism. *J. Geophys. Res.,* **79**, 1196-1206.

Marsh, B. D. and Kantha, L. H. (1978) On the heat and mass transfer from an ascending magma. *Earth Planet. Sci.Lett.,* **39**, 435-443.

Mason, B. (1966) *Principles of Geochemistry.* 310pp. Wiley, New York.

Masursky, H. Schaber, G. G., Soderblom, L. A. and Strom, R. G. (1979) Preliminary geological mapping of Io. Nature, *280*, 725-729

Masursky, H., Eliason, E., Ford, P. G., McGill, G. E., Pettengill, G. H., Schaber, G. G. and Schubert, G. (1980) Pioneer-Venus radar results: Geomorphology from imagery and altimetry. J. *Geophys. Res.* **85**, 8232-8260.

McBirney, A. R. and Murase, T. (1984) Rheological properties of magmas. *Ann. Rev. Earth. Planet. Sci.,* **12**, 337-357.

McBirney, A. R. and Noyes, R. M. (1979) Crystallisation and layering of the Skaergaard Intrusion. *J. Petrol.,* **20**, 487-554.

McCauley, J. F., Smith, B. A. and Soderblom, L. A. (1979) Erosional scarps on Io. *Nature,* **280**, 736-738.

McCord, T. B. and Adams, J. B. (1977) Use of ground-based telescopes in determining the composition of the surfaces of solar system objects. *NASA SP-370*, Part 2, 893 - 922.

McEwen, A. S. and Soderblom, L. A. (1983) Two classes of volcanic plumes on Io. *Icarus,* **55**, 191-217.

McEwen, E., Greeley, R., Nagel, E. and Belton, M. J. S. (1993) Lunar Impact Basins: New data for the western limb and far side from the first Galileo flyby. *J. Geophys. Res.,* **98**, 17149 - 17181.

McGee, K. P. and Head, J.W. (1995) The role of rifting in the generation of melt: Implications for the origin and evolution of the Lada Terra-Lavinia Planitia region of Venus. *J. Geophys. Res.*, **100**, 1527 - 1552.

McGetchin, T. R. and Smyth, J. R. (1978) The mantle of Mars: some possible geological implications of its high density. *Icarus,* **34**, 512-536.

McGetchin, T. R. and Ulrich, W. G., (1973) Xenoliths in maars and diatremes with inferences for the moon, Mars and Venus. *J. Geophys. Res.,* **78**, 1833-1853.

McGetchin, T. R., Pepin, R. O. and Phillips, R. J. (eds.) (1981) *Basaltic Volcanism on the Terrestrial Planets.* Pergamon, New York. 1246-1254.

McGill, G. E. (1994) Hotspot evolution and Venusian tectonic style. *J. Geophys. Res.,* **99**, 23,149 - 12,161.

McGovern, P. J. and Solomon, S. C. (1993) State of stress, faulting, and eruption characteristics of large volcanoes on Mars. *J. Geophys. Res.,* **98**, 23,553 - 23,579.

McKee, C. O., Cooke, R. J. S. and Wallace, D. A. (1976) 1974-75 eruptions at Karkar volcano, Papua New Guinea. In *Volcanism in Australasia*, R. W. Johnson (ed.), p. 173-190, Elsevier.

McKenzie, D., Ford, P. G., Fang Liu, Pettengill, G. H. (1992) Pancake-like domes on Venus. *J. Geophys. Res.,* **98**, 9113 - 76.

Meyer, B. (ed.) (1965) *Elemental sulfur, Chemistry and Physics.* Interscience, New York.

Meyer, J. D. amd Grolier, M. J. (1977) Geologic Map of the Syrtis Major quadrangle of Mars. USGS Map I-995.

Minakami, T. (1951) The O-shima basalts, Japan. *Bull. Earthq. Res. Inst.,* **13**, 629-644 and 790-800.

Mohr, P. (1992) Ethiopian flood basalt province. *Nature,* **303**, 577 - 584.

Moore, H. J., Boyce, J. M., Schaber, G. G. and Scott, D. H. (1980) *Lunar Remote Sensing and Measurements. Apollo 15-17 orbital investigations.* Geological Survey Professional Paper 1046-B, USGS Printing Office, Washington,.

Moore, H. J. and Schaber, G. G. (1975) An estimate of the yield strength of the Imbrium flows. *Proc. Lunar Sci. Conf. 6th,* 101-118.

Moore, J. G., Fleming, H. S. and Phillips, J. D. (1974) Preliminary model for extrusion and rifting at the axis of the Mid-Atlantic Ridge, 36°48 North. *Geology,* **2**, 437-440.

Moore, J. M. (1984) The tectonic and volcanic history of Dione. *Icarus,* 59, 205 - 220.

Moore, R. C. (1933) *Historical geology.* McGraw-Hill, New York, 128pp.

Morabito, L. A., Synnott, S. P., Kupferman, P. N. and Collins, S. A. (1979) Discovery of currently active extraterrestrial volcanism. *Science* **204**, 972.

Morrison, D. (1982) *The Satellites of Jupiter.* University of Arizona Press, Tucson.

Morrison, D. (1983) Outer planets satellites. *US Nat. Rept. 1979-82,* 151-159.

Mouginis-Mark, P. J., Wilson, L. and Head, J. W. (1982) Explosive volcanism on Hecates Tholus, Mars. *Proc., Lunar Planet. Sci. Conf.,12th.,* 1431-1447.

Murase, T. and McBirney, A. R. (1973) Properties of some common igneous rocks and their melts at high temperatures. *Bull. Geol. Soc. Amer.,* **84**, 3563-3592.

Murchie, S. L. and Head, J. W. (1988a) The evolution of volcanism on Ganymede: Possible importance of a low melting point volatile. *Lunar Planet. Sci.,19th*, 819-820.

Murchie, S. L. and Head, J. W. (1988b) Tectonic and volcanic evolution of Ganymede. *Lunar Planet. Sci., 19th*, 823-824.

Nelson, R. M., Pieri, D. C., Nash, D. and Baloga, S. M. (1982) Reflection spectrum of liquid sulphur and its implication for Io. *Rep. Planet. Geol. Prog. - 1982*, NASA TM-85127, 12-15.

Nixon, P. H. (ed.) (1973) *Lesotho Kimberlites*. Maseru: Lesotho National Development Corporation.

Nockolds, S. R. (1934) The contaminated tonalites of Loch Awe, Argyll. *Quart. J. Geol. Soc. Lond.,* **90**, 302-322.

Noe-Nygaard, A. (1974) Cenozoic to Recent volcanism in and around the North Atlantic Basin. In *The Ocean Basins and Margins, Vol. 2. The North Atlantic*, (A. E. M. Nairn and F. G. Stehli, eds.), 391-443. Plenum, New York.

O'Nions, R. K., Pankhurst, R. J. and Gronvold, K. (1976) Nature and development of basalt magma sources beneath Iceland and the Reykjanes Ridge. J. Petrol. *17*, 315-388.

Oversby, V. M. and Ringwood, A. E. (1971) Time of formation of the Earth's core. *Nature,* **234**, 463-465.

Oxburgh, E. R. and Turcotte, D. L. (1978) Mechanisms of continental drift. *Rep. Prog. Phys.,* **41**, 1249-1312.

Pai, S. I., Shu, Y. and O'Keefe, J. A. (1978) Similar explosive eruptions of lunar and terrestrial volcanoes. *Proc. Lunar Planet. Sci. Conf.,9th*, 1485-1508.

Papanastassiou, D. A. and Wasserburg, G. J. (1969) The determination of small differences in the formation of planetary objects. *Earth Planet. Sci. Lett,* **5**, 361-376.

Papike, J. J. and Bence, A. E. (1978) Lunar mare versus terrestrial mid-oceanic ridge basalts: Planetary constraints on basaltic volcanism. *Geophys. Res. Lett.,* **13**, 803-806.

Pavri, B., Head, J. W., Brennan Klose, K. and Wilson, L. (1992) Steep-sided domes on Venus: Characteristics, geologic setting, and eruption conditions from Magellan data. *J. Geophys. Res.*, **97**, 13,445 - 13,478.

Payne, C.H. (1925) *Proc. Nat. Acad. Sci.*, **11**, 192; Payne, C.H. (1925) *Stellar Atmospheres*, pp.56, 185, 188. Harvard Univ., Cambridge.

Peale, S. J., Cassen, P. and Reynolds, R. T. (1979) Melting of Io by tidal dissipation. *Science,* **203**, 892-894.

Phillips, R. J. and Malin, M. C. (1989) The interior of Venus and tectonic implications. In *Venus* (ed. D. M. Hunten etal.), p. 159 - 214. University of Arizona, Tucson.

Phillips, R. J., Banerdt, W. B., Sleep, N. L. and Saunders, R. S. (1981) Tharsis: Ten years later. Paper presented at 3rd International Mars Colloquium, NASA, Pasadena.

Phillips, R. J., Raubertas, R. F., Arvidson, R. E., Sarkar, I. C., Herrick, R. R., Izenberg, N., Grimm, R. E. (1992) Impact craters and Venus resurfacing history. *J. Geophys. Res.*, **97**, 15923 - 48.

Phillips, R. J. and Ivins, E. R. (1979) Geophysical observations pertaining to solid state convection in the the terrestrial planets. *Phys. Earth Planet. Int.* **19**, 107-148.

Pieri, D. C. and Baloga, S. M. (1984) Effusion rates, areas and lengths for some lava flows on Hawaii and Mt. Etna with planetary implications. *Rept. Planet. Geol. Geophys. Program-1983*, 141ff. NASA TM-86246.

Pieri, D. C. and Baloga, S. M. (1986) Eruption rate, area and length relationships of some Hawaiian lava flows. *J. Volcanol. Geotherm. Res,.* **30**, 29-45.

Pieri, D. C., Baloga, S. M., Nelson, R. M. and Sagan, C. (1984) The sulfur flows of Ra Patera, Io. *Icarus,* **60**, 685-700.

Pieters, C. M. and many authors (1986) The colour of the surface of Venus. *Science* **234**, 1379-1383.

Pieters, C. M., Head, J. W., Sunshine, J. M., Fischer, E. M., Murchie, S. L., Belton, M., McEwen, A., Gaddis, L., Greeley, R., Neukeum, G., Jaumann, R. and Hoffmann, H. (1993) Crustal Diversity of the Moon: Compositional analysis of Galileo solid state imaging data. *J. Geophys. Res.,* **98**, 17127 - 17148.

Pinkerton, H. and Sparks, R. S. J. (1976) The 1975 sub-terminal lavas, Mount Etna: A case history of the formation of a compound lava field. *J. Volcanol. Geotherm. Res.,* **1**, 167-182.

Pinkerton, H. and Sparks, R. S. J. (1978) Field measurement of the rheology of lava. *Nature,* **276**, 383-385.

Plescia, J. B. (1981) The Tempe volcanic province of Mars and comparisons wioth the Snake River Plains of Idaho. *Icarus* **45**, 586-601.

Prentice, A.J.R. (1978) Towards a modern Laplacian theory for the formation of the Solar System. In: Dermott, S.F. (ed) *The Origin of the Solar System.* Wiley, New York, pp 111 - 162.

Presnall, D. C. (1969) The geoemetrical analysis of partial fusion. *Am. J. Sci.,267*, 1178-1194.

Raitala, J. (1989) Development of the Alba Patera volcano on Mars. *Adv. Space Res.,* 9, 6143 - 6146.

Raitala, J. and Kauhanen, K. (1989) Magma chamber-related development of Alba Patera on Mars. *Earth, Moon and Planets,* **45**, 187 - 204.

Ramberg, H. (1967) Gravity deformation and the earth's crust as studied by centrifuged models. Academic Press, London and N.York.

Ramberg, H., (1970) Model studies in relation to intrusion of plutonic bodies, in: *Mechanics of Igneous Intrusion*, G. Newall and N. Rast (eds.) Gallery Press, Liverpool.

Reeves, H. (1978) The origin of the Solar System. In: Dermott, S.F. (ed) *The Origin of the Solar System.* Wiley, New York, pp 1 - 18.

Richey, J. E., Thomas, H. H. *et al.* (1930) The Geology of Ardnamurchan, North-west Mull and Coll. *Mem. Geol. Surv. G. B.*

Ridley, W. I. (1975) On high-alumina mare basalts. *Proc. Lun. Sci. Conf. 6th.,* 131-145. LPI, Houston.

Ringwood, A. E. (1960) Some aspects of the thermal evolution of the Earth. *Geochim. Cosmochim. Acta,* **20**, 241-259.

Ringwood, A. E. (1966) The Mineralogy of the Upper Mantle. In: P. M. Hurley (ed.) *Advances in Earth Science*, 357-399. MIT Press, Boston.

Ringwood, A. E. (1975) *Composition and Petrology of the Earth's Mantle*. McGraw-Hill, New York.

Ringwood, A. E. (1979) *The Origin of the Earth and Moon*. Springer-Verlag, New York, 26-162.

Ringwood, A. E. and Anderson, D. L. (1977) Earth and Venus: a comparative study. *Icarus,* **30**, 243-253.

Robson, G. R. (1967) The thickness of Etnean lavas. *Nature* **216**, 252-252.

Roedder, E. (1951) Low temperature liquid immiscibility in the system K_2O-FeO-Al_2O_3-SiO_2. *Am. Mineral.,* **36**, 282-286.

Roedder, E. (1979) Silicate liquid immiscibility in magmas. In: H. S. Yoder Jnr (ed.) *The Evolution of the Igneous Rocks: Fiftieth Anniversary Perspective*. 483-520. Princeton.

Ross Taylor, S. (1992) *Solar System Evolution*. Cambridge University Press.

Rothery, D. (1992) *Satellites of the Outer Planets*. Clarendon, Oxford.

Runcorn, S. K. (1980) An iron core in the moon generating an early magnetic field. *Lunar. Planet. Sci. Conf.,10th.*, 2325-2333.

Safronov, V. S. (1954) On the growth of planets in the protoplanetary cloud. *Astron. Zh.,* **31**, 499-510.

Safronov, V. S. (1972) *Evolution of the Protoplanetary Cloud and Formation of the Earth and Planets*. Translated from the Russian. Israel Program for Scientific translations, Tel Aviv.

Sakuma, S. (1954) Effect of thermal history on viscosity of Oosima lavas. Bull. Earthq. Res. Inst. *32*, 216-230.

Sato, M. (1978) Oxygen fugacity of basaltic magmas and the role of gas-forming elements. *Geophys. Res. Lett.,* **5**, 447-449.

Scarfe, C. M. (1973) Viscosities of basic magma at varying pressure. *Nature,* **241**, 101-102.

Schaber, G. .G. (1982) Syrtis Major: A low-relief volcanic shield. *J. Geophys. Res.* **87**, 9852-9866.

Schenk, P.M. (1991) Fluid volcanism on Miranda and Ariel: flow morphology and composition. *J. Geophys. Res.,* **96**, 1887 - 1906.

Schilling, J.-G., Unni, C. K. and Bender, M. L. (1978) Origin of chlorine and bromine in the oceans. *Nature* **273**, 631 - 636.

Schonfeld, E. (1979) Estimated viscosities of Arsia Mons lava flows. *Lunar Planet.*

Schubert, G. (1979) Subsolidus convection in the mantles of the terrestrial planets. *Ann. Rev. Earth Planet. Sci.* ,**7**, 289-342.

Schubert, G. ,Cassen, P. and Young, R. E. (1979) Subsolidus cooling histories of terrestrial planets. *Icarus,* **38**, 192-211.

Schultz, P. H. and Spudis, P. D. (1983) Beginning and end of lunar mare volcanism. *Nature,* **302**, 233 - 236.

Schultz, P. H. (1977) Lunar and martian floor-fractured craters. (Abstr.) In: *Basaltic Volcanism 2nd Inter-team Meeting*, 53-55. LPI, Houston.

Scott, D.H. (1969) The geology of the southern Pancake Range and Lunar crater volcanic field, Nye County, Nevada. PhD thesis, UCLA 1969.

Scott, D. H. (1982) Volcanoes and volcanic provinces: western hemisphere of Mars. *J. Geophys. Res.* **87**, 9839-9851.

Scott, D. H. and Carr, M. H. (1978) *Geologic Map of Mars*. USGS Map I-1083.

Scott, D. H. and Condit, C. D. (1977) Correlations: martian stratigraphy and crater density. NASA TM-X3511, 56-68.

Scott, D. R. and Stevenson, D. J. (1980) Magma ascent by porous flow. J. *Geophys. Res.*, **91**, B9, 9283-9296.

Scott, D. H. and Tanaka, K. L. (1982) Ignimbrites of Amazonis Planitia region of Mars. *J. Geophys. Res.* **87**, 1179-1190.

Scott, D. H. and Tanaka, K. H. (1986) *Geologic Map of the Western Equatorial Region of Mars*. USGS I-1802A.

Senske, D. A., Schaber, G. G. and Stofan, E. W. (1992) Regional topographic rises on Venus: geology of western Eistla Regio and comparison with Beta Regio and Atla Regio. *J. Geophys. Res.* **97**, 13395-13420.

Settle, M. (1978) Volcanic eruption clouds and thermal power output of explosive eruptions. *J. Volcanol., Geotherm. Res.*, **3**, 309-324.

Settle, M. (1979a) Lava rheology: Newtonian suspension model for lava flow behaviour at high shear rates. *Lun. Planet. Sci. 10th*, 1104-6.

Settle, M. (1979b) Lava rheology: thermal buffering produced by latent heat of crystallisation. *Lun. Planet. Sci. 10th*, 1107-9.

Sharpe, H. N. and Peltier, W. R. (1978) Parameterized mantle convection and the Earth's thermal history. *Geophys. Res. Lett.*, **5**, 737-40.

Sharpton, V. L. and Head, J. W. (1982) Stratigraphy and structural evolution of southern Mare Serenitatis: A reinterpretation based on Apollo Lunar Sounder Experiment. *J. Geophys. Res.* **87**, 10983-10998.

Shaw, H.R. (1980) The fracture mechanisms of magma transport from the mantle to the surface, in: *Physics of Magmatic Processes*. R. B. Hargreaves, (ed.), 201-264. Princeton University Press, N. Jersey.

Shaw, H. R. and Swanson, D. W. (1970) Eruption and flow rates of flood basalts, in: *Proc. 2nd. Columbia River Basalts Symposium*, E. H. Gilmour and D.Stradling (ed.) E. Washington State College Press, Cheney.

Shaw, H. R., Peck, D. L., Wright, T. L. and Okamura, R. (1968) The viscosity of basaltic magmas: an analysis of field measurements in Makaopuhi lava lake, Hawaii. *Amer. J. Sci.*, **226**, 225-264.

Shoemaker, E. M., Lucchitta, B. K., Plescia, J. B., Squyres, S. W. and Wilhelms, D. E. (1982) The geology of Ganymede. In: *Satellites of Jupiter*, D. Morrison (ed.), 435-520. University of Arizona Press, Tucson.

Sigvaldason, G. E. (1974) Basalts from the center of the assumed Icelandic mantle plume. *J. Petrol.*, **15**, 497-524.

Singer, R. B. (1980) The dark materials of Mars, II, new mineralogic interpretations from reflectance spectrosopy and petrologic implications. *Lunar Planet. Sci. 11th*, 1048-1050.

Sleep, N. H. (1994) Martian plate tectonics. *J. Geophys. Res.*, **99**, 5639 - 5656.

Slyuta, E. N., Nikolaeva, O. V., Kreslavsky, M. A. (1988) Distribution of small domes on Venus: Venera 15/16 radar data [in Russian]. *Astron. Vestnik.*, **22**, 287.

Smith, B. A., Shoemaker, E. M., Kieffer, S. W. and Cook, A. F. (1979) The role of SO$_2$ in volcanism on Io. *Nature* **280**, 738-743.

Smith, B. A. and 64 other authors (1989) Voyager 2 at Neptune: imaging science results. *Science*, **246**, 1422 - 1449.

Smrekar, S. E. and Phillips, R. J. (1991) Venusian highlands: geoid to topography ratios and their implications. *Earth Planet. Sci. Lett.*, **107**, 582-597

Solomon, S. C. (1975) Mare volcanism and lunar crustal structure. *Proc. Lunar Sci. Conf. ,6th*, 1021-1042.

Solomon, S. C. (1977) The relationship between crustal tectonics and internal evolution in the Moon and Mercury. *Phys. Earth Planet. Int.*, **15**, 135-145.

Solomon, S. C. (1979) Formation,history and energetics of cores in the terrestrial planets. *Phys. Earth Planet. Int.*, **19**, 168-182.

Solomon, S. C. (1980) Differentiation of crusts and cores of the terrestrial planets:Lessons for the early Earth? *Precamb. Res.*, **10**, 177-194.

Solomon, S. C. and Chaiken, J. (1976) Thermal expansion and thermal stress in the Moon and terrestrial planets: clues to early thermal history. *Proc. Lunar Sci. Conf. 7th*, 3229-3243.

Solomon, S. C. and Head, J. W. (1979) Vertical movement in mare basins: Relation to mare emplacement, basin tectonics and lunar thermal history. *J. Geophys. Res.* **84**, 1667-1802.

Sparks, R. S. J., Pinkerton, H. and Hulme, G. (1976) Classification and formation of lava levees on Mount Etna, Sicily. *Geology*, **4**, 269-271.

Sparks, R. S. L., Wilson, L. and Hulme, G. (1978) Theoretical modelling of the generation, movement and emplacement of pyroclastic flows by column collapse. *J. Geophys. Res.*, **83**, 1727 - 1739.

Spence, D. A. and Turcotte, D. L. (1985) Magma driven propagation of cracks. J. Geophys. Res., **90**, 575-580.

Spencer, J. (1987) *Icarus*, **69**, 297-313.

Spudis, P. D. and Greeley, R. (1978) Volcanism in the cratered uplands of Mars. *Eos* **58**, 1182.

Squyres, S. W., Jankowski, T. G., Simons, M., Solomon, S. C., Hager, B. H. and McGill, G. E. (1992) Plains tectonism on Venus: the deformation belts of Lavinia Planitia. *J. Geophys. Res.*, **97**, 13579-13599.

Steinbach, V. and Yuen, D. A. (1992) The effects of multiple phase transitions on Venusian mantle convection. *Geophys. Res. Lett.*, **19**, 2243 - 2246.

Stephenson, P. J. and Griffin, T. J. (1976) Some long basaltic lava flows in North Queensland. In: *Volcanism in Australasia*, R. W. Johnson (ed.), 41-51.

Stevenson, D. J. (1982) Volcanism and igneous processes in small icy satellites. *Nature*, **298**, 142 - 144.

Stofan, E. R. and Head, J. W. (1990) Coronae of Mnemosyne Regio: morphology and origin. *Icarus*, **83**, 216-243.

Stofan, E. R., Head, J. W. and Campbell, D. B. (1987) Geology of the southern Ishtar Terra/Guinevere Planitia region on Venus. *Earth, Moon and Planets*, **38**, 183-207.

Stofan, E. R., Head, J. W., Campbell, D. B., Zisk, S. H., Bogongolov, A. F., Rzhiga, O. M., Basilevsky, A. T. and Armand, N. (1989) Geology of a rift zone on Venus: Beta Regio and Devana Chasma. *Geol. Soc. Am. Bull.*, **101**, 143-156

Stofan, E. R., Sharpton, V. L., Schubert, G., Baer, G., Bindschadler, D. L., Janes, D. M. and Squyres, S. W. (1992) Global distribution and characteristics of coronae and related features on Venus: implications for origin and relation to mantle processes. *J. Geophys. Res.*, **97**, 13347-13378.

Stooke, P. J. (1989a) Volcanism on Callisto. *Lunar Planet. Sci., 20th,* 1073-1074.

Stooke, P. J. (1989b) Tethys: Volcanic and Structural Geology. *Lunar Planet. Sci., 20th,* 1071-1072.

Strom, R. G., Schaber, G. G. and Dawson, D. D. (1994) The global resurfacing of Venus. *J. Geophys. Res.*, **99**, 10899 - 10926.

Suppe, J. and Connors, C. (1992) Critical taper wedge mechanics of fold- and thrust-belts on Venus: initial results from Magellan. *J. Geophys. Res.*, **97**, 13545-13561.

Suppe, J., Powell, L. and Berry, R. (1975) Regional topo-graphy, seismicity, Quaternary volcanism, and the present-day tectonics of the western United States. *Amer. J. Sci.* **275A**, 397-436.

Surkov, Yu. A., Kirnizov, F. F., Khristianov, V. K., Glzov, B. N., Ivano, F. and Korchuganov, B. N. (1977) Investigations of the density of the venusian surface rock by Venera 10. *COSPAR Space Res.* **17**, 651-657.

Swanson, D. A. (1973) Pahoehoe flows from the 1969-71 Mauna Ulu eruption, Kilauea volcano, Hawaii. *Geol. Soc. Amer. Bull.*, **84**, 615-626.

Swanson, D. A., Duffield, W. A., Jackson, D. B. and Peterson, D. W. (1979) Chronological narrative of the 1969-1971 Mauna Ulu eruption, Kilauea volcano, Hawaii. *USGS. Prof. Paper 1056.*

Swanson, D. W., Wright, T. L. and Helz, R. T. (1975) Linear vent systems and estimated rates of magma production and eruption for the Yakima basalt of the Columbia River Plateau. *Am. J. Sci.*, **275**, 877 - 905.

Tanaka, K. L. and Chapman, M. G. (1992) Kasei Vallis Mars: Interpretation of canyon materials and flood sources. *Proc. Lunar Planet. Sci. Conf. 22*, 73 - 83.

Tatsumoto, M., Hedge, C. E. and Engle, A. E. J. (1965) K, Rb, Sr, Th, U and the ratio of Sr-87 to Sr-86 in oceanic tholeiitic basalts. *Science,* **150**, 886ff.

Taylor, S. R. (1992) *Solar System Evolution.* Cambridge, p.190.

Theilig, E. and Greeley, R. (1986) Lava flows on Mars: analysis of small surface features and comparison with terrestrial analogs. *J. Geophys. Res.,* **91**, E193-206.

Thomas, P. J., Squyres, S. W. and Carr, M.H. (1990) Flank tectonics of Martian volcanoes. *J. Geophys. Res.*, **95**, 14,345 - 14,355.

Thorarinsson, S. (1950) The eruption of Mt. Hekla, 1947-48. *Bull. Volc.,Ser. 2,* **10**, 157-168.

Thorarinsson, S. (1967) Some problems of volcanism in Iceland. *Geol. Rundschau,* **57**, 1-20.

Thorarinsson, S. and Sigvaldason, G. E. (1962) The eruption of Askja, 1961, a preliminary report. *Am. J. Sci.,* **260**, 641-651.)

Thorarinsson, S., Steinthorsson, S., Einarsson, Th., Kristmannsdottir, H. and Oskarsson, N. (1973) The eruption of Heimay, Iceland. *Nature,* **241**, 372-375.

Thornhill, G. D. (1993) Theoretical modelling of eruption plumes on Venus. *J. Geophys. Res.*, **98**, 9107 - 9111.

Thornhill, G. D., Rothery, D. A., Murray, J. B., Cook, A. C., Day, T., Muller, J. P. and Iliffe, J. C. (1993) Topography of Apollinaris Patera and Ma'adim Vallis: Atomated extraction of digital elevation models. *J. Geophys. Res.*, **98**, 32,581 - 23, 587.

Tozer, D. C. (1965) Thermal history of the Earth: 1. The formation of the core. *Geophys. J. R. Astron. Soc.* **9**, 95-112.

Turcotte, D. L. (1974) Membrane tectonics. *Geophys. J. Roy. Astron. Soc.* **42**, 33-42.

Turcotte, D. L. (1993) An episodic hypothesis for Venusian tectonics. *J. Geophys. Res.*, **98**, 17,061 - 17,083.

Turcotte, D. L. and Ahern, J. L. (1978) A porous flow model for magma mi8gration in the asthenosphere. *J. Geophys. Res.,* **83**, 767-772.

Turcotte, D. L. and Oxburgh, E. R. (1973) Mid-plate tectonics. *Nature,* **244**, 337-339.

Turekian, K. and Clark, S. P. (1969) Inhomogeneous accumulation of the Earth from the primitive solar nebula. *Earth Planet. Sci. Lett.,* **6**, 346-348.

Tyrell, G. W. (1937) Flood basalts and fissure eruptions. *Bull. Volc.* **1**, 89-111.

Uchupi, E. and Emery, K (1993) *Morphology of the Rocky Members of the Solar System.* Springer-Verlag.

Underwood, J. R. and Trask, N. J. (1978) Geologic Map of the Mare Acidalium region of Mars. USGS Map I-1048.

Urey, H. C. (1952) *The Planets.* Yale University Press, New Haven, Connecticut.

Urey, H. C. (1962) Evidence regarding the origin of the Earth. *Geochim. Cosmochim. Acta,* **26**, 1-13.

Veeder, G. J., Matson, D. L., Johnson, T. V., Blaney, D. L. and Goguen, J. D. (1994) Io's heat flow from infrared radiometry: 1983-1993. *J. Geophys. Res.*, **99**, 17,095 - 17,162.

Wadge, G. (1981) The variation of magma discharge during basaltic eruptions. *J. Volcanol. Geotherm. Res.,* **11**, 139-168.

Waff, H. S. and Bulau, J. R. (1979) Equilibrium fluid distribution in an ultramafic partial melt under hydrostatic stress conditions. *J. Geophys. Res.,* **84**, 6109-6114.

Walker, D., Stolper, E. M. and Hays, J. F. (1979) Basaltic volcanism: The importance of planet size. Proc. Lun. Planet. Sci. Conf.10th.,*Geochim. Cosmochim Acta, Suppl.,11*, 1995-2015.

Walker, F. and Poldervaart, A. (1949) Karroo dolerites of the Union of South Africa. *Geol. Soc. Amer. Bull.,* **60**, 591-706.

Walker, G. P. L. (1973) The lengths of lava flows. *Phil. Trans. R. Soc. London, Ser. A,* **274**, 107-118.

Walker, G. P. L. (1974) Volcanic hazards and the prediction of volcanic eruptions. In *The Prediction of Geological Hazards.* Misc.Paper 3, Geol. Surv. London, 23-41.

Walker, K. R. (1969) The Palisade Sill, New Jersey: a re-investigation. *Geol. Soc. Amer. Sp. Paper 111.*

Wasson, J. T. (1985) *Meteorites*. W. H. Freeman, New York.

Waters, A. C. (1962) Basaltic magma types and their tectonic associations: Pacific Northwest of the United States. *Amer. Geophys. Union Geophys. Mon.* **6**, 158-170.

Weertman, J. (1971) Theory of water-filled crevasases in glaciers applied to vertical magma transport beneath oceanic ridges. J. *Geophys. Res.,* **76**, 1171 - 1183.

White, R. and McKenzie, D. (1989) Magmatism at rift zones: the generation of volcanic continental margins and flood basalts. *J. Geophys. Res.,* **94**, 7685 - 7729.

Williams, H. and McBirney, A. R. (1979) *Volcanology.* 397pp. Freeman, San Francisco.

Wilson, L. and Head, J. W. (1981) Ascent and eruption of basaltic magma on the Earth and moon. *J. Geophys. Res.,* **80**, 2971-3001.

Wilson, L. and Head, J. W. (1983) A comparison of volcanic eruption processes on Earth, Moon, Mars, Io and Venus. *Nature,* **302**, 663-669.

Wilson, L. and Head, J. W. (1993) Mars: Review and analysis of volcanic eruption theory and relationships to observed landforms. Submitted to: *Rev. Geophys.*

Wilson, L., Sparks, R. S. J. and Walker, G. P. L. (1980) Relationships between pressure, volatile content and ejecta velocity in three types of volcanic explosion. *J. Volcanol. Geotherm. Res.,* **8**, 297-313.

Wise, D. U. (1979) Geologic Map of the Arcadia quadrangle of Mars. USGS Map I-1154.

Wood, C. A. (1984) Calderas - a planetary perspective. *J. Geophys. Res.* **89**, 8391-8406.

Wood, J. A., Dickey, J. S., Marvin, U. B. and Powell, B. N. (1970) Lunar anorthosites and a geophysical model of the Moon. *Proc. Lun. Planet. Sci. Conf. 11th.,* 965-988.

Yoder, H. S. (1976) *Generation of basaltic magmas.* Nat. Acad. Science Washington, DC.

Zimbelman, J. R. (1985) Estimates of Rheologic Properties for Flows on the Martian Volcano Ascraeus Mons. *Proc. Lunar Planet. Sci. Conf. 16th., J. Geophys. Res.,* **90**, D157-162.

Zisk, S. H., Mouginis-Mark, P. J., Goldspiel, J. M., Slade, M. A. and Jurgens, R. F. (1992) Valley systems on Tyrrhena Patera, Mars: Earth-based radar measurements of slopes. *Icarus,* **96**, 226 - 233.

Zuber, M. T. and Mouginis-Mark, P. J. (1992) Caldera subsidence and magma chamber depth of Olympus Mons Volcano, Mars. *J. Geophys. Res.,* **97**, 18,295 - 18,307.

Index

aa   70, 71, 150
accretion   10, 15, 16, 17, 22
adiabatic heating   17
Aeolis   176–177
Aino Platina   322
Alba Patera   73, 83,84, 85–93, 104, 107,
    108, 171, 174, 263, 264, 265, 266,
    267, 280, 285–294, 304, 305
Albor Tholus   264, 305
Alpha Regio   113, 140, 153, 160, 318, 340,
    343, 345
Amazonis Planitia   154, 176–177
ammonia   367, 368, 372, 379
Amphitrites Patera   298, 299, 305
Anala   334–335
andesite   71
anemone-type volcanoes   310–311
anorthosite   19, 22, 39, 180, 214, 215
ANT   180
Apennine Bench Formation   136, 185, 186
Aphrodite Terra   139, 153, 326, 340
Apis Tholus   357, 360
Apollinaris Patera   176, 264, 298
Apollo spacecraft   6, 20, 40, 179, 180
Arabia   166
Arachnoids   see Venus
Aramaiti   115, 322, 323
Archimedes   184, 186
Arecibo   138
Argyre   164
Ariel   375, 376
Aristarchus   83, 84, 85, 181, 186, 222,227,
    236
Aristarchus Plateau   223, 237
Arsia Mons   67, 84, 171, 177, 264, 266,
    267, 268–270, 274, 275, 283

Artemis Chasma   317, 321, 341, 343
Ascraeus Mons   82, 83, 264, 266, 270, 271,
    272, 275, 283
ash flows   132, 282, 290, 304–305, 325
Asteroids   7, 12
Atalanta Planitia   145, 320
Atla Regio   139, 140, 152, 153, 310, 314,
    315, 316, 325, 329, 331, 346

basalt   28, 29, 31, 97, 131–133, 137
    alkali   131, 132, 254, 255
    flood   48, 98, ,100, 117, 120–130, 340
    Fra Mauro   181–183, 188
    Hawaiian   254–256
    high-Ti   201, 210, 212, 213
    intermediate-Ti   211
    KREEP-rich   182–188, 203
    low-Ti   207, 212, 213
    lunar   32, 67
    mare   183
    MORB   38, 39, 40, 117, 119, 122
    oceanic island   38, 117, 119, 122
    terrestrial   67
    tholeiitic   122, 254, 255, 341
Bell Regio   325, 335–381, 343
Berenghinya Planitia   143
Beta Regio   139, 140, 144, 310, 315, 324,
    325, 326–328, 331
Bezymianny   50
Biblis Patera   266, 283
Big Glass Mountain   71, 72
Bingham fluids   61, 79, 325
Bishop Tuff   35
Boosaule Montes   354
Brito-Arctic province   117

calderas   106–109, 148, 168, 247, 253, 257–
    260, 269, 271, 275, 277, 283, 284,
    285, 287, 292, 295, 296, 299, 311, 333,
    338, 353, 357, 358
Callisto   367, 372, 378
Caloris basin   135, 136
canteloupe terrain   379, 380, 381
carbonaceous chondrites
Cayley Formation   189, 190
Ceraunius Fossae   171, 174, 281, 287
Ceraunius Tholus   280, 281, 282
channels   288–291, 296, 299–304
Chyrse   165
Clementine mission   see Galileo spacecraft
Colette   140
Columbia River Plateau   46, 48, 52, 84, 85,
    100, 103, 120, 125–128, 153
comets   12
complex ridged terrain (CRT)   330
cones   109, 110, 111
Copernicus   83, 194, 201
coronae   see Venus
cryovolcanism   368, 369, 377
cryptomare   see Moon

Dali Chasma   330
Danu Montes   338, 339
Deccan Traps   120, 121, 128, 129
degassing   3
Descartes Formation   189
Devana Chasma   326, 327, 328
Diana Chasma   330
diapirs   320, 321
diatremes   46
Dione   373, 374
domes   47, 112, 113, 143, 165, 234, 243
dykes   46, 47, 126, 156, 165, 168, 256, 260,
    370

Earth
    asthenosphere   46
    core   2, 16, 17, 18
    crust   1, 21, 27
    geothermal gradient   26, 29, 30
    lithosphere   2, 25
    LVZ   22, 25, 44
    mantle   2, 21, 24, 25, 26, 28, 30
    plate tectonics   2, 27

    subduction zones   26
    temperature   16, 26
East African Rift   131–133, 139, 310
East Pacific Rise   118
Eistla Regio   159, 315, 325, 328, 331–335,
    343, 347
elements   33, 34
    condensation of   10
    incompatible   34, 36, 46
    refractory   10
    trace   34, 36, 37, 181–183
    volatile   3, 11
Elysium   162, 165, 171, 265, 294–297, 307
Elysium Mons   264, 294, 295, 306
eruption   48–52
    clouds   54, 56
    fissure   47, 48
    Hawaiian   48, 61
    Peléan   50, 114
    phreatomagmatic   56, 303–305
    Plinian   49, 56, 93, 111, 114, 305, 325
    plumes   see eruption clouds
    rate   126–127
    Strombolian   48
    Surtseyan   56
    Vulcanian   48, 58
Etna   48, 60, 61, 77, 84, 85, 86, 87, 91
Europa   351, 367, 370

Faeroes   120, 121
FAMOUS project   118, 119
Fernandina   257–259
fire fountain   48, 148, 245
fissures   46, 51, 86, 102, 126, 172, 260
    width   52
flows   see lavas
fractional crystallization   2, 32, 33, 34
Fra Mauro Formation   181–183, 191, 193
Freyja Montes   339

Galápagos Islands   106, 107, 243, 245,
    257–259
Galileo mission   8, 203–208
Ganis Chasma   330
Ganymede   371, 372, 378
geyser   56, 57, 362
Gomera, Isla La   98
Great Bombardment   21

Greenland   120, 121, 123
Grimaldi   207, 208
Gruithuisen   192, 193, 194, 223
Guinevere Planitia   143, 145, 155
Gula Mons   315, 331, 332, 333, 335
Gunung Agung   56
Guor Linea   332, 333, 334

Hadley Rille   185, 213, 225
Hadriaca Patera   264, 299–301, 305
Halemaumau   259
Harbinger Montes   193, 196, 228, 234
Hathor Mons   311
Hawaii   44, 70, 77, 97, 245–257
   volcano ages   248–249
   propagation rates   249
Hawaiian Ridge   246, 248
Hebridean province   35, 46, 124
Hecate Chasma   316, 343, 346
Hecates Tholus   264, 281, 294, 296
Hekla   73
Hellas   164, 263, 298–305, 307
Heng-O   341, 343
Herigonius   209, 211, 227, 228
Hesperia Planum   164, 165
hot spots   249, 256, 340, 359
Hualalai   253

Ice Harbour Flows   126
Iceland   35, 57, 111, 119, 120, 121–123,
   128, 243, 245
ice volcanism   see cryovolcanism
ignimbrite   94, 112, 114, 325
Inachis Tholus   360
Innini Mons   311
Io   3, 4, 22, 43, 50, 56, 108, 351–356
   mountain massifs   353
   plains   353, 355–357
   surface composition   353
   vent materials   353, 357–362
   volcanic plumes   56, 351, 356, 361, 362
Iopolis Planum   356
iron   17, 18
Ishtar Terra   139, 142, 330, 341
Isidis   165, 168, 265

Japan   60, 66
Jovis Tholus   266

Jupiter11, 12, 369

Karroo basalts   120, 128
Kasei Vallis   174
Katmai   258
Kawah Mas   364
Kibero Patera   357, 358
Kilauea   44, 64, 126, 247, 252, 258, 293,
   359
   East Rift Zone   2, 259–260
kimberlite   46
Kohala   247
Komatiite   2, 341
Krakatoa   50
KREEP   180–188

Lacus Autumnii   237, 241
Lacus Veris   228, 237, 241
Lada Terra   139, 152, 342, 343, 345
lahar   60, 95, 111, 113, 114
Laki fissure   85, 100, 122, 323
Lakshmi Planum   338, 339
Lansberg   218, 219, 230
Laplace   9
Latona   323, 341
lava
   block   71
   channels   104, 105, 150, 151, 166, 267,
      311, 314, 338
   cooling   74–75, 148–149
   density   80
   dome   73
   effusion rate   78, 79, 80–86, 245, 325, 357
   Hawaiian   78, 245
   lakes   255, 300, 359
   laminar flow   62
   lunar   67, 91, 179–242
   Martian   76, 86, 93
   mass eruption rate   82
   rheology   60–73, 77–93
   tubes   75, 76, 104, 105, 153, 174, 223–
      235, 285–288
   Venusian   75, 93
   viscosity   61, 63, 64, 74, 81, 82, 285
   yield strength   61, 63, 64, 75, 81, 83, 84,
      91, 285
Lavinia Planitia   140, 143, 145, 147, 150,
   153, 311, 316, 343, 344, 345

levees   63, 79, 80, 82, 285
liquid immiscibility   35
lithosphere   22, 23, 118
lithospheric plates   5, 117
Loihi Seamount   247, 248, 249
Loki   362
Luna spacecraft   6, 200
Lunae Planum   164, 174, 264

maar   111, 116
Maasaw Patera   356, 358
Maat Mons   328, 329, 330
magma   2, 24–28
    ascent   44–47
    chamber   255, 277, 293, 347–348
    diapiric rise   45–46
    eruption rate   50–52, 293–294
    production rate   50–52, 293–294
    rheology   60–96
    viscosity   63–67
    yield strength   61, 63, 64, 81, 82
Magellan spacecraft   7, 139
Makaopuhi   76
Malea Planum   171, 176
mantle   21, 22
    convection   22, 24
    Earth's   2, 21
    plumes   2
    segregation   22
Mare Cognitum   229
Mare Crisium   199, 204
Mare Frigoris   187
Mare Humorum   203, 209, 211, 227, 229,
    236
Mare Imbrium   99, 181, 184, 203, 209, 216,
    217, 223, 233
Mare Insularum   213
Mare Nubium   227, 229, 230
Mare Orientale   206, 216, 227, 228, 237–
    242
Mare Serenitatis   195, 199, 203, 204, 210,
    211
Mare Tranquillitatis   201, 204, 209, 213
Mare Vaporum   209
maria   see Moon
Mariner spacecraft   6, 265
Marius Hills   228
Mars   4, 22, 41, 46, 55, 57, 80, 160–178,

    257, 261–308
    ash eruptions   53, 55
    core   20
    elastic lithosphere   308
    flow plains   171–174
    mantle   20, 92, 307
    Northern Plains   177
    paterae   263, 285–294, 297–305
    plateau plains   161–164
    ridged plains   164–171
    shields   174, 261–284
    volcanic plains   21, 58
mascons   198
Mauna Kea   247, 253
Mauna Loa   44, 64, 70, 79, 83, 84, 85, 244,
    247, 253, 256, 271, 293
    Makaopuhi   260
    Mokuaweoweo   244
    South Pit   244
    Southwest Rift   363
    sulphur flow   363
Mauna Ulu   105
Maunder Formation   239, 241
Maxwell Montes   140
Memnonia   171
Mercury   4, 7, 11, 20, 134–138
    core   20, 21
    intercrater plains   135–137
    light plains   135, 136
meteorites   11, 12, 38, 39
    chondrites   12, 13, 14, 38
    SNC   39, 41
methane   367, 372, 379
mid-Atlantic Ridge   118
mid-oceanic ridges   26, 27, 97, 118, 119,
    341
Milichius   227
Mimas   372, 373
Miranda   376, 377
Moho   28
Mont Pelée   50, 112
Montes Harbinger   see Harbinger Montes
Montes Rook   237, 238
Moon   22, 39–41, 98, 179–242
    agglutinates   201
    basins   20, 40
    cones   229–230
    core   19

cryptomare   195
crust   2
dark halo deposits   58
dark mantling deposits   212, 235–237
domes   190, 192, 193, 194, 198, 225–229, 241
flood volcanism   159
glass   212, 236
internal structure   39
highland crust   19, 39, 179
KREEP-rich rocks   182–188
light plains   188, 189, 190
magma ocean   2, 39, 180, 188
mare basalt   20, 40, 41, 52, 64, 215–242
mare arches   231–236
maria   159, 179, 201, 215–242
regolith   201–212
Mount Meru   84, 110, 111, 112, 113, 114
Mount St. Helens   58, 95, 112, 114
mudflows   see Lahars
Mylitta Fluctus   152, 153

Navka Planitia   311
Nectoris basin   190
Nemea Planum   355
Neptune   11
neutral buoyancy zone   310, 347–348
New Zealand   57
Niobe Planitia   156, 157
nitrogen   367
Noachis   163
norite   180
Northern Queensland   46, 76
novae   see Venus
nuées ardentes   122

Oceanus Procellarum   39, 193, 195, 197, 200, 203, 204, 219, 228, 229
Oldonyo Lengai   84, 97
olivine   27, 28, 29, 34, 122, 213, 214
Olympus Mons   64, 83, 84, 85, 86, 177, 264, 266, 267, 275–280, 307
aureole   173, 276, 279, 280
basal scarp   276, 278
O-shima   66
Ozza Mons   152, 329, 330

pahoehoe   69, 71, 150

Palus Putredinis   185, 186, 236
pancake dome   see Venus
Paraná Basin   120, 121, 128, 129
Parga Chasma   316, 319, 346
partial melting   25, 28, 30, 31, 43
partition coefficient   36
paterae   107, 108, 243, 263, 357, 358
Pavonis Mons   264, 266, 267, 270, 272
Pele   361, 362
peridotite   28, 29, 341
Phoebe Regio   139, 153, 315, 335–339
phonolite   112, 117, 130, 132
Pioneer-Venus spacecraft   7, 139
plagioclase   119, 180, 212
plains volcanism   100, 101, 102
pressure ridges   71, 72, 73
Prinz   195, 219, 222, 227, 228
Procellarum Basin   181, 184, 187
pyroclast   50, 51, 55, 56
pyroclastic deposits   98, 177, 212, 228, 242, 278, 289, 297, 298, 299, 365
pyroclastic flows   53, 60, 93–95, 114
pyroxene   33, 34, 119, 214

Ra Patera   357, 358
radioactive nuclides   3, 16, 17
Ranger spacecraft   6
Rayleigh-Taylor instabilities   44, 45
remote sensing   6–8
Rhea Mons   139, 327, 328
rheology   see Magma
Reyjkanes Ridge   121, 123
Rhea Mons   327, 328
rhyolite   130, 159, 255, 305
Riccioli   208
rifts   46, 130–135, 153
Rio Grande Rift   132, 133

Sacajawea   140
Santorini   258
Sapas Mons   151, 153, 315, 330
Sappho   333, 334
Saturn   368, 372
Schickard   207
Schiller-Zucchius   207
Sedna Planitia   140, 142, 143, 153
seismic discontinuities   28, 29
Selu   318–320, 343

shields   100, 105–107, 228–229, 243–285,
   310–315
  breakout   358, 361
  fields   *see* shield fields
  Galápagos   106, 107, 243
  Hawaiian   105, 106, 246–257
  Martian   107, 261–284
  Venusian   107, 310–314
shield fields   107, 140, 155–157
Shorty Crater
Sif Mons   315, 331, 332, 333, 335
sills   232
sinuous rilles   67, 68, 192, 218, 219, 222,
   228
Sinus Aestuum   209
Snake River Plains   100–103, 105, 150,
   174, 175, 219, 223, 227, 243, 292, 365
solar nebula   9–12
South Pole – Aitken basin   206–207
spatter ridges   165, 168, 292
spectral reflectance   202, 203
strato-volcanoes   109
Stromboli   48
subduction zones   26, 118
sulphur   22, 56, 67, 95, 96, 353, 357, 362
  flows   67, 357–361, 363–366
Sun   9, 11, 12
Supernovae   9
Surveyor spacecraft   200
Syrtis Major Planum   164, 165, 168, 263,
   305

Tamfana   317, 343
Taurus-Littrow   201, 212, 214, 236
Tellus Regio   325
Tempe Terra   164, 174, 175, 264
Tepev Mons   337
tesserae   *see* complex ridged terrain (CRT)
Tharsis   160, 171, 263, 265
  bulge   160, 164, 168, 261, 265, 284, 306,
   307
  Montes   174, 263, 271, 283, 306, 307
  Tholus   264, 266, 283
Theia Mons   327, 328
Themis Regio   310, 316
Thetis Regio
tholeiite   118, 122
Tibesti   97, 294

Tinatin Planitia   324
Titan
Titania   368, 376
troctolite   180
trachyte   117, 130, 132
Triton   43, 349, 379, 380, 381
Tycho   83, 85, 194, 201
Tyrrhena Patera   263, 264, 301–304, 304,
   305

Ulysses Patera   264
Ulysses Tholus   283
Uranius Tholus   171, 264, 266, 280, 281,
   284
Uranius Patera   171, 172, 263, 264, 266,
   280, 281, 284
Uranus   11, 368, 372
Ushas Mons   311

Valles Marineris   164, 171
Vega landers   141
Venera spacecraft   6, 138, 139, 141
Venus   4, 7, 11, 20, 41, 46, 54, 138–160
  arachnoids   116, 143, 148, 312, 315, 323
  CRT (complex ridged terrain)
  core   20
  coronae   115, 116, 143, 144, 145, 148 ,149,
   244, 309, 315–323, 343
  domes   148, 157–160, 309, 320, 323–325,
   330
  fracture belts   143, 341
  gravity data   139
  heat structure   341–350
  highlands   138, 328
  lithosphere   320, 341–347
  lowlands   138
  mantle   343–350
  novae   116, 148, 149, 315
  pancake domes   158–159, 323–324
  plains   140, 141–160, 348–350
  shield volcanoes   148, 149, 154, 155, 325,
   336
  surface composition   141, 148
  tesserae   *see* CRT (complex ridged terrain)
  upland rolling plains   138
  volcanic flows   148, 149, 150–153
  volcanic resurfacing   348–350
Vesta Rupes   338–339

Vesuvius   244
Viking spacecraft   265
volatiles   52–58, 66, 149, 290–292, 303–
    305
volcanic plains   100, 309, 310
volcanic rise   309, 310, 325–339, 345
Voyager spacecraft   351, 357, 362, 379

welded deposits   112
wrinkle ridges   99, 145, 147, 163, 166–170,
    231–236, 240

Yakima Basalt   100, 122, 126–128
    Roza Member   126–128
yardangs   176

zone melting   46

WILEY-PRAXIS SERIES IN ASTRONOMY AND ASTROPHYSICS
Forthcoming titles

THE VICTORIAN AMATEUR ASTRONOMER: Independent Astronomical Research in Britain 1820–1920
Allan Chapman, Wadham College, University of Oxford, UK

TOWARDS THE EDGE OF THE UNIVERSE: A Review of Modern Cosmology
Stuart G. Clark, Lecturer in Astronomy, University of Hertfordshire

LARGE-SCALE STRUCTURES IN THE UNIVERSE
Anthony P. Fairall, Professor of Astronomy, University of Cape Town, South Africa

MARS AND THE DEVELOPMENT OF LIFE, Second edition
Anders Hansson, Ph.D.

ASTEROIDS: Their Nature and Utilization, Second edition
Charles T. Kowal, Computer Sciences Corp., Space Telescope Science Institute, Baltimore, Maryland, USA

ACTIVE GALACTIC NUCLEI
Ian Robson, Director, James Clerk Maxwell Telescope, Head Joint Astronomy Centre, Hawaii, USA

ASTRONOMICAL OBSERVATIONS OF ANCIENT EAST ASIA
Richard Stephenson, Department of Physics, University of Durham, UK; Zhentao Xu, Purple Mountain Observatory, Academia Sinica, Nanjing, China; Yaotiao Tiang, Department of Astronomy, Nanjing University, China

EXPLORATION OF TERRESTRIAL PLANETS FROM SPACECRAFT, Second edition
Yuri Surkov, Chief of the Planetary Exploration Laboratory, Russian Academy of Sciences, Moscow, Russia